HUMAN IMPLICATIONS OF SCIENTIFIC ADVANCE

Proceedings of the
XVth International Congress
of the History of Science
Edinburgh 10-15 August 1977

edited by

E.G.FORBES

Edinburgh

© 1978

Edinburgh University Press
22 George Square, Edinburgh

ISBN 0 85224 323 5

Printed in Great Britain by
The Scolar Press, Ilkley, Yorkshire

CONTENTS

Symposium 1.   SCIENCE AND HUMAN VALUES

J.R.Ravetz

INTRODUCTION TO SYMPOSIUM 1:

SCIENCE AND HUMAN VALUES

This Symposium, the first of the Congress, was designed to provoke discussion of the themes which make the history of science relevant to the concerns of people of all nations. The two sessions looked at science from complementary perspectives. In the first, the focus was on European civilisation; for it is here that our modern science had its effective origins, and here that its technological applications are most developed for good or ill. In the second session, the perspective shifted; the main speakers looked at European science from the outside. They offered criticisms, based on their scholarly researches, of some important assumptions about European science that have contributed to our image of it.

We were fortunate that Dr. Joseph Needham, a great scholar and humanist, could introduce and chair the first session. In his person we have someone who sees European civilisation and its science from both within and without: And he has done more than any other person to make Europeans aware of their debt to the other great civilisations. In each of the sessions the concluding remarks were made by scholars from the Socialist part of Europe; in the morning Prof. G. Dobrov (U.S.S.R.), currently at the International Institute of Applied Systems Analysis; and in the afternoon Prof. R. Richta from Czechoslovakia. Their optimistic presentations reminded us that the problems of European science can also be seen not so much as a worry as a challenge, and that there are some communities where this is the consensus.

SCIENCE AS A CULTURAL SYMBOL

Allow me to begin by quoting from a recent statement by Jean Houston, on the re-mythologisation of science.

"This was brought home to me [she writes] when I questioned a group of Catholic girls. It was in a Catholic girls' college where I was teaching, and I said: "Young ladies, I am going to name some images, and I would like you to tell me what reaches you, what turns you on." First I said to them 'the Mother of God" and there was no reaction, and then I said "the Sacred Heart of Jesus" and 'the Annunciation of Mary' and there was no reaction. At this stage I said 'DNA" and many girls tensed up, and when I followed it with "RNA" the breathing was perceptibly faster. Then I said 'Inter-galactic nebula", and there was an echoing chorus of "Oo-wow" - and as you may know "Oo-wow" is the modern liturgical equivalent of "Amen". If you want the whole response, it's 'Oo-wow, far out'. Well, such a controlled experiment shows you the mythological matrix of our time."*

Now this is only one aspect of the popular attitudes to science today. It long ago became a commonplace to say that scientists have come to occupy the position of the mediaeval clergy as the sages, magi and guides of mankind. The movement has reached the point when whole bookshops are entirely devoted to science fiction, often the only reading of many people, often a mirror-image of ourselves and our fears and hopes, yet so often, especially in the Western capitalist world, extremely frightening, and intended to frighten. Hence the reaction of many millions who have come to abominate science with all the dictatorial and unfeeling high technology which springs from it, not only with regard to the future, but in the here and now. Hence the anti-science movement and the "counter-culture".

* In "Beyond the Mechanical Mind", ed. P. Fry & M. Long, Australian Broadcasting Commission, Sydney, 1977, p.132.

Some of the most telling formulations of the disenchant-
ment of the young with science have been presented by Theodore
Roszak in his books, "The Making of a Counter-Culture",[*] and
"Where the Wasteland Ends".[†] He and the young are against
modern science because they feel that it has had evil, totalitarian and inhuman social consequences. They are not content
to put this down merely to misapplied technology; their criticism
of science itself goes deep. They attack "the myth of objective
consciousness", detesting that "alienative dichotomy", which
separates the observing self from the phenomena in Nature, and
set up what they call an "invidious hierarchy", which raises
the observer to an inquisitorial level, free to torment Nature,
living or dead, in whatever way will bring intellectual light.
They feel too that science encourages a "mechanistic imperative",
that is to say, an urge to apply every piece of knowledge, in
every possible way, whether or not its application is health-
giving for human beings, or preservative of the non-human world
in which they have to live. The scientific world-view is thus
accused of a cerebral and egocentric mode of consciousness,
completely heartless in its activity. It is not as if scientific
methods of control were applied only to non-human nature; the
"scientisation of culture" is calculated to enslave man himself.
There are many techniques of human control, such as the behavioural

[*] "The Making of a Counter-Culture; Reflections on the
Technocratic Society and its Youthful Opposition" (New
York, 1968; London, 1971).

[†] "Where the Wasteland Ends; Politics and Transcendence in
Post-Industrial Society" (New York and London, 1972-3).

5

and management sciences, systems analysis, control of information, administration of personnel, market and motivational
research, and the mathematisation of human persons and human
society.[*]   In a word, technocracy is rampant, and the more
complete the domination of Nature, the more fully does it
become possible for ruling élites to increase their control
of individual human behaviour.[†]

Since Francis Bacon's time the essence of the scientific
method has been alienation, in the sense of an absolute distinction between the observer and the external world, with which he
can have "no sense of fellowship nor personal intimacy, nor
any strong belonging".   Nothing inhibits the ability to understand, and after understanding to manipulate and exploit to the
full.   There is much that can be said against the "callousness"
of science, yet without what one might call clinical detachment
no scientific medicine would ever have come into being.   In the
same way the opponents of science cannot deny its pragmatic
value in that pharmacological knowledge does lead to the relief,
or cure, of disease, and that flight would be impossible without
a knowledge of aerodynamics and thermodynamics.   The anti-
science movement is in rather a quandary here, for it can hardly
wish mankind to return to the infancy of pre-scientific ignorance,
while at the same time it is justifiably uneasy, indeed outraged,
at the uses which are constantly made of scientific knowledge;
and filled with fears for the future.[§]

[*]   Cf. J. Ellul, "The Technological Society" (New York, 1964).
[†]   Cf. W. Leiss, "The Domination of Nature" (New York, 1972).
[§]   This whole subject has been the theme of C.H. Waddington's
        Bernal Lecture at the Royal Society: "New Atlantis
        Revisited" (Proc. Roy. Soc. 1975).

I am inclined to think that the real meaning behind the anti-science movement is the conviction that science should not be taken as the only valid form of human experience. Actually philosophers have been calling this in question for many years past, and the forms of human experience - religion, aesthetics, history and philosophy, as well as science - have been delineated in many integrated surveys.   Roszak himself hints this, as when he denies that scientific objectivity can be "the only authentic source of truth", or when he says that "we must be prepared to see the truth as a multi-dimensional experience".

The whole anti-science movement has arisen because of two characteristics of our Western civilisation: on the one hand the conviction that the scientific method is the only valid way of understanding and apprehending the universe, and on the other hand the belief that it is quite proper for the results of this science to be applied in a rapacious technology often at the service of private capitalist profit.   The first of these convictions is held as a semi-conscious assumption by a great many working scientists, though formulated clearly only by a small number; at the same time it spreads widely through the populations, often leading to great callousness and insensitivity in personal relationships, quite beyond the power of the traditional codes of religion and ethics to modify. Similarly, the mass-production technology of the capitalist world, so freely paralleled and imitated in the Soviet Union and the Socialist Republics of Eastern Europe, has indeed supplied the peoples of the developed world with a vast wealth of material goods, but only at the cost of debauching their aspirations, limiting their freedoms, and imposing controls every day more insidious and unhealthy.

Let us turn now to an entirely different aspect of the situation. Even in a properly balanced human society, where the natural sciences were counter-balanced all the time by what used to be called in Cambridge the "moral sciences", and other forms of human experience such as the religious and the aesthetic, there would still be great difficulty in dealing with the all but intolerable ethical choices which applied science places before mankind and will increasingly place as time goes on. The young people of the counter-culture are revolted by the necessity of making such choices, but neither they nor we can go back to the "bliss of ignorance" as in primitive times. Actually it never was bliss, of course, because the very mission of science was to lead us out of the wilderness of ancient fears, taboos and superstitions. But the promised land will never be won by science alone. The control of applied science is probably the greatest single problem for humanity today, and one might even go so far as to wonder whether the most penetrating social critiques, such as the theory of the class struggle, and historical materialism, are not simply aspects of this basic question.

No doubt man has been facing it ever since the discovery of fire, but today it threatens his very existence. Everyone knows about nuclear power and the devastating possibilities of nuclear weapons, but such apparently simple problems as the disposal of the radio-active waste from nuclear power stations are nightmares to those who worry about the social responsibilities of science. Nowadays mathematical engineering is almost as dangerous, and the possibilities of "artificial intelligence", and the vast computing machines which can and will be built, with their fabulous information-storage and retrieval, are quite breath-taking. The privacy of the individual is now endangered, the rights of children to be taught by living teachers, and the safety of whole populations exposed to the danger of some electrical or mechanical fault when computers are harnessed to "defence".

The possibilities of biology and medicine are at least as challenging. My own professional background has made it natural for me to follow such developments. One of the largest fields in which they arise is that of generation, for this is the first time in human history that man is on the point of acquiring absolute control both of reproduction and infertility. All too soon we shall be in possession of means for controlling the sex of the human embryo. After this the sterilisation of whole groups might become a live issue. Ethical controversies have raged for years round contraception and abortion, but problems are also raised by the new foetal medicine, which can detect grave abnormalities long before birth, and by artificial insemination, which is only by convention attached to infertile marriages. Legal considerations and changes are lagging far behind the actual possibilities, such as spermatozoa banks, maintained from donors outstanding for physical or intellectual brilliance,

and possibly several generations older than the receiving
womb.[*]

Again, now that we know the chemical structure and
coding of the semantophore molecules of desoxyribonucleic
acid (DNA), which carry the instructions for making each new
human organism, infinite possibilities are open for interfering
with this hereditary material.[†] That would be biological
engineering applied at the molecular level; one could envisage
the insertion of an entirely new piece of chromosome, or the
removal of another.   Or one could produce hitherto unheard-of
hybrids by substituting a mixed-cell nucleus for the original
one of the fertilised egg itself.   These may seem distant
prospects, requiring enormous expenditure of money; but there
has already been unexpected success in transferring genes (the
hereditary units) from one lot of bacteria to another.   Certain
viruses can pick up genes and put them into the bacterial nuclear
systems.   What if an antibiotic-resistant strain of bacteria
were produced, which quickly spread all over the world like
wild-fire, and decimated human populations?   That this is a

* On all these questions see "Our Future Inheritance; Choice
  or Chance?", a study by a British Association Working Party,
  ed. A. Jones & W.F. Bodmer (Oxford, 1974).

+ See the Trueman Wood Lecture of C.H. Waddington, "Genetic
  Engineering", Journ. Roy.Soc.Arts, 1975, 123, 262.   My old
  friend and collaborator takes a refreshingly cool view of
  the dangers before us because of the immense expense which
  researches in such embryology and genetics involve, and
  the consequent certainty of public scrutiny.   I am not
  so optimistic, for two reasons:(a) the possibile activities
  of totalitarian States, and (b) the doubt that public scrutiny - or debate - necessarily leads to right ethical policy.
  As Waddington himself says: "One wonders whether we are
  intellectually, emotionally, or morally, prepared to face
  such choices....".

real danger has been shown by a self-denying ordinance achieved very recently in California, where the scientists working in these fields agreed to establish a moratorium on such experiments, at least until more laboratories with adequate safety equipment and security become available. Here there is one very tempting possibility open to mankind, namely the possible insertion into plants of genes favouring the symbiosis of nitrogen-fixing bacteria, as happens in the leguminous plants today. If that could be arranged for the staple crop cereals it would be a gift to humanity almost as great as the gift of fire. What effect would this immeasurable increase in food production have upon the human race?

Medicine is also confronting humanity with almost insoluble problems.[*] The conquest of transplantation intolerance has already led to a great proliferation of organ transplants, and no doubt the surgeons in time to come will have access to whole banks of spare parts for human beings. But transplantation studies go much further, for it is now possible to make chimaeras between animal species, since certain killed viruses makes the tissues stick together, and this could be used to unite human with animal tissues. What is going to come of that? Ethical problems also arise in all cases where the treatment may be very expensive, needing elaborate machinery - for example the kidney machines which dialyse the blood, and can keep a person going even though his or her kidneys are only able to function very ineffect-

---

[*] One of the most obvious, and widely debated, is of vourse that of euthanasia. On all these questions, see G. Leach, "The Biocrats", (London, 1970).

ively.   Who is going to choose who gets the advantage of restricted techniques in short supply?

Again, much work is being done on the fertilisation and
cultivation of human eggs _in vitro_ up to the blastocyst stage
before their reimplantation into a uterus to go on developing
until term.   Aldous Huxley in his famous novel "Brave New
World" visualised the isolation of totipotent blastomeres so
as to reproduce many identical copies of low-grade human beings,
and this is not at all impossible.   But there are other ways
of effecting such "cloning".   For example, nuclei from adult
cells can take the place of the egg's own nucleus itself, so
that a whole regiment of individuals with identical genetic
material could be created.   The question would then arise:
do all human beings have an inalienable right to individuality?
Such is the fix that Faust has got himself into, and the young
suspect that they know why.

I could go on a lot longer, but it must be evident that
humanity has never hitherto had to face anything like the tremendous ethical problems posed by the physico-chemical and biological sciences.   Now it is not at all obvious that the traditional ethics of the Western world, even with all its tomes of
moral theology and casuistry, is the best equipped to deal with
these problems, and certainly not on its own.   Even within the
sciences it is not obvious that the traditional modes of thinking
of Western philosophy are the most adequate for the extraordinary
and incredible events which go on in the world of sub-atomic
particles;   and indeed there are those, such as Odagiri Mizuho,

who are showing that Buddhist philosophy may give a good deal of help to the nuclear physicist which could not come from Western ideas alone.

A similar line of thought has been pursued of late by Fritjof Capra, another nuclear physicist, especially in his book "The Tao of Physics". Essentially his argument is that modern sub-atomic physics has made it quite clear that reality completely transcends all ordinary language, and that this was seen intuitively by the Taoist and Buddhist thinkers of ancient China and India. In the sub-atomic world the concepts of space and time, the idea of separable material objects, and the usual understanding of cause and effect, have all lost their meaning. Mass and energy are interconvertible, radiation is "not exactly" waves and "not exactly" particles, time does not uniformly flow, changes always include the observer in an essential way, and no precise prediction is possible. Polar opposites are complement-ary rather than antagonistic, particles are both destructible and indestructible, matter both continuous and discontinuous, and objects are relational events rather than substances, spontaneous dynamic patterns in a perpetual dance. Reality is beyond exist-ence and non-existence. It is hardly surprising therefore that many minds, especially of the younger generation, are attracted to the thought of Lao Tzu and Chuang Chou, the strange system-atisation of the I Ching (Book of Changes), and the insights of Tantrism and Chhan (Zen) Buddhism. The only unanswered question is how it came about that the ancient and mediaeval thinkers of India and China came to conclusions so close to those we have now arrived at with a great deal of trouble, building gigantic cyclotrons and following laboriously the traces of hadrons, electrons, photons and the like in bubble-chambers. Also it was one thing to guess it, and quite another thing to prove it. Without modern science that could not have been done.

What needs saying is that most of those who have worried quite properly about the control of applied science in the West have so far failed to realise that there is a great culture in the East, which for two thousand years has upheld a powerful ethical system never supported by supernatural sanctions. This is where Chinese culture may have, I think, an invaluable gift to make to the world. Nearly all the great philosophers of China have agreed in seeing human nature as fundamentally good, and considering justice and righteousness as arising directly out of it by the action of what we in the West might call the "inner light". The Johannine light, perhaps, "which lighteth every man that cometh into the world". Let men and women have proper training in youth, the right ideals, and a classless society which will bring out the best elements potentially within them.

For the Chinese, then, ethics was accepted as internally generated, intrinsic and immanent, not imposed by any divine fiat, like the tables of the law delivered to Moses on the mountain. I should go so far as to say that never have the Chinese been more faithful to this doctrine, interpreting it in terms of selfless service to others, to people, than they are at the present day. Wei jen min fu wu! This is my second fundamental point. If the world is searching for an ethic firmly based on the nature of man, a humanist ethic which could justify resistance to every dehumanising invention of social control, an ethic in the light of which mankind could judge dispassionately what the best course to take will be in the face of the multitude of alarming options raised by the ever-growing powers the natural sciences give us, then let it listen to the sages of Confucianism and Mohism, the philosophers of Taoism and Legalism. Obviously we must not expect from them exact advice on choices arising from techniques which they would never have been able to imagine. Obviously, also, we are in no way bound down to the formulations they gave to their ideas in ancient feudal, or mediaeval feudal-bureaucratic, society - time marches on. But what matters is their spirit, their undying faith in the basic goodness of human nature, free from all transcendental elements and capable of leading to a more and more perfect organisation of human society.

J-J.Salomon

LE MOUVEMENT DE PENDULE SCIENTIFIQUE:
UNE CRISE DU PROGRES

Rien n'est plus significatif de la "crise de civilisation" que nous vivons que le procès fait à la science,
et que ses procureurs viennent de l'intérieur aussi bien que
de l'extérieur de la communauté scientifique. Naguère
encore, au lendemain de la deuxième guerre mondiale -
malgré Hiroshima et l'impossibilité de revenir à la paix -
la science apparaissait comme un facteur et davantage
comme une garantie de progrès. Si le monde avait des ennuis, c'est parce qu'il y avait trop peu de science ou un
type inadéquat de science ou parce qu'on ne savait pas
bien l'appliquer. Tout semblait possible grâce à la
science : des moyens d'action nouveaux, toujours promis à
se multiplier et à s'étendre dans la conquête de l'infiniment grand et de l'infiniment petit, et le transfert de
ces succès de la rationalité du domaine de la matière à
celui de la vie sociale. Vous voulez la lune ? Nous l'a-
vons eue : le désenchantement a succédé à l'euphorie.

L'inquiétude écologique, le malaise universitaire, la
révolte étudiante, la critique du "complexe militaro-
industriel" et aussi l'indifférence, quand ce ne sont pas
les sarcasmes, qui a suivi la répétition presque routi-
nière des promenades sur la lune : "Comment, vous inves-
tissez tant d'efforts et d'argent pour en ramener des cail-
loux, mais vous n'êtes pas capables de résoudre les problèmes de la terre !" - autant de signes précurseurs d'un
mouvement qui n'a cessé de s'étendre et qui, en mettant en
cause les objectifs et les résultats des politiques de la
science, a fini par mettre en question les normes internes
de la science et jusqu'à son statut épistémologique.

Ce réquisitoire, s'il n'était qu'une manière littéraire de renouer avec l'anti-intellectualisme, ne prête-
rait pas à plus de conséquences que du temps où la science
était la cible de critiques mystiques ou romantiques tels
que Blake, Keats, Ruskin, poètes, artistes, romanciers ou
philosophes dont le discours n'avait aucune prise sur le
cours de la science. Mais il est difficile de ne pas se
souvenir que l'anti-intellectualisme a nourri aussi les
pires mouvements politiques du XXème siècle - et ce qu'il
a coûté en termes de régression scientifique aux régimes
qui en ont fait leur doctrine, en termes de vies humaines

aux pays qui en ont subi le fanatisme.

Ce qui, en tout cas, constitue la nouveauté de la situation actuelle, c'est que la critique de la science ne vient plus seulement de l'extérieur, mais aussi de l'intérieur. De plus, cette critique ne s'arrête pas aux cercles restreints des intellectuels, elle se manifeste dans le grand public. Parmi ceux qui s'interrogent, on trouve des scientifiques et non des moindres ; parmi ceux qui s'inquiètent, il y a l'homme de la rue. Pour la première fois dans l'histoire du rationalisme, le bilan de la science n'apparaît pas comme nécessairement "positif" ni la vocation du chercheur comme nécessairement "heureuse". Si les signes de cette crise se sont multipliés depuis la fin des années 60, il est clair qu'il faut remonter plus haut - et plus profond - pour en comprendre le sens.

Il y a des crises de la science et des crises pour et par la science. De toute évidence, on ne parle pas de la même chose dans le premier cas que dans le second. Mais peut-on dire qu'il n'y a aucun rapport entre celles-ci et celles-là ? Et n'y a-t-il pas des situations où il est difficile, sinon impossible d'isoler les unes des autres ? C'est tout le fond du débat qui oppose les "internalistes" aux "externalistes" en histoire des sciences. Un marxiste ne se fera pas faute de souligner qu'à aucun moment la science envisagée comme superstructure n'est un sous-système autonome par rapport au système économique et social. Ce type d'explication peut aller jusqu'à l'absurde, par exemple lorsque Hessen rapportait les Principes de Newton à la lutte des classes sous la Révolution anglaise et à l'essor de la "bourgeoisie capitaliste" (1). Mais point n'est besoin d'être marxiste pour montrer comme Schmookler ou Merton, que le cours de la science répond aussi aux demandes de la société et aux enjeux dont elle est le théâtre (2).

Dans les rapports entre science et société, la conjonction "et" ne renvoie pas plus à une simple juxtaposition de deux domaines hétérogènes qu'à une subordination étroite de l'un à l'autre. Il y a des crises de la science qui n'intéressent d'aucune façon les milieux extérieurs à la communauté scientifique ; il en est d'autres qui, si éloigné des préoccupations de la société qu'en soit l'enjeu théorique, ne sont ni indifférentes ni étrangères aux changements sociaux. L'histoire des sciences n'a sans doute pas connu de plus "grande crise" que celle qui déboucha sur la révolution scientifique du XVIIème siècle : rupture épistémologique, mutation des esprits, mais aussi des moeurs, des institutions et des structures sociales, la révolution galiléenne peut apparaître à la fois comme l'effet et la cause des grandes transformations de la Renaissance (3). En ce sens, crise de la science et crise de la société ne sont jamais que deux aspects du même phénomène.

Pourtant, on trouve une différence sensible d'appréciation entre les crises sociales (en particulier les révolutions politiques, grandes accoucheuses de l'Histoire) et les crises scientifiques : si les premières, comme les maladies, peuvent avoir une issue heureuse ou malheureuse et donc se conclure parfois sur une régression, les secondes apparaissent toujours comme débouchant sur un progrès. Telle du moins que les scientifiques la pensent et que, jusqu'à présent, les historiens des sciences l'ont écrite, l'histoire des sciences se distingue de toutes les autres histoires en ce qu'elle ne connaît pas d'issue négative. Il n'est pas surprenant que toutes les philosophies de l'histoire, de Vico à Marx en passant par Hegel et Comte, aient vu dans le progrès des sciences un modèle de la nécessité qu'elles assignent aux différentes étapes du devenir humain pour trouver son aboutissement.

Il y a, bien sûr, des problèmes scientifiques sans solution. Mais alors de deux choses l'une : il s'agit soit de faux problèmes (la quadrature du cercle, le mouvement perpétuel), soit de problèmes mal posés dont la solution est reportée ou refoulée jusqu'au mûrissement de la crise qui conduit à les aborder dans des termes nouveaux (ainsi le phlogistique et la génération spontanée auxquels Lavoisier et Pasteur ont permis une fois pour toutes de tourner le dos). Autrement dit, toute crise majeure en histoire des sciences - toute révolution - est source de progrès.

Tel est précisément le titre du dernier chapitre du livre de Kuhn, qui fonde sur ce caractère cumulatif et progressif des sciences de la nature sa conception fameuse du paradigme et des révolutions scientifiques. Il faut reconnaître, si l'on s'en tient à l'histoire interne des sciences, qu'il y a quelque chose d'irréfutable dans cette conception. En période de "science normale", quand le paradigme auquel se réfère une communauté scientifique n'est pas remis en question, les spécialistes constituent "un instrument extrêmement efficace pour résoudre les problèmes ou les puzzles que définit le paradigme. Et le résultat de cette efficacité doit inévitablement être un progrès" (4). En période de crise, le passage du paradigme ancien au paradigme nouveau se traduit à son tour par un progrès. Kuhn en donne deux sortes de preuves : l'une est de caractère intellectuel, car le nouveau paradigme permet de "résoudre les problèmes qui ont conduit à la crise" (5), l'autre est matérielle et pédagogique, puisque les spécialistes renoncent aux livres et aux articles fondés sur le paradigme ancien (6).

Irréfutable si cette positivité s'applique exclusivement aux crises internes de la science, on peut s'interroger sur cette conception quand les crises de la science sont aussi des crises pour et par la science. On peut même aller plus loin et se demander si, de nos jours, les cri-

ses de la science peuvent encore être strictement res-
treintes aux débats internes de la communauté scientifique.
En ce cas, la positivité spécifique des révolutions scientifiques va-t-elle encore de soi ? Kuhn lui-même en a conscience, puisqu'il note - en passant - que "le bilan d'une
révolution scientifique comporte des pertes aussi bien que
des gains et (que) les scientifiques ont tendance à se montrer particulièrement aveugles à l'égard des premières".
Je veux bien que Kuhn ne pense ici qu'aux pertes que subit
le savoir ou l'institution scientifique en tant que telle(1).
Mais il me semble, au contraire, que tout l'enjeu du soupçon qui pèse sur la fonction sociale de la science tient
dans cette question : s'il n'y a pas que des gains, qu'en
est-il de ces pertes ?

Née au XVIIème siècle, la science moderne a été immédiatement associée à l'idée de progrès. Par la méthode expérimentale, le savoir agit sur la nature et la transfor-
me : savoir opérationnel, qui non seulement renouvelle le
champ de la connaissance, mais aussi se prolonge en instruments d'action. De ce point de vue déjà, la démarche scientifique moderne est par elle-même tournée vers l'avenir, en
elle-même source de progrès : la théorie est soumise à l'é-
preuve pratique de résultats qui s'additionnent, se cumu-
lent et se multiplient.

On peut discuter à loisir sur la réalité du progrès
en morale ou du progrès en esthétique : inépuisable sujet
de dissertation pour potaches, sur lequel le moindre bon
sens interdira toujours de conclure. L'histoire contemporaine, individuelle et surtout collective, est si "pleine
de bruit et de fureur", si riche en défis ou en dénis de
ce qui passe pour être le bien, que ce serait folie d'attribuer plus de vertu à l'homme du XXème siècle qu'à celui
des générations qui l'ont précédé.

Du point de vue de l'art, les monuments des civilisations disparues ou des sociétés transformées sont là pour
témoigner que le progrès est affaire de perspective autant
que de goût, mais jamais de cumul : notion absurde, en vérité, qui tendrait à reconnaître une "moindre beauté" à
l'art rupestre ou grec qu'à celui du XVIIème ou du XXème
siècle, aux tableaux de Breughel ou de Franz Hals qu'à ceux
de Picasso ou de Soulages. Les Musées se remplissent
d'oeuvres nouvelles, mais elles n'annulent pas celles du
passé.

La science, au contraire, est ce domaine de la culture qui, par l'accumulation et le renouvellement, rend ca-
duques, obsolètes comme l'on dit aujourd'hui, des pans en-
tiers de connaissances du passé (et, de plus en plus du passé le plus récent). Dans tous les autres domaines de la culture, on tourne les pages d'un même livre ; en science,
comme l'a souligné Kuhn, on change de livre : "Il n'y a

rien dans la formation scientifique qui soit l'équivalent
du musée artistique ou de la bibliothèque des classiques"(7).

Sur la lancée des succès remportés par la méthode expérimentale, les liens entre l'idée de progrès et la science
sont apparus si étroits, que celle-ci est devenue le modèle
de ce qu'il fallait à la fois penser et faire pour permettre à la société d'accéder au "plus" : bonheur, mieux-être,
progrès social et, pourquoi pas, progrès moral ? Ce modèle
qu'offre la science, c'est essentiellement, ne l'oublions
pas, celui des sciences de la nature : tout le siècle des
Lumières s'en inspire et en tout, attendant de leur "grand
exemple", dit Cassirer, "le renouvellement des sciences
morales, une vision approfondie de l'esprit des lois, de
l'esprit de la société, de la politique, de l'art poétique
même" (8). Présentant le savoir comme pouvoir - celui-là
même qui doit, suivant le mot de Descartes, rendre l'homme
"maître et possesseur de la nature" -, les sciences de la
nature substituent aux évidences du monde vécu l'ordre du
monde soumis au calcul et à la mesure.

La tentation est grande non seulement d'invoquer le
modèle dans la pratique sociale, mais encore de l'y transposer : le siècle des Lumières voit dans le progrès des
connaissances l'accès au progrès du genre humain ; il suf-
fira donc d'appliquer à la conduite des hommes les méthodes,
la démarche, les principes de ces sciences qui ont tant fait
(et ne vont pas cesser de faire) leurs preuves dans le domaine des choses, pour tracer tout droit, dans les sillons
désordonnés de l'histoire, la route du progrès. Comme l'en-
fant grandit, comme Zazie mûrit, plus il y a de science,
plus l'humanité doit irrésistiblement marcher vers le mieux.

Pour les sociétés modernes, marquées de part en part,
jusque dans leur vie quotidienne, par les succès pratiques,
tangibles, incessants des sciences de la nature, le rationalisme joue le rôle d'un institut d'émission qui a toujours
frappé même monnaie : côté pile, la science ; côté face, le
progrès. On peut imaginer la pièce ainsi gravée : sur l'avers,
"République des savants" ; sur le revers, la **corne** d'abon-
dance des découvertes et des innovations. Il y a des hauts
et des bas, des périodes de déflation et d'inflation, avec
une circulation plus ou moins grande de papier monnaie.

Quand il y en a peu en circulation, le cours est au plus
haut, la monnaie inspire confiance, elle est même prestigieuse : équation simple, la science c'est le progrès et le
progrès c'est la science. Ainsi dans l'Europe du XVIIIème
siècle, où la science promettait plus qu'elle ne pouvait tenir et venait à peine de conquérir sa légitimité : inflation. De même aujourd'hui, dans les pays peu développés où
traditions et structures scientifiques sont inexistantes,
la science apparaît comme la voie royale de la modernité,
instrument et garantie du processus accéléré d'industrialisation : surévaluation. En revanche, dans les sociétés in-

dustrialisées : dévaluation.

Un des signes de la modernité est que le thème de la croissance a pris le relais de l'idée de progrès : celui-ci ne se mesure pas, et se révèle, faute d'expression quantitative, une idée de philosophe ou un mythe ; celle-là, au contraire, n'a de sens qu'au niveau des comptabilités nationales dont le volume des biens et des services, et non pas le "mieux-être", donne la mesure.

Une des causes du malaise que vivent nos sociétés depuis les années 60 tient assurément à la prise de conscience des dommages qu'entraîne le processus de croissance tel qu'il a été jusqu'à présent conçu. Ce n'est pas la seule, loin de là. Mais cet exemple est particulièrement significatif du point de vue qui nous occupe ici, et à un double titre. D'abord, il souligne que la croissance n'est pas sans limites, et que les sociétés industrialisées sont tenues de prendre en compte, au bilan des gains et des pertes, les unes tout autant que les autres : dans ce processus de "destruction créatrice" suivant la formule de Schumpeter, on ne peut plus tirer un trait aussi léger que naguère sur ce qui est détruit. Ensuite, cet exemple signale de la part de la société une attitude tout-à-fait nouvelle à l'égard de la science, un véritable retournement.

Dans la mesure même où la science n'est pas étrangère à ces dommages, son image cesse de coïncider avec l'image du progrès. Hier encore, l'activité scientifique pouvait s'épanouir, sans risque de se révéler coupable ou complice de conséquences désastreuses ; aujourd'hui, on n'est pas loin de lui demander de faire à l'avance la preuve de son innocence. Tel est bien le sens profond du <u>technology assessment</u> : contrôle, évaluation de la technologie, sans doute ; mais la formule renvoie, étymologiquement, en français comme en anglais, à l'idée d'un tribunal. Or, à ce tribunal, ce n'est pas seulement la technologie qui est sur la sellette, mais la science elle aussi.

En 1970, quand j'ai publié <u>Science et Politique</u> - dont le sous-titre devait être "Essai sur la situation des scientifiques dans le monde moderne" -, la contestation de l'intérieur de la communauté scientifique commençait à peine. Il n'était pas besoin d'être prophète pour prévoir que, venue des Etats-Unis pour des raisons évidentes, elle s'étendrait vite à l'Europe. De fait, les livres critiques de la science se sont multipliés (9), tout comme les mouvements plus ou moins organisés de scientifiques "contestataires". Mouvements marginaux, à première vue, dont un décompte statistique montrerait qu'ils ne regroupent qu'une très petite fraction de la population des chercheurs. Mais, outre ceux qui participent activement à la <u>critique</u> de l'institution scientifique, le nombre des chercheurs qui se sentent en <u>crise</u> n'a cessé de grandir, comme s'il y avait de moins en moins de chercheurs "heureux" au sens où Michel Serres, s'indignant de l'instinct de mort qui s'est emparé de la science, parle justement du bonheur et du plaisir de la recherche (10).

Quelque chose s'est joué dans l'histoire de l'institution scientifique qui la retient désormais d'invoquer avec la même force de conviction les valeurs non moins que les joies dont elle se réclamait depuis l'origine. On peut dater ce tournant de la deuxième Guerre mondiale en fonction d'événements-symboles tels que le Manhattan Project, la première explosion atomique d'Alamogordo ou celle d'Hiroshima et de Nagasaki, mais on peut tout aussi bien remonter à la première Guerre mondiale dont André Malraux a dit qu'elle révélait pour la première fois, à cause de l'usage des gaz asphyxiants, "du négatif au bilan de la science". Depuis la seconde Guerre mondiale, et désormais dans le domaine civil non moins que dans le domaine militaire, les événements-symboles se sont accru et accéléré qui légitiment d'autant plus le dossier de la critique et de la contestation à dénoncer dans ce tournant une trahison (Minimata, Torrey Canyon, Seveso, Ekofisk, etc.).

Si le Manhattan District Project est l'exemple le plus significatif de l'association délibérée entre la science et le pouvoir - association si étroite que le soutien à grande échelle de l'un doit impliquer les résultats à grande échelle de l'autre -, cet exemple est loin d'être le seul : il illustre une situation qui va se généraliser à la plupart des activités de recherche. L'ère de la "grande science" a commencé qui fait du savoir scientifique un travail organisé, souvent industrialisé, parfois bureaucratisé, en tout cas tributaire d'investissements massifs et par suite de décisions politiques. Mais le même savoir, par la diffusion de ses résultats et l'échelle de ses conséquences, affecte à son tour le théâtre des décisions politiques, et le scientifique devient conseiller, stratège, diplomate, agent parmi d'autres au service de l'Etat, quand il ne passe pour prêtre d'une religion dont le culte est lié au pouvoir. Le mouvement qui, depuis les débuts de la science moderne, a fait de la recherche scientifique une profession sécularisée, débouche avec les politiques de la science sur sa métamorphose en institution, quand ce n'est pas en religion d'Etat.

L'idéologie de la science conçue comme pure aventure de l'esprit, recherche désintéressée, institution autonome dans le système social, apparaît du même coup illusoire, sinon mystificatrice face aux réalités de la pratique de la recherche. Mais, précisément, la situation du scientifique est tout entière inscrite dans cette ambiguïté inévitable qui veut que la science se donne comme une fin en soi, alors qu'elle n'est reconnue ni soutenue qu'en fonction de son instrumentalité. Même la science qu'on dit pure ne peut échapper aux critères d'utilité auxquels se subordonne le système de production : non pas seulement parce que la science pure se voit attribuer des valeurs étrangères à ses fins, mais aussi parce que ses propres valeurs se dissolvent toujours davantage dans les fonctions qu'elle remplit comme technique parmi d'autres.

Pour qui s'interroge sur le sens philosophique de ce qui s'est joué, il est impossible de ne pas se référer aux conférences dans lesquelles Husserl, il y a plus de quarante ans, s'interrogeait déjà sur la "crise des sciences européennes", crise de "l'humanité européenne" comme il disait, où se préparaient tous les événements dont nous avons été et sommes toujours les témoins (11). Textes testament, écrits alors que la barbarie s'étendait déjà sur l'Europe, et qui dénonçaient dans "l'échec apparent du rationalisme" toutes les racines d'une crise dont la seule nouveauté aujourd'hui est précisément qu'elle soit perçue et vécue à leur tour par les scientifiques eux-mêmes.

La crise de la science ne date pas d'aujourd'hui, même si c'est effectivement depuis la fin de la deuxième Guerre mondiale qu'elle est devenue plus aiguë dans le sillage des crises renouvelées que vivent nos sociétés. C'est qu'il y a un rapport étroit entre la science réduite à l'efficacité de ses résultats et la "faillite de l'humanisme" sur lequel Husserl s'interrogeait dès 1935-36. Plus la science est efficace, moins elle apporte de réponse aux questions du sens ou du non-sens de l'existence humaine ; plus elle est "payante", moins elle semble servir l'humanisme. Le constat dressé par Husserl ne s'applique pas moins à l'ère atomique, spatiale ou électronique, qu'à celle qui les a précédées. Et l'on s'en veut de ne pas citer tout ce passage en entier, tant il semble aborder de front la question même que je voulais soulever :

"Nous prendrons notre point de départ dans un renversement qui eut lieu au tournant du siècle dernier dans l'attitude à l'égard des sciences. Ce renversement concerne la façon générale d'_estimer_ les sciences. Il ne vise pas leur scientificité, il vise ce que les sciences, ce que la science en général avait signifié et peut signifier pour l'existence humaine. La façon exclusive dont la vision globale du Monde qui est celle de l'homme moderne s'est laissée, dans la deuxième moitié du XIXème siècle, déterminer et aveugler par les sciences positives et par la "prosperity" qu'on leur devait, signifiait que l'on se détournait avec indifférence des questions qui pour une humanité authentique sont les questions décisives. De simples sciences de faits forment une simple humanité de fait. Ce renversement dans la façon d'estimer publiquement les sciences était en particulier inévitable après la guerre et, comme nous le savons, elle est devenue peu à peu dans les jeunes générations une sorte de sentiment d'hostilité. Dans la détresse de notre vie, - c'est ce que nous entendons partout - cette science n'a rien à nous dire. Les questions qu'elle exclut par principe sont précisément les questions qui sont les plus brûlantes à notre époque malheureuse pour une humanité abandonnée aux bouleversements du destin" (12)...

Assurément, "la crise européenne s'enracine dans l'erreur d'un certain rationalisme" (13), celui qui a mis "hors circuit", pour parler encore le langage de Husserl, "le monde de la vie". Peu importe ici que la solution préconisée par Husserl, le retour au sujet par la phénoménologie transcendantale, se révèle à nos yeux dérisoire pour remonter la pente ou redresser la route. Ce qui importe, au contraire, c'est l'angoisse qu'exprimait ce texte à la veille de la deuxième Guerre mondiale et le refus, aussi, qu'il opposait à la barbarie : les jeux ne lui paraissaient pas joués, il restait une marge d'espoir et d'action pour surmonter ce qu'il appelait "l'échec apparent du rationalisme" (14).

La critique de l'institution scientifique n'est pas un avatar parmi d'autres de la critique sociale (ou de l'autocritique) à laquelle la pensée européenne s'est toujours livrée. Jusqu'au XIXème siècle, c'est-à-dire jusqu'au moment où les sciences sociales se sont constituées en tant que sciences indépendantes et de la philosophie et des sciences de la nature, la critique sociale s'est en fait appuyée sur les acquis intellectuels et pratiques de l'institution scientifique pour dénoncer les injustices et les erreurs de l'ordre existant. Aujourd'hui, ce sont précisément ces acquis qui inclinent à contester l'institution scientifique elle-même, et c'est à ce titre que cette contestation est à la fois la plus révélatrice et la plus contradictoire de la crise européenne : la plus révélatrice, parce que l'institution qui incarne avec le plus d'éclat la réussite de la rationalité européenne tend à se récuser elle-même ; la plus contradictoire, parce qu'on ne voit pas comment cette crise pourra être surmontée si elle engage à tourner le dos aux possibilités d'action que l'institution scientifique, même contestée, ne cesse d'offrir.

En tant que sous-système du système social, la science et la technologie en sont venues à reproduire les contradictions - insatisfactions et incertitudes - de la société dans son ensemble. Nul n'aura l'illusion de penser qu'il suffirait d'agir sur l'institution scientifique et de la réformer d'une manière ou d'une autre pour que les raisons de la crise s'évanouissent. Même si la science et la technologie passent pour les agents les plus efficaces du changement social, ce ne sont jamais les seuls, et c'est encore leur accorder une fonction autonome dans le système social que de ne pas voir que leurs effets sont toujours médiatisés par d'autres agents que les chercheurs, d'autres institutions que les laboratoires.

Nous ne sommes pas loin ici de la politique de la science. Le point que je veux souligner, c'est que la crise de la science reflète manifestement celle de sociétés qui payent un tribut trop lourd à ce que Husserl appelait "l'objectivisme". Et nous ne pouvons plus demeurer ni aveugles ni sourds aux "pertes" résultant du processus scientifique tel qu'il est associé aux objectifs de croissance des sociétés industrialisées. Les liens entre la technologie et la science sont aujourd'hui si étroits que les objectifs scientifiques des chercheurs sont nécessairement influencés par les objectifs technologiques des "décideurs".

Non seulement les chercheurs vont là où sont les crédits, mais encore la nature des champs de la recherche à défricher et les conditions dans lesquelles ils sont creusés sont définies en grande partie par la "demande sociale". Mais parler d'une "demande sociale" est encore trop vague : tout comme la notion de marché renvoie à des intérêts précis, la notion de "demande sociale" renvoie à des options et à des structures de pouvoir données au sein de la société en général et de la communauté scientifique en particulier. La "grande science" du XXème siècle, avec ses instruments et ses "crash-programs" aux budgets géants, n'est pas seulement à l'image des ambitions que se donnent les chercheurs, elle dénonce aussi les choix de sociétés dont les orientations économiques et politiques légitiment et façonnent les orientations de leurs activités de recherche.

L'enjeu véritable de la crise dont témoigne l'institution scientifique, c'est la découverte d'un ou de paradigmes nouveaux non pour la science, mais pour la société. Tel est, me semble-t-il, le sens profond de cette prise de conscience nouvelle : le bilan du progrès scientifique n'est plus nécessairement positif. S'il y a aussi du négatif, il importe que les politiques de la science se préoccupent toujours davantage de répondre à la question suivante : comment contrôler - sans l'asphyxier - un système créatif de la recherche qui implique néanmoins une telle menace pour l'humanité ?

Il serait aisé de démontrer comment "l'objectivisme" dénoncé par Husserl s'est traduit dans les politiques de la science de tous les pays industrialisés, quels que soient leurs opinions, leur idéologie ou leur régime, par des pertes qu'on ne peut plus minimiser ni négliger en postulant que science signifie toujours progrès. Pour les études sur la politique de la science, comme pour les responsables chargés de cette politique, la tâche n'est plus seulement de prêter attention au processus cumulatif des gains ; il s'agit aussi d'éviter ou de réduire les conséquences négatives d'un procès de changement dans lequel les activités scientifiques et technologiques jouent l'un des rôles les plus importants.

NOTES AND REFERENCES

1. B. Hessen, "The Social and Economic Roots of Newton's
   Principia" dans Science at the Crossroad, re-édition
   des contributions soviétiques au deuxième Congrès International d'Histoire de la Science de 1931, Cass,
   Londres, 1971, pp. 151-212. L'ensemble de ces contributions, introduit par Boukharine, eut une influence
   considérable sur les scientifiques et historiens des
   sciences anglais, notamment Bernal et Needham.

2. Jacob Schmookler, "Catastrophe and Utilitarianism in the
   Development of Basic Science", dans Economics of Research
   and Development, édité par R.A. Tybout, Ohio State
   University Press, 1965, pp. 19-33 ; Robert K. Merton,
   en particulier, Science, Technology and Society in
   17th Century England, 1938, re-édité par Harper
   Torchbooks, New York, 1970 et Social Theory and Social
   Structure, Free Press, New York, 1957.

3. Le chapitre consacré par le Père Lenoble aux "Origines
   de la pensée scientifique moderne" constitue une admirable démonstration de ces liens entre révolution scientifique et changements sociaux, dans Histoire de la
   Science, sous la direction de Maurice Daumas, La Pléïade,
   Gallimard, 1957, pp. 370-534.

4. Thomas S. Kuhn, La structure des révolutions scientifiques, Flammarion, Paris, 1970, p. 197. La traduction
   française est si mauvaise - community devient "groupe"
   et puzzle "énigme" - que je préfère revenir à l'original.

5. Idem, p. 183.

6. Idem, p. 198.

7. Thomas S. Kuhn, op.cité, p. 198.

8. Ernst Cassirer, La Philosophie des Lumières, Fayard,
   Paris, 1966, pp. 76-77.

9. En particulier, J.R. Ravetz, Scientific Knowledge and
   its Social Problems, Oxford University Press, 1971. Du
   même auteur, voir "Criticisms of Science", qui montre
   fort bien l'évolution de cette littérature, dans Science,
   Technology and Society, édité par D.J. de Solla Price
   et I. Spiegel-Rösing, Sage, Londres, 1977. Aussi, les
   deux livres édités par Hilary and Steven Rose, The Radi-
   calisation of Science et The Political Economy of
   Science, MacMillan, Londres, 1976.

10. Michel Serres, "Trahison : la thanatocratie", Hermès III - La Traduction, Editions de Minuit, Paris, 1974, pp.73-104. Ainsi : "Le savoir naît heureux. Il se partage, heureux sans se pouvoir diviser, il multiplie, de soi, les fruits de la réjouissance. Il faut n'avoir jamais reçu de piqûre aiguë, délectable d'une solution ou d'une idée, n'avoir jamais évalué de fait son pouvoir thaumaturge, son buissonnement à foison dès lors qu'on la donne, pour se dessécher sur pied, comme tant et tant, parmi un métier qui n'a cependant de rapport exact qu'au rire et à l'éros..." (p. 74).

11. Edmond Husserl, La crise des sciences européennes et la phénoménologie transcendantale, Gallimard, 1976, pp. 61-62.

12. Op.cité, p. 10.

13. Ibid., p. 371.

14. Ibid., p. 382.

J. R. Ravetz

THE SYMBOL OF SCIENCE IN EUROPEAN THOUGHT

My contribution to this Symposium is intended to be an historical
supplement to the analysis given by Jean-Jacques Salomon. I am not
merely giving a scholarly background to the present crisis, interesting
as that might be. Rather, I hope that by using history we might
plumb the depth of this crisis, always in a tentative and qualitative
fashion. For, depending on how deep are its various roots, we may get
some estimate of how radical must be its resolution.

I think that my contribution will also be of interest to colleagues
who come from non-European cultures. For there cannot be a crisis of
European science without a crisis of European culture; the culture has
been symbolized by the science for so long that it now depends on it
for a large share of its content and meaning. Also, at the present
time other powerful cultural institutions are falling into decline, as
for example organized religion and élitist academic education. In
consequence, "science" as a stock of achievements in knowledge and
power, as well as a method and style, must take an increasing share of
the content and meaning of European culture. A crisis of the scientific
part then becomes one of the whole.

THE PROBLEM AS DEFINED HERE

I shall argue that in these philosophers there is a sense of commitment to science, with an optimism that rises continuously, from the
early seventeenth to the late nineteenth century. Then quite suddenly
the lights begin to go out all over Europe, in certainty about science
as much as in political optimism, at the beginning of this century.
And while political democracy seems just now to be a more hardy growth
than many dared hope, a resurgence of scientific optimism has yet to
occur.

It is therefore of some practical concern, to examine this cultural
symbol, science, and to see which aspects are essential to what we now
consider to be its real and vital meaning. Of course we cannot simply
dismember a cultural symbol, keeping the nice parts and discarding the
nasty. But within Europe and even more, outside, conscious decisions
do influence the evolution of this symbol in practical and institutional
terms. And so we do well to become as familiar as possible with this
symbol and its problems as they have developed.

THE EARLY PROPHETS

Since prophecy, rather than experiment, was Francis Bacon's chosen
path, we may well start with him. His Novum Organum is about a method,
and an attitude, deriving more from a Puritan morality than from the
practice of the arts and sciences of his time. Indeed, as I have
argued elsewhere, he saw the reform of natural science as only a part,
though an essential and inseparable part, of the Reform of man that
would usher in the Christian millenium.

Bacon's promises for science come only at the end of Book I of the
Novum Organum; he was careful to lay the groundwork before he made any
claims on behalf of his method. In Aphorism 129 he described a variety

of benefits; for brevity I will mention only some. Discoveries and inventions are actually superior to civil and political deeds, he claims, for they benefit all men for all times, and "Moreover the reformation of a state in civil matters is seldom brought in without violence and confusion; but discoveries carry blessings with them, and confer benefits without causing harm in any way". Then, comparing the condition of peoples in Europe and "the wildest and most barbarous districts of New India", he recalls how "man is a god to man", and explains, "this difference comes not from soil, not from climate, not from race, but from the arts". He cites also the three great inventions, printing, gunpowder and the magnetic compass; such that "no empire, no sect, no star seems to have exerted greater power and influence in human affairs than these discoveries". And finally he lists his three grades of ambition: that for self (which is degenerate), for country (still covetous), and, best and noblest, to "establish and extend the power and dominion of the human race itself over the universe".

A striking feature of the Scientific Revolution is the way that the same themes emerge in the writings of philosophers whose intellectual formation and personal quests were really very different. Descartes was engaged on a very personal search for the Good and the True. Disillusioned with the world of academic learning in a fashion that closely anticipates our present drop-outs, he found certainty in his own reason, operating on a natural world denuded of all real properties except those studied in geometry. Finding intellectual security in those famous "long chains of reasoning, all simple and easy" of the geometers, he eventually broadened his perspectives to include the techniques of the artisans and (in spite of himself) the dreams of the alchemists.

Speaking of his duty to publish his work, (in Part 6 of the <u>Discourse on Method</u>) he utters the classic "Baconian" manifesto.

This form of denigration of the literary disciplines, for the credit of those mathematical, had direct ancestors   (and probably sources) in sixteenth-century philosophical propaganda for mathematical architecture. Galileo also went on to try a theological argument for the perfection of human knowledge of mathematics. Agreeing that God's Knowledge is infinite <u>extensively</u>, and therefore unattainable by man, he argues that man's <u>intensive</u> understanding of particular mathematical truths can actually be perfect. Thus he believes that "its knowledge equals the Divine in objective certainty, for here it (the human mind) succeeds in understanding necessity, beyond which there can be no greater sureness". What Galileo understood by mathematics in this connection is not entirely clear; while he makes no claim in this particular place for conclusions about Nature, his praise of geometry in connection with "disciplined experience and necessary demonstration", as a means of achieving certainty, is well known.

So our philosophical prophets were all sure they possessed a method leading to truth, both in contemplation and in action. What of the Good? Could not evil come from the new science? Bacon was the most acutely aware of the problem, and his answers were not all consistent.

On this problem Galileo is silent, and Descartes produces a solution which, though limited to a personal commitment, is of classic simplicity for a "scientists' Hippocratic oath." He would not be able to work on any project "That can be useful to some only be being harmful to others".

LATER YEARS OF OPTIMISM

The faith in science of later generations could not become more intense concerning the object itself; but it could extend the domain of application of science and of its method. Newton saw the task in what we might call "pre-Enlightenment" terms. At the very conclusion of the Opticks he anticipates a moral benefit from his method of experimental analysis and theoretical synthesis.

Later visions were, of course, more bold. We have Condorcet, while suffering from some errors and distortions of the Revolution he did so much to create, still sure that "all errors in politics and morals, have their basis in errors in philosophy, themselves resulting from errors in natural science".

Those whom we might call the Victorian English philosophes, Huxley, Tyndall and their colleagues, were equally certain of the source of factual truth and intellectual virtue. For them "science" was above all a method, replacing credulity and intellectual indolence in the studies of Nature and of man alike. The Higher Criticism of the Bible was a scientific endeavour parallel to that of the elucidation of the Descent of Man.

Huxley's praise of science in his famous lecture of 1866 on Natural Knowledge, breathes a Victorian optimism that is no less high minded in its attack on the religious style of thought, than was the prevailing Christian consciousness itself.

He speaks of "intellectual ethics" of man, and describes the "moral convictions most fondly held by barbarous and semi-barbarous people. They are the convictions that authority is the soundest basis of belief; that merit attaches to a readiness to believe; that the doubting disposition is a bad one, and scepticism a sin; that when good authority has pronounced what is to be believed, and faith has accepted it, reason has no further duty. There are many excellent persons who yet hold by these principles, and it is not my present business, or intention, to discuss their views. All I wish to bring clearly before your minds is the unquestionable fact, that the improvement of natural knowledge is effected by methods which directly give the lie to all these convictions, and assume the exact reverse of each to be true." And then, listing the intellectual virtues of science, he concludes with the neat adaptation of a theological motto, "The man of science has learned to believe in justification, not by faith, but by verification."

We must keep in mind that Huxley spoke in the context of a period when science was triumphantly vindicating its traditional promises: it was indeed overcoming dogma and superstition, it was correcting old errors and discovering new truths; and it was supplying new powers for the transformation of the material situation of mankind. The problems of what to do when science produced something other than the Good and the True were scarcely recogniseable, at least for those whose basic commitment was to the reality that their science described. And the others, romantics, mystics, eccentrics, survived on the fringe of intellectual respectability.

THE ONSET OF DOUBT

The problem of error in science, perhaps the most sensitive one for its ideological image of itself, has only recently attracted the attention of historians and philosophers of science. For most earlier periods, in modern times, the work of science was largely new construction or the disposal of pre-scientific beliefs. Those occasional erroneous views which emerged from apparently genuine scientific research could be explained away on the basis of incomplete knowledge, or, if necessary, prejudice. Such were the concepts of phlogiston, and caloric, and (at any time) the previous theory of the nature of light. Palpable errors by the masters themselves, were conveniently forgotten. We might recall the classic among these, Lavoisier's insistence on the respirable air that supports combustion by its supposedly characteristic property of acid-formation, thus oxygène, or the German Sauerstoff, a most misleading term that has survived to this day.

The reasons why Einstein's theories caused such a popular stir, a bit more than half a century ago, and were widely interpreted as a refutation of Newton, are assuredly complex and beyond my present concern. Certainly, one may view the history of physics since Descartes as a series of grand research programmes which yielded many viable and lasting positive results in the course of complete failures on their stated metaphysical objectives. But there was a dramatic quality to "relativity", particularly in its being announced to the media at the end of the Great War with all its disillusions.

As Sir Karl Popper tells it, Einstein's announcement of his predictions for the solar eclipse, with the request for a test that could falsify his theory, clearly showed that idealistic young man the difference between genuine science and pseudo-science. This was pre-eminently a moral perception; and so it is no wonder that Popper's ethical message, the call for self-criticism, rings so much more clearly than his epistemological studies on knowledge defined by falsifiability.

It is not usually emphasized that Popper actually rescued the Good in theoretical scientific knowledge only by sacrificing the True. It is not easy to describe the sort of epistemological status that may be possessed by knowledge which is defined by its falsifiability. But we can be certain that such knowledge is likely to be proved erroneous in some way or other at some time, and therefore almost certainly not "True and Necessary" in the sense of Galileo.

Thus it was truly an historic joke that the positivism of the Vienna Circle, basing its destruction of metaphysics on the supposed verifiability of scientific assertions, was obsolete on technical grounds nearly from the time of its conception, and then remained so through the several decades of its influence on philosophers and scientists alike.

Attempts to refine Popper's insights and to make them the basis of explicit criteria of rational choice in science were made, quite heroically, by Imre Lakatos. He participated fully and self-consciously in the great tradition of seeing natural science as the example and guarantor of the intellectual and therefore political virtues. But he was perhaps too wise and too realistic to engage successfully in philosophizing on the grand scale.

Much of Lakatos' later work was in response to the challenge made by Thomas Kuhn, in his Structure of Scientific Revolutions. Now, Kuhn was not seeking a debate with philosophers of science on this ideological plane; in many ways he represents the end of ideology. But, as Lakatos correctly sensed, this made Kuhn's doctrines particularly dangerous. We should recall Huxley, and also Popper, when we consider Kuhn's description of the behaviour of "normal" scientists. Thus we learn that "Normal science, for example, often suppresses fundamental novelties because they are necessarily subversive of its basic commitment."

But anomalies do occur, and "normal science repeatedly goes astray. And when it does (I continue quoting) – when, that is, the profession can no longer evade anomalies that subvert the existing tradition of scientific practice – then begin the extraordinary investigations . . ." that yield a new paradigm. Elsewhere Kuhn speaks of the dogmatism that is necessary for scientific research; and he also questions, rather casually, whether the ideal of Truth is really relevant for science.

Kuhn's may well be a more accurate picture of the behaviour and styles of scientists, ordinary and great, than is Popper's Einstein who calmly asks the world to try to refute his theory. But it then leaves open the question of the relevance of Kuhn's sort of science to the True and the Good. Indeed, Paul Feyerabend was not being totally mischievous when he propounded the following paradox: since organized crime has both puzzle-solving within paradigms and scientific revolutions, how is science different? Kuhn answered that he had never claimed his description to apply uniquely to science; and he thereby emptied his idea of "science" of all its positive meaning as a cultural symbol in European thought.

I hope I have indicated here some evidence for the thesis that studies of the epistemology of science have in the present century not been effective in continuing the tradition of promoting natural science as the bearer of the Good and the True. My sample is biassed, I know; there are other contemporary philosophers who still try to re-establish the certainties of earlier centuries. I apologise to them for my neglect; it is only that to me, those I have cited are the most interesting and significant failures.

TAINTED POWER
As Jean-Jacques Salomon has shown, the idea of "pure science" is really an ideology conceived by professors in quite recent times, and quite inconsistent with the essence of European Science since the scientific revolution.
One important aspect of the Scientific Revolution was that it adopted the aspirations of magic, while rejecting its means and its reality. In the Discours, Descartes quickly dismisses the "false sciences" as the very last and very worst of the bad lot he studied in his youth. Galileo jeered at the alchemists and soothsayers for their deceptions, and also at the astrologers for their silly belief that the moon influences the tides.

Bacon was more sober; being closer to the traditions of popular science and natural magic, he did not dismiss them out of hand. Indeed, his recommendations for an empirical astrology are now being vindicated, with the discovery of the correlation of climatic changes (from solar activity) with alignments of planets and variations of the earth's axis.

But Bacon could satisfy himself on the fundamental implausibility of the occult arts, for they all depended on the production of large effects from small causes. Thus in speaking of alchemy, he says it is "far more probable" that a detailed scientific study of metals could eventually produce gold, "than that a few grains of an elixir should in a few moments of time" be able to do it.

Now here, I submit, is an avenue by which the science of our times has brought itself back towards the categories of magic. Of course a nuclear bomb releases cataclysmic forces; but what of a thimblefull of liquid (petrol or gasoline or benzine), driving a heavy metal wagon for many metres along a road? Further, when information takes on power, using matter only as coding systems of one form or another, the ratio of "effect" to "cause" becomes very large. With the development of biological engineering, where the information is even self-replicating, we are indeed in a new realm of powers and responsibilities. Or perhaps the realm is a very old one, and Norbert Weiner's prophetic warnings about "sorcery" and "simony" must be heeded very seriously indeed.

Those who worry about such problems are not merely eccentrics and "anti-science" prophets. Now the most distinguished and responsible scientists recognise that our technology can create problems that may be beyond human ability to solve: thus the dangers of a "Plutonium economy", in the problem disposal of long-lived wastes and of terrorism, recently identified by Sir Brian Flowers.

Truth and error, good and evil, are now inextricably entwined in any realistic view of science in the later twentieth century. The bright vision of the early prophets, the assured confidence of their successors, are no longer justified or even plausible. As a symbol of European culture, science has become troubled and ambivalent.

CONCLUSION

Where do we go from here? What can be done to repair or replace that bruised symbol of science? This is a question that goes far beyond the bounds of scholarship; it is prophecy, and whatever the history of science might be, it is not that.

But I should try to say something useful, and to do so I will first comment on a distinction that is just now very popular among certain American scientists. Those who promote Recombinant DNA research make a sharp separation between the acquisition of knowledge and its application. The latter lies in the realm of social responsibility, and should be properly controlled. But (they say) to acquire knowledge is a fundamental right and drive in man, and provided that research is conducted safely and ethically, there is no reason in morality, and no possibility in practice, of stopping it.

Regardless of the obvious criticisms of such a doctrine as a rationalisation of the practices of an interest-group, we may consider what it presupposes. There is the implicit assumption that safety questions are straightforward; I happen to disagree. But more seriously, there is the implicit shifting of the burden of proof of dangers, onto the opponent of the research. Since the field of genetic engineering is in its infancy, any such dangers are very speculative; therefore they may be disregarded until such time as they are real. Presumably then the active social conscience of the scientific-industrial establishment will ensure their rapid containment.

The sincerity with which such arguments are advanced by highly intelligent and socially conscious scientists tells us something about their picture of the world. Even though they know that technology can be evil, they retain the faith of Descartes and Galileo that scientific truth is, in itself, harmless and benign.

In this way they (and with them the consensus of scientists every-where) keep to themselves the right to create a "potential force",

leaving to others the responsibility for its eventual control should
it become actual.

I hope that many of my audience now are asking, what is wrong with
that? For that question would show how deeply we are all still
committed to the innocence of knowledge, that we cannot imagine otherwise. But of the three prophets Bacon was wiser; in his famous prayer
he referred to a _lust_ for knowledge, that caused the fall of man, just
as a lust for power caused the fall of the angels. To these he counter-
posed charity, "of which there can be no excess."

So, Bacon knew how deeply evil and good were related; indeed, I
believe that his real solution for the ills of science as he found them
was nothing less than a millenarian reform. But he was the exception;
generally the prophets of science could see only one side, and that was
the good.

It seems to me that the classic simple optimism of science, which
might now be quite a dangerous thing, may relate to a stylistic feature
of European thought. With this speculation, offered for the many non-
European colleagues here, I will conclude my remarks.

A STRATEGY FOR ORGANICAL TECHNOLOGY

In his contribution Prof. Dobrov discussed work now in progress
at the International Institute of Applied Systems Analysis, Vienna.
One part of this is the analysis of technological systems into
a hierarchy of types. These include the "simple system", "process
of activity", "family of technologies" and "technology-as-a-whole".
Different sorts of policy-making techniques are appropriate to the
various levels; distinguishing among them will thereby prevent confusion of analyses and policies.

At the most general level, the criteria for judging technologies
can be described by the acronym WEIMM -
Water (and more generally environment); Energy (production and
utilization); land (use and improvement); Materials (raw and man-
made); and Manpower (employment and productivity). This approach
is to be combined with existing methods for the theory and practice
of "organized technology".

A most significant quantitative characteristic of contemporary
technological innovation was discussed. This is the tendency for
new technologies to be displaced after an ever-decreasing interval
of time. This interval appears to decay exponentially, halving
every twenty years, and is now some 7 years.

Since we cannot have "instant institution" of innovations, we may
either stabilize at a very quick turnover; or we may possibly
experience an increase in substitution time, with its important
implications for the R & D effort.

For solving the increasingly large and complex problems of developing new technologies, it has been realised that traditional
"hardware" and "software" need to be supplemented. The new component is "orgware", which ranges from the systems for servicing and
maintaining equipment, over to interactions with the social-economic-
political structures of the country wherever they affect the
technology.

The task of analysing new technologies has led to many systems in
many nations. In the U.S.S.R. they have developed a "Systems
Assessment of New Technologies" (SANT).

"The problem is to find systematic means of using a selection of
methods which mutually compensate for each other's weaknesses.
This is the same as the problem of creating a reliable system from
relatively unreliable parts. We know of at least one example of
the successful solution of this problem (Homo sapiens!) and this
inspires hope..."

ALTERNATIVE SCIENCES OF THE EAST

Modern European science has shed an ever stronger light over the
natural world, while it has also inevitably created shadows. It has
proven most successful in the mechanico-materialistic aspects of
Nature, while relatively speaking other approaches have been overshadowed, neglected and often put into an inferior status of human
inquiry. With the object of counterbalancing, at least, the overwhelming conventional approach, and adding a fresh dimension in the
established framework of thought, it should be worthwhile to look upon
and examine some of the entirely different, nevertheless sufficiently
rationalistic, approaches to Nature, that have been overshadowed in
the light of modern science. Located in the farthest east from the
centre of activity in the West, China and Japan had been fairly well
isolated up until modern times, say the eighteenth century. In spite
of occasional infiltration they have maintained, in goals as well as
in approach, the major characteristics of traditional science, the
basic framework of which was well formulated in China around Christ's
time. We shall try to pay due attention to their goal of science,
avoiding as much as possible the application of criteria of Western
history of science.

Science of the Changeable and Irregular
Celestial bodies attract people's attention, mainly because of their
changing aspect. Eternity and regularity appear to be dull, trivial,
stereotyped routine, unless they are linked to some specific intellectual
values.

Celestial portents were usually considered bad rather than good. The
people of the ancient world, having no control over natural calamities,
always reacted to any sign of change with wonder, awe and fear. The
portents were, in ancient China, interpreted as the warnings of the
heavens given to the rulers for their misconduct. In order to pacify
the wrath of the heavens, the ruler appointed astrologers to observe
the celestial changes as quickly as possible and to advise him on the
methods of pacifying the angry heavens. The ancient rulers acted
according to the advice of their court astrologers.

One of the ways to escape the fear caused by celestial changes was to
find regularity in them, and thereby turn them into non-sensational,
predictable phenomena. Thus, lunar eclipses, which were earlier considered unpredictable portents, lost significance astrologically as
early as Christ's time, whereas solar eclipses, chiefly because of the
technical difficulty of prediction, continued to be considered as
uncontrollable portents up to medieval times. When a solar eclipse
was predicted by a court astronomer but did not occur, the non-
appearance was often ascribed by the court to the ruler's virtue, and
he was simply happy. Astronomically, however, it created a crisis in
prediction theory and technique. In a case where discrepancy between
a theory and an observation was inexplicably great, astronomers often
labelled it "irregular", and considered that "the moon moved erratically".

It was not the astronomers' fault, but the heavens were beyond control.

This attitude towards Nature offers an easy way out, but it does not provide a solution for the crisis. It actually never generated a crisis between theory and observations, as Chinese astronomers were easy-going and they were satisfied simply to classify phenomena as regular or irregular without analysing deeply the cause of discrepancy. Nevertheless, it has certain merits in its own right.

In Western Platonic tradtion, we take it for granted that there exists regularity in Nature, and thereby exclude any other possibilities of science. A notorious outcome of this conviction was the Western rejection of listing celestial portents such as novae and cometary phenomena that could not be explained by contemporary theory, giving them scant attention as irregular and hence insignificant earthly phenomena not worth noting. On the other hand, the Chinese believed that the ultimate texture of reality was too subtle to be fully measured or comprehended by human intellect. Although their science too assumed regularities to some extent, they showed keener curiosity about the particular and the evanescent. In the "irregular"-oriented East, extraordinary phenomena were keenly observed and carefully classified and recorded. The incomparable mass of carefully dated astrological portents that thus accumulated has proved an indispensable heritage to scientific analysts in modern time.

One might say that in the classical Western tradition there was an urge to put every phenomenon into a single box of regularity; those that could not be assigned to the box of regularity were rejected. In the Eastern tradition, in addition to the box in which all the regular pieces were assembled, another box to hold all the irregularities was also provided. In practice, the Chinese provided two chapters for their treatment of celestial phenomena in successive official dynastic histories, i.e. a "calendrical science" chapter for regularities and an "astrology" chapter for irregularities, both of equal status.

If science is defined, as Europeans conventionally do, as the pursuit of natural regularities, then the Eastern tradition is bound to appear weak because it lacked analytical rigour and theoretical consistency. Judged in its own terms, however, there is some merit in the Chinese relatively catholic and unprejudiced interest in everything that happened in Nature.

This double-box approach may be seen in the writings of a Japanese astronomer, Shibukawa Haruni (1639-1715). In his own work of mathematical astronomy Shibukawa remained thoroughly positivistic, and he also left a somewhat problematic astrological treatise (1698). In this work he often expressed the scepticism towards astrological interpretations that one might expect of a practical astronomer.

Shibukawa believed that a professional astronomer must be thoroughly competent in both the major branches of celestial studies, calendrical science and portent astrology. His calendar reform provided a box for regularities. It was no less important to furnish the means by which astrological portents might be classified. He was convinced that the heavens could not be fully comprehended only through mathematical regularity. The sky was a unity of such depth that the tools of no single discipline could plumb it. Although he found astrological interpretations to be often equivocal, the vast historical accumulation of omen records suggested that it had to be taken seriously. There must have been, he thought, justified passion and reason behind that tireless activities of the ancients. His keen sense for the changing nature of the heavens was expressed in a form more distinct than can be found among his Chinese contemporaries. Criticising his contemporary Chinese astronomers who, from their astronomical points of view, distrusted the dates in the oldest Chinese chronicle attributed to Confucius he stated: "This error is due to their commitment to regularity-based mathematical astronomy, so that they do not admit that extraordinary celestial change happened in the sky ... Celestial changes do in fact take place in the heavens. We should not doubt the authority of Confucius' sacred writing brush".

Admitting that regular motion was too limited an assumption, Shibukawa could easily conceive, for example, that astronomical parameters could vary from century to century. In the official Chinese calendar of the thirteenth century, the discrepancy between ancient records and recent observations was explained by a secular variation in tropical year length. Shibukawa revived this variation in his Japanese calendar reform (c. 1690) in spite of his Chinese contemporaries' rejection.

The variation terms used in Chinese and Japanese astronomy were too large to survive empirical testing, and were eventually discarded. Whenever the Aristotelian notion of an unalterable universe was followed rigorously, any change in the celestial course was simply inconceivable. In the West, the first systematic study of variations in basic astronomical parameters was delayed until the time of Simon Laplace (late eighteenth century). It is significant for the history of ideas that in China and Japan there was no reason to resist such variations.

Shibukawa's concern for celestial change and irregularity was backed by Ogin Sorai (1666-1728), the most influential person among all the Japanese Confucian philosophers. Commenting on the variation of astronomical parameters, he stated that since the heavens are imbued with vital force, the length of the year can change freely and constancy is not to be expected in the sky. He said that "Sky and earth, sun and moon, are all living bodies". Indeed, only a dead universe could be governed by law and regularity. As Ogiu was interested only in the vital aspects of nature, he always remained an agnostic in physical cosmology. His anarchic and dynamic cosmology was bathed in historicism. "All scholarship should finally converge in historical studies", said Ogiu.

Looking back at the main current of academic traditions, while the West remained centered upon philosophical and logical enquiries in the Platonic and Aristotelian traditions, Eastern scholarship definitely inclined to history, with the Shi-chi (c. 100 BC) of Ssu-ma Ch'ien, the Imperial astrologer-historian, as the prototype.

The energetic approach to the human body

Before the introduction of Western anatomy, the Chinese and Japanese
did not believe that thought takes place in the head. The brain was
considered to be merely mud, and there was no specific function
assigned to it. When asked about the seat of the mind, they might have
preferred to locate it in the heart or chest. It is well to remember
that the interrelation between brain and mental process was not
established on the history of science; it was the history of a belief
rather than a scientifically proven fact. If there had been any differences
in the way of thinking between those who believed thinking took place in
the head and those who believed it took place in the chest, this would
be a very challenging theme for the history of ideas. Suffice it to say
here that the former tends to be analytical, as mind is separated from
heart, while the latter is akin to organismic thinking in which mind
and body are inseparably integrated.

What seemed to be more important, here, was that the Chinese did not
seriously try to locate thinking power in any particular place in the
body. In fact, Confucian treatises lacked vocabulary to describe mental
process. As it was concerned solely with mundane affairs of socio-
ethical nature, the Confucian tradition had nothing introspective to
yield elaborate epistemological terms. Such terms were often borrowed
from the vocabulary of the Buddhists, whose goal was personal enlightenment.

In Confucian tradition, there was no Cartesian dualism. Even in the
speculative school of Neo-Confucianism, the characteristic approach of
Ko-wu (the investigation of things) was not built on the dualism of
self vis-a-vis the outer world but had the pan-naturalistic concept
that there was li (an organising or harmonising principle) behind everything in the universe, while the phenomenological world was explained
in terms of the most fundamental protyle, ch'i, which spreads throughout
the universe and also permeates the human body.

In order to permeate the universe, ch'i must be modelled on a fluid,
most possibly a gas. Dynamical ch'i moves and functions. In the Eastern
way of looking at Nature, dynamical fluid is always more appreciated
than dull immobile material solid. Spiritual and evanescent fluid that
permeates indefinitely and infinitely is more essential than finite solid
matter limited by form and shape. The latter is merely a contamination
of the former. The main application of the ch'i concept was made in
the medical field. The fundamental of Chinese pathology was to maintain
the harmony of ch'i in microcosm in accordance with the ch'i in macrocosm,
and thus health would be preserved. Once this harmony is upset, the
health would be damaged.

In the history of Western medicine, there is a well-known rivalry between
solidists and humoralists. While the former tried to localise the seat
of disease in a solid part of the body, such as the stomach or the brain,
and gave importance to anatomical research, the latter believed in the
significance of maintaining the balance of all humour circulating all
over the human body. The Eastern physiology and pathology are closer
to humouralist theory in its claim of the balance of ch'i. The Chinese
never tried to localise the cause at a particular organ, and no

physiological function was associated to solid matter.

Although there seems to be a certain corporeal notion attached to ch'i, either gaseous or liquid, ch'i can be translated into incorporeal and invisible "energy" as often maintained by Western sinologists, and it works out nicely, especially in medical usage. Thus, the Chinese ch'i pathology may be called a system of energetics that explains everything in terms of invisible energy, as opposed to Western traditions of solidists and atomists.

From the energetics point of view, a heart causing a pulse beat is much more important than a brain, a mere container of inactive muds. It is also clearly understandable, in this connection, that in the actual practice of Eastern medicine, pulse diagnosis is more important than the conditions of internal organs, and also that the holistic function of the entire body is more important than the anatomical view of localised organs. This viewpoint has led to a logical conclusion that the maintenance of harmony of the entire body is more essential for medical care than intensive treatment of the specific affected parts. In view of the Chinese way of non-localised thinking, and also in view of the circulation of dynamic ch'i, it was reasonable enough to find the correlations between the heart and the ears and between the liver and the eyes - the action-at-distance - as appeared in the celebrated medical classic, the Nei-ching.

The anatomical charts passed down from ancient times in the East were quite simple and crude. One might wonder why the Chinese, who had great empirical knowledge of medical treatment, had been satisfied with them. It is because the Chinese chart had an entirely different purpose from that of the Western counterpart. Since the Chinese had no solidist ideas and were interested rather in the general system of physical function, they made a general layout omitting insignificant details, like a diagram of an electric circuit. Hence even when con-fronted with elaborated Western anatomical charts, their conceptual scheme could not be jeopardised.

It is interesting to note, in this connection, a criticism of Sano Antei, an eighteenth-century Japanese physician, on anatomy. In his Hi Zoshi (A refutation of the anatomical charts, 1760) he said, "what internal organs really signify is not a matter of morphology; they are the containers to store vital energy with various functions. Lacking that energy, the internal organs become only emptied containers". In other words, what characterises the internal organs is not their morphology but the difference in their functions defined by the energy they store. Nothing can be learned by dissecting a cadaver, since it lacks this vital energy.

In the Chinese natural philosophy of energetics, the operation and function of invisible ch'i circulating around the universe was con-sidered to be most essential of all, and morphological observation of figure and appearance was thought to be superficial and inferior in value. Observation per se was merely a childish way of investigating things. "The observation of too obvious facts is", said Sano, "of much less value than groping speculation ... even a child is as good an observer as an adult". A scholar who refrains from tracing speculatively the connections between form and function is no better than a child.

One merit of the ch'i physiology and pathology may be the emphasis on the meteorological and climatic factors in diagnosis. As ch'i is basically the cosmic protyle, it is responsible for bringing up the linkage between macrocosm and microcosm.

Non-localising, energetic concerns of Chinese medicine precluded the naming of diseases associated with specific organs. Instead, the emphasis was made on the phase of energetic dynamism; that is to say, even though the apparent symptoms or syndromes were similar, different diagnoses and treatments were applied according to the upward or downward phase of the disease trend. Doctors were, like a strategist, always concerned with the disposition of the frontier in the war against disease.

The Playful World of Mathematicians
There is a belief that disinterested pursuit was something characteristic of Western scientific tradition, but only a brief look at the behaviour of Japanese mathematicians of the Tokugawa period (17th to 19th century) immediately disproves the above thesis. Mathematicians in Japan formed private groups and gathered together purely for mathematical activity. While in such a group as the Royal Society of London the activities of the amateurs gradually declined, Japanese groups expanded up to the nineteenth century, recruiting members from the increasingly lower strata of society. The secret of the vigour of mathematical activity lies in its enjoyable leisurely nature, like crossword puzzles. In fact, modern historians of Japanese mathematics have commonly observed that traditional Japanese mathematics is more of an art form than a field of scholarly inquiry.

I am not attempting here to define scholarship and art but merely trying to say that scholarship is some sort of publicly admitted function and legitimate activity and art is a kind of private indulgence.

As an example of ideological legitimation, Confucian scholars were often asked to write prefaces to the mathematical texts to show that contemporary people should respect mathematics simply because of its high status in antiquity. But the mathematicians of the Tokugawa period themselves were entirely emancipated from classics-oriented authoritarian Confucianism, being socially marginal curiosity seekers outside the Confucian regime, and thus they did not care whether mathematics had been an old art or not. Although applied mathematics such as calendrical science and land-surveying had firm positions within the Tokugawa bureaucracy, mathematics itself was not formally recognised in the governmental structure and its tradition existed entirely in the private sector.

The popular image of mathematics was that of the abacus, but daily use of the abacus did not require anything like elaborate mathematical technique. In merchant families, studying mathematics to a higher degree in preference to business was generally forbidden. Thus isolated, mathematics did not establish itself as a socially respectable vocation. The status of a mathematics teacher had no economic implications. There were mathematicians who travelled from one town to another and received support from country amateur groups, but their living was not different from travelling tumblers.

Generally mathematicians had their greatest achievements in the early years of their life, but the mathematicians of the Tokugawa period were most active in the later part of their life. This was mainly due to the fact that the time and money required to assemble the necessary books came only with the more leisurely pace of life at old age. In particular, civil servants (samurai) had scruples about participating in such activities and for the most part only published mathematical works after retiring from office.

With the lack of ideological, as well as social and practical legitimation, mathematicians could not defend the scholarly aspects of their work effectively. Nevertheless, they showed greater signs of activity in certain respects than the Royal Society did.

The artistic character of Japanese mathematics helps to explain this fact. In a publisher's catalogue of the Tokugawa period, mathematical books were classified in the same category as books on tea ceremony and flower arrangement, which indicated that mathematics was mostly viewed as a popular art. During the Tokugawa period there were several pursuits that were not scholarly occupations and had no academic prestige, games like go and Japanese chess; it was probably these activities that, in conjunction with haiku and waka (short poetry), encouraged the development of Japanese mathematics.

What was it then that distinguished Japanese mathematics from poetry? It definitely came closer to the discipline of modern science than any other field of inquiry existing during the Tokugawa period, for it was an activity following a well-established paradigm in Kuhn's sense.

Japanese mathematics developed, in its formative period, a unique system of posing mathematical problems that was called idai (bequeathal questions). A mathematician would pose scores of problems of several kinds in the latter part of his treatise and then publish it. Another mathematician would supply answers to these problems and present his own in the same manner. According to convention, a third mathematician would supply answers to the second set of problems and issue his own in relay fashion. This interest in mathematical puzzles greatly stimulated the formation of mathematical-connoisseur groups. As idai passed on among the enthusiastic puzzle-solvers, they developed a trend towards problems of a purely intellectual or entertainment character uninhibited by utilitarian constraints. But as problems became more complex, impossible problems appeared and mathematicians began to spend a greater part of their energy with but little effect. At this time of confusion, Seki Takakazu (1642-1708) appeared on the scene. What really gives Seki his enormous reputation, in consequence, was his creation of the basic paradigm both for posing questions and for answering them in an intellectual setting where almost total confusion had prevailed earlier.

The artistic spirit of Japanese mathematicians seems to embody the essential ethos of mathematics; and in contrast one suspects that the

Western institution under Platonic tradition gave mathematics an excessively authoritarian aura that may not be essential to its nature.

During the nineteenth century, abstract mathematics in Europe tried to detach itself from various practical applications. We see a typical example of this in the career of Karl Weierstrass. Rigorous questioning of the logical basis of mathematics was the effective way to elevate its status from the handmaiden of physics to the queen of the sciences. We may be assured that students who decided to take mathematics at the university and make their career in it were grateful to Weierstrass, and evaluated his work highly. The emergence of scholarly elitism, especially that of German-type universities must have reinforced the legitimacy of the quest for fundamentals. Since no tradition of university scholarship existed in Tokugawa Japan, it is unlikely that any Japanese mathematics supporters would have rejoiced at the approach of Weierstrass. Only if a student was being trained as an expert at a university would he be likely to tolerate any questioning of fundamentals.

It would appear, therefore, that the difference between Western mathematics and Japanese mathematics was due to the difference in patronage: the former was recognised and given legitimacy in a university system while the latter was appreciated only by private citizens as a significant art form.

Final Remarks
Our present value-free attitude towards Nature was actually produced in the nineteenth-century German university. Furthermore, as the disciplines separated out of the ancient unity of science, each professional learned to view Nature from the viewpoint of his own field. The fragmented conceptions of and assumptions about Nature, centred in academic specialities and heavily coloured by their prejudices, will be referred to as professional views of Nature.

Prior to the nineteenth century, intellectual activity pursued goals whose scope transcended the pettiness of modern value-free activity. In earlier times, science in the West was pursued on the assumption that its investigations would demonstrate the glory of God, whereas in the East the ideology of science was derived from Confucianism, which emphasises individual moral cultivation and social pacification. Morality was the basis of law; laws of nature conformed to and were necessarily subordinate to it. Thus, astronomy and medicine had to subordinate themselves to an essentially Confucian order of priorities in order to guarantee respect for their status as disciplines. Value-free pursuits like Japanese mathematics, which diverged from moral values or had nothing to do with them, became isolated from the Confucian framework. In its liberation from moral values Japanese mathematics was perhaps the closest to modern science of all disciplines in the traditional East.

Modern science does not have such grandiose and far-reaching objectives as alchemy, which tried to prolong life indefinitely, or astrology, which sought to predict the future course of nature and human affairs. It rather tries to achieve such objectives as are immediately possible.

Modern scientists, unlike Shibukawa Haruni, avoid using historical
changes in the celestial movements and tried to restrain astronomy
within the framework of a mechanical clock. In coping with disease they
emphasise solidistic explanations rather than presupposing a relationship between the human body and the vicissitudes of time, as in the
Heaven-Earth-Nature conception of Chinese medicine. It has analysed
the phenomena objectively and dispassionately and rigorously excluded
value-laden admixtures from its purview. But modern scientists today
may not fully explain why they are pursuing the set conventional
objectives.

Joseph Needham has stressed China's worldwide priority in the history of
discovery and invention. There is sufficient basis for his claims of
priority; but when he compares and evaluates the past scientific and
technical contributions of China with those of the West, he uses
proximity to "today's standard" as the basis for evaluation. "Today's
standard", however, is quite simply a Westerner's standard of evaluation.
It may well be that to persuade ignorant Westerners of China's great
contributions, Needham had no choice but to use this standard. But
his plea for comparison or "titulation" seems to imply that Chinese
and Western science had the same objective and universal goal to be
synthetised in future.

There is a serious question, however, as to whether this was actually
the case. Would Chinese science or that of Japan in the Tokugawa period
have developed in the same direction as Western science in the absence
of influence on them from the latter? Needham thinks that both were
groping toward the same point of sublation, but can we reasonably
assume the existence of such a point a priori? I do not believe it
exists. It seems to me that the two are actually diverging in different
directions. The intellectual framework differed from one culture to the
other, and for that reason it would seem reasonable for the goals of
science to have differed as well. One need not, from this point of view,
be concerned with the problem of priority since even if Eastern science
is set over against that of the West, the two were not following
parallel courses of development toward some fixed, predetermined end.

It might be that the Easterner created the science of change and
irregularity on the basis of their acute consciousness of historicism.
They might, as well, have created the science of energetics, the field
of continuous ch'i. And the whole science might have been created in
a joyful world in which laymen can participate.

R. RASHED

LA NOTION DE SCIENCE OCCIDENTALE

La science classique est européenne et ses origines sont
directement lisibles dans la science et dans la philosophie
grecques: cette doctrine - une fois n'est pas coutume en
histoire de la philosophie et de la science - a survécu
intégralement à tous les conflits d'interprétation, pourtant multiples au cours des deux derniers siècles. Les philosophes, sans exception ou presque, l'ont admise comme
postulat destiné à caractériser globalement la Raison Classique. Kant aussi bien que Comte, les néo-Kantiens aussi
bien que les néo-positivistes, Hegel aussi bien que Husserl,
les hegeliens et les phénoménologues aussi bien que les
marxistes, tous reconnaissent en ce postulat la base de
leurs interprétations de la modernité classique. On cite,
jusqu'à nos jours encore, les noms de Bacon, Descartes et
Galilée, dont on retranche parfois le premier, auxquels,
selon les cas, on en ajoute bien d'autres, comme autant
d'étapes dans la reprise d'une marche interrompue par les
siècles de décadence, comme autant de marques sur le chemin
d'un retour révolutionnaire à la science et à la philosophie
grecques. Retour que tous ont compris à la fois comme la
recherche d'un modèle et la redécouverte d'un idéal, ainsi
qu'en témoignent les métaphores platonicienne et archimé-
dienne par lesquelles un Brunschvicg et un Koyré caracté-
risent les modes d'existence de la science classique. On
pourrait imputer cette unanimité des philosophes à leur
démarche de dépassement des données historiques immédiates,
à leur souci de radicalité, à leur effort pour saisir, selon la description de Husserl, "le phénomène originel (Ur-
phänomen) qui caractérise l'Europe au point de vue spiri-
tuel"; et, par conséquent, on pourrait s'attendre que la
situation de ceux qui sont aux prises directes avec les
faits de l'histoire des sciences soit tout autre. Or il n'en
est rien: ce même postulat, les historiens des sciences l'a-
doptent comme point de départ de leurs travaux et, surtout,
de leurs interprétations. A cet égard, entre Poggendorff,
Rosenberg, Dühring, Gerland d'une part et Duhem de l'autre,
en histoire de la physique, les différences sont infimes;
de même qu'entre Tannery, Cantor et Bourbaki en histoire
des mathématiques. Qu'ils fassent, donc, de l'avènement de
la science classique le produit d'une rupture avec le Moyen-
Age, qu'ils défendent, à l'opposé, la thèse d'une continuité
sans rupture ni coupure, qu'ils adoptent - c'est le cas le
plus fréquent - une position éclectique, la majorité des
historiens s'accordent pour reconnaître plus ou moins implicitement le même postulat.

Aujourd'hui, en dépit des travaux de Woepcke, Suter, Wiedmann, Luckey ..., pour l'histoire de la science arabe, et, tout récemment, du Dictionary of Scientific Biography, en dépit de l'oeuvre de Needham pour l'histoire de la science chinoise, les travaux effectifs des historiens reposent sur une conception fondamentalement identique. Bien plus: alors que, depuis peu, le concept même d'histoire des sciences et les méthodes de la discipline sont devenus objets de controverse et de critique, internalistes et externalistes, continuistes et discontinuistes, sociologues de la science et analystes de concepts conviennent tacitement pour laisser la précédente doctrine hors du débat et, par suite, à l'abri du doute. On retrouve donc la même représentation: dans sa modernité aussi bien que dans son historicité, la science classique apparaît finalement comme l'oeuvre de la seule humanité européenne; plus encore, c'est par elle essentiellement que l'on définit cette humanité. Seules les activités scientifiques de l'humanité européenne, en effet, sont objet d'histoire. Il arrive, il est vrai, que l'on reconnaisse aux autres humanités une certaine pratique scientifique; celle-ci demeure néanmoins en dehors de l'histoire, ou bien y est intégrée au seul titre de ses apports aux sciences essentiellement européennes; ces apports n'étant que des suppléments techniques qui ne modifient nullement la configuration intellectuelle ou l'esprit de ces dernières. L'image donnée de la science arabe constitue une excellente illustration de cette démarche: il s'agit d'un conservatoire du patrimoine grec, transmis tel quel, ou enrichi de quelques innovations techniques, aux héritiers légitimes de la science antique. Dans tous les cas, mal intégrée à l'histoire des sciences, l'activité scientifique extérieure à l'Europe fait l'objet d'une certaine ethnographie de la science dont la traduction universitaire n'est autre que l'orientalisme.

I

La notion de science européenne est déjà présente dans les travaux des historiens et des philosophes du XVIIIe siècle. Elle assure alors deux fonctions différentes, mais qui ne sont pas sans lien: moyen de définir la modernité dans un débat dogmatique qui se prolonge au XVIIIe siècle, elle est aussi un élément constitutif d'une diachronie naïve dont les visées demeurent polémiques et critiques. Dans le débat sur les Anciens et les Modernes engagé auparavant, savants et philosophes, en effet, se référaient pour définir la modernité à la science où on combine raisonnement et expérience: la préface du Traité du Vide et, d'une certaine manière, De la Recherche de la Vérité, tentaient ainsi de démontrer la supériorité des modernes dès le XVIIe siècle (1). L'induction historique, ou prétendue telle, entendait donner à ce débat dogmatique sa détermination concrète, et rendre ainsi contraignante la supériorité des Modernes. C'est d'ailleurs là une des raisons, et non des moindres, pour lesquelles l'histoire des sciences fut introduite sur la scène du XVIIIe siècle. Mais déjà on assimilait l'Occident à l'Europe, déjà on opposait à la "sagesse orientale" la philosophie naturelle de l'Occident post-newtonien: tel Montesquieu dans les Lettres Persanes (1721) (2).

Outre ce rôle critique et polémique dans un débat continu et à rebondissements, la notion de science occidentale assumait alors une fonction dans l'élaboration de l'histoire comme diachronie de l'esprit humain. Elle intervenait pour marquer une étape de son mouvement progressif, mouvement réglé à la fois par un ordre cumulatif et par une perte continue des erreurs acquises. Telle est, schématiquement décrite, la représentation d'un Fontenelle, d'un d'Alembert, ou d'un Condorcet. Lorsque ce dernier par exemple, comme bien d'autres ensuite, désigne la modernité en avançant les noms de Bacon, de Galilée et de Descartes, c'est pour marquer le passage de la Huitième Epoque à la Neuvième Epoque du Tableau historique (3) d'une humanité dont l'avenir se confond avec un devenir indéfini des Lumières. La science classique est européenne et occidentale dans la seule mesure où elle représente une étape dans la succession continue et normée d'une seule et même individualité: l'humanité. Pour un Fontenelle, un d'Alembert ou un Condorcet, il serait donc absurde de lire dans les seules science et philosophie grecques les origines de la science classique, dont la qualification d'Européenne ne renvoie à aucune anthropologie, mais simplement à la coïncidence d'une histoire empirique et d'une histoire idéale, vérité de la première. De cette conception, le Discours Préliminaire de l'Abbé Bossut à l'Encyclopédie Méthodique offre une illustration, limitée, certes, à l'histoire des sciences. Le postulat de départ de ce tableau historique du progrès des sciences exactes, divisé en trois périodes, où se mêlent conjectures, prétendus faits, et faits, est que "tous les peuples considérables de l'Ancien Monde ont aimé et cultivé les Mathématiques. Les plus distingués en ce genre sont les Chaldéens, les Egyptiens, les Chinois, les Indiens, les Grecs, les Romains, les Arabes etc.; dans les temps modernes, les nations occidentales d'Europe" (4). La science classique est européenne et occidentale parce que, écrit l'Abbé Bossut, "les progrès que les nations occidentales de l'Europe ont faits dans les sciences, depuis le XVIe siècle jusqu'à nos jours, effacent tellement ceux des autres peuples" (5).

Ainsi formulée au XVIIIe siècle, la notion de science occidentale changea de nature et d'extension au tournant du XIXe siècle. En un mot, avec ce qu'au siècle dernier Edgar Quinet appelait "Renaissance orientale"(6), c'est-à-dire l'orientalisme, la notion se trouvait complétée de la dimension anthropologique qui jusqu'alors lui manquait. Cette renaissance orientale a en effet fini par jeter sur la science en Orient un discrédit auquel l'histoire par les langues donnait un support prétendûment scientifique.

S'il est vrai que la conception du XVIIIe siècle survivait encore çà et là, et particulièrement chez les historiens de l'astronomie, dès les premières années du XIXe

siècle les matériaux et les notions de l'orientalisme ont
le plus contribué à la constitution des thèmes historiques
des différentes philosophies. En Allemagne aussi bien qu'en
France en effet, les philosophes de tendances diverses mi-
sent sur l'orientalisme, pour des motifs différents, certes,
mais selon une représentation identique: l'Orient et l'Occident ne s'opposent pas comme lieux géographiques, mais comme
positivités historiques; cette opposition ne se limite pas
à une période de l'histoire, mais renvoie pour ainsi dire
à l'essence de chaque terme. On peut rappeler à cet égard
les Leçons sur la Philosophie de l'Histoire, ainsi que d'autres ouvrages de Hegel (7), comme on peut également évoquer
le du Pape (8) de Joseph de Maistre. A cette époque encore,
comme on le voit avec le philosophe de la Restauration aussi
bien que plus tard chez les Saint-Simoniens, apparaissent
les thèmes d'"appel de l'Orient", de "Retour à l'Orient",
qui traduisent la réaction contre la science et plus généralement contre le Rationalisme. Mais c'est avec l'avènement et l'essor de l'école philologique allemande que l'on
a cru doter la notion de science occidentale du support
scientifique, et non plus seulement philosophique, qui jus-
que-là lui faisait défaut.

On connaît, en effet, l'importance de cette école pour
l'ensemble des disciplines historiques; on sait moins bien,
pour l'heure, comment s'est opérée son influence sur l'histoire des sciences; tout indique cependant que cette influence ne fut pas seulement directe, mais indirecte, grâce à
l'extension de cette école en mythologie et en études reli-
gieuses. En tout cas, dès le départ, les travaux de Friedrich von Schlegel (9) et de F.Bopp, notamment, ont placé
l'historien dans une situation nouvelle: son objet constitue désormais une totalité irréductible, quant au mode
d'être et à la nature de ses éléments; sa méthode lui impose à présent de comparer entre totalités analogues quant à
leurs structures et à la fonction qu'elles assurent.

Le but déclaré d'Ernest Renan consistait à faire "pour
les langues sémitiques ce que Bopp a fait pour les langues
indo-européennes" (10). Sa tâche revenait en réalité à
mettre à profit toute la littérature consacrée à la philo-
logie et à la mythologie comparées, pour parvenir à une
description pour ainsi dire éidétique de l'esprit sémitique
et de ses manifestations dans l'histoire. Or, pour Renan
comme pour Lassen (11), Aryens et Sémites se partagent, à
eux seuls, la civilisation: l'historien n'a donc plus qu'à
apprécier d'une manière différentielle et comparative
leurs contributions respectives. C'est donc désormais la
notion de race qui constitue le support de l'historiogra-
phie. Mais par "race", on entendait simplement l'ensemble
des aptitudes et des instincts reconnaissables seulement
grâce à la linguistique et à l'histoire des religions (12).
C'est donc en dernière analyse pour des raisons qui tien-
nent aux langues sémitiques que les Sémites, contrairement
aux Indo-Européens, n'avaient, et ne pouvaient avoir, ni
philosophie, ni science. "La race sémitique, écrit Renan,

48

ne se reconnaît presqu'exclusivement qu'à des caractères
négatifs: elle n'a ni mythologie, ni épopée, ni science,
ni philosophie, ni fiction, ni art plastique, ni vie ci-
vile" (13). Les Aryens, quelle que soit leur origine, défi-
nissent à la fois l'Occident et l'Europe. Dans un tel contexte, Renan, qui par ailleurs a combattu tous les miracles, en a quand même conservé un: le "Miracle Grec" (14).
Quant à la science arabe, elle est, écrit Renan, "un
reflet de la Grèce, combiné avec des influences de la
Perse et de l'Inde" (15); en bref, la science arabe est un
reflet aryen.

Les conséquences de cette notion de science occidentale
étaient élaborées au XVIIIe siècle comme élément d'une
simple diachronie, fondée au XIXe siècle sur une anthropologie. S'il arrive que leurs origines soient oubliées, ces
conséquences demeurent encore vivaces dans les travaux des
philosophes et des historiens, et particulièrement lorsqu'il s'agit de la science classique. A cette idéologie,
nous n'allons pas en opposer une autre. Nous nous proposons
simplement de confronter certains de ses éléments aux
faits de l'histoire des sciences, en commençant par l'algèbre pour conclure sur le problème crucial des rapports
entre mathématiques et expérimentation.

II
Pas plus que les autres sciences arabes l'algèbre n'a
échappé aux précédentes caractérisations: visées pratiques,
allure calculatoire, absence d'exigence de rigueur. C'est
précisément ce qui permettait à Tannery d'écrire que cette
algèbre n'est pas parvenue au niveau atteint par Diophante.
C'est aussi, semble-t-il, ce qui, encore récemment, a auto-
risé Bourbaki à exclure de fait la période arabe lorsqu'il
retrace l'évolution de l'algèbre. Il va de soi que nous n'a-
borderons pas ici la discussion de thèses litigieuses - et
pour nous erronées - telles que l'existence d'une théorie
algébrique dans les Arithmétiques de Diophante, ou encore
l'existence d'une algèbre géométrique, reconnue comme telle,
chez les Grecs; nous limiterons donc notre étude à la question de l'occidentalité de l'algèbre classique. N'a-t-on
pas souvent affirmé, depuis Condorcet et Montucla, jusqu'à
Bourbaki, en passant par Nesselman, Zeuthen, Tannery, Klein,
pour ne citer que quelques noms, que l'algèbre classique
est l'oeuvre de l'Ecole Italienne, achevée par Viète et
par Descartes? Ne continue-t-on pas, hier Milhaud, Dieudonné
aujourd'hui, à faire remonter à Descartes le commencement
de l'histoire de la géométrie algébrique (16)? La rédaction
historique du mathématicien est, à cet égard, significative:
entre la préhistoire grecque de la géométrie algébrique et
Descartes, Dieudonné ne trouve qu'un vide qui, loin de faire
peur, est idéologiquement rassurant. En dehors de ces cas
exemplaires, Bourbaki et Dieudonné par exemple, il arrive,
certes, que des historiens citent al - Khwārizmī, sa définition de l'algèbre et sa résolution de l'équation quadrati-
que, mais c'est généralement pour réduire l'algèbre arabe

à son initiateur. Or, cette restriction est grave et fait
tort à l'histoire de l'algèbre. Celle-ci en effet ne se
présente pas comme une simple extension de l'algèbre d'al-
Khwārizmī, mais principalement comme une tentative de dépassement théorique et technique de cette dernière. Plus
encore, ce dépassement n'est pas la résultante d'une somme
d'oeuvres individuelles, mais pour ainsi dire l'effet de
véritable traditions, à l'époque actives. La première de
ces traditions avait conçu le projet précis d'arithmétiser
l'algèbre héritée d'al-Khwārizmī et de ses successeurs im-
médiats; la seconde, pour surmonter l'obstacle de la résolution par radicaux des équations du 3ème et du 4ème degré, avait dans une première étape, formulé pour la première fois une théorie géométrique des équations, pour changer
ensuite de point de vue et étudier les courbes connues au
moyen de leurs équations, autrement dit pour engager explicitement les premières recherches en géométrie algébrique.
Le schéma traditionnel de l'histoire de l'algèbre ne serait
donc, dans ces conditions, qu'un simple mythe historique;
la preuve en sera fournie par le rappel de quelques faits.

La première tradition, nous l'avons dit, avait pour projet d'arithmétiser l'algèbre héritée. Ce programme théorique
fut inauguré à la fin du Xe siècle par al-Karajī, et se
trouve ainsi résumé par l'un de ses successeurs, as-Samaw'
al (1176): "Opérer sur les inconnues comme les arithméticiens opèrent sur les connues" (17).

Les travaux de cette tradition des algébristes ont, de
plus, ouvert la voie à de nouvelles recherches en théorie
des nombres et en analyse numérique (18). Si l'on s'arrête
à l'analyse numérique, par exemple, on peut affirmer qu'a-
près avoir renouvelé l'algèbre par l'arithmétique, les
mathématiciens des XIe et XIIe siècles ont aussi opéré un
mouvement de retour à l'arithmétique pour trouver, dans
certains chapitres de celle-ci, le prolongement appliqué
de la nouvelle algèbre. Il est vrai que les arithméticiens
prédécesseurs des algébristes des XIe et XIIe siècles
extrayaient les racines carrées et cubiques et disposaient
des formules d'approximation pour les mêmes puissances. Mais,
à défaut du calcul algébrique abstrait, ils ne pouvaient
généraliser ni leurs résultats, ni leurs méthodes, ni leurs
algorithmes. Avec la nouvelle algèbre, la généralité du calcul algébrique devenait constituante d'un chapitre d'analyse
numérique qui n'était jusqu'alors qu'une somme de procédés,
sinon de recettes. C'est au cours de ce double mouvement
qui s'établit entre l'algèbre et l'arithmétique que les mathématiciens des XIe et XIIe siècles ont atteint des résultats encore attribués à tort aux mathématiciens des XVe et
XVIe siècles. Ainsi, la méthode dite de Viète pour la résolution des équations numériques, la méthode dite de Ruffini-
Horner, des méthodes générales d'approximation et en particulier celle que D.T. Whiteside désigne sous le nom d'al-
Kāshī-Newton, enfin, la théorie des fractions décimales.
En plus des méthodes, qu'ils voulaient itératives et susceptibles de conduire d'une manière récursive aux approxima-

tions, les mathématiciens des XIe et XIIe siècles formulè-
rent encore de nouveaux procédés de démonstration comme
l'induction complète, telle qu'on la trouve toujours au
XVIIe siècle. Ils engagèrent également de nouveaux débats
logico-philosophiques concernant par exemple la classification des propositions algébriques et le statut de l'algèbre par rapport à la géométrie. Ce sont leurs successeurs,
enfin, qui posèrent plus tard le problème du symbolisme.

C'est dire que des conceptions, des méthodes et des
résultats attribués à Chuquet, Stifel, Faulhaber, Scheubel,
Viète, Stevin..., sont bien l'oeuvre de cette tradition de
l'école d'al-Karajī, qui du reste était connue des mathématiciens latins et hébreux.

Nous venons de voir que parmi les concepts élaborés par
les algébristes arithméticiens dès la fin du Xe siècle se
trouve celui de polynôme. Cette tradition de l'algèbre
comme "arithmétique des inconnues", selon l'expression même
de l'époque a, par conséquent, frayé la voie à une autre
tradition algébrique, inaugurée par al-Khayyam (XIe s.) (19),
et renouvelée à la fin du XIIe siècle par Sharaf al-Din al-
Tūsī. Alors que le premier a formulé pour la première fois
une théorie géométrique des équations, le deuxième a de
manière décisive, marqué les débuts de la géométrie algébrique.

Or, exclure ces deux traditions, celle des arithméticiens et celle des géomètres, et aussi analystes avant la
lettre, de l'histoire de l'algèbre, les isoler de cette
histoire par une mise à l'écart que l'on tente arbitrairement de justifier en invoquant les visées pratiques et cal-
culatoires, l'absence d'exigence de rigueur, voilà ce qui
permet d'écrire l'histoire de l'algèbre classique comme
l'oeuvre de la Renaissance, aboutissant, selon l'expression
de Tannery, à la "révolution cartésienne". L'occidentalité
de l'algèbre apparaît donc comme le résultat d'une interprétation oblique ou d'une histoire tronquée, parfois même
des deux à la fois.

III
Parmi les disciplines mathématiques, l'algèbre n'est
ni un cas unique, ni un exemple privilégié. A des degrés
divers, trigonométrie, géométrie, déterminations infinité-
simales, sont autant d'illustrations de la précédente analyse. Plus généralement, l'optique, la statique, la géographie mathématique et l'astronomie n'y font pas non plus
exception. Ainsi en histoire de l'astronomie les travaux
récents, parfois en cours, rendent manifestement caduque,
sinon erronée, la manière dont Tannery comprend les astronomes arabes et les interprétations qu'il en donne (20).
Mais comme nous nous sommes fixé la tâche d'examiner la
doctrine de l'occidentalité de la science classique, nous
nous en tiendrons à une composante essentielle de cette
doctrine: l'expérimentation. N'a-t-on pas en effet souvent
marqué le clivage entre les deux moments de la science
occidentale - la période grecque et la Renaissance - par

l'introduction des normes expérimentales? Sans doute l'accord massif des philosophes, historiens et sociologues de la science s'arrête-t-il là; et de fait les divergences ne tardent pas à apparaître dès qu'ils s'efforcent de définir le sens, la portée et les origines de ces normes expérimentales. Ainsi les origines sont-elles liées dans un cas au courant du platonisme augustinien, dans un autre à la tradition chrétienne, et particulièrement au dogme de l'incarnation (21), dans un troisième aux ingénieurs de la Renaissance, dans un quatrième au Novum Organum de F. Bacon, dans un cinquième enfin à Gilbert, Harvey, Kepler, Galilée. Quelques attitudes parmi d'autres, qui se superposent, s'enchevêtrent et se contredisent, mais convergent toutes sur un point: l'occidentalité des nouvelles normes. De cette position dominante, quelques historiens et philosophes s'écartent néanmoins, et déjà au XIXe siècle, pour ramener à la période arabe les origines de l'expérimentation: tels Alexandre von Humboldt en Allemagne, et Cournot en France (22).

Mais il fallait attendre que s'établissent de nouveaux rapports entre mathématiques et physique pour qu'une telle notion, diffuse, de l'expérimentation, se vît conférer la dimension qui la détermine: une composante à la fois systématique et réglée de la preuve. C'est essentiellement en optique, avec Alhazen, que l'on constate l'émergence de cette nouvelle dimension. Chacun sait qu'avec Alhazen nous avons définitivement rompu avec l'optique comme géométrie de la vision ou de la lumière. On sait également que l'expérimentation est bien une catégorie de la preuve. Il faut à présent nous demander ce qu'Alhazen entend par expérimentation. Nous trouverons chez Alhazen autant de sens à ce mot et autant de fonctions assurées par l'expérimentation que de rapports entre les mathématiques et la physique. La simple fréquentation des textes d'Alhazen nous indique que le vocable et ses dérivés - expérimenter, expérimentation, expérimentateur - appartiennent à plusieurs systèmes superposés, que la seule analyse philologique risque de ne pas discerner. Mais, si l'on s'attache d'abord au contenu plutôt qu'à la forme lexicale, on distinguera plusieurs types de rapports entre mathématiques et physique, qui permettent de repérer les fonctions correspondantes de la notion d'expérimentation. Les rapports entre mathématiques et physique s'établissent en effet selon plusieurs modes qui, s'ils ne sont pas thématisés par Alhazen, sont sous-jacents à son oeuvre et en permettent l'analyse (23).

Un type d'expérimentation, non pratiqué par Alhazen lui-même mais rendu possible par sa propre réforme et par ses découvertes en optique, apparaît au début du XIVe siècle, chez son successeur al-Fārisī. Les rapports instaurés entre mathématiques et physique visent, dans ce cas, à construire un modèle; et par conséquent à réduire systématiquement et au moyen de la géométrie la propagation de la lumière dans un objet naturel à sa propagation dans un objet fabriqué. Il s'agit donc de définir, pour la propagation, entre l'objet naturel et l'objet fabriqué, des correspondances analogiques véritablement assurées d'un statut mathématique. Ainsi le modèle de la sphère massive en verre, remplie d'eau, pour l'explication de l'arc-en-ciel. L'expérimentation a donc ici pour fonction de réaliser les conditions physiques d'un phénomène que l'on ne peut étudier ni directement, ni complètement.

IV

Pour conclure, rappelons quelques points:

(1) La doctrine de l'occidentalité de la science classique, dégagée au XVIIIe siècle pour édifier une diachronie de la Raison Universelle, doit à l'Orientalisme du XIXe siècle l'image que nous lui connaissons aujourd'hui: on pensait alors pouvoir déduire d'une anthropologie que la science classique est européenne, et que ses origines sont directement lisibles dans la science et dans la philosophie grecques.

(2) L'opposition entre l'Orient et l'Occident, d'une part sous-tend la critique de la science et du rationalisme en général; d'autre part, écarte, en droit comme en fait, de l'histoire des sciences, la production scientifique en Orient. Pour la science écrite en arabe, on a invoqué son absence de rigueur, son allure calculatoire et ses visées pratiques pour justifier cette non-intégration effective à l'histoire des sciences. De plus, dépendant étroitement de la science grecque, incapables, enfin, d'introduire les normes expérimentales, les savants de cette période se trouvent finalement relégués au rôle de gardiens consciencieux du musée hellénistique. Cette image de la science arabe, si elle s'est atténuée au cours de ce siècle, et particulièrement ces vingt dernières années, persiste cependant dans l'idéologie de l'historien.

(3) Confrontée aux faits, cette doctrine révèle son mépris des données historiques et sa fertilité en interprétations idéologiques: sont en effet admises comme évidences des notions qui soulèvent bien plus de problèmes qu'elles n'en résolvent. Ainsi la notion de Renaissance scientifique, alors que dans plusieurs disciplines tout indique qu'il n'y eut, tout au plus, qu'une réactivation. Ces pseudo-évidences ne tardent pas à devenir les bases conceptuelles d'une philosophie ou d'une sociologie de la science, ainsi que le point de départ d'élaborations théoriques en histoire des sciences, comme le montrent des tentatives très récentes.

NOTES

1. Pascal: <u>Oeuvres complètes</u>, Paris, 1963, p. 231.
   Malebranche: <u>De la Recherche de la Vérité</u>, Paris, 1910, vol. I, p. 139.
2. Montesquieu: <u>Oeuvres complètes</u>, Paris, 1964. Voir les Lettres 104 et 135, et plus particulièrement la Lettre 97.
3. Condorcet: <u>Esquisse d'un tableau historique des progrès de l'esprit humain</u>, Paris, 1966, p. 201.
4. <u>Encyclopédie méthodique</u>, Paris, 1784, p. III.
5. Op. cit.
6. Titre donné par E.Quinet à un chapitre du <u>Génie des religions</u>, 1841.

7. Hegel: <u>Leçons sur la Philosophie de l'Histoire</u>, trad.
   fr. par J.Gibelin, Paris, 1963, pp.82 sqq.
   Hegel: <u>Leçons sur l'Histoire de la Philosophie</u>, trad.
   fr. par J.Gibelin, Paris, Coll. Idées, vol.II, pp.19-21.
8. Joseph de Maistre, <u>Du Pape</u>, Lyon, 1884, pp.487 sqq., par
   exemple.
9. Fr. Schlegel: <u>Essai sur la langue et la philosophie des</u>
   <u>Indiens</u>, trad. fr. de A.Mazure, Paris, 1837. Rappelons
   que les deux classes flexionnelle et non flexionnelle,
   selon Schlegel, "épuisent complètement tout le domaine
   du langage" (p.51). D'après Schlegel, les langues sémitiques ne sont pas flexionnelles, car la structure
   flexionnelle, à partir des racines est, selon l'auteur,
   empruntée par ces langues (pp.54-61).
   Quant aux langues indo-germaniques, "il fallait
   l'intelligence la plus claire et la plus pénétrante, car
   elles expriment "les plus hautes notions de la pensée
   pure et universelle, ainsi que l'entier linéament de la
   conscience" (p.79).
10. E.Renan: <u>Histoire générale et système comparé des lan-</u>
    <u>gues sémitiques</u>, Paris, 1863, p.ix. Cf. pp.18,22.
11. Ch. Lassen: <u>Indische Alterthumskunde</u>, Leipzig, 1847,
    vol.I, pp.494 sqq.
12. Renan, <u>op. cit.</u> pp.490-1.
13. Renan, <u>op. cit.</u> p.16.
14. G.Milhaud: <u>Leçons sur les origines de la science antique</u>,
    Paris, 1893, p.306. Voir aussi: E.Renan: <u>Souvenirs</u>
    <u>d'enfance et de jeunesse</u>, Paris, 1883, p.59.
15. E.Renan: <u>Nouvelles considérations sur le caractère géné-</u>
    <u>ral des peuples sémitiques</u>, Paris, 1859, p.89.
16. Il s'agit de G.Milhaud: <u>Descartes savant</u>, Paris, 1921 et
    de J.Dieudonné: <u>Cours de géométrie algébrique</u>, Paris
    1974, vol.I.
17. As-Samaw'al: <u>al-Bāhir en Algèbre</u>, ed. par S.Ahmad et
    R.Rashed.
18. R.Rashed: L'extraction de la racine n[ième] et l'invention
    des fractions décimales, in <u>Archive for the History of</u>
    <u>Exact Sciences</u> (à paraître).
19. F.Woepcke: <u>L'algèbre d'Omar Alkhayyâmī</u>, Paris, 1851.
20. Il s'agit en particulier de la traduction par Carra de
    Vaux du texte <u>Les sphères célestes selon Nasū Eddīn</u>
    <u>Attūsī</u>, intégrée par Tannery aux <u>Recherches sur l'his-</u>
    <u>toire de l'astronomie ancienne</u>, Paris, 1883, App. VI,
    pp.337-61.
21. Cette position est illustrée par l'hegelien Alexandre
    Kojève: L'origine chrétienne de la science moderne, in
    <u>Mélanges Alexandre Koyré</u>, vol.II, Paris, Hermann, 1964,
    pp.295-306.
22. A.A.Cournot: <u>Considérations sur la marche des idées et</u>
    <u>des événements dans les temps modernes</u>, Paris, Vrin,
    1973, pp.42-3
23. Voir les travaux de Wiedmann, M.Nazīf Schramm, Sabra, et
    les nôtres sur Alhazen et al-Farīsī. G.Sarton: <u>The in-</u>
    <u>cubation of western culture in the Middle East</u>,
    Washington, 1951, pp.27-9.

TOWARDS A NEW SCIENCE

In his contribution Prof. Richta described the elements of a "new science", which, in the context of a new and better social order, would be a more highly developed form of cognition and practice, more truly serving human development.

First (and immediately relevant to the work of this Congress) the new science would be self-conscious, in a critical and historical way, unlike in most present contexts. Its social functions would also be vastly increased, transforming the means of production through mass involvement of a highly educated working people. It would not be an instrument for control of human society by an external force, but would be the means whereby society could control itself and its direction of development - towards goals chosen for their worth and feasibility.

In its handling of the natural world the new science would steer between the alienated reduction of nature to a mere object on the one hand, and the magical or anthropomorphic enchantment of Nature on the other. Care and respect for non-human objects can be well-founded on an appreciation of the essential needs of humanity as inhabitants of the earth.

The self-awareness of the new science would not be based on the "value-free" rationality on which "pure science" traditionally rested; this has shown itself to be too prone to abuse. There would need to be a humanistic commitment; and the establishment of appropriate theories and methods for that is a challenging task for our time.

Symposium 2.   INTERNAL AND EXTERNAL CAUSATION OF SCIENTIFIC IDEAS

Symposium 2: <u>Internal and External Causation of Scientific</u>
            <u>Ideas</u>

<u>Introduction</u>

It is now commonly accepted that the two approaches to
the history of science labelled "internal" and "external" are
complimentary and not contradictory, and that any so-called
conflicts between them are pseudo-conflicts. Both approaches
have in fact reached out from their original bases to embrace
each other in a more genuinely and comprehensively historical
stance. On the one hand studies in the Koyré tradition, and
those of his more recent successors Pagel, Yates, Foucault,
and others, have shown how the "internal rationality" of
science has to be widened to include the ideology of belief
systems of all kinds - magic, mythology, alchemy, religious
sectarianism, etc. - in other words all those cultural
factors which in more Whiggish history were thought to be the
antithesis of science. On the other hand, "external" or
social history of science is no longer content only to study
peripheral social factors in the provenance of science - for
example the educational, financial, or class-based determinants of scientific work, or even the sociology and psychology
of the scientific sub-culture itself in the manner of Kuhn.
Social history of science is increasingly, and most
interestingly, taken to mean study of the social conditioning
of the <u>theoretical</u> belief systems of science - in other words
sociology of science has become a branch of sociology of
knowledge.
     In his book <u>Knowledge and Social Imagery</u> (London, 1976),
David Bloor has described as the "strong programme in the
sociology of knowledge" the thesis that social explanation is
required not only of the errors and blind alleys of the
history of science, but also of its successes, and of the
scientifically rational inferences of scientists. This thesis
contrasts sharply with that of one of the founding fathers of
sociology of knowledge, Karl Mannheim, whose programme
explicitly <u>excluded</u> sociological and ideological explanations
of the so-called rational disciplines of the natural sciences,
logic, and mathematics. Recent philosophy of science, however,
has made it possible to understand sociology of science in a
fashion which does not impugn the "rationality" of science,
for rationality here must be understood to include an openness
of theory-choice. Scientific theories are not logically and
uniquely determined by data - they are <u>under</u>determined, to
use Quine's phrase. It follows that of all the theories that
might have been acceptable on purely empirical and rational
grounds at any given time and in any given state of the
evidence, there is still a historical problem of explaining

why that one was chosen that was in fact chosen. Even more
fundamentally, recent historical and anthropological studies
have indicated that what are taken to be empirical and rational grounds have varied in different societies. There are
social and historical explanations of why western society has,
roughly since the 17th century, understood and practised
science in the way it has.

Discovery of social causes need not conflict with the
fact that certain rational and empirical norms of scientific
discovery are accepted in our society. Neither does it imply
any kind of rigid determinism of the development of scientific
knowledge. It is indeed a general principle of the philosophy
of science that causal explanation does not entail causal
determinism. Determinism is the characteristic of a theory
according to which from a complete description of the present
and perhaps some past states of a system, all future states
can be precisely and uniquely calculated. No natural science
has ever shown the world to be deterministic in this sense,
not even Newtonian mechanics if this had turned out to be a
permanently acceptable theory, if only because the complete
information required to test such determinism can never be
available. In any case Newtonian mechanics is not an acceptable theory, and modern physics describes the world as in
principle indeterministic, and gives explanations and
predictions only in terms of statistical laws. Moreover there
are other natural sciences, such as cosmology, geology, and
evolutionary biology, that contain even more loosely specified
causal correlations and trends. In some cases even significant
individual events postulated in a theory, such as individual
particle disintegrations or individual mutations, are wholly
unpredictable. All of this does not prevent us speaking of
causal laws and explanations in natural science, and at least
the same variety of non-deterministic causality must be
expected in social theories in general, and in sociology of
knowledge in particular.

A question does remain, however, about the _kind_ of causal
relations that may be expected among the varied conditions
associated with the development of science. Let me introduce
a distinction which cuts across so-called external and
internal factors, and speak of an _extrinsic_ and _intrinsic_
approach to causality in the history of science. In the
extrinsic approach, to put it crudely, one expects to find
generalizable correlations between, say, social class of
origin and propensities to particular kinds of theory -
positivism or idealism, atomism or continuum theories,
catastrophism or uniformitarianism, individualism or holism.
But it is evident that at different periods of history and
under different circumstances such line-ups of allegiance
change. For example, Priestley, the marginal man and religious
dissenter of the 18th century, favoured dynamic theories of
matter against the pure mechanism then adopted by the
religious establishment, but in the early 17th century it had
been the radical secularists who had on the whole been
associated with mechanical atomism against the then scholastic
establishment with their multiplicities of essential forms and

qualities of matter. Another example occurs in the first paper of our Symposium. Professor Rossi points to a correlation between occultism and elitism and the pre-scientific mind on the one hand, and the openness and "democratic" character of the new 17th-century science on the other. But in the 20th century we have a radical and "participatory" reaction against orthodox science, which has itself become elitist and opaque to the uninitiated mind. There are no simple correlations in history, and not enough particular cases on which to ground complex correlations.

The intensive approach to historical causation, on the other hand, is exemplified in all the better examples of sociology of scientific ideas. For example, in their study of the Pearson-Bateson controversy between biometric and Mendelian theories of heredity,* Donald MacKenzie and Barry Barnes illustrate its principles well. They show a general correlation of the professional middle-class Fabian gradualism of Pearson's social group with their interest in eugenic "engineering", and their consequent positivist and precautionary aversion to speculative and ill-grounded theory. On the other hand, Bateson's social background was the Cambridge academic elite, which led him to an almost romantic conservative affinity for non-mechanical, holistic theories, to an aversion to social engineering, and to rejection not only of social Darwinism, but also of the natural gradualism apparently implied by Darwin's theory. The historical method here is not general induction, for no claim is made that these social characteristics and these scientific propensities will be found correlated at all times and places. There are indeed in the MacKenzie and Barnes paper some attempts at Baconian-type arguments about presence and absence of factors in different particular cases: for example, scientific socialization and training are rejected as sufficient causes of differences of subsequent scientific style, on the grounds that some individuals cross over from one group to the other. On the other hand, single individuals are not allowed to be counter-examples to the main thesis, as when the authors remark that Pearson's collaborator Weldon did not reveal any commitments in politics, and that in general there are no necessities in an individual's chain of affiliations. The analysis involves rather "the ongoing practice, ideology and institutional structure" of "coherent social groups" - where the social groups are evidently ideal types in Weber's sense.

All this suggests that the causality sought in such cases has more to do with perceptions of rational relevance and irrelevance than with mere external correlations of factors in particular cases. For example, MacKenzie and Barnes judge that features of scientific training are possibly relevant to subsequent scientific style, and that Fabian affiliations are possibly relevant to social and natural Darwinism. In a case as close to us historically and culturally as this one, where we share conceptions of rationality, the historian can be

---

* "Biometrician versus Mendelian: a controversy and its explanation", Kölner Zeits. fur Sociologie und Socialpsychologie, 1975.

guided by his own "understanding" of the relevance judgments
and inferences of his actors. In cases more remote in time
and space, such as the debate about hermetic and alchemical
influences on early 17th-century science, this kind of
historical understanding becomes more problematic. The
intensive approach to the history of science seems to demand
that the historical methods of hermeneutics and <u>Verstehen</u>
should be given at least as much analytic attention by
philosophers of science as causality has received, and
certainly much more than is generally the case at present in
methodological studies of history of science.

The papers of the Symposium reflected various of these
concerns. Professor Rossi, as already mentioned, finds one
of the distinctive changes associated with 17th-century
science to be its open character: it "levels men's wits", in
contrast to the hidden and secret character of the magical
cults. He briefly suggests that this feature reflects the
new democritization of politics. However, as Dr. Ravetz
pointed out in discussion, this suggestion does not seem to be
confirmed by the mid-17th-century alliance between Paracelsians
and the politically radical Puritan sects in England, where
the newly established mechanical science begins rather to
look like the ideology of the rising bourgeoisie. Clearly the
historical ramifications here are complex, and need extended
and (in the sense explained above) intensive study.

Professor Ben David develops the theme of historical
causation of the sociological systems of the 19th century in
France and Germany, in contrast to non-systematic approaches
to sociology in England. He finds an explanation, not in the
direct influence of the French revolution, but in the decline
of the 18th-century legacy of social science for political
reasons during the Napoleonic era, and the consequent need to
re-found the discipline on the basis of individual systems
such as those of Saint-Simon, Fourier, Comte, and Marx.

That there are sociological reasons for the development
and style of sociology itself as a professional discipline is
to be expected, but not yet sufficiently understood in any
period of its history including our own. Professor Thackray
brings sociological analysis even nearer home by asking
questions about the origins of the history of science itself
as a discipline, in the U.S.A. in the 20th century. He
describes the characteristics of the individuals, scientists,
and administrators who forwarded this development, and refers
to the cultural needs which these people saw the history of
science as fulfilling. Again, further questions for research
arise about latent as well as manifest functions. For example,
how have distinctive interests in social and intellectual
history of science been related in American studies, and why
has post-Second World War history of science in America been
almost exclusively "internalist"?

Professor Mikulinsky turns to the general issue of
internalism versus externalism, and finds it a pseudo-problem.
Neither approach understood exclusively can be adequate because
science and society are organically related, and there are
multiple causative links between the intellectual and socio-

economic structures of society. He illustrates his own approach by the example of causative influences in the work of Charles Darwin. Professor Kröber illustrates a similar thesis by considering the influence of the symposium of Soviet papers entitled "Science at the Cross Roads", which were delivered to the Second International Congress of the History of Science in London in 1931. He suggests that Boris Hessen's classic paper in that symposium, "The Social and Economic Roots of Newton's 'Principia'", should not be understood in terms of a "vulgar Marxist" approach,which neglects all causative factors but the socio-economic.

These last two papers stimulated a lively discussion in which it was remarked how historical-materialist and western interpretations of the history of science seemed to be near consensus on historical method. Professor Needham expressed the hope that the term "externalist", which had originally been coined to describe the socio-economic emphasis of Marxist historiography, would not now be used solely to refer to a narrow and exclusive socio-economic approach. It seemed that not only in the 1970's, but also in the 1930's, there was happy consensus regarding the nature of good historiography of science. Some doubts about this were also expressed, however, and indeed if one considers the bulk of the work in the discipline in the last 40 years, one may be forgiven for doubting whether all historians of science had actually received news of the consensus. At all events, it is unquestionable that the message has now been widely received and understood, and as historians widen and deepen their study of science as an aspect of the general history of men and society, it is to be expected that consensus about method will lead to greater maturity of the discipline, to a greater community of interest with general historians, and perhaps also to fruitful debate and controversy about interpretations from diverse ideological standpoints.

MAGIC, SCIENCE, AND EQUALITY OF MEN

Becoming a magician - either in the sphere of natural magic or in that of demonic magic - is not the same as becoming an accountant, a biology professor or a theoretical physicist, for a very simple reason: science and truth have a fundamental characteristic in the universe of magic: they are not accessible to all men, neither in fact nor in principle.

The term, <u>initiation</u>, has always been used and is still used (and not by chance) in reference to magic. In order to know the truths of magic and in order to practice magic, something must be added to natural man, to his innate capacities and his common sense. Man must embrace a principle greater than human nature and he must assume a role which is qualitatively different from that which men have by nature. A nature which is in some way divine must be superimposed on his original nature. He must confer upon himself a new and almost divine way of being which allows him to carry out admirable or miraculous actions. St. Thomas' definition of grace, <u>quaedam similitudo divinitatis partecipata in homine</u>, could - if removed from its context - be inserted in many natural magic texts. The series of activities which we usually designate by the term, magic, have in fact the very strong tendency to become difficult to distinguish from other activities usually designated by other names. As D. P. Walker has written, magic is always on the point of turning into art, science, applied psychology and, mainly, religion.

Magical techniques are, together, a way to operate on the world and a process of religious regeneration. Magic is also salvation. The domination of nature presupposes the achievement of individual perfection, and the process which permits the achievement of perfection coincides with that which leads to the domination of nature. Not all men can achieve perfection. Consequently, not all men can know reality and operate on it. The ascetic discipline, withdrawl from the world, hearing the word of the master with the aim of illumination, the capacity to raise oneself to a level unattainable by other men, are some of the elements which constitute a magical type of knowledge. This mode of thinking gives rise to several closely bound themes which reappear in innumerable texts, which are taken up and repeated by several authors and which take the form, to a certain degree, of constants. These are, 1) the secret and reserved character of a knowledge whose divulgation would have ill-omened consequences; 2) the extreme difficulty and complication of procedures and rituals which permit one to approach the truth and operate on the world; 3) the distinction between the narrow group of sages or "true men" and the <u>promiscuum hominum genus</u> or the mass of profane; 4) the extraordinary character of the figure or personality of the magician who can accomplish impressive feats and who has reached a level of knowledge and perfection which sets him apart.

These, as we have said, are closely linked themes among which distinctions have merely a function of convenience. Does not the extraordinary character of the feats realized by the magician prove that he belongs to the group of the elect? And does the distinction between the elect and the common people not imply the necessity of secrecy regarding the patrimony of ideas, in which profound truths must be veiled so as to appear unrecognizable? Do the reservedness and extreme difficulty of knowledge not result from the incapacity of the majority of men to approach it? Do the variety, ambiguity and allusiveness of magical terminology not depend on one hand on the complication of the procedures and on the other hand on the necessity of reserving knowledge to the few and of hiding its essence? Is understanding the truth notwithstanding the language used rather than through that language not a way of verifying that one belongs to the group of the enlightened?

The fourteen treatises of the Corpus Hermeticum which Marsilio Ficino translated from the Greek between 1463 and 1464 circulated widely as a manuscript, and sixteen editions were printed between 1471 and the end of the XVIth century. Those texts - which date to the second century after Christ, but which throughout the sixteenth century were attributed to the legendary Hermes Trismegistus, contemporary of Moses and indirect teacher of Pythagoras and Plato - are at the basis of the great rebirth of magic in the XVth and XVIth centuries, and operated strongly on the culture of the ages of Ficino and Kepler. The text of the Picatrix latinus was also widely read in the XVth and XVIth centuries in a latin version derived from the Arabic through a lost Spanish version. The Arabic text probably dates to the XIth century, but it is an essential treatise for understanding the philosophical and artistic culture of the Renaissance. Its extraordinary importance, Eugenio Garin wrote, consists "in the insertion in a theoretical context, which is Neoplatonic on one hand and hermetic on the other, of all the vast antique and medieval magical-astrological patrimony, in terms which are surprisingly close to the fifteenth century Platonic tendency, sometimes so close that the proximity cannot be accidental".

Ad laudem et gloriam altissimi et omnipotentis Dei, cuius est revelare suis praedestinatis secreta scientiarum: the theme of secrecy appears in the first lines of the Picatrix and reappeares time and time again.

The "secret" which the book intends to partially reveal cannot be acquired if knowledge is not acquired first. The secret "cannot be possessed if not by the sage and by he who studies science in an orderly way". Science is divided into two parts, one of which is manifest and the other, hidden. The hidden part is deep, but he who explores it will attain what he desires and will be able to draw from it what he wants. The occult will open up to him. The words which refer to the ordering of the world are the same ones which Adam received from God, and they can be understood only by those sages who dedicate themselves constantly to the sciences and who have understood the essence of the Being by drawing on the truth.   In the hermetic-magical tradition, the themes of a distinction between two types of man, of secrecy and of difficulty, seem inextricably bound up with one another. Around 1330 Bono da Ferrara wrote that the art of alchemy is extremely difficult because of equivocation, allegory and metaphor. The oratio of alchemical philosophers is written "in alien and enigmatic terms, with extraneous and impossible figures". The philosophers speak a language which only they know. This allows them to understand one another and exclude all others (loquela extranea ab omnibus aliis et nota solis eis):

"I pray and beseech all those who understand these things and in whose"
"hands this precious gem will arrive, to communicate it to men who ded-"
"icate themselves to these problems, who ardently desire this art and"
"who are learned in natural principles. They must instead hide it from"
"the profane and from children, for they are undeserving."

The authority of rank, the merit of sainthood and of doctrine and the dignity of nature are elements which constitute the figure of the sage for Cornelio Agrippa (who writes two centuries later). Agrippa presents the secrecy of truth and of the processes which permit man to attain it as closely linked to the distinction between divine and common mortal men. "Secrets must be communicated verbally only, through a small group of" "sages and the sacred arcana must be guarded by a small number of elect..." "Every experiment of magic abhors the public, wants to be hidden, grows" "stronger in silence and is destroyed when revealed."

This arcanum, writes Paracelsus in a text of 1584, had already been considered occult by the ancient Fathers, so as not to be reached by the hands of undeserving men. We beseech you, in imitation of those Fathers, "to deal with and conserve this divine mystery in secret". Sublimated mercury, the great German physician affirms in another text, becomes gold, silver, copper and iron; it appears malleable like wax, liquifies in the sun like snow and, finally, returns to its primitive state. This is a secret which must be kept well hidden and which must not be revealed to those who are unworthy of it: "A goose will prefer a turnip to a gem; therefore the common people are not worthy of knowing this secret, and God has expressly forbidden throwing pearls before swine". We have placed very few examples before the eyes of the reader, Giovambattista Della Porta writes in Magia naturalis. But these and what can be derived from them must be kept with a faithful heart, so that they are not degraded by reaching the hands of profane men who belong to the herd (ne passim per manus ignari et gregarii hominis pervenientia vilescant). Even Robert Fludd referred to the herd of the profane, in the XVIIth century, in one of his many answers to Kepler, who had accused him of "dabbling in tenebrous enigmas". All the great philosophers - Hermes and Apuleus of Madaura, Pithagoras, Solomon and even Christ himself - have acclaimed "silence and the concealment of the secret signs". Rightly, heavenly wisdom "has been hidden from the non-elect".

"Tous les Sages de nom de tous les siècles vieux/ Sans lettres ont tousjours par parelle interprète/ D'âme en âme entonné leur doctrine secrète."

For all the exponents of magic and alchemistic culture, the texts of ancient wisdom take the form of somehow sacred texts which include secrets that only a few men can decipher. The truth is hidden in the past and in the profound. It must be searched for and individuated beyond the expedients which were wisely used to hide it from the unworthy. Like when dealing with sacred texts, it is necessary to continuously go <u>beyond the letter</u>, in search of a message which is more and more hidden. That message expresses a truth which is always the same. History is only apparently varried. It contains a single, immutable <u>sapientia</u>: "when Aristotle wrote of the <u>prima materia</u>, Plato of the <u>hyle</u>, Hermes of the <u>umbra orrenda</u>, Pythagoras of the Symbolical unity and Hippocrates of the deformed chaos, they were all writing in reality of the darkness or dark abyss of Moses".

A very ancient theme reappears in Fludd's insistence on the difference between regenerated men and those who have only human form and figure. Assuming different meanings and values, this theme had crossed both the "pious" and "psychiatric" currents of Renaissance magic as well as those currents linked to demonic and necromantic prospectives. Robert Fludd belongs to the very numerous and - in the course of the XVIIth century - belligerent group of so-called "reactionary hermetists". In the years in which he wrote, the theme of the distinction between sages and common men had undergone an extremely harsh criticism.

Only he who has not underestimated the historical significance of magic, alchemy and astrology in Renaissance and XVIIth century culture, and who clearly knows "how slight was the armour with which many philosophers advanced to attack the prevailing magical view of nature" (Yates), can realize the meaning of and reasons for the numerous polemic and "extremist" statements presented by many contemporaries of Bacon and Mersenne. Convinctions concerning the sources and processes of knowledge, on the nature of man, and on the roads to truth, multiply and are expressed with a disconcerting radicality. They are born within different perspectives, with different philosophical presuppositions at their core. Sometimes they even have very different meanings, but independently from their origins, they contribute to the affirmation and consolidation of an image of knowledge which can effectively contrast the hermetic-magical tradition. They pose themselves as alternative to this tradition and, later, they exercise a hegemonic function on the various forms and manifestations of culture.

Those truths which are called common notions, Descartes writes in the <u>Principia</u>, can be known by many with all clarity and distinction. Nevertheless, to some people, these truths are not sufficiently evident. This does not depend in any way on the fact that "la faculté de connaître qui est en quelques hommes s'étende plus loin que celle qui est communément en tous". It depends only on the prejudices acquired in childhood, from which it is very difficult to free oneself. Let us briefly recall the famous beginning of the <u>Discours sur la methôde</u>, which affirms that "good sense is of all things in the world the most equitably distributed". The power of judging well and of distinguishing between the true and the false (which is what is called reason) "is by nature equal in all men". Furthermore, reason or good sense, which distinguishes us from the animals, "is complete in each one of us". The diversity of opinions "is not due to some men being endowed with a larger share of reason than others", but only to the fact that our thoughts proceed along different paths, and that we are not attending to the same things.

Every man, according to Hobbes, "brought Philosophy", that is, Natural Reason, "into the world with him", and errors and deviations depend only on the lack of the right method, "as it were for want of sowing and planting". The recta ratio is present in all men and makes them equal: differences in understanding derive from the passions, that is to say, from the body and from customs. Reason, he affirms in another text, "is no less of the nature of man than passion, and is the same in all men". Even in the first lines of the De Corpore, Hobbes had opposed his philosophy to that "which is found in the metaphysic codes", and to that "which makes philosophers' stones". All knowledge which "is acquired by Divine inspiration" or which "comes to us not by reason but by Divine grace in an instant and, as it were, by some supernatural sense", is strictly excluded from philosophy.

Hobbes thinks that most men, in regard to science, "are like children, that having no thought of generation, are made to believe by the women that their brothers and sisters are not born, but found in the garden". Notwithstanding this, they who have no science "are in better and nobler condition with their naturall prudence, than men that by mis-reasoning or by trusting them that reason wrong, fall upon false and absurd generall rules". The situation is such that "they who content themselves with daily experience" and "either reject or not much regard philosophy" are "men of sounder judgement" than those who are full of refined opinions but full of uncertainty and who do nothing but dispute and wrangle.

In the De corpore Hobbes insists on the necessity of explaining "the few and first elements of philosophy", which are similar to "seeds from which pure and true philosophy may hereafter spring up little and little". Those seeds or first fundations of science appeared to him "not only not beautiful, but poor, arid and, in appearence, deformed". Philosophy, he had said in the Author's Epistle to the Courteous Reader, "the child of the world and of your own mind, is within yourself". The method which Hobbes followed and theorized and which leads to science and truth, is made for all men: "if it like you, you may use the same".

The elements of continuity (which are doubtless present and on which many scholars have rightly insisted) in the so-called hermetic tradition and natural philosophy of the XVIIth century, cannot lead one to neglect the differences - which are really very great - between the new image of the man of science and that of the magician-priest theorized by Ficino or of the magician as titanic and extraordinary being, presented by Agrippa.

To theorize that truth is accessible and that men are equal in principle in regard to its attainment meant also to open the door to new ways of understanding "political" relationships. The distinction between lords and servants, which for Hobbes is artificial and unnatural, <u>does not derive from a difference in intelligence</u>: philosophers and sages were the ones who transformed a factual difference into an ontological one: "The inequallity that now is, has bin introduced by the lawes civill. I know that Aristotle...maketh men by Nature, some more worthy to command, meaning the wiser sort (such as he thought himselfe to be for his philosophy); others to serve (meaning those that had strong bodies, but were not philosophers as he); as if Master and Servant were not introduced by consent of men, but by difference of Wit: which is not only against reason, but also against experience. For there are very few so foolish, that had not rather governe themselves, than be governed by others".

The desire to widen the control of all the human race over nature, to build a universal method and to teach everyone everything (as Comenius wanted), marked the end of the gnostic-hermetic distinction between two types of men, between the initiates, the elect, the "regenerate" on one hand and common mortals on the other. After the age of Bacon, Descartes, Hobbes and Galilei, every form of knowledge which theorized secrecy, calling on its inaccessibility, which conceived the difficulties encountered along the way of knowledge as "superhuman", which affirmed the initiatic character of access to <u>episteme</u>, appeared structurally connected to the political thesis according to which men were not capable of governing themselves and - as Pietro Pomponazzi says - are similar to children who, "in order to lead them to good and deliver them from evil", must be told fables and stories of miracles, angels and devils.

It is well known that Girolamo Cardano enjoyed certain success as a mathematician and that Giovambattista Della Porta occupies a non-negligible place in the history of optics. The calculations of many astrologers are much less questionable than the mathematical ramblings of Hobbes and Paracelsus is much less "scholastic" than Descartes. The so-called <u>internal history</u> is often inadequate to explain the great changes or turning points of history. Modern science was not born from the rediscovery of classical scientific texts alone, nor from simply reconsidering technical problems that had remained unsolved. To understand the "scientific revolution" we must move not only on the level of "calculations", but also on that of general problems: conceptions of man, nature and history, and the construction of a new image of science and of the natural philosoper. Is the problem of the relationships between scientists and other men part of the history of science?

J. Ben-David

THE STRUCTURE AND FUNCTION
OF NINETEENTH-CENTURY SOCIAL SCIENCE

In the nineteenth century much of social science assumed characteristics that set it apart from its eighteenth century antecedents and twentieth century consequents. The founders of this kind of social science were Saint-Simon, Comte, and to some extent also Fourier, but he was much less concerned with science than the other two (Bourgin; Gouhier, 1933-41; Gouhier, 1965; Manuel, 1956; Manuel 1962).

These thinkers claimed to have discovered comprehensive "systems" with laws that explained how society had developed until their times, and predicted its course in the future. Furthermore, they were not content to publish their discoveries, but also set out to organize mankind for action to realize the goal of history as perceived by them. In addition to their claim of being scientists, they claimed to be prophetic leaders and were adopted as such by ideological-religious movements and sects. This type of social scientist cum prophet was almost completely absent in England. Herbert Spencer's system, for example, had many of the characteristics described above, but he had no prophetic pretensions. But on the continent this new intellectual type became widespread. His best known representative was Marx, and Freud also belonged to it to some extent.

This was a drastic change compared to the eighteenth century. Turgot, D'Alambert, Condorcet and even the idéologues attempted to explain social development and to apply their theories to the changing of society (Manuel, 1962; Hankins, Baker). However, their theories were not as comprehensive as those of their nineteenth-century successors, nor did they regard their theories as closed systems that one either accepted or rejected as a whole. Eighteenth-century thinkers consciously resisted system building, and did not found movements. Their theories were regarded as contributions to a collective effort for the advancement of knowledge and the improvement of human life (Manuel, 1962, p. 17; Hahn, 32-34). The validation of ideas was considered as a long drawn out process of criticism and accumulation of evidence in which the verdict would be pronounced by the "Republic of Letters" (Hahn, pp. 36-45).

In view of these antecedents, the rise of the all embracing historicistic systems in the nineteenth century strikes one as a step backward. The pretension to explain on the basis of a few metaphysical principles the entire history of mankind, and also to predict and actually bring about its future, was incompatible with all scientific tradition. Furthermore, these closed systems which pretended to determine scientifically the cognitive structure (religion, metaphysics, science) and morality appropriate for every age, introduced into social science an element of scientific and moral relativism which, carried to its logical conclusion, eventually led to the denial of the possibility of finding increasingly universalistic ways of establishing truth and right, and - by implication - to the justification of intellectual and moral coercion. Ironically, social scientists who set out to complete the work of the Scientific Revolution by adding the science of man and society as the last block to the edifice of science, and to complete the work of Enlightenment by the creation of scientific religions or movements in order to replace all religion and prejudice, ended up with weakening the scientific endeavor and reinforcing the intolerance and dogmatism they set out to destroy.

This development may be considered by some as inevitable and desirable, and by others as a deplorable result of faulty thinking and reliance on insufficient evidence. In either case there is a tendency to view this whole development as an unfolding of the inner logic (including the self-contradictions) of these systems (Mill, Hayek, Popper II., Habermas). In contrast, the present paper tries to relate the nineteenth-century systems to their social background. Without denying their weaknesses, it will be shown that they were genuine scientific efforts, and that the effects attributed to them were due to the setting in which the systems were propagated rather than to their far from compelling internal logic.

II. The explanation of the emergence of the social science cum prophecy pattern lies in the changes that occurred in the structure and function of social science in France during and after the Napoleonic period. The changes in France served as a model for the rest of continental Europe. England was influenced by these changes in France but, as will be shown below, the scientific institutions of early eighteenth-century England were significantly different from those of France, and unsuitable for the implantation of the French model.

These changes had intellectual, as well as social and political roots. Social thought in the eighteenth century developed in a symbiotic relationship with natural science. Its function was to interpret the significance of the scientific revolution to the non-expert public and to apply the scientific approach to the solution of economic, social and political problems (Hankins, 74-95, 109-118, 143). It was - as social science still so often is - a combination of philosophy of science and social thought. The rules of scientific thought had to be abstracted from their original context in the natural sciences in order to apply them to social problems. In order to undertake the task one had to have scientific competence as had D'Alambert, Condorcet and many of their contemporaries. One did not have to abandon one's scientific role while devoting oneself to this new enterprise. A large part of natural science - such as chemistry, electricity, magnetism, biology - was still not much more than an approach. Therefore, applying the same approach to the problems of society could be regarded as a legitimate and feasible part of science.

With the rise of scientific disciplines at the end of the eighteenth century the image of a general "science" became blurred and unattractive. "Science" was on its way to being replaced by discrete "sciences" each with a distinct tradition and set of practices (Hahn, 275-276; 304-307). Each discipline was, or strove to be, a "system" with a logical structure and laws of its own. The general scientific approach to the investigation of a problem was no longer adequate.

There was a parallel development towards specialized disciplines based on empirical philological enquiry in humanistic learning. History, philology and linguistics took their place as empirical sciences next to experimental and exact science (Schnabel, 31-152). This discredited the social scientific tradition which - as has been pointed out - was closely linked to a philosophy of general science. Social science appeared as lacking proper subject matter and method, compared to the new natural science and humanistic learning. Thus, instead of appearing as a pioneering extension of scientific thought, as it was in the eighteenth century, social science came to be regarded as an anachronistic anomaly among the new scientific disciplines.

This problem was aggravated by the disappearance of the need for the social scientists as publicity agents for science. The work of the philosophes who founded social science in the eighteenth century furthered the cause of modern science and scholarship in general. By interpreting the educational implications of science they helped to destroy the scholastic traditions and church authority in education, thus paving the way to the establishment of institutions of higher education devoted to the study of the new scientific and scholarly disciplines. Once these disciplines were recognized and officially supported, they did not need the spokesmanship of social scientists any longer. In fact that spokesmanship was occasionally embarrassing, in view of the propensity of social sciences to incur the enmity of the rulers.

There was a similar loss of sympathy for social science among the upper bourgeoisie, and the professional classes who also attained their goals in the late revolutionary and Napoleonic period. Their newly acquired stake in the political and social status quo (except for the interval of Restoration between 1815-1830), and their experience with the social experiments of the revolutionary period, made them skeptical about the utility of the attempts to apply a scientific approach to the interpretation and reform of society (Manuel, 1956, 189-207).

This change in the fortunes of the social scientists coincided with the dissolution of the second class (moral sciences) of the Institut, the elimination of the social sciences from the curricula of upper secondary and higher education and of the social scientists from educational policy-making, which they had monopolized between 1794 and about 1800 (Manuel, 1956, 60-61, 80-81, 382; Moravia, 564-607). This was due in large part to political reasons - Napoleonic autocracy could not tolerate free enquiry in social affairs - but it was facilitated by the declining support of social science among its erstwhile clients. Thus social scientists, who spearheaded the successful movement for the recognition of science and modern learning in education and academic institutions throughout the eighteenth century, were excluded from partaking of the fruits of their victory.

This did not lead to the elimination of social science. The question of what the meaning of science is for the understanding of man and society was too interesting and the intellectual traditions of enquiry into these matters were too strong to be abandoned. But while other intellectual fields now possessed an elaborate institutional framework, social science lost its academic status and even the semi-institutionalized status that it had attained before the Revolution. There was no official support for social science, and there was no recognized scientific community within the field that could exert informal authority by withholding or granting recognition to contributions.

In consequence, social science was practiced by different people

in a variety of careers. Some people with academic appointments in
the sciences and humanities such as Laplace, Pinel, Clapeyron, Duhamel,
Lamé, Andrieux, and many others were part time or occasional social
scientists (Gouhier I., 133-166). And private scholars engaged in
other careers, such as de Tocqueville, could still do outstanding
work.

But those who wanted to practice social science programmatically
were not accommodated by any institutional framework in France (or
Germany), and they had to contend with prejudice and disrepute among
the academic circles and to face the suspicion and sometimes persecution of the authorities.

III. This was the background of the emergence of comprehensive
systems, and the formation of scientific sects in social science. The
propensity to create systems had always existed in France (as well as
in Germany, in natural, as well as social science) but it was kept in
check by the leading academic scientists (Manuel, 1956, 382; Hankins,
81; Hahn, 32-34; Olson, 37). These continued to oppose the "spirit
of systems" in the early nineteenth century (Biot, 109-16) but at the
same time respectable scientists were busy writing books which attempted to systematize and codify broad fields of knowledge. The great
advances in physics and chemistry, and the demand for textbooks in the
newly established institutions of higher scientific study, such as the
Polytechnique, gave rise to attempts at the presentation of different
fields of science as logically coherent systems, based on certain premises and ruled by a limited number of principles (Gouhier, I , 135-
139). These were much more careful and much less pretentious efforts
than those of Saint-Simon, Comte, and the other builders of historicistic systems. But since the existence of a "system" of knowledge
now became the hallmark of a discipline, social scientists were left
with the choice of either abandoning the claim that social science was
a scientific discipline, or attempting to discover a system of their
own. Thus while the actual state of knowledge did not allow the
establishment of a social scientific system, the tradition to apply contemporary modes of scientific thought to society provided great motivation to create such systems. This conflict is reflected in Comte's
attempt at developing a complete philosophy of all the positive sciences, and then discovering the laws of history and society at a single
stroke. This also explains why - in spite of the strictures against
system building in general, and the shortcomings of all the proposed
systems in particular - Saint-Simon and Comte succeeded in getting the
attention of many people of outstanding education and ability (Manuel,
1956, 344-47; Gouhier, 1941, 72-79; Simon, 13-18). Only a few of
these remained permanent followers, but the lively interest aroused by
the work of Saint-Simon and Comte shows that their attempt to create
a social science through the discovery of a system appeared in the
first half of the nineteenth century at least as a potentially useful
exercise.

This suggests that the emergence of systems was much more a continuation of eighteenth-century traditions than it appears. It was
an attempt to interpret the social implications of the new disciplinary state of science. But there was a serious departure from tradition in the forms of the propagation of social science. As a result
of its de-institutionalization, the pursuit of social science was

forced into a peculiar framework of social and political movements.

IV. It is usually assumed that this form of organization was a
consequence of the comprehensive systems. Since these systems pre-
tend to determine scientifically the appropriate moral order for every
age, they have an implicit or explicit claim that people adhere to
them. But while movements usually need comprehensive systems to justify their far-reaching claims on their members, the systems themselves do not necessarily generate movements. World views do not
automatically translate themselves into social action. The philosophical systems of Germany in the early nineteenth century, such as
the different versions of idealism, and Naturphilosophie, or Spen-
ciarian evolutionism that had much following at the end of the nineteenth or early twentieth century, did not lead to the emergence of
movements.

By the 1830s - when there was a chance of re-institu-
tionalizing social science - there was already a pattern of social
movements exclusively devoted to a single master and his system.
Comte, Marx, and Freud all tried to escape this fate, but were caught
up in it. The movements were a semi-institutionalized framework for
the pursuit of social science. They organized lectures, study groups
and publications, and were frequently supported by people who were
not adherents (Simon, 73-93). The movement pattern was even adopted
by social scientists who were not system founders, such as LePlay
(Clark, 104-11).

The fact that the new kind of social science now became practiced by such movements had fateful potentialities. As social scientists who claimed to have discovered the laws that predict the future
course of history, lack of success to obtain wide recognition was a
particularly disturbing thing for the inventors of systems. It not
only meant penury, but also a challenge to their theories, since in
their expectations mankind, and first and foremost scientists, should
have recognized the power of their ideas and rallied around them
instantly. This problem of the "blindness" of mankind and even of
scientists to what was claimed to be palpable truth was a central
motive in their thought. The ways to solve this problem were set down
by Saint-Simon. He eventually assumed that his theories should appeal
first and foremost to the industrial class that included both managers
and workers, because their occupational interests coincided with the
direction of historical development discovered by him (Manuel, 277-
280, 288-294). As those actually carrying out the historical mission,
this class was bound to recognize the truth of his insight. This was
an idea with political potentialities, since under the Restoration the
leaders of business were locked in a fight against the resurgent aris-
tocratic reaction. The idea that women and working people would be
more capable of accepting their doctrine than the better educated also
appears in positivism (Simon, 11); but the original meaning of the idea
was simply to justify and promote the search of support from the
industrial, or working classes.

Even more important than explaining the lack of general support
for their ideas, was the need of the system builders to explain the
rejection of their ideas by the bulk of scientific and scholarly
opinion. Again, the patterns of dealing with this problem were laid
down by Saint-Simon and his less scientific contemporary, Fourier.

It consisted of rejecting all the philosophy of the past and the present - with few exceptions - as unscientific (OSSE XXXIX, 17-18, 29-31; OSS VI, 117; Bourgin, 241-42, 250-54; Manuel, 1956, 137-60; Manuel, 1962, 214-15, 237). This - at least in principle - neutralized the most important and consistent source of doubt and opposition. Since experimental and exact scientists did not actively engage in polemics against social scientists, their sporadic opposition was less of a problem. This opposition was dealt with by pointing out the lack of relevance of the physical sciences to the study of man which could explain the rejection of the overtures of Saint-Simon, Fourier, and Comte to the leading scientific figures of their age who were physicists, by accusing scientists of excessive specialization and lack of interest in large issues; and by viciously attacking some of the scientific leaders. But in spite of much bitterness against scientists, science was never rejected, and its validity never denied. Only the validity of philosophy (except, of course, their own) was denied on the ground that it was based on a historically dated false way of thinking.

The internal logic of the nineteenth century historicistic systems has been loose enough to allow more than one kind of interpretation. Old and new Hegelians reached diametrically opposed political conclusions from the doctrine about the identity of reason and reality; Littré believed that his interpretation of positivism that led him to liberal views was the correct conclusion of positivism, and that the late teachings of Comte were an aberration; and Marx' expressed views on scientific relativism have been - as pointed out - contradictory (Berlin, 64 ; Charleton, 55-71; Seliger, 30-57). This is not to say that the historicistic systems have been valuable to social science, but only that the causes of relativism and intolerance are to be sought in the sectarian spirit of movements using scientific doctrines for purposes of domination. When such movements got hold of scientifically much more valuable ideas than historicism, such as heredity, they could turn it to equally authoritarian uses and justify it logically (Rosenberg, 89-97).

The second observation supporting the present interpretation is the difference between the nineteenth-century history of social science in France as against its history in England. English and French social scientists in the early nineteenth century shared a common tradition, and worked along parallel lines. French ideas had to be translated into a different philosophical idiom before they had an influence in Germany, but no such translation was needed in England. Comte was more rapidly appreciated in England than in France (Simon, 12). Yet sectarian organization of social science spread to the continent and was rejected in England (except by some very peripheral groups). The early and illustrious followers of Comte in England, such as John Stuart Mill, refused to follow Comte into his sectarian phase (Mill). Later, at the end of the nineteenth century, the presence of Marx in London had little influence on the development of English socialist thought that remained (at least at that time) pragmatic, empirical and tolerant. And psychoanalysis, which developed as a closed and ideologically tinged sect on the Continent of Europe, became in England and the United States an accepted part of psychology.

The reason for this difference was that the development of disciplinary science and specialized academic institutions was much slower in England than in France or Germany. Science in England remained a relatively unified movement struggling for official support and recognition throughout much of the nineteenth century. As a result, English social scientists – such as Mill, Bagehot, and Lewes – were accepted as part of the intellectual élite of the country, much as their eighteenth-century predecessors. No de-institutionalization and drastic change in the standing and function of social science occurred in England as in France, nor was social thought eventually excluded from the academic framework as in Germany. Therefore, in England social movements did not become the principal framework for the pursuit of social science with the result that discourse in the field continued to be pursued in the traditional manner of people submitting their contributions to the criticism of their peers, rather than to their followers, observing the rules of a game based on the assumption that logic and evidence are things shared by all the participants, and accepting the fact that final judgment in scientific matters is within the competence of an impersonal scientific community and not the monopoly of a particular group or movement.

Bibliography

N. G. Annan, The Intellectual Aristocracy, in T. H. Plumb (ed.), Studies in Social History: A Tribute to G. M. Trevelyan, London, Longmans, Green & Co., 1955, pp. 243-87.

K. M. Baker, Condorcet: From National Philosophy to Social Mathematics, Chicago, Univ. of Chicago Press, 1975.

I. Berlin, Karl Marx: His Life and Environment, 3rd ed., New York, Oxford U.P., 1963.

T. B. Biot, Mélanges scientifiques et litteraires, II., Paris, 1858.

H. Bourgin, Fourier:contribution a l'etude du socialisme francais, Paris, 1905.

D. G. Charleton, Positivist Thought in France During the Second Empire 1852-1870, Oxford, Clarendon, 1959.

T. N. Clark, Prophets and Patrons: The French University and the Emergence of the Social Sciences, Cambridge, Harvard, 1973.

A. Comte, Cours de philosophie positive, 6 vols., Paris, 1830-42, (4th ed. 1877) (Cours).

A. Comte, Système de politique positive, 3rd ed., Paris, 1890.

H. Gouhier, La jeunesse d'August Comte et la formation du positivisme, 3 vols., Paris, Vrin, 1933-41.

H. Gouhier, La vie d'Auguste Comte (2nd ed.,), Paris, Vrin, 1865.

J. Habermas, Knowledge and Human Interests, Boston, Beacon, 1971.

R. I. Hahn, The Anatomy of a Scientific Institution: The Paris Academy of Sciences, 1666-1803, Berkeley, U. of California, 1971.

T. L. Hankins, Jean d'Alambert: Science and the Enlightenment, Oxford, Clarendon, 1970.

F. A. Hayek, The Counter-Revolution of Science: Studies on the Abuse of Reason, Glencoe, Free Press, 1952.

F. E. Manuel, The New World of Henn Saint-Simon, Cambridge, Harvard, 1956. Also The Prophets of Paris, Cambridge, Harvard, 1962.

J. S. Mill, <u>Auguste Comte and Positivism, 1865</u>, in <u>The Collected Works of John Stuart Mill</u> (J. M. Robson, ed.) Vol. X, Toronto, 1969, pp. 261-368.

S. Moravia, <u>Il tramonto dell'illuminismo: Filosofia e politica nella societa francese (1770-1810)</u>, Bari, Laterza, 1968.

R. Olson, <u>Scottish Philosophy and British Physics 1750-1880</u>, Princeton, Princeton U.P., 1975.

K. R. Popper, <u>The Open Society and Its Enemies</u>, 2 vols. (6th ed.), New York, Harper, 1967.

C.-H. de Saint-Simon, <u>Oeuvres de Saint-Simon et Enfantin</u> (E. Dentu ed.), 47 vols., Paris, 1865-78 (<u>OSSE</u>)

C.-H. de Saint-Simon; <u>Oeuvres: Groupant les ecrits de Saint-Simon non inclus daus les Oeuvres de Saint-Simon et Enfantin</u>, Vol. 6, Paris, Anthropos, 1966 (<u>OSS</u>).

C. E. Rosenberg, <u>No Other Gods: On Science and American Social Thought</u>, Baltimore, Johns Hopkins, 1976.

F. Schnabel, <u>Deutsche Geschichte im neunzehnten Jahrhundert</u>, Vol. III, Freiburg, Herder, 1934.

M. Seliger, <u>The Marxist Conception of Ideology</u>, Cambridge, Cambridge U.P., 1977.

W. M. Simon, <u>European Positivism in the Nineteenth Century</u>, Ithaca, Cornell, 1963.

Professor Arnold Thackray

SCIENTIFIC IDEAS AND SOCIAL CAUSATION: AN AMERICAN EXAMPLE

THE THEORETICAL CONTEXT

We are by now familiar with the proposition that scientific ideas are socially constructed. Thomas S. Kuhn has stated that "scientific knowledge, like language, is intrinsically the common property of a group or else nothing at all." Mary Hesse has elegantly explored the formal, philosophical characteristics of this common property while David Bloor has undertaken an articulation and defense of "the strong programme in the sociology of knowledge." That programme seeks to explain "knowledge, including scientific knowledge, purely as a natural phenomenon." To put the matter in other terms, the Ding an sich remains unknown. Instead we deal in socially constructed representations of reality.[1]

Less controversial but more-often overlooked is the proposition that historical ideas are also socially constructed. The point is familiar. It was made most pithily by E. H. Carr when he said, "Before you study the history, study the historian.... Before you study the historian, study his historical and social environment."[2] Like nature, the past can be known to us only in mediated forms. The history of science is thus subject to a double modifier. Through it we obtain socially-constructed (historical) accounts of a socially-constructed (scientific) world.

Recognition of this situation provides no sufficient reason for a despairing Pyrrhonism or an anxious flight to authority. It rather opens the way for the continuous cognitive work of constructing new historical accounts more congruent with our shifting sense of intellectually rewarding problems. It also enables us to understand why previous historians of science found satisfaction in perceptions quite different from those we currently employ. The double modifier offers us a double opportunity. We need not see past science as simply true or false. No more need we agree or disagree with past historical enquiry. Instead of condemning outmoded sensibilities, we may explore the utilities afforded by previous categories.

Guidelines for such an exploration have been well formulated by Barry Barnes. He postulates that both scientific and historical ideas are "tools with which social groups may seek to achieve their purposes." Ideas suit purposes not because of any logical relationship but because of their fit to particular problems in particular circumstances. Accordingly, "beliefs which 'work' in one situation may be quite inappropriate in another. The connection between ideas and interests is contextually mediated."[3]

To find examples of such contextual mediation we need only look back at the subjects which have been considered within these congresses over the past half century. The 1930s discussion of "The Social and Economic Roots of Newton's Principia" comes immediately to mind.* So too do later attempts to demarcate internal from external history and to define their proper spheres of influence. Those attempts illustrate how ideas may be used as tools. Our present unease with the very categories of "internal"

---

*It forms the subject of Professor Kröber's paper, in this symposium.

and "external" history also shows how beliefs which served well in one historical situation may make poor implements in another.

The example on which I wish to focus is however slightly more distant. It is the development of the history of science as a recognized academic activity, in the United States in the early years of this century.

THE CASE-STUDY

The Emergence of the American University

The American university, as an institution devoted to scientific and scholarly discovery and teaching and training in that spirit, was a product of the last quarter of the nineteenth century. 4 The founding of the Johns Hopkins provides one benchmark. The establishment of Clark University in 1889, Stanford University in 1891 and the University of Chicago, also in 1891, confirmed and accelerated a trend apparent in such older, slower-moving institutions as Columbia, Harvard and Pennsylvania. In the five years 1876-1880 alone, universities in the United States awarded more Ph.D. degrees than they had given in their whole previous history (Table 1). By the mid-nineties a "first generation" of American doctors of philosophy had emerged to the world. Members of that generation were to participate in, be moulded by and profit from the massive reorganization of intellectual life of which their own degrees were one earlier indicator.

University teaching became an attractive career for intellectual, ambitious young persons. It offered prospects of advancement and it permitted concentration on special field of interest. Universities increasingly recognized achievements in research and provided facilities and resources for that research. The greater size of the student bodies, both at the undergraduate and the graduate levels, required more teachers and larger departments. At Harvard, for example, the number of post-graduates went from none in 1870 to 429 in 1908, while the number of professors increased from 45 in 1870 to 194 in 1908. Of especial significance was the expansion in "other teachers and research fellows" from 14 at the beginning to 416 at the end of the period (Table II).

The profession of science and the number of scientific societies grew proportionately. Intellectual specialization and social differentiation proceeded apace. Until 1881 the American Association for the Advancement of Science was divided into but two sections. By 1916 its vastly expanded program was jointly sponsored by forty-three specialist societies. The size and structure of the American Chemical Society indicates the transformations occurring, in the case of a single discipline. In 1889 its membership was 204, in 1917 over 10,000. In the former year the society was a unitary organization, in the latter it conducted its business through seven intellectually specialized divisions. In the latter year also, more doctoral degrees were granted in chemistry alone than in all academic subjects combined some three decades earlier. 5

This rate of academic growth was without precedent in the Western world. It formed the setting for the "heroic age" of American academic science. The possibilities inherent in that heroic age were profoundly to influence the careers fashioned out by the scientific members of the first generation of Ph.Ds. By the time the United States was drawn into World War I, a small elite from among them had become famous and influential, not only in pure science but also in scientific medicine, in "governmental science," in the administration of new and old universities, in the creation of learned societies, and in the development of philanthropic institutions.

The History of Science as Intellectual Program

In the 1880s and 1890s the major universities conducted their first systematic classes for graduate students. The obvious danger that these seminars might breed a race of narrow "technicians" was perhaps most keenly perceived by the scientists who taught them. The scientific disciplines were still somewhat defensive in the face of the criticisms, explicit and unspoken, coming from the older academic disciplines such as classics, philology, and moral philosophy. University scientists increasingly sought to provide a rationale that could both meet these criticisms and help to establish the higher justification for science. And this, it appears, helped lead to regular courses on the history of the various scientific disciplines, sometimes for graduate students and sometimes for advanced undergraduates.

Appropriately enough, it was the leader in technical education—M.I.T.—which took the initiative. In 1887 Sedgwick began lecturing on the history of biology and Cross on the history of the physical sciences. By 1905 their two courses had been combined into a regular series on the history of science. A similar pattern was manifest at Harvard, where T. W. Richard's lectures on the history of chemistry—first given in 1890—led to L. J. Henderson's launching in 1911 of his survey of the history of the physical and biological sciences. Stanford from its very beginning provided teaching in the history of chemistry, while historical courses in one or other fields of science were also given at the Universities of California, Illinois, Michigan, North Carolina and Pennsylvania, and at Chicago, Cornell, and Northwestern. 6

By the early 1900s, it became common to discuss explicitly the aims of such teaching. The purpose most often avowed was "the much desired humanizing of science." That aim was simply and unself-consciously linked with the recruitment and induction into science of undecided students, the development of bold claims regarding the place of natural science and "scientific method" in modern culture, and the desire to reintegrate the increasingly differentiated scientific specialties. The educational needs of students of science and the moral and cultural concerns of scientists together defined the proper criteria that should govern the growth of this subject.

The outbreak of European war brought to the front other themes in the developing historiography. Science was international. It was universal. It was "the arbiter of the future of man and of peace." It gave life, stability, truth and wealth. Knowledge of its history would "enhance in the breast of every young man or woman, faith in human progress and goodwill to all mankind." 7

The war drew together the elite of the first generation. It placed them in command of new ventures. It made vivid to them the reality of the national scientific community. Their experiences also made them aware of growing disciplinary differences in science and of the extent to which the political realities of American life remained obstinately rooted in local, ethnic and class attachments. The variety of those attachments and the pluralistic, dispersed and loosely federated forms of organization to which they pointed, together mandated strenuous new administrative roles for the leaders of the scientific community.

The war in Europe also led to the arrival in America of George Sarton, an idealistic young refugee committed to a secular faith in science as the guiding thread of history. Behind him in Belgium, Sarton had left his fledgling journal, Isis: Revue consacrée à l'histoire de la science. His troubles with Isis - primarily financial - were to serve as the catalyst to a social transformation, whereby the elite of the first generation would render explicit and institutionalized their commitment to the intellectual program of the history of science, as an expression of their common values. 8

The Scientific Elite and New Legitimations for Science
The decade following the war was one of exultant self-confidence and callow optimism, coupled with apprehension about the new phase of America's existence. There was uneasiness about having departed from the "faith of the fathers," about the perils of an urbanized life, amidst entanglements with other countries and other cultures. A small and generally unregarded group of literary intellectuals, radical and liberal, denied the value of the prosperity, the hedonism, and the still persisting piety of American society and culture. Their criticism, made from the periphery of the circles in which the leadership of the country moved, added, nonetheless, to a sense of need to restate the rightness of America's course. One ready answer lay in the affirmation of the value of science. In this affirmation science was seen not only as sustaining the technology which makes America progressive and thus promises to relieve mankind of the evils of poverty and the pain of illness. Science would also guide the country in the realms of ideals, goals and values.

The secular faith in science found its spokesmen among the elite of the "first generation." By the 1920s that elite was at the peak of its collective power. They found the continuing expansion of academic science a not unmixed blessing. It implied new financial pressures and unfamiliar organizational problems. As leaders they had to seek out wealthy and powerful benefactors. As spokesmen, they had to enlarge the public appreciation of science as a distinctive strand in modern culture. To them, the history of science became valuable in a variety of ways at once wider and more focussed than those known to the rank-and-file members of their generation who also drew upon the subject. The new interest displayed by the scientific elite was to prove decisive in facilitating the forms, and shaping the content, of the history of science as an organized activity.

The History of Science Society
The history of science very easily achieved the kind of "imperfect institutionalization" which was to characterize it between the Wars. In 1920 a "History of Science Subsection" was established in Section L of the American Association for the Advancement of Science. Sufficient communal energy was available not only to sustain this sub-section but also to engender the creation of the History of Chemistry Division of the American Chemical Society (1923) and the American Association for the History of Medicine (1924), in addition to the History of Science Society itself (1924).

The forty-seven men who constituted the organizing group of the History of Science Society may quite naturally be divided into three subgroups (Table III). 9 The largest subgroup, containing an extraordinary combination of talent and power, was that of the "patrons". Its twenty-six members--all scientific administrators and researchers from the "first generation"--were in their mid-fifties or older. They had extensive European experience as students--21 of 26--strong and explicit commitments to peace and internationalism--10 of 26--and enormous influence within and outside the academic community. Thirteen of them were in the National Academy of Sciences. America's only Nobel-laureate was one. The thirty-four presidencies of national scientific societies they had severally held included the American Association for the Advancement of Science, the National Academy of Sciences, the American Chemical, Mathematical, Philosophical, Physical, and Psychological Societies, the American Societies of Anatomists, Naturalists, and Orientalists, the American Neurological Association, the American Council of Learned Societies, the National Research Council, and the Royal Society of Canada.

A clear difference in age, prestige and cumulative advantage separates this sub-group from the next, that of the nine "proto-historians" of science. The "proto-historians" of science were relatively young--five of the nine were in their forties--, they had less experience of Europe and they did not enjoy eminence in learned societies. The more or less exiguous academic positions they enjoyed owed more to their energetic exploitation of the pre-war tradition of history of science as an appropriate subject for science students than to any deep commitment to them by the patrons. The social space between patrons and practitioners-- exemplified in the distance between, say, Frederick Barry and Edgar Fahs Smith--helps to account for the contrast between the lofty expectations of the former sub-group and the difficulties of the latter in establishing their subject in the universities. The final subgroup--the "sympathizers"--consisted of physicians and a remarkable camaraderie of young historians associated with James Harvey Robinson and the call for a "new history." Had the links between these historians and the history of science proved stronger, developments might have followed a different course. However the "new history" had battles enough of its own to fight, in disciplinary and departmental contexts quite remote from those of either patrons or practitioners of the history of science.

THE EXECUTION OF THE PROGRAM

Sarton was of course the most active among the various propagandists on behalf of the history of science, as he was also the most tireless organizer of its necessary intellectual apparatus. He created Isis, the journal which the History of Science Society sustained. He produced the long succession of "critical bibliographies" in the subject. He worked on what was designed to be the definitive Introduction to the History of Science. He produced a stream of arguments for his chosen field. His formulations of the importance of the subject, and of the moral worth of science, took on power and resonance to the extent that they coincided with the ideology pervasive among the patrons of the activity. This was true when he argued for the importance of science and of encyclopaedic knowledge, or for the special merits of the Anglo-Saxon tradition. Just as resonance between the intellectual orientations of patrons and practitioners helped shape the work of the latter through patterned approbation and applause, so a lack of resonance could be powerful in shaping by indifference and neglect. By way of example, one of Sarton's ambitions was "to carry on simultaneously research on ancient science and on nineteenth century science." His first intention blossomed while the second quietly withered.10

The history of science itself made only slow progress toward an independent academic status, but remained a quiet source of support to that positive, progressive view of science which was formed in the

heroic age. Only in the aftermath of another World War and in the context of new concerns about the place of science in American society was
the subject to gain a full-fledged professional identity.   A still
further generation was to elapse before historians of science would feel
the need to analyze anew their relations both to the scientific estate
and also to the historical profession.

TABLES

TABLE I

Ph.D. (or Equivalent) Degrees Conferred: Five Year Cumulations, 1871-
1896

Period	Number of Degrees Conferred
1871-75	89
1876-80	192
1881-85	276
1886-90	574
1891-95	1150

Source: U. S. Bureau of the Census, Historical Statistics of the
United States. Colonial Times to 1970. (Government Printing
Office, Washington, D. C., 1976), Series H 751-765. In this
and subsequent tables, statistics will be given to three
significant figures.

TABLE II: INSTITUTIONS OF HIGHER EDUCATION—INCOME AND SIZE

Size and Income—All Institutions

Year	Total Faculty	Enrollment	Income: All Sources, in Millions of Dollars
1880	11,600	116,000	not available
1890	15,800	157,000	21
1900	23,900	238,000	35
1910	36,500	355,000	76

Statistics for Certain Universities

Year	Registration—all Departments			Graduate School Enrollment		Teachers of Professional Grade	Other Teachers and Research Fellows
	Harvard	Yale	Columbia	Harvard	Johns Hopkins	Harvard+	Harvard+
1870	1,320	755	776	-	-	45	14
1880	1,370	1,300	1,530	43	102	70	65
1890	2,270	1,650	1,670	132	233	90	108
1900	4,290	2,540	3,180	353	168	134	277
1910	4,120	3,280	5,120	429*	187*	194	416

* Figures for 1908   +Figures for 1868, 1878, 1888, 1898 and 1908

Sources: A. Historical Statistics, H 689-699, H 700-715, H 716-727.

B. Henry James, Charles W. Eliot (2 vols, Houghton, Mifflin, Boston, 1930) 2, appendices C and D.

84

TABLE III

The Organizing Group of the History of Science Society

Subgroup	No.	Age Range in 1924	No. with Foreign Birth or Educational Experience	No. of members of Nat. Academy of Science	Presidencies of Other National Societies, pre-1924	Active Commitment to Internationalist Politics
Patrons						
Administrator	9	55-74	8	7	24	6
Researcher	17	54-73	13	6	10	4
Total	26	54-74	21	13	34	10
"Proto-historians of Science"	9	40-57	5	0	3	0
"Sympathizers"	12	35-55	6	2	2	0
Total	47	35-74	32 (68%)	15 (32%).	39	10

Sources: Information derived from the DAB, The Biographical Memoirs of the National Academy of Sciences, American Men of Science, the National Cyclopaedia of American Biography, obituary notices in Isis, etc.

NOTES AND REFERENCES

This present paper presents the results of an extensive investigation, in abbreviated form. I am indebted to many colleagues for helpful conversation, comments and criticism.

1. See Thomas S. Kuhn, The Structure of Scientific Revolutions (2nd edition, University of Chicago Press, Chicago, 1970: quotation from p. 210); Mary Hesse, The Structure of Scientific Inference (Macmillan, London, 1974); David Bloor, Knowledge and Social Imagery (Routledge and Kegan Paul, London, 1976: quotation from p. 2).

2. E.H. Carr, What is History? (Vintage Books, New York, 1967 [original edition, 1961]: quotation from p. 54). The same point, in other words, may be found in much recent historiographical writing. See for example p. xxiv et seq. in G.N. Clark's "General Introduction" to Volume 1 of The New Cambridge Modern History (Cambridge University Press, Cambridge, 1961).

3. Barry Barnes, Scientific Knowledge and Sociological Theory (Routledge and Kegan Paul, London, 1974: quotation from p. 116).

4. The standard account is Lawrence R. Veysey's The Emergence of the American University (University of Chicago Press, Chicago, 1965). See also Christopher Jencks and David Riesman, The Academic Revolution (Doubleday, New York, 1968). Among the many excellent studies of the broader context, I am most indebted to Robert H. Wiebe, The Search for Order. 1877-1920 (Hill and Wang, New York, 1967).

5. See Herman Skolnick and Kenneth M. Reese, A Century of Chemistry (American Chemical Society, Washington, D.C., 1976). Statistics on degree conferrals are taken from Douglas L. Adkins, The Great American Degree Machine (Carnegie Commission on Higher Education, Berkeley, 1975).

6. The best guide to these early developments is F.E. Brasch, "The Teaching of the History of Science," Science, 1915 $\underline{42}$, 746-760. The M.I.T. course of Sedgwick and Tyler led to their Short History of Science (MacMillan, New York, 1917), a revised edition of which reached its seventh printing in 1958. Sedgwick, Cross and Tyler were all "first generation" Ph.Ds. Sedgwick (1855-1921) is memorialized in the DAB. Cross was educated at, and headed the physics department of M.I.T. Their colleague H.W. Tyler (b.1863) was also educated at M.I.T., and at Erlangen (Ph.D., 1889). He became in turn Secretary of the Institute, Chairman of its maths department and (in 1916) General Secretary of the American Association of University Professors. See Isis, 1940 $\underline{31}$, 60-64.

7. On the future-affirming character of modern science, see Charles C. Gillispie, "Remarks on Social Selection as a Factor in the Progressivism of Science," American Scientist, 1968 $\underline{56}$, 434-450 (especially p. 447). For a characteristic examination of the utilities offered by the subject, see C.R. Mann, "The History of Science—an Interpretation," Popular Science Monthly, 1908 $\underline{72}$, 313-22. Quotation from p. 315.

8. These developments are more fully documented and discussed in my contribution to the Fiftieth Anniversary Celebration of the [History of Science] Society, entitled "Five Phases of Pre-History, Depicted from Diverse Documents," Isis, 1975 $\underline{66}$, 445-453 and in Arnold Thackray and Robert K. Merton, "On Discipline-Building: the Paradoxes of George Sarton," Isis, 1972 $\underline{63}$, 473-495.

9.   The organizing committee of the History of Science Society con-
     sisted of thirty-seven men (their names are listed in _Isis_, 1924
     _6_, 3-7). If to their number we add the ten officers appointed in
     the first five years who were not among the original committee,
     we arrive at an "organizing group" of forty-seven people.

10. See Sarton's characteristic editorials in _Isis_, 1919 _2_, 1-7 and
     1924 _6_, 2-42. The quotation is from a 10 April 1918 letter of his
     to Robert S. Woodward (Sarton archives, Carnegie Institution,
     Washington, D.C.).

Prof. Semen R.Mikulinsky (USSR)

INTERNALISM-EXTERNALISM CONTROVERSY AS A PHONY PROBLEM

Both in my report to the XIVth Congress on the History of Science in Tokyo and in my earlier works I have already voiced my conviction to the effect that neither internalism nor externalism can serve as a theoretical foundation for elucidation of the history of science.

Internalism as well as externalism represent simplifications, limited and one-sided schemes; therefore one cannot speak of the choice between them as there can be no choice between vitalism and mechanicism.

Since both these theoretical constructions are wrong, their synthesis cannot be fruitful; this is similar to the fact of impossibility of the synthesis between mechanicism and vitalism. The situation can be resolved only by rejecting these two constructions and developing an alternative theory.

First of all, it is necessary to provide a rigorous definition of internalism and externalism.

It has been already about forty years that these trends struggle with each other, but the question is still far from being clear, especially with respect to externalism. The overwhelming evidence points to the fact that at present this issue is even more confused and less clear than forty years ago.

According to the internalist conception, the development of science is governed by the immanent laws of its own. Internalism is frequently ascribed the rejection of any influence of economic and social factors on the development of science. This is not quite correct. Their influence is recognized by internalists, but only in the sense that these factors can either promote the development of science or slow it down. They can exercise no influence, however, on the structure of knowledge, its nature and the direction of scientific development. They are but a stage for the show. It can impede or favour the show but can have no effect on the course of events depicted in the play.

The situation with externalism is more complex. As will be shown below, the term "externalism" is actually used to denote two completely different approaches to the analysis of the development of science; in this process the Marxist and the purely externalist are mixed, the latter meaning the reduction of science to a mere epiphenomenon of the economic and social conditions, its reduction to the direct expression of the economic and social relations of a given epoch.

INTERNALISM AND EXTERNALISM AND THE PROBLEM OF CAUSAL EXPLANATION IN THE HISTORY OF SCIENCE

Any science steps into the phase of its maturity at the time when a certain theoretical framework of the field is being built and the necessity to provide explanation of the phenomena of the field is realized. The most ancient eposes of India, Mahabharata and Ramayana contain the most refined description of the way of life of many animal and plant species; the ecology, however, appeared only after many centuries when the causal relation between the environment and the way of life had been established.

The history of science, if it desires to be a science, should not just answer the questions what, when, where and who, but analyze the law-governed movement of scientific knowledge and the historical conditionality of events in the history of science, their relation to specific historical conditions as well.

If a conception claims the capability of a theoretical interpretation of the history of science, of being able to play the role of methodology of historico-scientific research, it should elucidate the development of science as a regular process, identify its motive forces.

Let us make an attempt to evaluate internalism and externalism from the point of view of this criterion. Such a formulation does not cancel the problems produced by these trends in the course of their history, of their struggle with each other. None of them was completely fruitless. Each trend has identified the problems previously were not paid attention by the historians of science. Over the last quarter of the century, however, the history of science as a scientific field witnessed considerable changes in connection with large-scale shifts in the science and the change of its social role, its place in the contemporary society, with the new aspect of the old question concerning the social value of scientific cognition. The historiography of science became mature and turned into an independent science with its composition of empirical and theoretical knowledge. Instead of being a purely descriptive field it becomes an explaining science. It is this fact that produces the need for the new interpretation of its theoretical grounds.

We can and must now, having in mind the above changes in the position and role of internalism and externalism, deal with the controversy on a new level, and first of all, to adopt a point of view determining to what extent these trends are capable of identifying the regular nature of development of the scientific knowledge, and give it a causal explanation.

One cannot ignore the fact to-day that the internalist conception results in that the intellectual aspects in the development of science are totally opposed to socio-historical ones.

This conception, even if we take into consideration the merits of its proponents in the reconstruction of the history of the scientific ideas, actually failed to solve the problem of the explanation of the history of science as it interpreted as absolute the thesis that the development of scientific knowledge can be explained only from itself and restricted in principle the analysis of the scientific development by the development of the scientific concepts inside themselves.

For instance, Alexandre Koyré demonstrated brilliantly that the scientific revolution of the XVIIth century was associated with the radical changes of the Antique and Medieval worldview, the destruction of the picture of the Universe that was characteristic for these epochs. But how can one explain this very destruction, what caused it and why it occurred exactly in the XVIIth century and not substantially earlier? Why the Scientific Revolution that led to the formation of the New Science took place in Europe only and why science understood from the modern point of view did not emerge in countries of ancient Orient civilizations (China, India) in spite of the high cultural level in those countries and in spite of the fact that in the Medieval times they were considerably abead of Europe in many areas of knowledge and its practical application? Internalism, restricting itself to the sphere of "pure" thought only, neither does nor can answer these questions. On the other hand, if one explains the above phenomena by a random intellectual "mutation", one rejects any explanation since such an "explanation" can be offered with respect to any new event in science and to any epoch; such explanation, therefore, has no cognitive value whatsoever.

Externalism, quite contrary to internalism, appeared from the very beginning promoting the explanation of the historic causality of the development of science. The externalists placed on causality the essence of the problem, worked for it and are still struggling with internalism for about half a century. This undoubtedly constitutes a merit of externalists and determines their role in the historiography of science. Externalism, however, adopted the straightforward and simplified form of the Marxist thesis concerning the determining role of the social and historical practice in the development of society, and, therefore, in the spiritual life. As a result, externalism, contrary to the true Marxist view on the development of science, employed vulgar sociology and economic determinism that were alien to Marxism and tried to deduce the explanation of the development of science directly to the social and economical conditions. It led, naturally, only to a rough scheme since the most important link was missing - the interaction between the material and the spiritual spheres, which was fine, complex, multilevel, multimediated and by no means straightforward, like mechanistic determinism.

We come across a kind of such simplification and reduction to a rough scheme in the works of Kinosuke Ogura, the Japanese historian of mathematics, the author of "Arithmetic and the Class Society" (1929), in the works of M.Leroy, F.Borkenau, B.Gessen, E.Zilsel, and most recently in A.Sohn-

Rethel's monograph "Geistige und körperliche Arbeit" (Frankfurt, 1972).

In order to arrive at true explanation of the history of science it is necessary to trace, in a concrete way, the manner in which transformation of scientific concepts and theories, the problems and structure of science took place in science itself; to analyse, how, through what mediated links the social and economic conditions and needs realize in science. This analysis should be carried out not in general but specifically in each case and take into account the transformation of the above concepts, in correspondence with the science's own laws, into its specialized forms, being, so to say, translated into the language of science and incorporated into its tissues. This cannot be done if one ignores the specific nature of science as "spiritual production" (K.Marx). One can obtain an appearance of the explanation of the historic causality, but not a true causal explanation.

But the greatest irony is in the fact that the externalism, with all the outward show of its being opposite to internalism, implies that the social and historic conditions appear as purely external with respect to science, which is exactly the case with internalist conception. But externalism and internalism ignore the problem of identifying the mechanisms and the methods of their transformation into the form, specific for science. The interaction of social and economic conditions with science is considered by mechanistic-causal (Laplacian) determinism as a direct, immediate influence. In this case the economic and social conditions indeed appear with respect to science as purely external, and it becomes impossible to identify their influence on the inner status of science; the reason for this is that science by its nature differs in principle from the material production, economics, social relations, the political history and for all their enormous influence on science it cannot be reduced to them.

Thus, since neither internalism nor externalism can explain the development of science, we call the externalism-internalism controversy a phony problem. It is imaginary for us as although internalism for the first sight is opposed to externalism, no solution can be found in the frame of this controversy.

We have already mentioned that externalism reduces science to a mere epiphenomenon of the economic and social conditions, neglects the specificity and the relative independence of science. Unlike externalism, Marxism is alien to the direct reduction of science to economics and social conditions. It does not ignore the specificity of science, the activity and relative independence of thinking, the effect of the available system of knowledge, of the ideas and concepts on the subsequent development of science. These two approaches to science are radically different and it would mean to quarrel with logic if one does not differentiate them, mixes them with each other. At the same time, this is frequently exactly the case, and the term "externalism" is being used for denoting the fact that in order to explain the development of science it is not sufficient to restrict oneself to the analysis of the movement of scientific ideas, that the material conditions of the life of society exercise a determining influence over the set of problems, the nature and the direction of the scientific development for all its relative independence and the presence of specific regularities of development of the scientific knowledge. If we neglect the material conditions of the social life, it becomes impossible to understand and to give an adequate elucidation of the history of science. But in this case we are dealing already not with the internalism-externalism controversy but with that between both and Marxism; here we come across a quite real and essential problem, instead of an imaginary one. The problem does not consist in that internalism has as its basis the presence of specific, inner laws of the development of scientific ideas, concepts and theories. Beginning from the times of K.Marx, the Marxist conception of the scientific development not just used it as its foundation but stressed this aspect continuously. There is nothing new in principle in internalism as far as the above is concerned. The object of divergence is different. It lies in a completely different plane. Internalism limits in principle the analysis of the scientific development by that of the movement of scientific ideas, theories, methods; it negates the internal connection, between the nature, structure and the trends of development of the scientific knowledge, on the one hand, and the material and spiritual life of society, the socio-historical practice, on the other. From its point of view science represents an isolated, completely autonomous sphere which does not depend on economic and social conditions. This constitutes the essence of internalism and of the radical divergence between it and the Marxist concept of the development of science, rather than the acceptance of the inner laws of movement of the scientific ideas, concepts, theories and methods.

There is absolutely no doubt about internalism being opposed to the Marxist theory of the development of science. But it would be as absurd to call this theory externalism, as, for instance, to classify Shakespear, Balzac and Leo Tolstoi as naturalists because they depicted extremely accurately the life of people, or to call them modernists because each of them emerged as an innovator when compared to his predecessors and created a style of his own reflecting his view of the world.

We believe that it would not be justified to say that such scientists as J.Bernal, and all those historians of science who identify the connection between the development of science and the economical and social structure of the society and realize clearly the presence of specific laws of development of science, without permitting to reduce the history of science to the economic and social conditions, are externalists. It would be more correct to call the trend represented by such historians of science historico-monistic if we have in mind the internalist absolute division of the history of science into "external" and "internal" being opposed to.

THE SOCIAL NATURE OF SCIENCE AND THE MECHANISMS OF ITS DETERMINATION

When dealing with the question of social determination of the development of science we consider essentially three levels of analysis. The first one consists in the identification of the social nature of science.

The second level of analysis is associated with the study of the science belonging to certain large-scale epochs being historically conditioned in its history under the conditions of various civilizations and cultures. The third level represents an analysis of determination of development of individual scientific theories, conceptions, notions, etc.

As the analysis proceeds from one level to the next the problem usually becomes ever more complex. It should be concerned not so much with the already obtained knowledge but with the genesis of new one, if it is to be fruitful. As we see it, the theoretical difficulties of the history of science as a scientific field of the last half-century are associated precisely with the fact that the study of genesis of the new knowledge, being influenced by positivism, was not paid sufficient attention. We think that the future of the history of science is related exactly to the analysis of the mechanisms of establishing the new knowledge.

The question of the motive forces of the development of science represents a key problem for the theory of the development of science and the methodology of historico-scientific research. This question can be formulated in a more narrow form as that of the relation between the cognitive (epistemological) and the socio-historic (meaning the set of socio-economic, cultural-historic, sociological, ideological, psychological, etc.) factors in the development of science.

One cannot obtain an answer to this question by applying the philosophy of science based on positivism, inductionism, or conventionalism. Neither has it been developed in the history of science.

In order to approach this problem, it is necessary first of all to take into account the fact that science is not just a set of various forms of knowledge, but a product of certain activity, some kind of work.

This activity is of a specific nature, characteristic only of itself; it is directed at the production of new knowledge. The scholar in the process of this work forms special relations with all the researches existing previously and at present, as well as with nature. This generates the specific features of the scientific work.

According to K.Marx, the scientific work, unlike other forms of work, should be called "universal labour". Regardless of who and where the knowledge has been obtained, it becomes the property of all humanity. At any time of its history science should be viewed as a result of the efforts of all previous generations, the cooperation of contemporaries and the foundation for its future development. It is clear even from the above that although the scientific work is intellectual and creative and its results become associated with the name of a certain scientist, it is distinctly social in nature.

Like any social work, it is caused by the needs of society, and its existence is conditioned on its being able to satisfy the society needs; it is therefore determined by the latter. Although the scientific work is predominantly intellectual and creative, it cannot be realized in the frame of pure thought. It requires certain skills, techniques, means of research, tools; all those result from the effert of many people engaged in material production. The social practice the needs of which at the same time stimulate the development of science, varifies the product of the scientific work the new knowledge. Finally, and this should be kept in mind, the cognitive activity itself is always being carried out in certain forms that are historically conditioned and is governed by definite norms and criteria chosen by the society. The result of the scientist's work, in order to be recognized and accepted in science, should meet the norms and criteria which are conventional for the scientific community. Therefore, the scientific work is associated with society so closely that although a scientific discovery is arrived at by a concrete scientist or a group of scientists, the subject of cognition is actually the society itself.

The very nature of science is social, and its social characteristics, instead of being external with respect to it, are inherent for science.

It is always important to have in mind when one carries out the analysis of development of science that the latter exists as a certain abstraction in our consciousness only, that it appears over the course of history in historically different forms. It is not just the contents of science but its approach to nature as well, the method of thinking and even the very concept of science, its perception by the society that varies from one historical epoch to the other.

As far as the nature and structure of science of each period are concerned, it is essential to grasp that of its components which is dominant for a given epoch. The key concepts gradually become prevalent in the whole scientific community. Each period creates its own picture of the world which is formed by complex interweaving of the chief scientific, philosophical, ideological and, up to a certain time, religious notions and represents a reflection of the real being of a given time period, its material and spiritual life that has been transformed in the people's mind.

This image reflects the world view which is special for every epoch and prevailing in the people's consciousness, gradually acquiring the force of a prejudice. The achievements of the scientific thought are being projected on it. It determines, in its turn, the characteristic features of science of a given time period. The influence of the economic and social factors exercised over the structure and the nature of science is mainly transferred through this picture as a mediating link. It becomes all the more possible since these factors have been already transformed in the world image and appear in it as included in the intellectual context rather than in their pure form.

An expression - science and society - is frequently used. It can be accepted only conditionally. Science is created by society and constitutes an organic part of it. There is no science outside society. It does not have an existence separate from society, science represents a subsystem of society, its culture. Therefore the absolute division of the history of science into the intellectual (the internal) theory and the social (the external) one which is characteristic of internalism does not agree with the real situation in science. We can and do speak of two forms in which science appears - of science as a system of scientific knowledge and of science as a special form of activity, or the social institute. But those are just different crossections of science which we consider, and not two independent forms of reality. Science regarded as a social institute or a special form of activity without its results - the system of knowledge - and divorced from its object content becomes meaningless and turn into fiction and nonsense. The forms of the organization of science, the social psychology of scientific communities, etc. do change; it is accompanied and influenced by the changes in the object contents, the structure and the social functions of science and cannot be completely understood outside of them.

Science has no two histories independent from each other - the internal and the external ones. If the process of development of science is to be reflected in the same way as it occurs in reality rather than to be represented by abstract schemes, we can only consider the histories in their unity. It is a different matter that it is possible to single out one or another aspect of science for the purposes of research, and to consider its history. It is being done when the task is to investigate deeply and in a special manner the process of institutionalization of science, the history of scientific organizations, communications in science, forms of its organizations etc. or the history of scientific ideas, especially the development of the scientific knowledge.

But also on the same ground it is not possible to recognize the reasonable character of absolute division and counterposition of external and internal factors influencing the development of science, which tendency was supported by internalism, surely, one must admit that both types of factors are distinguished. But the absolute counterposition of them is wrong. One cannot once for ever put some factors in the category of external ones, other of those of internals. Here each case requires a concrete analysis, as the factors, which in one case play the role of external ones, in other situation could be recognized as internal ones and vice versa.

The economic, social, psychological, logical, etc. factors are external inasmuch as they remain neutral with respect to the contents of science; they become transformed into internal, however, as soon as they start interacting with the object contents of science. Here one can establish an analogy with organisms. The environment can influence the phenotype, the ontogenesis of organisms, promote their growth and development and leave their heredity unchanged. But if they affect the hereditary basis, the genetical apparatus, the heredity is changed. The external factors turn into internal ones.

The history of science clearly reveals the relative nature of the notions "external" and "internal" and their mutual transitions.

As is well known, the founder of the scientific embryology Karl Baer had arrived at the recognition of the evolution of organisms even before Darwin and independently of him. He presented in his work of 1834 "The Fundamental Law of every Development in Nature" a large set of data in favor of the acceptance of evolution - the variability of domesticated animals and plants, the geographic distribution of plants and animals over the Earth Globe, the paleontological evidence of evolution. All these groups of facts complemented later by the data on triple parallelism were extremely important in Ch.Darwin's system of evolution proofs. Only one step was missing in Baer's theory - to find the explanation of evolution. He literally was holding a key for such an explanation. According to his statement of 1850, some find it, in that the animals who are poorly adapted to the environment become extinct. But this explanation was rejected by Baer. And later, when Ch.Darwin's "Origins of Species" appeared, Baer, although he demonstrated in detail 25 years ago that there existed the evolution of the organic world, spoke against Ch.Darwin's theory. How could these facts be explained if one accepts only the logic of the development of the evolution idea? The logic of development of the evolution idea can prompt nothing here. Baer's course of thinking contradicts it. At the same time, the above facts can be explained quite obviously. Baer was a convinced idealist; he believed that "goal-orientedness" (Zielstrebigkeit) was inherently present in nature, and evolution represents a realization of the immanent law of nature - the movement towards the greater dominance of spirit over matter. In 1850 and later, when Ch.Darwin's theory was published, Baer rejected the idea of selection because it contradicted his philosophy and, according to his own statement, led to materialism in both cases. What, then, was the

structure-forming nucleus of Baer's conception? There can
be no two opinions here: Baer's ideology, his philosophical
convictions, in other words, a factor which was external
from the point of view of a consistent internalism. This
external factor, however, turned into the nucleus of Baer's
evolution conception, determined its structure, i.e., became the internal one.

As followed from the subsequent development of biology,
Baer's conception concerning evolution was wrong. The external which became the internal and determined the contents
of Baer's approach to evolution was realized as contradictory to the scientific facts, in other words, external to
it and incompatible with it. Thus the external became the
internal, to be transformed later in the course of development of science into the external again.

Gerald Holton's conception is of considerable interest
which was formulated in his book "Thematic Origins of Scientific Thought. Kepler to Einstein" (Cambridge, Mass., 1973)
and which represents a definite step forward as compared
to the traditional internalist approach. G.Holton applies
the thematic analysis to the study of the new knowledge in
its "nascent moment", the influence of the peculiar features of an individual scientist activity and personality on
the resultst of his work and thus attempts to construct a
model of the growth of science which would be free from
being "far too blunt for laying bare the fine structure"
and "popular caricatures" which could only be obtained by
the internalist and the externalist approaches.

Holton considers the difference between two types of activity - the activity of an individual scientist and that
of the scientific community on the whole "The former aspect
does not have to remain tacit, nor should it be discredited
as inaccessible to rational study" (p.15).

One may consider two types of activity in Holton's works
as the ontogenesis and philogenesis of science. While philogenesis is not a sum of ontogeneses, and ontogenesis does
not represent a repetition of philogenesis, they interact
closely with each other and carry each other's traces. In
this respect they are erased, transformed, do not appear in
their pure form, and nevertheless it is exactly the interaction between ontogenesis and philogenesis that expresses
the continuity and the historic development of one or another evolution branch. If this biological phenomenon is
used as a model, it would be easier to understand the mechanism of inclusion and transformation of the individual
and the external into the internal which constitutes the
tissue of science.

When the problem of interaction between external and internal factors in the development of science is analyzed in
connection with the activity of an individual scientist,
the experience of other sciences in the study of the cognitive process, psychology in particular, deserves close attention of the historians of science. Psychology employed
experimental and quantitative methods to reveal a number of

important laws of the consciousness formation in the ontogenesis. Among these mechanisms that were traced in the
works of the outstanding psychologists Piaget (Switzerland).
Vygotsky (USSR) and others one can single out the establishing of determination dependence between the psychological
functions, on the one hand, and the process of actual communication of the individual with the external world and the
social environment, on the other. Before a mental act becomes an inherent property of the subject, it has been formed
in the objective system of relations, in the object external actions. Later those external actions become transformed into the internal ones (become interiorized).

The theory of interiorization, naturally, cannot be immediately applied to the activity of the scientist; the
depth mechanisms of the cognitive processes revealed by it,
however, can frequently prompt the historian of science and
protect him from the long existing notions of spontaneity
and independence of the intellectual processes from everything external.

ATTEMPT TO RECONSTRUCT THE GENESIS OF A SCIENTIFIC THEORY

I am far from adopting a naive conception to the effect
that an example or even a sum of examples can disprove internalism. We know only too well that a single example can
be opposed by ten others - the opposite ones. Thus, it is
not the individual examples that we are being concerned
with but the juxtaposition of internalism to the whole experience of the scientific development considered in its
wholeness as a social cognitive activity.

The historian of science, however, always deals in his
work with a concrete phenomenon - the history of a certain
science, a theory, a discovery, etc.,the activity of a certain scientist. And if his general theoretical or methodological conception suggested for the interpretation of the
development of science is true, it should help the historian of science to arrive at a deeper and better understanding of the concrete phenomenon under study. Therefore, I
would like to suggest considering the genesis of Ch.Darwin's
evolution theory not as an example but rather as a model of
historic reconstruction and verify whether this model corresponds to internalism.

As is well known, the idea of evolution of species was
arrived at by Darwin as early as in the beginning of the
fourth decade of the XIXth century when he was influenced
by the observations performed by him during his round-the-
Earth journey on board of Beagle. Ch.Darwin's evolutionary
point of view became so well shaped by the middle of 1837
that he began writing his famous Notebooks on Transmutations of Species. The real turning point, however, came when he conceived
of natural selection; thus it is important to seek an explanation of
how this idea originated.

There have been numerous attempts to explain the appearance of Darwin's theory by the influence exercised on Darwin by Malthus' "Essay on Population". It was Darwin him-

self who was to some extent responsible for this version. Those who advanced it, however, did not pay attention to the fact that after having mentioned in the Autobiography the book of Malthus, Darwin immediately made a statement that he "was well prepared to appreciate the struggle for existence which everywhere goes on". We owe to Prof. Gavin de Beer who drew the attention to Darwin's Notebook of 1837 the fact that we know now that even in 1837, i.e., fifteen months before he read Malthus' book Darwin arrived at the thought that those not adapted to the conditions of existence become extinct and only the adapted ones survive Species "die if they do not change ... it is a generation of species", as Darwin wrote in his Notebook in 1837.

At the end of the XVIIIth - the first half of the XIXth centuries there appeared very many writings on the fierce struggle for existence in nature. I would like to refer the reader to the statement of the Moscow University professor Ch. F. Rouiller (1814-1858). His publications were mainly in Russian, and unfortunately nobody in the West knows about him. Rouiller was a convinced and consistent evolutionist. The development and advancement of the evolution approach to the organic world was the most important occupation of his life. Even before Darwin he created a school of zoologists-evolutionists.

The experience of English animal breeding and gardening served for Darwin as a key to the construction of his evolution theory. It was actually this idea and not that of the struggle for existence that was the main, structure-forming nucleus, the pivot of Darwin's theory. His practical experience was not merely an impetus, a stimulus for the development of a theory, not just a factor that promoted the clarification of the needs and thus prompted the choice of the direction of research and stimulated them, which is frequently the case in the history of science (for instance, the development of thermodynamics in connection with the development of steam engines. It was extremely important for the very structure of the theory. The idea of selection borrowed from practice determined the nature of Darwin's theory.

As to the idea of competion and struggle for life which was (regardless of how accurately this factor was reflected in Darwin's book) of considerable importance for the construction of the theory, it was not difficult to establish that it was produced by the social relations in England of the time of flourishing capitalism. The idea was "in the air" of England of that period. At that time the competition, the struggle for a place under the sun which was accompanied by enrichment of some at the expense of others, the struggle for existence were recognized as a natural and indisputable law of the human society. One did not need Malthus' book in order to perceive this. The above was the opinion widely spread at that time.

The ideas of competition and the struggle for existence arose outside the realm of science. Barry Gale (Isis, 1972, vol.63, No.218, pp.321-344) correctly writes that "... in some sense the language of the Origin was the language of the Victorian age... Within the context of the Victorian era Darwin seems to be merely another in a long line of Victorians emphasizing struggle" (p.343).

The influence of practical experience and the social relations on the nature of a scientific theory that was revealed in such a distinct and clear form represents a rather rare case. It does not become less convincing, however, because of this. The fact that one does not come across such cases frequently just speaks of the complexity of relations between science and social practice, the economic, social and spiritual conditions of the life of society.

Although scores of thousands of people knew about the variability of animal and plants when domesticated, thousands applied it in practice in various countries, hundreds of thousands had some idea about the struggle for existence and even lived under the conviction that it represents a natural law of the existence of the world and the human society, it was only Ch.Darwin who used these concepts to create his theory of evolution.

The notorious story about the falling apple that led to the emergence of the theory of gravitation is known by everybody and can serve as an example of simplification and vulgarization of the problem of explaining the genesis of a theory. Scores of millions of people observed the free falling of bodies but only Newton created the theory of gravitation.

We are facing here necessity to study the personality of a scientist, the unreproduceable peculiarities of his mind, ideology, his perception of the world and, correspondingly, the problem situations in science to which he is devoted.

When tracing the formation and the peculiarities of the personality and the creativity of an individual scientist, reconstructing the path of his research, the course of his thoughts, we can, if allowed by the available material, penetrate into the laboratory of the scientific creativity. And here it becomes extremely obvious visually what can be by no means always extracted from studying the final product of to scientist's work - the scientific text, since, according to the norm established in science, the scientist thoroughly purifies it from everything which is not directly related to the object of research.

Speaking about this, I have in mind something greater than the creation of the biographies of scientists which represent a very important and independent kind of historic scientific studies. I have in mind one of the means the combination of which with other methods makes it possible to approach closer the process of origination and creation of new knowledge and the identification of invariants which allow to perceive the mechanisms and laws of the development of science.

Both the morphologist who begins to study the embryonal stages of an organism development by certain features which disapper when the embryo reaches the adult age and establishes the connections between a given species and others clarifying the philogenesis of species and the historian of science who includes into his sphere of analysis not just the final results in the form of scientific texts can identify those elements which cannot be located in the scientific text.

When the creative process is studied in such manner one frequently identifies very clearly the effect of the philosophical views of a scientist, his ideology, his seeing of the world, his motives, the direct and indirect stimuli of his creative work, the role of the general and immediate social environment, the conditions under which he has been formed, lives and works, on the choice of direction and the course of research.

CONCLUSION

Thus, the mechanisms and laws of the development of science which in the present context means first and foremost the formation of new knowledge can be explained not through the fundamental limitation of research by the sphere of logical unwinding of the scientific concepts (this is what internalism calls us to), not through reduction of the explanation of the history of science to social and economic conditions exclusively (it is what externalists try to do unsuccessfully)but by means of an analysis of interaction of the object contents of science, the socio-economic and cultural-historical conditions and the personality factors in the realization and identification of their dialectic unity and the determining influence of the social-historical practice on the development of this interaction.

When internalism substitutes the concept of logical in place of that of rationality and calls everything which cannot be directly deduced logically from the previous knowledge, any appeal to the social, economic, cultural-historical conditions, the psychological peculiarities of the scientist's personality or the social psychology of a scientific community for the purpose of explaining the genesis of new problem situations and appearance of new elements in science the irrationalism, it dooms the historians of science to eternal life on an isolated island among a territory that is unknown to and closed for them, a territory which is called the living human world, the world of developing culture where the process of mastering nature takes place continuously, and in the course of this process a person hot only enriches its knowledge but transforms its thinking as well.

Günter Kröber

"Science at the Cross Roads" - Voraussetzungen und Folgen

"Science at the Cross Roads"

Von den 15 Internationalen Kongressen für Wissenschaftsgeschichte, die in den verflossenen 50 Jahren in verschiedenen Teilen der Welt stattgefunden haben, hat keiner soviel Aufsehen erregt wie der II. Kongreß 1931 in London. Das herausragende Ereignis, dem er seinen nachhaltigen Eindruck und seine bleibende Bedeutung verdankt, waren die Beiträge der sowjetischen Wissenschaftshistoriker, insbesondere der Aufsatz Boris Hessens über "Die sozialökonomischen Wurzeln von Newtons 'Principia'"[1]. Teilnehmer des Kongresses, wie Joseph Needham, erinnern sich seiner als eines "epochemachenden Textes"[2]. Neuere Reflektionen über den II. Kongreß werten Hessens Beitrag noch immer als eine "Sensation"[3].

Was war der Grund jener Faszination, die von den sowjetischen Beiträgen und besonders von der Arbeit Hessens ausging? Worauf beruhte die Kraft ihrer Ausstrahlung, die die Organisatoren des Kongresses bewog, sie sofort nach dem Kongreß gesondert zu veröffentlichen?[4]

Noch ganz unter dem Eindruck der sowjetischen Vorträge spricht J. D. Bernal eine Woche nach dem Kongreß als von einem "Kongreß von Ideen, geboren durch die Revolution"[5], später - offensichtlich mit dem Blick auf Hessens Beitrag - vom "Ausgangspunkt für eine Neubewertung der Geschichte der Wissenschaft"[6]. Needham nennt Hessens Aufsatz eine "klassische Darlegung der marxistischen Wissenschaftshistoriographie"[7], und P. Weingart sieht in ihm "eines der anspruchsvollsten ... Beispiele marxistischer Wissenschaftsanalyse"[8]. Die Reihe solcher Einschätzungen ließe sich fortsetzen. Was die Sensation ausmachte, war also nicht, daß eine repräsentative Delegation sowjetischer Wissenschaftler erstmalig an einem Internationalen Kongreß für Wissenschaftsgeschichte teilnahm. Vielmehr war dies die theoretische Konzeption der Wissenschaftshistoriographie, die im Beitrag Hessens ihren konzentrierten Ausdruck fand.

Als Sensation konnten der Beitrag Hessens und anderer sowjetischer Kongreßteilnehmer deshalb erscheinen, weil im westlichen Teil der Welt die seit 1917, seit der Großen Sozialistischen Oktoberrevolution in Sowjetrußland und später in der Sowjetunion eingesetzte Entwicklung wissenschaftswissenschaftlicher und wissenschaftshistoriographischer konzeptioneller Ideen weitgehend unbekannt geblieben war. Diese Entwicklung gründete sich ihrerseits auf die bereits von Marx und Engels ausgearbeitete und praktizierte historisch-materialistische Betrachtungsweise der Geschichte der Wissenschaft.

Voraussetzungen

Deren Kerngedanke ist der von der sozialen Natur und

der sozialökonomischen Determiniertheit der Wissenschaft. Er besagt, daß wissenschaftliche Tätigkeit sich nicht nur schlechthin unter bestimmten gesellschaftlichen Verhältnissen vollzieht, die gleichsam als äußeres Milieu für die Wissenschaftsentwicklung auftreten. Voraussetzungen, Ausführung und Zwecke wissenschaftlicher Tätigkeit sind durch die Produktionsverhältnisse der betreffenden Gesellschaft und die aus ihnen entspringenden sozialen, ökonomischen, politischen, kulturellen und wissenschaftlichen Erfordernisse und Interessen ebenso determiniert wie die Bestimmung ihrer Inhalte und ihres Umfangs. Es gehört zum Wesen historisch-materialistischer Geschichtsauffassung, die Entwicklung der Wissenschaft aus den materiellen Existenz- und Entwicklungsbedingungen und -bedürfnissen der menschlichen Gesellschaft, genauer: der jeweiligen ökonomischen Gesellschaftsformation als Ganzes zu verstehen. Wissenschaft ist aus der Sicht des historischen Materialismus das historisch determinierte Produkt und ein aktuell notwendiges Glied der gesellschaftlichen Arbeitsteilung innerhalb einer historisch-konkreten ökonomischen Gesellschaftsformation. "Wird die materielle Produktion selbst nicht in ihrer spezifischen historischen Form gefaßt, so ist es unmöglich, das Bestimmte an der ihr entsprechenden geistigen Produktion und die Wechselwirkung beider aufzufassen", charakterisierte Marx diesen Zusammenhang (9). Engels hat dies ausführlich für die antike Astronomie, Mathematik und Mechanik (10) sowie für die Wende vom scholastischen Wissenschaftsbetrieb zur neueren Naturwissenschaft im 15.–16. Jahrhundert (11), aber auch für spätere Perioden der Entwicklung der Natur- und Gesellschaftswissenschaften belegt.

Der historische Materialismus faßt wissenschaftliche Tätigkeit als "geistige Produktion" (Marx). Er stellt damit zugleich die spezifischen Besonderheiten in ihrer Entwicklung in Rechnung, die sie von anderen Arten gesellschaftlicher Tätigkeit, insbesondere der materiellen Produktion, unterscheiden. Diese bewirken, daß die historische Entwicklung der Wissenschaft spezifischen Gesetzmäßigkeiten unterliegt, die gewöhnlich auch als "innere Logik" der Wissenschaftsentwicklung bezeichnet werden. Die relative Selbständigkeit der Entwicklung der Wissenschaft vollzieht sich auf der Basis ihrer generellen sozialökonomischen Determiniertheit, existiert nicht außerhalb dieser, sondern ist Ausdruck und Bestandteil derselben.

Die historisch-materialistische Sichtweise der Wissenschaft und ihrer Geschichte ist für viele Arbeiten sowjetischer Wissenschaftler und Wissenschaftshistoriker in den 20-er und 30-er Jahren kennzeichnend. Seit Lenins "Entwurf eines Plans wissenschaftlich-technischer Arbeiten" im April 1918 an die Russische Akademie der Wissenschaften übergeben worden war, spielten Fragen der Entwicklung der Wissenschaft in Sowjetrußland und in der UdSSR, der Stellung und Funktion der Wissenschaft in der Gesellschaft generell, des Verhältnisses von Wissen-

schaft und gesellschaftlicher Praxis, von Grundlagenforschung und angewandter Forschung, Prinzipien der Wissen-
schaftsplanung, der Leitung und Methodologie der Forschung,
des wissenschaftlichen Schöpfertums, aber auch des Aufbaus
einer allgemeinen Theorie der Wissenschaftsentwicklung eine
zunehmende Rolle in der Diskussion wissenschafts- und for-
schungspolitischer Probleme. Die Namen, die uns hier be-
gegnen, sind vor allem die führender Naturwissenschaftler,
weitsichtiger Gesellschaftswissenschaftler und hervorragen-
der Organisatoren der Wissenschaft jener Zeit.

V. I. Vernadskij - Geologe, Mineraloge, Wissenschaftshistoriker, Vorsitzender der Kommission Wissenschaftsgeschichte der AdW der UdSSR in den zwanziger Jahren und Mit-
begründer der Kommission zur Erforschung der natürlichen
Produktivkräfte Rußlands (1915) - betont in seinen Werken
den Gedanken des engen Zusammenhangs zwischen Wissenschaftsentwicklung und gesellschaftlichen Bedürfnissen, die Bedeutung der Wissenschaft für den gesellschaftlichen Fortschritt und die Verbindung der Wissenschaft mit den Volksmassen (12). S. F. Ol'denburg - Orientalist und Ständiger
Sekretär der Russischen und später der AdW der UdSSR von
1919-1929 - beschäftigt sich in seinen Reden und Aufsätzen
der zwanziger Jahre wiederholt mit dem Verhältnis von
Grundlagenforschung und angewandter Forschung (13). Der
Ökonom S. G. Strumilin wirft Anfang der dreißiger Jahre
die Frage nach einer Methodologie der quantitativen Erfas-
sung wissenschaftlicher Arbeit auf (14). Der Historiker M.
N. Pokrovskij, seinerzeit Direktor der Kommunistischen Akademie, wendet sich gegen eine einseitige utilitaristische
Auffassung von der Funktion der Wissenschaft in der Gesellschaft (15). T. I. Rajnov, Wissenschaftshistoriker und Mit-
arbeiter in der Kommunistischen Akademie, erörtert die neu-
en gesellschaftlichen Anforderungen an den Wissenschaftler
(16). Der Physiker A. F. Ioffe umreißt in verschiedenen
Beiträgen die Grundprinzipien, Quellen und Triebkräfte der
Entwicklung der Physik und der Naturwissenschaft über-
haupt (17).

Die Aufzählung ließe sich fortsetzen. Erwähnt seien lediglich noch der Philosoph und Logiker I. A. Boričevskij
und der Mathematiker G. A. Gruzincev. Boričevskij entwarf
bereits 1926 ein Programm für den Aufbau einer als Wissen-
schaftswissenschaft verstandenen Theorie der Wissenschaftsentwicklung, in dem - ganz im Sinne des heutigen wissen-
schaftstheoretischen Konzepts - Wissenschaft als eine spezifische Art gesellschaftlicher Tätigkeit verstanden wurde,
für deren Erforschung er die Soziologie und deren theoretisches Fundament, den historischen Materialismus, für un-
erläßlich hielt (18). Auch Gruzincev geht von der Auffassung der Forschung als einer gesellschaftlichen Tätigkeit
aus, entwickelt die Wissenschaftstheorie aber mehr in
Richtung einer Methodologie der wissenschaftlichen Arbeit (19).

Es trifft also keineswegs zu - hier stimme ich mit F.
G. Werskey überein -, daß der tiefe Eindruck, den die so-

wjetischen Beiträge auf und nach dem II. Kongreß hinterließen, etwa dadurch bedingt war, "daß - die ökonomische
Depression gegeben - sich die Aufmerksamkeit im Jahre 1931
natürlicherweise auf ökonomische Fragen konzentrierte, und
daß diese Vorbelastung der spezifisch marxistischen Doktrin,
derzufolge jeglicher wissenschaftliche Fortschritt real
durch ökonomische Ursachen determiniert ist, Vorschub lei-
stete"[20]. Die Spuren, die diese Beiträge auf dem wissen-
schaftshistoriographischen Terrain der westlichen Länder
hinterließen, waren vielmehr Ausdruck einer beginnenden
neuen Etappe der Wissenschaftsgeschichtsschreibung, deren
allgemeines theoretisches Fundament der historische Materialismus und nicht ein vulgärer ökonomischer Materialismus war und ist.

Folgen

Am ausgeprägtesten ist dieser Einfluß bei J. D. Bernal
sichtbar. Noch ganz unter dem Eindruck der sowjetischen
Beiträge schreibt er: "Die alte Konzeption der Wissenschaftsgeschichte - die einfache Aufzählung von Entdeckun-
gen und Erfindungen, die Darstellung des Lebens und Wir-
kens großer Leute, die Aufstellung der genealogischen Folge
des zeitgenössischen Wissens - erscheint heute als partiel-
le, wenngleich notwendige Grundlage für die Erforschung
der Wechselwirkung der Wissenschaft mit Ökonomie und Poli-
tik, Kunst und Industrie während des gesamten Geschichts-
verlaufs"[21]. Ein anderer Grundzug der historisch-materialistischen Betrachtungsweise der Geschichte der Wissenschaft, den er in den sowjetischen Beiträgen sofort ak-
zeptiert, war, daß die Wissenschaftsgeschichte für die sowjetischen Gelehrten "offensichtlich lebenswichtig war:
Sie war nicht lediglich akademische Forschung, sondern Anleitung zum Handeln"[22]. Beide Gesichtspunkte - die Wissenschaftsgeschichte als Bestandteil der Geschichte der Gesellschaft und in Wechselwirkung mit deren einzelnen Kom-
ponenten zu sehen und es als Aufgabe wissenschaftshistori-
scher Forschung zu begreifen, durch das Studium der Wissenschaftsgeschichte Erkenntnisse zu gewinnen, die der praktischen Gestaltung gegenwärtiger Entwicklungsprozesse der
Wissenschaft und der Prognose künftiger dienen - finden
wir in den späteren Werken Bernals als konzeptionelle Leit-
gedanken realisiert. Seine "Social Function of Science"
(1939) nennt er "den Versuch einer Analyse" dessen, wie die
Arbeiten, die Wissenschaftler ausführen, "mit den sozialen
und ökonomischen Entwicklungen zusammenhängen, die in ihrer
Umgebung vor sich gehen"[23]. Und seine "Science in History" (1954) schreibt er in der Absicht: "Um herauszufin-
den, wie man die Schwierigkeiten, vor denen wir stehen,
überwinden und die neuen Kräfte der Wissenschaft zum allgemeinen Wohl statt zur Zerstörung verwenden kann, ist es
notwendig, aufs neue zu untersuchen, wie die gegenwärtige
Situation entstanden ist"[24].
    Bernal hat diese Grundhaltung zeit seines Lebens be-
wahrt. 25 Jahre nach Erscheinen der "Social Function of

Science" kann er feststellen, daß die grundlegenden Prognosen über die Wissenschaftsentwicklung und das Verhältnis von Wissenschaft und Gesellschaft, die er auf Grund seiner historisch-materialistischen Analyse 1939 getroffen hatte, sich inzwischen erfüllt haben; dies gilt insbesondere für die Voraussage der in der Mitte des 20. Jahrhunderts einsetzenden wissenschaftlich-technischen Revolution (25). 1957 bekräftigt er auf einer Konferenz in Moskau nochmals seine Auffassung vom Wechselverhältnis zwischen Wissenschaft und Gesellschaft als einem Grundanliegen der Wissenschaftsgeschichtsschreibung: "Ich meine, man kann die Entwicklung der Naturwissenschaft nicht begreifen, ohne zu untersuchen, wie sie auf gesellschaftliche Anschauungen reagiert hat, und wie ihrerseits die Entwicklung der Naturwissenschaft die Gesellschaftswissenschaften - direkt, unter dem Einfluß von Ideen, oder indirekt, über die Veränderung der ökonomischen Struktur der Gesellschaft und damit über die Produktionsverhältnisse deren Ideologie - beeinflußt hat"(26).

Die in "Science at the Cross Roads" vereinigten Beiträge fanden bekanntlich neben Bernal Widerhall auch bei anderen, vor allem britischen Naturwissenschaftlern. Genannt seien J. B. S. Haldane, Lancelot Hogben, Hyman Levy, J. G. Crowther and Joseph Needham (27). In ihren Arbeiten greifen sie wichtige Grundgedanken der marxistischen Konzeption der Wissenschaftsgeschichtsschreibung auf und liefern - von ihnen ausgehend - Darstellungen bestimmter Aspekte der historischen Entwicklung der Wissenschaft, bestimmter ihrer Perioden, regionaler oder disziplinärer Entwicklungen im Bereich der Wissenschaft, die bis heute ihre Bedeutung nicht verloren haben (28). Dennoch verzeichnen wir nach dem II. Kongreß keine sich allgemein durchsetzende Tendenz einer marxistischen Wissenschaftsgeschichtsschreibung in den westlichen Ländern. Vorbehalte philosophisch-weltanschaulicher wie politischer Natur hielten viele der Wissenschaftler, die geneigt waren, die heuristischen Potenzen der historisch-materialistischen Konzeption der Wissenschaft und ihrer Geschichte zu akzeptieren und im eigenen Schaffen zu erproben, davon ab, zu jener notwendigen Komplexität der marxistischen Betrachtungsweise fortzuschreiten, die für Bernals Werk so kennzeichnend ist, und die Konsequenzen aus der Anerkennung des historischen Materialismus für ihr wissenschaftliches, gesellschaftliches und politisches Wirken zu ziehen, die Bernal in so beeindruckender Weise für sich gezogen hat.

Bernals Werk - selbst inspiriert durch die Entwicklung der neuen Gesellschaft und der Wissenschaft in der UdSSR in den zwanziger und dreißiger Jahren - fand in den Jahren nach dem 2. Weltkrieg in der Sowjetunion und den sozialistischen Ländern größtes Interesse und hohe Würdigung(29). Es widerspiegelt sich darin ein besonders in den fünfziger Jahren sich verstärkendes Interesse für Probleme der Entwicklung der Wissenschaft und ihrer Geschichte, dessen soziale Grundlage die qualitativ neue gesellschaftliche Wer-tigkeit der Wissenschaft in der sozialistischen Gesellschaft ist. Die sich in diesen Ländern in den sechziger und siebziger Jahren entwickelnde marxistisch-leninistische Wissenschaftstheorie und Wissenschaftsgeschichtsschreibung (30) baut auf dem Marxschen Verständnis der wissenschaftlichen Tätigkeit als "allgemeine Arbeit" (31) auf. Sie versteht Wissenschaft als das "Produkt der allgemeinen geschichtlichen Entwicklung in ihrer abstrakten Quintessenz" (32), als sozialökonomisch determinierte spezifische Art gesellschaftlicher Tätigkeit, die in der Produktion, Reproduktion, Verbreitung und praktischen Nutzung von objektiv wahrem Wissen über Gegebenheiten der Natur- und Gesellschaft besteht (33).

Marxismus versus Internalismus und Externalismus

Die historisch-materialistische These von der sozialökonomischen Determiniertheit der Wissenschaft reduziert sich keineswegs auf die externalistische Behauptung, Entstehung und Entwicklung der Wissenschaft seien einzig und allein durch die Bedürfnisse der Produktion und der Entwicklung der Produktivkräfte der Gesellschaft bedingt. Den marxistischen Standpunkt so zu interpretieren hieße, ihn gründlich mißzuverstehen. Freilich ist er in dieser simplifizierten Form einfacher zu bekämpfen (34). Die historisch-materialistische Betrachtungsweise der Geschichte der Wissenschaft hat jedoch nichts mit Externalismus zu tun. Dafür gibt es zwei wesentliche Gründe:
1. Für den Marxismus sind Gesellschaft und Wissenschaft nicht zwei sich äußerlich zueinander verhaltende Wesenheiten, von denen die eine die Entwicklung der anderen - womöglich noch linear und unmittelbar - determiniert. Wissenschaft wird selbst als gesellschaftliches Phänomen verstanden, Wissenschaftsentwicklung als untrennbarer Bestandteil gesellschaftlicher Entwicklung. In der wissenschaftlichen Tätigkeit realisieren und reproduzieren sich die Produktionsverhältnisse der jeweiligen Gesellschaftsordnung, und Wissenschaft hat selbst eine Produktivkraftfunktion, zumindest seit ihrer Nutzung im Dienste des Kapitals. Für die marxistische Konzeption der Wissenschaftshistoriographie ist es deshalb fundamental, die Wissenschaft als einen Teil des Lebensprozesses realer, d. h. historisch-konkreter Gesellschaftsordnungen zu verstehen. Durch konsequente Realisierung ihrer jeweiligen gesellschaftlichen Funktionen provoziert die Wissenschaft - über die Entwicklung der ganzen Gesellschaft vermittelt - ihre eigene Höherentwicklung als soziales System. Wird von dieser determinierenden Vermittlung über die Totalität des Gesellschaftsganzen abstrahiert, so wird der soziale Entwicklungsmechanismus der Wissenschaft unverständlich. Die Determination des Entwicklungstyps der Wissenschaft (ihres Entwicklungsmechanismus, ihrer Entwicklungsdynamik und ihrer Entwicklungsrichtungen) durch die materielle Produktionsweise der Gesellschaft ist für die historisch-materialistische Sichtweise der Wissenschaftsgeschichte wesentlich.

Sie ist fundamentaler als die Determination des Musters
der jeweiligen Erkenntnisinhalte der Forschung durch kon-
krete Anforderungen der gesellschaftlichen Praxis, die für
den Externalismus kennzeichnend ist.
2. Die Totalität des Gesellschaftsganzen, über die die so-
zial-ökonomische Determination der Wissenschaft vermittelt
ist, schließt einen bestimmten Entwicklungsstand der Wissenschaft selbst ein, der somit seinerseits zu einem determinierenden Faktor weiteren Wissenschaftsfortschritts
wird. Engels deutet dies an, wenn er feststellt: "Die Wissenschaft schreitet fort im Verhältnis zu der Masse der
Erkenntnis, die ihr von der vorhergehenden Generation hin-
terlassen wurde" (35). Es versteht sich, daß auch diese
These - die auf die "innere Logik" der Wissenschaftsentwicklung abzielt - nichts mit Externalismus zu tun hat, wie
sie andererseits im Gesamtkontext der marxistischen Konzeption der Wissenschaftsentwicklung auch nicht für Internalismus spricht.

Die marxistisch-leninistische Wissenschaftstheorie er-
weist den Gegensatz von Externalismus und Internalismus somit als Scheinalternative. Die theoretischen Voraussetzungen in den Werken von Marx, Engels und Lenin, von denen
die Beiträge in "Science at the Cross Roads" ausgingen,
und deren Folgen in Gestalt der heutigen Arbeiten zur Wissenschaftstheorie und -geschichte in den sozialistischen
Ländern weisen dies noch deutlicher aus als "Science at
the Cross Roads" selbst.

Fußnoten
1 In den vorliegenden englischen und deutschen Übersetz-
  ungen der Arbeit von Hessen wird deren Titel wiederge-
  geben als "Die sozialen und ökonomischen Wurzeln...". Bei
  Hessen ist jedoch von "sozialökonomischen Wurzeln" die
  Rede (vgl.: B. Gessen: Social'no-ekonomičeskije korni
  mechaniki Njutona. Izdanije vtoroje. Moskau-Leningrad
  1934). Der Unterschied ist wesentlich, da der Terminus
  "sozialökonomisch" in der marxistischen Theorie auf die
  Totalität einer Gesellschaftsordnung - in der Einheit
  ihrer materiellen Produktivkräfte und Produktionsverhältnisse - abzielt und infolgedessen etwas anderes bezeichnet als "sozial und ökonomisch".

2 J. Needham: Foreword. In: Science at the Cross Roads.
  London 1971. S. VIII

3 Vgl. z. B.: P. G. Werskey: Introduction. In: Ebenda. S.
  XV

4 Vgl.: Science at the Cross Roads. Published by Frank Case
  and Company Limited. London 1931

5 J. D. Bernal: Science and Society. In: J. D. Bernal: Freedom of Nesessity. London 1949. S. 334

6 J. D. Bernal: The Social Function of Science (London
  1939; reprinted Cambridge, Mass. 1967). S. 406

7 J. Needham: Foreword. In: A. a. O. S. VIII

8 P. Weingart: Vorwort. In: Wissenschaftssoziologie II. Determinanten wissenschaftlicher Entwicklung (Hrsg. v. P. Weingart). Frankfurt/M. 1974. S. 8

9 K. Marx: Theorien über den Mehrwert. Teil I. In: K. Marx/F. Engels: Werke Bd. 26.1. Berlin 1965. S. 257

10 Vgl.: F. Engels: Dialektik der Natur. In: K. Marx/F. Engels: Werke Bd. 20. Berlin 1962. S. 456

11 Vgl.: F. Engels: Einleitung zur englischen Ausgabe der "Entwicklung des Sozialismus von der Utopie zur Wissenschaft". In: K. Marx/F. Engels: Werke Bd. 22. Berlin 1963. S. 299; F. Engels: Dialektik der Natur. In: A. a. O. S. 457; Brief von Engels an W. Borgius vom 25. Januar 1894. In: K. Marx/F. Engels: Werke Bd. 39, Berlin 1968. S. 205

12 Vgl.: Vernadskij, V. I.: O zadsčach i organizacii prikladnoj raboty, Leningrad 1927. Vernadskij hatte bereits 1903 eine Arbeit zum Thema "Wissenschaftsfortschritt und Volksmassen" konzipiert, die jedoch unvollendet geblieben ist (vgl.: Močalov, I. I.: Vernadskij, V. I.: Čelovek i myslitel', Moskau 1970, S. 151f).

13 Vgl. insbesondere: Ol'denburg, S. F.: Polozenije našej nauki aredi nauki mirovoj. In: Nauka i technika SSSR za 10 let (1917-1927). Moskau-Leningrad 1927

14 Strumilin, S. G.: K Metodologii učeta naučnogo truda. Leningrad 1932

15 Vgl.: Pokrovskij, M. N.: O dejatel'nosti Kommunističeskoj akademii. In: Vestnik Kommunističeskoj akademii. 1927, kn. XXII, S. 7

16 Vgl.: Rajnov, T. I.: O tipe raznostoronnego učenogo. In: Sorena, Nr. 9/1934

17 Vgl. z. B.: Ioffe, A. F.: Peredelka mira. In: Front nauki i techniki. Nr. 6/1931; Ioffe, A. F.: Moja žizn' i rabota. Avtobiografičeskij očerk. Leningrad 1932

18 Vgl.: Boričevskij, I. A.: Naukovedenije kak točnaja nauka. In: Vestnik znanija. Nr. 12/1926

19 Vgl.: Gruzincev, G. A.: Očerki po teorii nauki. In: Zapiski Dnipropetrovs'kogo institutu narodnoi osvitu. 1928, Bd. III

20 Werskey, P. G.: Introduction. In: Science at the Cross Roads. A. a. O. S. XII

21 Bernal, J. D.: Science and Society. A. a. O.

22 Ebenda.

23 Bernal, J. D.: The Social Function of Science. A. a. O. S. XIII

24 Bernal, J. D.: Vorwort zur ersten Auflage. In: Die Wissenschaft in der Geschichte. 3. bearbeitete Auflage. Berlin
1967. S. XI

25 Vgl.: Bernal, J. D.: After Twenty-Five Years. In: The Social
Function of Science, A. a. O. XVII

26 Naučnaja konferencija po obsuždeniju nekotorych teoretičes-
kich problem istorii estestvoznanija i techniki 11.-12.11.
1957. In: Voprosy istorii estestvoznanija i techniki. Vyp.
6, 1958. S. 143

27 P. G. Werskey hat diesen Einfluß detailliert beschrieben in
seiner Einführung zur 2. Auflage von "Science at the Cross
Roads", London 1971

28 Vgl. z. B.: Crowther, J. G.: British Scientists of the Nine-
teanth Century. London 1935; Needham, J.: A History of Embryology, Cambridge 1934

29 Z. B. veranstalteten die Sowjetische Vereinigung der Wissen-
schafts- und Technikhistoriker und das Institut für Ge-
schichte der Naturwissenschaft und Technik der Akademie der
Wissenschaften der UdSSR im November 1957 eine speziell der
Diskussion von Bernals Buch "Die Wissenschaft in der Ge-
schichte" gewidmete Konferenz, an der Bernal selbst aktiven
Anteil nahm. Vgl.: Voprosy istorii estestvoznanija i techniki
Vyp. 6, 1958

30 Vgl.: Mikulinskij, S. R.: Naukovedenije: problemy i issledo-
vanija 70-x godov. In: Voprosy filosofii, H. 7/1975; Aktual'-
nyje problemy sovetskogo naukovedenija. Knigi 1966 - 1976 gg.
Katalog vystavki. Moskau 1977

31 Vgl.: Marx, K.: Das Kapital. Dritter Band. In: Marx/Engels:
Werke Bd. 25, S. 113/114

32 Vgl.: Marx, K.: Theorien über den Mehrwert. Erster Teil. In:
Marx/Engels. Bd. 26.1. S. 367

33 Vgl.: Wissenschaft - Stellung. Funktion und Organisation in
der entwickelten sozialistischen Gesellschaft (Hrsg. von G.
Kröber und H. Laitko). Berlin 1975. Kap. 2.3.

34 So unterstellt z. B. Gerard Radnitzky: "Das marxist-pragma-
tische (?-G.K.) (instrumentalistische) Wissenschaftskonzept
faßt praktische Anwendbarkeit als Kriterium für wissenschaftliche Gültigkeit auf, verwechselt Nützlichkeit mit wissen-
schaftlichem Wert (etwa Bewährtheit einer Theorie)... Das,
und die Überbetonung der Praxis, führt schließlich dazu,
daß das sogenannte technische Erkenntnisinteresse als das
für die Naturwissenschaft allein sinnkonstitutive Interesse
aufgefaßt wird." (G. Radnitzky: Prinzipielle Problemstel-
lungen der Forschungspolitik. In: Zeitschrift für allgemeine Wissenschaftstheorie. Band VII. Heft 2/1976. S. 394).
Solche und ähnliche Entstellungen der marxistischen Wissen-
schaftskonzeption dienen Radnitzky dazu, "in ihrer Kritik
eine wichtige praktische Aufgabe zu sehen" (Ebenda. S. 401).

35 Engels, F.: Umrisse zu einer Kritik der Nationalökonomie.
In: K. Marx/F. Engels: Werke Bd. 1, Berlin 1972, S. 521

Symposium 3.   INTERNATIONAL CO-OPERATION AND DIFFUSION IN SCIENCE

Prof. Maurice Crosland

INTRODUCTION

In choosing the subjects of the international organisation of science and the international diffusion of science for this symposium, I wish to suggest not only that this is a useful area of study but that it is one which is particularly appropriate for an international congress of the history of science. By inviting speakers from different countries and different continents, by ensuring that North America and Asia are represented as well as Europe and by having an audience even more widely representative of different countries of the world, we have made a beginning on truly international lines. There are of course difficulties of language,[1] translation sometimes creates problems and there are potential conflicts of political ideology, but we must be prepared to pay such a price for a multinational symposium.

The symposium falls into two parts. In the first part speakers were asked to consider some of the origins and development of international collaboration and international scientific meetings. Considering the influence of Humboldt in encouraging international collaboration in science in the nineteenth century, he probably deserves a special study. For the twentieth century Professor Schroeder has provided a useful sociological and political analysis of the problems of international scientific co-operation. In the second part of the symposium speakers, looking at the science of three different countries of major importance in the modern world, discussed the extent to which science might be considered a national product and the extent to which it related to science in other parts of the world. Does it make sense to speak of "European science", "American science", or "Japanese science"? Of course scholars have been investigating for some time the development of science in different countries: science in France, in Britain, in Russia etc.[2] But by considering such questions within the framework of a symposium on international science we are hoping to turn research in another direction. I trust the contributors to this symposium will forgive me if I suggest that they have not been able to provide a final answer. Some argue for national schools, for "western science", etc., but the question of the precise difference between science in different parts of the world is too large to be answered in a one-day meeting.

The symposium is presented,therefore,as a contribution to a problem which I expect will continue to exercise historians and sociologists of science in the 1980s and even beyond. Many of the speakers are recognised authorities on their respective subjects. My own contribution can be explained as a historical introduction which I thought useful but, because of the large time span and enormous documentation, I found no one else willing to contribute. The way forward obviously lies in more limited studies which can on a later occasion be brought together to provide a useful synthesis.

1. For those who do not read German, Dr Cawood's commentary may be useful in giving Professor Biermann's argument as well as criticism.
2. For a recent contribution see M.P. Crosland, "History of Science in a National Context", Brit. Jour. Hist. Sci., 10 (1977), 95-113.

Prof. M.P. Crosland

ASPECTS OF INTERNATIONAL SCIENTIFIC COLLABORATION AND ORGANISATION
BEFORE 1900

INTRODUCTION

Most of the work that has been done on international science has been
focussed on the twentieth century by scholars whose primary interests are
political and sociological. Historians of science have not shown much
general interest in this area, although occasionally a historian concerned
with the history of an individual science has looked at some international
collaboration within that particular discipline. Unfortunately the
separate study of the history of individual sciences precludes the appreciation of general patterns. There are more similarities between problems
of quite different sciences than one would at first imagine. I shall be
referring in the course of this paper to the general question of agreement
on a common language. I could also point out that expeditions concerned
with one science, say astronomy, have benefitted other areas of science
like natural history, and vice-versa. No science therefore stands in
isolation, whether considered methodologically or from the point of view
of organisation.

Because this early period is not an area in which much research has
been done, one of my principal objectives will be to present basic
information and a few perspectives. What follows therefore, is of the
nature of a general survey. There is, of course, a vast documentation,
but I hope later to be able to explore more fully some aspects which I
only touch upon here or even omit entirely. This paper presents some
preliminary thoughts on a general problem which I think deserves attention
in any comprehensive study of the organisation of science.

In the literature of scientific conferences, largely drawn up by
librarians, compilers have delighted in searching out some particularly
early date. Thus a recent compilation of scientific periodicals carries
the title Scientific Conference Proceedings, 1644-1972. I might mention
that the date 1644, predating the foundation of the Royal Society of
London, refers to a meeting of falconers at Rouen - hardly a serious
challenge to the established views on early scientific societies! It
could be argued that such antiquarian tit-bits do no harm. My own view
is that a title which implies an activity of an unchanging nature over
three centuries and comprehensive treatment of that activity is grossly
misleading. Another publication is called: International Congresses ...
Full List. Here the claim for completeness is explicit. A third
compilation claims an exhaustive enumeration of 1,978 international
organisations founded since 1815. Any such claim to be complete is very
dangerous, particularly in the earlier period. The scientific conferences
of the twentieth century are well documented and can therefore be
discussed without challenging the data. Before 1900 the data requires
careful historical study. In superficial research it is all too easy to
be misled by titles. The Actes du 1er Congrès International de Botanique,
which refers to a meeting held in Paris in 1900, provides an example of
such a title. It represents the vanity of its secretary, a certain Emile
Perrot, and ignores the fact of international botanical meetings over the
previous 40 years, some of which had made significant contributions to
international collaboration in botany.

We might distinguish three kinds of international activity, each with their own development, although having implications for the others. The first and most basic kind of contact is the free flow of scientific information by correspondence. In the seventeenth century those indefatigable correspondants Mersenne and Oldenburg exchanged letters on a large scale and they passed on information to natural philosophers in all parts of Europe. This aspect of international collaboration continued to be important in science and from the nineteenth century had spread from Europe to all continents of the world, although of course by then scientific journals had to a large extent superceded private correspondence. Translations, reviews and abstracts have all contributed to the dissemination of information on an international scale.

Travel is a second kind of transnational activity, often allowing an individual to discuss his work with other interested parties far from home. But scientists and particularly astronomers have also needed to make investigations beyond their national frontiers and they have sometimes required a guarantee of safe conduct from the government through whose territory they were to pass. From the beginning the claim was made that the study of nature transcended all political and national considerations, except perhaps the pursuit of glory. This level of co-operation obviously involved a certain amount of negotiation at government level. I shall be focussing attention in this paper on the collaboration of scientists rather than the collaboration of governments but this is obviously a dimension that requires detailed study.

Whereas correspondence and travel were aspects of international scientific activity well established in the seventeenth and eighteenth centuries, the third aspect, the international meeting, involved a greater infrastructure and was slower to emerge. Nevertheless by the early nineteenth century meetings which had begun on a local or national scale gradually came to take on an international aspect. The presence of foreign visitors at meetings helped prepare the way for a multi-national conference in which representatives of different countries met on equal terms. Arising out of international meetings there came into being in the second half of the nineteenth century a number of permanent international scientific organisations. Thus all the main features of modern international science had appeared before 1900.

NATIONAL SOCIETIES

One might tend to think of a national meeting as being the antithesis of an international meeting. In fact the growth of scientific societies organised on national lines greatly helped the development of international contacts and prepared the way for international gatherings. Already in the seventeenth century the Royal Society had accepted foreign members and in the eighteenth century foreign members represented a large fraction of the total membership. The Paris Academy of Sciences and the Berlin Academy each had a number of foreign associates. Although such a position was largely honorary, it did help to promote international awareness. In the nineteenth century the Gesellschaft Deutscher Naturforscher und Ärzte (founded in 1822), although explicitly a German organisation, attracted attention outside the German states and its meetings came to be attended by an increasing number of foreign scientists. The multinational character of the meetings became a source of pride to the organisers so that at the 1828 meeting in Berlin a map of Europe was printed showing the towns from which participants came and a table of countries was attached showing that, in addition to the German-speaking states, there were several representatives from each of the following countries: Sweden, Denmark, Poland, Holland as well as one each from Britain, France and Russia. The Scandinavian participation in the German meetings was particularly striking and this encouraged the Scandinavian scientists to organise their own congresses. From 1839 there were regular meetings of the Congress of Scandinavian scientists (Scandinaviske Naturforskeres Møde) which met every two or three years in the larger cities of Sweden, Denmark and Norway. These meetings attracted a limited amount of interest outside Scandinavia but a few visitors from neighbouring countries (notably Prussia, Finland and Russia) attended later meetings. The German society had important implications for the organisation of science in Britain and France as well as in Scandinavia and it was not long before the question arose of coordinating meetings of the various national societies.

The best known influence of the Deutsche Naturforscher is that in Britain, where it served as a model for the British Association for the Advancement of Science. The British Association played a significant part in international relations, thus challenging the authority of the Royal Society to represent British science on the international scene. The first international congress on terrestrial magnetism was held in Cambridge in 1845 in conjunction with the meeting of the British Association there. The idea of such a congress in Britain had come from Kupffer, director of the Russian program of magnetic and meteorological observations. Although British physicists constituted the majority at the meeting there were half a dozen important representatives from the Continent, including Erman from Berlin and Kreil from Prague. Weber, Gauss and Humboldt did not attend but sent letters to the organisers of the congress. Another magnetic congress in conjunction with the British Association took place in 1898. Meanwhile, however, there had been many proposals at meetings of the British Association which had implications for international collaboration. Joseph Henry, wrote to the 1855 meeting in Glasgow, to point out the desirability of an international catalogue of scientific literature, and in the 1860s there were several discussions on the rationalisation of electrical units.

On the political front the Congress of Vienna in 1815 was a landmark in the history of international collaboration and it helped to establish a system of international consultation. States were brought together in another way by the growth of railways in the nineteenth century and the introduction of the electric telegraph. Improving communications also brought new problems, some associated with the growth of national pride. However there were obvious mutual benefits to be obtained by agreement on collective measures to improve trade and to check the growth of disease. International River Commissions were established in the early nineteenth century. The spread of epidemics brought about diplomatic conferences although the first international health conference in 1851 in Paris achieved very little. The coming of the railway required collaboration at national frontiers and it was not long before there were demands to send telegraphic messages across national boundaries. In 1865 twenty European states met in Paris to work out an international convention on the telegraph. The Austro-German Postal Union of 1850 showed the advantages of a unified postal administration but the Universal Postal Union came into being only in 1874.

All of these international conferences involved minor concessions of national sovereignty. They paved the way for international collaboration in technological areas. The interaction of scientists for technological purposes in turn provided a good precedent for the organisation of pure science on international lines. Political and economic collaboration were not therefore the _cause_ of scientific collaboration but they reinforced the trend and made it easier for scientists to cross frontiers. The political suspicion with which, for example, some of the earlier meetings of the Deutsche Naturforscher had been met came to be replaced by an acceptance and, by the end of the nineteenth century, a positive fashion for international conferences of all kinds.

It is inevitable that most of the discussion on international collaboration before 1900 should be focussed on Europe because that was the centre of activity. The great powers in the eighteenth century were clearly Britain and France, whose political influence was rivalled only by the major contributions to science emanating from these two countries. Although French political power declined after the defeat of Napoleon, the prestige and high quality of French science ensured a continuing French authority in the nineteenth century. After 1800 German scientists too began to make some contribution: Humboldt and Gauss in geomagnetism, Bessel in astronomy and Weltzien in chemistry. After several important French contributions to geodesy it was the Prussian government which took the main initiative in the mid-nineteenth century in earth measurement. The first meeting of the Geodesic Association of Central Europe was held in Berlin in 1864. It was from this association that the International Geodesic Association sprang.

An important international venture which began as a German collaboration was the foundation of the International Association of Academies. In 1893 the Academies of Science of Vienna, Munich, Leipzig and Goettingen founded a cartel of German Academies. On the initiative of the Vienna Academy overtures were made to the Royal Society of London and the Paris Academy to suggest wider co-operation. This co-operation fitted in well with a plan of the Royal Society of London for collaboration to compile an international catalogue of scientific literature. The International Association of Academies was thus founded quite quickly after a meeting at Wiesbaden in 1899.

INDIVIDUAL AND COLLECTIVE AUTHORITY

A few outstanding individuals played a major role in the calling of international conferences. Although no single scientist in the second half of the nineteenth century could assume he had the authority to make decisions affecting his particular science, it was often the initiative taken by an individual which finally brought men together in an international congress. A good example of such authority and initiative is provided by the Belgian, Adolphe Quetelet. Although Quetelet is remembered for his pioneering work in statistics, he had begun as a geometer and astronomer. His appointment in 1828 as astronomer at the Brussels Royal Observatory and in 1834 as permanent secretary to the Brussels Academy helped to extend his contacts with scientists in other countries and in the 1830s he was actively engaged in international collaboration in astronomy, meteorology and geophysics. His monograph of 1835 on social statistics gave him an international reputation and if anyone was to call an international conference of statistics he was the obvious candidate and Brussels was the obvious place.

Yet,although an individual might help to launch a congress, he could not usually expect to dictate to his colleagues. The growth of science and the parallel between the development of political democracy and the working of the scientific community was expressed as follows by a group of British zoologists writing in 1842:

> "The world of science is no longer a monarchy, obedient to the ordinances, however just, of an Aristotle or a Linnaeus. She has now assumed the form of a republic, and although this revolution may have increased the vigour and zeal of her followers, yet it has destroyed much of her former order and regularity of government."

This then was a nineteenth-century conservative British view of an indirect influence of the French revolution on the organisation of science. The zoologists went on to suggest that order could be imposed by general consent, which was more likely to be obtained by basing any rules of nomenclature on clear rational principles.

It is interesting that the group, which included Richard Owen and Charles Darwin, did not go as far as to propose an international meeting. Starting from the premiss that Britain was a world power they felt that the British Association would have sufficient international authority. What they claimed publicly, however, was the authority of rationality:

> "All that is wanted then is that some plain and simple regulation, founded on justice and sound reason should be drawn up by a competent body of persons [i.e. themselves] and then be distributed throughout the zoological world."

Thus, despite the metaphor of science as a republic, they did not yet see it as an international community of equal partners. The idea of equal participation of many countries developed only slowly; a definite change can be seen in the international botanical congress held in Paris in 1867, i.e. a quarter of a century after the meeting of British zoologists. Alphonse de Candolle remarked ruefully [perhaps he was thinking of himself] that "Nowadays no-one likes to submit to the will even of a man of genius...." He thought that many scientists would rather accept a decision arrived at by a majority of their colleagues, as expressed for example at an international scientific congress. One had to be careful, however, that any such vote really was representative of the growing international community of that particular branch of science. In so many international congresses the host country was so well represented that any majority decision would have effectively been a decision agreeable only to the scientists of one country.

In a subject in which national and local advantages were constantly sought it is pleasing to record at least one example of national abnegation in favour of a truly international spirit. The idea of an international congress of physiology came from the British Society of Physiology but, far from insisting on London as the venue, they proposed Switzerland. It was thus that a group of physiologists representing Britain, France, Germany, Italy and Switzerland planned the first international physiological congress which took place in Basle in 1889. No doubt arguments for location at the centre of Europe were advanced for the Swiss meeting, arguments which competed with considerations of ease of communications. Perhaps one should also remember that in the nineteenth century there was a growing appreciation of Alpine scenery, not to mention the physiological problems posed by high altitudes.

Paris was chosen for the location of a large number of nineteenth-century scientific congresses for reasons geographical, political, linguistic and economic. In so far as it was the scene of several major international exhibitions which involved science and technology, it was appropriate to extend the occasion to bring together men in one particular scientific discipline. The supreme example of this is the way the

International Exhibition of 1900 was used to mount international
scientific congresses in mathematics, physics, meteorology, chemistry,
botany, geology, psychology and many branches of applied science. It was
also the occasion of the first international congress of the history of
science.

THE STANDARDISATION OF LANGUAGE

One of the most important motives for calling an international congress
was to obtain authority to introduce certain conventions in a particular
science. Most basic was language, particularly in the biological sciences,
geology and chemistry. Generally in physical science agreement was
necessary on units of measurement, with the metric system receiving
increasing support throughout the nineteenth century. The branch of
physics in which there was most discussion about units was electricity, a
science which in the nineteenth century, with the development of current
electricity, the telegraph and electric power, had a growing industrial
relevance.

In the eighteenth century Linnaeus had introduced into botany a
binomial nomenclature, which provided a welcome solution to the confusion
in the naming of plants at the time. Because it had been well thought
out, because it made use of an international language (Latin) and because
of the increasing stature of Linnaeus in natural history in the mid-
eighteenth century, the nomenclature was widely accepted. The circumstances of the reform of chemical nomenclature in 1787 were rather
different in so far as the main authority came from Lavoisier but the
ground work had been done by Guyton de Morveau. A collaboration between
these two and two other leading French chemists led to the publication of
a joint work which came to be accepted not only within France but also in
other countries. This can be explained by the dominant position of French
chemistry at the time and by the good sense of the proposals. Although
the new chemical names were French it was not too difficult to make minor
changes to adapt them to the English language. Although the historian
may give these purely practical reasons for the acceptance of the new
chemical language, Lavoisier would have claimed that, in following the
sensationalist philosophy of Condillac, he was proposing a system related
to nature. In so far as the terms derived from the Greek,he was tapping
a universally recognised spring of civilization.

In the early nineteenth century the Swedish chemist Berzelius was able
to introduce the symbols used in modern chemistry based on the initial
letters of the Latin names of the elements. Once again a Scandinavian
scientist had been able to gain acceptance by using an international
language. Yet we are now arriving at the end of a tradition. By the
nineteenth century both the growth of national consciousness and the
growth of science made it unthinkable that an individual could legislate
for scientists in other countries. At the most he might name a new
substance or define a unit and leave to a representative group of his
colleagues the task of rationalisation.

A major achievement in the second half of the nineteenth century was
the holding of international conferences in many different disciplines
with the object of agreeing on a common language. The fact that they did
not take place until the second half of the nineteenth century suggests
that language was not the first reason for international collaboration.
However it was an important motive for holding international meetings.
It may surprise some historians to learn how wide a range of international
conferences were inspired by the goal of rationalisation of nomenclature.

119

They range from statistics to geology and from chemistry to meteorology.
Several subjects will be briefly reviewed in the chronological order of
their first official international conference as examples of the importance
of language in promoting international meetings.

It is worth citing the remarks of Quetelet at the first international
statistical congress to show that the philosophy of the eighteenth-century
French philosopher Condillac was not dead in 1855:

"Le moyen le plus sur de faire progresser les sciences, c'est
d'en perfectionner le langage et d'adopter des notations uniformes
qui permettent de résumer plus facilement un grand nombre d'idées
et de rapprocher plus de faits pour en saisir les rapports et les
lois."

In statistics the problem was to persuade different governments to produce
data of population, resources, mortality, etc., in ways which were
strictly comparable. In chemistry a major problem was to persuade
chemists, whether in the same country or not, to agree on a common system
of notation in which the atomic symbols of Berzelius would stand for an
agreed atomic weight. Some accepted atomic weights twice the value
accepted by others, while others again had taken refuge in equivalents.
The Karlsruhe congress called by Kekulé, Weltzien and Wurtz in 1860 and
attended by many leading chemists from most European countries discussed
the problem of nomenclature among other questions. Nothing was settled,
but a valuable precedent was established for further international
meetings.

The subject of nomenclature was not the first problem of the botanical
congresses of the 1860s but it was a problem which did not take long to
emerge. An important part of the 1867 congress was devoted to the
question which had recently become more pressing because of the use by
horticulturists for simple cultivated varieties of the Latin names used
by botanists for the genuine species. Alphonse de Candolle headed a
commission to examine the problem and make recommendations.

Turning to the subject of meteorology, which held its first international congress in Vienna in 1873, there is an obvious parallel with
statistics in so far as the problem of language is concerned. Meteorology
with its agricultural and military implications was also of interest to
governments. But in so far as meteorology was a science like geomagnetism,
transcending national frontiers, it is surprising that it had to wait
until 1873 for its first international congress. Perhaps this merely
illustrates the difference between international cooperation and the
calling of a formal meeting. The conference was concerned to arrange for
observations to be made in different countries at the same local time and
with instruments so constructed and placed as to give results strictly
comparable with each other. But the concept of a common language was
also explicit in discussion of symbols which at that time were often
abbreviations of words in the different vernacular languages of Europe.
A committee recommended the substitution of symbols for letters but for
wind directions it chose the initial letters in the English language.

The first international congress of geology in 1878 had some points in
common with both meteorology and botany. Indeed the botanists' concern in
Paris in 1867 with nomenclature had led at least one geologist (the
Spaniard Vilanova) to suggest a discussion of geological nomenclature.
But the case of geology differs from that of the other sciences in that
most of the concern for an international meeting came not from a European
country but from the United States. American geologists had encountered
strata which had no name in the old world; in different American states

they had given different names to the same thing and had finally settled for a pragmatic nomenclature relating to place, leaving the problem of rationalisation to the future. James Hall took the initiative at a meeting of the American Association for the Advancement of Science in Buffalo in 1876 to suggest the calling of an international congress of geologists in Paris at the time of the 1878 Paris exhibition. Although many other subjects beside nomenclature were discussed, this was one of the principal motives for the meeting. Also disucssed was the question of agreement on the choice of colour for geological maps. It was not long before the question of a unified nomenclature for palaeontology was raised at these meetings (Berlin, 1885). Thus after much early confusion geologists made a major and continuing use of international conferences to achieve greater uniformity in language and cartographic conventions.

A concluding comment on the whole question of standardisation of nomenclature is that the importance of agreement on names and units tended to override considerations of national prestige and priority. It provided a strong incentive for scientists to meet and work out some agreement. When scientists had become accustomed to such international meetings they found that there were other advantages both intellectual and social in such meetings and they therefore tended to become permanently established.

THE METRIC SYSTEM

International collaboration in the metric system is sometimes considered to date from the conference of 1875 known as the Conférence diplomatique du mètre. Such a view gives a misleading perspective since the events of the 1870s were really the culmination of activities which had been going on for nearly a century. The possibility of standardising weights and measures had been raised in the British House of Commons in 1790 and, as similar ideas had been current in France, Talleyrand, French minister of foreign affairs, suggested collaboration between the two countries through their principal scientific societies, the Académie des Sciences and the Royal Society. Unfortunately political events in France made collaboration more and more difficult and it was the Académie des Sciences alone which examined the problem in detail. The Academy recommended a new unit of length, the metre, related to the size of the earth. The metre was thus international in theory. In practice it related to the meridian through Paris. The practical geodesic measurements and the construction of standards occupied several years. When the time came to make desisions on a definitive metre and a definitive kilogram some scientists in the Institute persuaded the French government to invite scientists from other countries to collaborate in the final steps. The arrival of representatives from nine allied and neutral countries in Paris in September 1798 has some claim to mark the beginning of the first international scientific conference. The presence of the foreign scientists may have been little more than a gesture but it did help to provide a little more authority for the metric system than a uniquely French determination would have done.

Nevertheless the metric system was usually considered in other countries to be French and the initiative for further action in the propagation of the system lay therefore with that country. In 1841 the French government sent copies of metre standards to a large number of European countries and metric standards were also sent as far as the United States, South America and Japan. It was, however, international exhibitions which did most to propagate the advantages of the metric system. The 1851 exhibition revived an interest in Britain in an international system, and at the 1855 Paris exhibition members of the panels of judges representing many different countries undertook to make representations to their respective governments of the advantages of a universal system of weights and measures. This raised the obvious question of whether the metric system could be accepted as that universal system. The metric system had already been accepted in several countries of Europe and Central and South America but the spread of the system emphasised the need for uniformity of standards. There was a very real danger in the mid-nineteenth centure that different countries would use metric standards which were slightly different. An international conference was therefore called in Paris and 24 foreign countries agreed to send representatives. The date chosen for the conference, August 1870, was unfortunate, because by the time the conference met, France was at war with Prussia. The meeting nevertheless took place and did some useful preliminary work, which prepared the way for the Conférence diplomatique du mètre and the establishment of an International Bureau of weights and measures.

ASTRONOMY

One could single out astronomy as an example of a science which has had a long and distinguished history of international collaboration. As an observational science on a large scale, astronomy came to involve the whole earth as a base. Most European astronomers of the seventeenth century were content with observation of the stars in northern latitudes but the Englishman Edmond Halley was able with royal support to secure a passage to St. Helena in 1677 to enable him to draw up the first catalogue of stars of the southern sky. Three quarters of a century later Lacaille with French government support was able to extend his star catalogue of the southern hemisphere from 350 stars to nearly 10,000.

A second motive for astronomical expeditions arose out of the two transits of Venus predicted for 1761 and 1769. In order to use this rare event to calculate the distance of the sun, different groups of astronomers needed to plan to observe from different parts of the world the exact moment of ingress and egress of the planet across the sun's disc. Le Gentil managed to obtain French government approval to observe the transit of Venus in India and the Paris Academy of Sciences responded to an invitation from the St. Petersburg Academy by sending Chappe to head a Russian party in Siberia. The most ambitious of the British plans was the famous voyage of Captain Cook which combined an observation from Tahiti with exploration of the South Pacific. International science here overlapped with colonial interests. The expeditions were planned by colonial powers and in so far as there was any collaboration it was between European astronomers.

Although the arrangements made to observe the eighteenth-century transits of Venus established something of a precedent, the expeditions were very exceptional. In the 1820s John Herschel and Friedrich Bessel each put forward proposals for a general survey of the stars but it was the Berlin Academy which took responsibility for most of the work. The first international cooperation in the nineteenth century took place in solar rather than in sidereal astronomy. Improvement in communications made it feasible to make plans on a world-wide scale for the observation of solar eclipses, a phenomenon which was of course strictly local so that the vantage point was crucial. One of the first solar eclipses to attract attention on an international scale was that which crossed central and southern Europe on 8 July 1842. The early eclipse expeditions were often a matter of individual decision and observers later compared results. They did not collaborate so much as share a common purpose. However, considering that the important phenomenon of a solar corona hardly lasted more than one minute, that was an obvious incentive to make the best use of those precious seconds. In the solar eclipse of 17 May 1882, when astronomers from many countries converged on Sohay in Upper Egypt, each observer was assigned a special task and the advantages of a strict division of labour became evident.

The new technique of photography provided both a permanent record of astronomical observations and an incentive for further international collaboration. The British astronomer, David Gill, who held the position of Her Majesty's astronomer at the Cape of Good Hope, was the first to chart accurately and measure star position by means of photography. His Cape Photographic Durchmeisterung, which included nearly half a million southern stars, was to be the forerunner of a general photographic survey of the heavens. Gill corresponded with astronomers in Paris and the result was an international congress in 1887 with representatives from 17 countries who were to plan the great Carte du ciel. By the early twentieth century the need had been felt for a Permanent Commission of the Carte du ciel. The institutional response to the need to centralise a speedy service of astronomical information had been met in 1881 by the creation of a European central office for astronomical telegrams. Another permanent institution was the Commission Internationale de l'Heure proposed in 1906 with a bureau in Paris.

NEW SCIENCES

A branch of science which claimed increasing attention in the late nineteenth century was psychology. The propagation of physiological psychology through the work of Wundt in Germany brought the subject under the banner of experimental science. The president of the first international congress was the psychologist Ribot, who as editor of the Revue philosophique had championed the cause of experimental psychology in France. Moreover, as the author of works on the state of psychology in Germany and Britain, Ribot was fully in touch with the latest developments outside his own country. When Ribot opened the first international congress of physiological psychology in Paris in 1889, he explained that up till then all psychologists had had in common had been vague aspirations and limited correspondence. By holding such a congress they were giving their subject some institutional unity. They could not hope at their first meeting to resolve problems but they could decide on the questions they were to ask. Although they had first favoured the term 'physiological' psychology, many people, including the British, preferred to use the wider term 'experimental' and the 1889 Congress agreed that the following Congress to be held in London in 1892 would be known as the Second International Congress of Experimental Psychology.

Another branch of knowledge with a claim to be scientific which developed in the nineteenth century was sociology. I mention this briefly as an example of the institutionalisation of a new area of study self-consciously related to the natural sciences. One of the key figures, René Worms, a young entrepreneur who in 1893 at the age of 24 founded both a journal which he called the Revue internationale de sociologie and the Institut International de sociologie. The use of the term international was justified by the participation of a few sociologists from Britain, Germany, Austria and the U S A. Both the journal and the institute were under the control of an editorial board of which Worms was secretary.

Within a year of their foundation they announced the first of a series of international congresses to be held in Paris: 1894, 1895, 1897, 1900. Although there were no more than twenty members at the early congresses, papers were sent in from outside and members of the public were admitted to fill up the meetings. The whole enterprise was financed by members but Worms aimed to obtain both government recognition and financial support. Official status was achieved by 1909 although they had to wait until 1931 for a government subsidy. The claim of these early sociological congresses to international status was in tune with the aspirations of the age. By the beginning of the twentieth century the international congress had become a new fashion.

CONCLUSION

One of the benefits of studying science from an international viewpoint is that it provides a truly wide perspective. It is always useful to look at a problem from more than one point of view and, although science is often the product of research in different national contexts, the growth in communication led to the rapid diffusion of science on a global scale.

The transmission of information and skills from one country to another is still a largely unexamined problem. Although I have not chosen to develop the connection, there is an obvious relevance to scientific publications. Many of the leading journals in the nineteenth century became more explicitly international, setting up editorial boards with representatives from different countries. International negotiation in scientific matters may have less dramatic consequences than negotiation on political issues but it is important that scientists should all use the same technical language and the same units. The history of the metre is a part of the history of international science. I have given the importance of a standard nomenclature and the precedents of diplomatic and technological meetings as factors in the growth of international science.

There are still historians of science who feel that the internal exploration of a branch of science is sufficient without any regard to context. But others recognise that institutions can encourage, guide, direct (though not determine) the course of science. If this is so, the historian does not have to stop with the study of a local scientific society or a national institution. He should be prepared to consider the institutional structure of science on an international scale.

By 1900 the international scientific conference had arrived. The study of twentieth-century international cooperation is therefore a rather different sort of activity more readily amenable, for example, to sociological analysis. For the nineteenth century such an approach would be premature. We still need to study the conditions which brought scientists together before such contact became a well-established routine. We need to investigate the part played by the leading scientific societies and we need to look at the organisation of individual sciences.

GENERAL BIBLIOGRAPHY

There is not room here to list the various proceedings of international conferences which have been used as a basis for this preliminary sketch. Apart from these and the Annual Reports of the British Association for the Advancement of Science a few works of general relevance are listed below.

Blaisdell, D.C.	International organisation, New York, 1966.
Crosland, M.P.	"The Congress on definitive metric standards, 1798-1799 : the first international scientific conference?" Isis, 60 (1969), 226-231.
Eijkman, P.H.	L'internationalisme scientifique, The Hague, 1911.
Gregory, W. (Ed.)	International Congresses and Conferences 1840-1937, New York, 1938.
Lyons, F.S.L.	Internationalism in Europe, 1815-1914, London, 1963.
Mangone, G.J.	A short history of international organisation, New York, 1954.
Schroeder, Brigitte	"Caractéristiques des relations scientifiques internationales, 1870-1914", Journal of World History, 10 (1966-67), 161-177.
Speeckaert, G.P.	The 1,978 International Organizations founded since the Congress of Vienna, Brussels, 1957.
Stratton, F.J.M.	"International Co-operation in Astronomy". Monthly Notices of the Royal Astronomical Society, Vol. 94. No.4. Feb. 1934, pp.361-372.
Union des Associations Internationales	Les Congrès Internationaux de 1681 à 1899. Liste complète, Brussels, 1960.
University of Cambridge,	Union Catalogue of Scientific Libraries Scientific Conference Proceedings 1644-1972, 1975.

Kurt-R. Biermann, Berlin (DDR)

ALEXANDER VON HUMBOLDT ALS INITIATOR UND ORGANISATOR INTERNATIONALER
ZUSAMMENARBEIT AUF GEOPHYSIKALISCHEM GEBIET

Bei der Vorbereitung und Durchführung des Internationaler
Geophysikalischen Jahres 1957/58 wurde das geistige Patronat des
grossen Naturforschers und Humanisten Alexander von Humboldt (1769-
1859) in Anspruch genommen; seither wurde es bei naturwissenschaft-
lichen Unternehmungen unter internationaler Beteiligung wiederholt
angerufen. Worin ist die Ursache hierfür zu erblicken?

Diese Thematik ist in neuerer Zeit bereits von Frau Kellner, London,
sowie den Herren Balmer, damals Basel, Chapman, seinerzeit in Ann Arbor,
Körber, DDR, und Otto Schneider in Argentinien behandelt worden. Ich
glaube indessen, man kann ihr einige neue Aspekte abgewinnen, wenn die
gestellte Frage in mehrere Teilfragen zerlegt wird: Warum bildete sich
gerade auf geophysikalischem Gebiet so frühzeitig eine Kooperation über
Grenzen und Meere hinweg heraus, warum geschah dies zuerst Ende des 18.,
Anfang des 19. Jahrhunderts, was zog Humboldt am Erdmagnetismus
besonders an und inwiefern war er Initiator und Organisator seiner
Erforschung in internationaler Zusammenarbeit.

Die Teilfrage, warum sich auf geophysikalischen Arbeitsgebieten wie
Meteorologie, Ozeanographie und Untersuchung des Erdmagnetismus sehr
fruh eine weltweite Zusammenarbeit herausbildete, ist relativ einfach
zu beantworten: Nur die Durchführung gleichzeitiger gründlicher
Beobachtungen nach einheitlichen Methoden und möglichst unter Benutzung
einheitlicher Messinstrumente an verschiedenen Orten des Erdballs kann
zur Erkenntnis der wirkenden Gesetzmässigkeiten mit höchster Bedeutung
für die Praxis der Schiffahrt, des Handels, des Verkehrs und der Land-
wirtschaft führen. So ist es kein Zufall, dass es gerade eine
meteorologische Gesellschaft, die Mannheimer Societas Meteorologica
Palatina, gewesen ist, die ein ausgedehntes Stationsnetz schon in den
Jahren 1780 bis 1792 von nicht weniger als 39 Beobachtungspunkten mit
einheitlichen Instrumenten, Instruktionen und Beobachtungsformularen in
Deutschland, Böhmen, Ungarn, in der Schweiz, in Italien, Frankreich,
Russland, Belgien, in den Niederlanden, in Norwegen, Danemark und
Schweden, auf Grönland und in den USA organisierte, sodass Ergebnisse
vom Ural bis Massachusetts, vom hohen Norden bis zum Mittelmeer erzielt
wurden. Dieser verheissungsvolle Beginn war nur von relativ kurzer
Dauer; 1795 erschien der letzte Band mit den Beobachtungen von 1792.[2]
Karl Theodor, Kurfürst von Pfalz-Bayern, in seiner Abkehr von der Auf-
klärung durch die Revolution in Frankreich bestärkt, stellte seinen
Zuschuss zu den Ephemeriden ein.

Im Prozess der revolutionären Ablösung der feudalistischen durch die
kapitalistische Produktionsweise nach der Grossen Franzosischen Revolution erlangte die Wissenschaft als Produktivkraft in der Entwicklung
der materiellen Produktion zunehmend grössere Bedeutung. Zunächst
nahm in Frankreich selbst die Naturwissenschaft und Mathematik nach
Beseitigung der anachronistischen Zwänge einen gewaltigen Aufschwung.
An den Beratungen zur Regulierung der Masse nach der Revolution und
zur Einführung des Dezimalsystems nahmen auch Wissenschaftler aus
Holland, aus der Schweiz, aus Italien, Spanien, Dänemark und Deutsch-

land teil, die nach Paris kamen, um das neue Mass kennenzulernenund es
mit dem in ihren Herkunftsländern geltenden zu vergleichen.[3]   Das war
gewissermassen der Beginn wissenschaftlicher Zusammenarbeit nach der
Revolution.   Dieser Austausch stellte etwas Neues dar, und es ist
nicht verwunderlich, dass er den Repräsentanten der alten Ordnung ebenso
suspekt erschien wie das Astronomentreffen auf dem Seeberg bei Gotha im
Jahre 1798 unter Beteiligung von französischen, schweizerischen und
deutschen Fachleuten.   So wurde denn auch in diesem Vorläufer späterer
internationaler Kongresse eine revolutionäre Manifestation erblickt.[4]
Die Zusammenkünfte in Paris und Gotha sind ebenso wie die sich verstär-
kende Neigung zur Begründung naturwissenschaftlicher Fachzeitschriften
unter internationaler Beteiligung Beweise für das Erstarken des
Bürgertums.

Gehen wir nun zu Humboldt und seiner Rolle über.

Betrachtet man Humboldts Forschungs- und Studienreisen, so fällt auf,
dass er stets von anderen Personen begleitet worden ist. 1789 in der
Pfalz, 1790 in den Niederlanden und in England, 1791 in Böhmen, 1795 in
Tirol und Oberitalien, 1797 in den Salzburger Alpen, auf der grossen
amerikanischen Forschungsreise 1799/1804, 1805 in Italien, 1811 auf der
Fahrt nach Wien, auf seinen Londonreisen 1817/18, 1822 wiederum in
Italien, 1824 in der Bretagne, 1829 auf der russischsibirischen Reise,
1845 in der Eifel - immer war Humboldt in Gesellschaft.   Im Verlauf
von 5 1/2 Jahrzehnten hat er auf seinen Reisen Teilung der Aufgaben
und Beobachtungen praktiziert und gegenseitige Konsultation gesucht und
gefunden.   Es hat, wie er selbst einmal konstatierte, kaum je einen
Menschen gegeben, der so viel wie er gefragt hat.   Sein "Kosmos" ist
ein bleibender Beweis für die Richtigkeit dieser Feststellung.   Der
englische Geophysiker Sabine, der Berliner Astronom Encke, der Physiker
und Astronom Arago in Paris, der Maler Albert Borg, der Russlandreisende
Abich, der Japanreisende Siebold und viele, viele andere Natur- und
Gesellschaftswissenschaftler, Künstler und Reisende des In- und
Auslandes haben auf Befragen durch bereitwillig erteilte Auskünfte, ja
ganze Abhandlungen zum "Kosmos"[5] beigetragen, sodass dies Werk durchaus
als Frucht interdisziplinärer und internationaler Zusammenarbeit betrachtet worden kann.   In den Akademien von Paris und Berlin hat er in
starkem Masse auf die Zuwahl führender ausländischer Mitglieder bzw.
Korrespondenten eingewirkt.

Wir halten fest: Von früher Jugend an bis zur Vollendung seines
neunten Lebensjahrzehnts hat sich Humboldt auf seinen Reisen und in
seinen wissenschaftlich-literarischen Schaffen über alle fachlichen und
Ländergrenzen hinweg in höchstem Masse kooperativ erwiesen.

Weiter ist ein für Humboldt charakteristischer Grundsatz in diesem
Zusammenhang hervorzuheben: Er war der Meinung, dass geachtete Privat-
personen verpflichtet wären, energisch ihren moralischen Einfluss auf
Regierungen zur Förderung der Wissenschaften geltend zu machen.[6]   Dies
Prinzip hat Humboldt sein Leben lang mit grossem Erfolg praktiziert,
gerade auch, wie noch näher zu belegen sein wird, auf dom Gebiet der
Erforschung des Geomagnetismus.   Allerdings muss einschränkend bemerkt
werden, dass er nicht immer den richtigen Blick für die politischen
Realitäten gehabt hat und deshalb gelegentlich nicht erkannte, dass der
Hebel seines Einflusses für die Beseitigung der Hindernisse unzuläng-
lich war.   Ich erinnere nur an seine geplante Reise nach Indien. 1818
war er der Meinung, seine guten Beziehungen zum Prinzregenten und zu
Ministern in London würden ausreichen, um den Widerstand der East India

Company zu brechen. Er irrte sich. Er blieb wegen seiner freimütigen Berichterstattung über koloniale Missstände in Lateinamerika eine persona non grata; die Reise wurde nicht gestattet. Übrigens war die Ursache damals kein Geheimnis. 1849 sprach der englische Ingenieuroffizier und Forschungsreisende Richard Strachey es in Kalkutta öffentlich aus, wer Humboldts Indienreise verhindert habe: "Englishmen alone need remember that he was prevented by them."[7] Es wäre sehr zu wünschen, dass die näheren Umstände der Verweigerung der Reiseerlaubnis durch britische Quellenforscher aus dem Dunkel der Archive an das Licht der Öffentlichkeit gebracht würden.

Ich gehe nun zu Humboldts geomagnetischen Aktivitäten über. Hatte er 1790 auf der Rückreise von England in Paris den Enthusiasmus des Volkes für die Ideen der französischen Revolution kennen- und bewundern gelernt und selbst Sand zum Bau des noch unvollendeten Freiheitstempels herbeizuschaffen geholfen, so gehörte er später nicht zu des zahlreichen Gleichgesinnten, deren Begeisterung rasch verrauschte. Sein erneuter Aufenthalt in Paris im Sommer 1798 brachte ihm den Zugang zur Pariser Akademie, und zwar zur Klasse der physikalischen und mathematischen Wissenschaften. Hier lernte er nicht nur den Elan einer der alten Beschränkungen ledig gewordenen Gemeinschaft hochbegabter und begeisterter Naturforscher unter bürgerlich-demokratischen Bedingungen kennen, sondern er erhielt auch ganz spezielle Anregungen , die für unser Thema von besonderer Wichtigkeit sind. Er mass nämlich in Paris mit und unter Anleitung von Jean Charles Borda die magnetische Deklination und wurde von ihm zu Intensitätsmessungen aufgefordert.[8] So wurden diese dann auf der amerikanischen Forschungsreise eine der Hauptaufgaben.[9] Warum war die Anregung Bordas bei Humboldt auf so fruchtbaren Boden gefallen? Hierfür waren mehrere Gründe bestimmend: Seit Humboldt 1796 im Fichtelgebirge Gesteinsmagnetismus entdeckt hatte,[10] hatte der Erdmagnetismus seine Aufmerksamkeit erregt. Das im "Kosmos" niedergelegte Bekenntnis zur Erlangung "mittlerer Zahlenwerte" als "Ausdruck physischer Gesetze" und damit als "letzter Zweck"[11] hat von Anfang an Humboldts Tätigkeit als Naturforscher bestimmt; bekanntlich trug ihm dies Schillers verständnislose Kritik ein, er wolle "die Natur schamlos ausmessen". Gerade der Geomagnetismus nun bot sich dazu an, durch "Bestimmung numerischer Mittelwerte" das "Quantitative der Gesetze" zu ergründen.[12] Hinzu kommt ein Spezifikum der Manifestationen des Erdmagnetismus, das besonders Humboldts ausgeprägtes historisches Interesse ansprach. Die Einbeziehung der Vergangenheit in die Analysierung des gegenwärtigen Zustandes war ein Grundelement Humboldtscher Betrachtungsweise. "Das Sein wird in seinem Umfang und inneren Sein erste als ein Gewordenes erkannt";[13] Diesem seinem Leitsatz folgte er nicht nur in der Darstellung naturwissenschaftlicher Sachverhalte und des Ganges ihrer Erforschung im Wandel der Zeiten, wie etwa in seinen "Kritischen Untersuchungen über die historische Entwicklung der geographischen Kenntnisse von der Neuen Welt und die Fortschritte der nautischen Astronomie in dem 15. und 16. Jahrhunderts";[14] er verwirklichte sein historisches Interesse ebenso in seinen Vorschlägen für die Erhöhung der Rentabilität von Gruben durch Auswertung alter Bergwerksakten oder in seinen Anregungen für eine Münzreform durch Rückgriff auf die Münzgeschichte, bei der Darstellung der Lage der Indianer in Lateinamerika oder etwa in der Schilderung des Reflexes der Natur in deren dichterischer Beschreibung in früheren Epochen. Die erste selbständige (nicht erhalten gebliebene) Abhandlung des Zwanzigjährigen betraf den Webstuhl bei den Griechen und Römern, als junger Oberberg-

meister in Franken studierte er bergbauliche Archivalien, in Amerika
suchte er die Archive auf und ging auf die Quellen zurück, wie er dies
drei Jahrzehnte später auch in Russland tat.   Wie sehr musste einen
Forscher mit einem so ausgeprägten Sinn für die Genese ein naturwissenschaftlicher Forschungsbereich anziehen, in dem nicht nur lange
zurückreichende Beobachtungen einen Erkenntniszuwachs gebracht hatten,
wie dies für andere Gebiete auch zutraf, sondern in dem das Phänomen
selbst sekularen Veränderungen unterworfen war.   Nach seiner Rückkehr
aus Amerika formulierte Humboldt am 17.12.1804, wiederum in der Pariser
Akademie, gemeinsam mit Biot, auf der Grundlage seiner Reisebeobach-
tungen[15] das Gesetz der Zunahme der magnetischen Intensität gegen die
Pole.[16]   1805/06 setzte er gemeinsam mit Gay-Lussac[17] in Frankreich,
Italien, der Schweiz und in Deutschland die Intensitätsmessungen fort.[18]
Nach der Rückkehr nach Berlin ging Humboldt einen Schritt weiter: Er
bildete ein Team, ausser ihm selbst bestehend aus einem Astronomen,
einem Techniker, einem Geographen, einem Geologen und einem Mechaniker,[19]
das unter seiner und des Astronomen Leitung von Mitte Mai 1806 bis Ende
Juni 1807 in Berlin 6000 Beobachtungen der stündlichen Variationen der
Deklination und der wiederkehrenden Perturbationen durchführte.[20]
Schon damals entstanden erste Pläne, durch gleichzeitige Beobachtungen
an mehreren Orten Zusammenhänge etwa zwischen magnetischer Perturbation
und dem Auftreten eines Nordlichts zu klären,[21] aber in dieser Zeit
napoleonischer Kriege fehlten dafür die Voraussetzungen.   Während
seiner Pariser Jahre (Ende 1807 bis April 1827) wurde Humboldts
Beschäftigung mit dem Geomagnetismus vor allem durch den bereits
genannten Francois Arago beeinflusst, mit dem er z.B. 1810 Inklination-
smessungen ausführte und der in seiner Gegenwart im November 1817 am
Greenwicher Hügel das 1825 als Rotationsmagnetismus erklärte Phänomen
entdeckte, welches seine Erklärung 1831 durch die von Faraday gefundene
Induktion erhielt.[22]   "Durch Arago", sagte Humboldt gelegentlich,
"begann eine glänzende Epoche für die Erforschung des tellurischen
Magnetismus.   Die auf der Pariser Sternwarte regelmässig zu bestimmten
Stunden gemachten Beobachtungen über die täglichen Veränderungen der
Abweichung umfassen eine grössere Periode von Jahren, als je diesem
Zweige der messenden Physik gewidmet worden sind."[23]   In Paris war es
auch, wo Humboldt aus in Kazań und Paris gleichzeitig angestellten
Beobachtungen lernte, was man aus korrespondierenden Messungen gewinnen
könne.   Als er 1827 in seine Heimatstadt zurückkehrte, gehörte es
daher erklärtermassen zu seinem wissenschaftsorganisatorischen Programm,
nicht nur die Arbeiten von 1806/07 wieder aufzunehmen und die neuen Daten
mit den früheren zu vergleichen,[24] sondern darüberhinaus "in- und
ausserhalb Europas einen regelmässigen Cursus correspondierender Beobachtungen des tellurischen Magnetismus zu begründen; Gleichmässigkeit
der Apparate und der Methoden, verständige Auswahl der Beobachtungsorte,
steter Verkehr zwischen den geübten Beobachtern und Sicherung des
Antheils gelehrter Corporationen, damit das van meinen Freunden und mir
gegründete Institut permanent bleibe, waren wesentliche Erfordernisse."[25]
Am 2.4.1829 stellte Humboldt in einem Vortrag über die Ergründung von
Phänomenen des tellurischen Magnetismus von den Mitgliedern der
Berliner Akademie fest: "Die naturwissenschaftliche Zivilisation der
Welt reicht kaum über jene glänzende Epoche hinaus, wo in dem Zeitalter
von Galilei, Huygens und Fermat gleichsam neue Organe geschaffen wurden,
neue Mittel, den Menschen (beschauend und wissend) in einen innigeren
Kontakt mit der Aussenwelt zu setzen, Fernrohr, Thermometer, Barometer,

die Pendeluhr und, ein Werkzeug von allgemeinerem Gebrauche, der Inifinitesimal-Kalkül. Wäre die alexandrinische Schule oder wären die Araber, eine Nation, die den im Altertum leider unbekannten praktischen Sinn des Beobachtens, Messens und Experimentierens zeigte, im Besitz der Hilfsmittel gewesen, die uns gegenwärtig zu Gebote stehen, so würden wir wissen, ob der Druck der Atmosphäre, die aus der Luft sich niederschlagende Regenmenge, die relative Frequenz vorherrschender Winde, die Richtung der isothermischen Linien, wie die Verteilung des Magnetismus auf dem Erdkörper sekulären Veränderungen unterworfen sind. Bei dem Mangel an numerischen Daten", fügte er hinzu, "bei dem Mangel an Dokumenten zur physischen Geschichte des Erdkörpers und des Luftkreises, ist unser Zeitalter berufen, den kommenden Geschlechtern die Lösung jener wichtigen Probleme vorzubereiten."[26] Bereits am 26.9.1830 konnte er erklären: Fast dürfen wir schon sagen, dass unsere Linie magnetischer Stationen sich jetzt von Südamerika quer durch Europa bis [zur russischen Mission in] Peking erstreckt."[27] Zu dieser Bilanz hat nicht nur die von ihm entfaltete briefliche Aktivität beigetragen, die z.B. zur Beteiligung des Freiberger Physikers Ferdinand Reich mit Messungen unter Tage, zur Teilnahme des französischen Naturforschers Boussingault mit Beobachtungen in Kolumbien und zu Messungen Wrangels in Alaska, zu Beobachtungen auf Kuba, auf der Krim und auf Island führte, sondern vor allem auch seine russisch-sibirische Reise von 1829. In Berlin liess er seine von ihm als "Zentralanstalt" bezeichnete eisenfreie Beobachtungsstation unter der Obhut des Meteorologen Dove, des Physikers Riess und einer Reihe anderer junger Naturforscher verschiedener Disziplinen, während er sein Ansehen und seinen Einfluss in Russland benutzte, um magnetische und meteorologische Stationen in den "verschiedensten klimatischen Zonen" des Reiches wärmstens zu befürworten. Dass er auf seiner Reise auch wieder selbst zahlreiche magnetische Messungen zusammen mit geographischen Ortsbestimmungen vernahm, versteht sich.[28] Seine Pläne, das russische Reich in grossem Massstabe an korrespondierenden Beobachtungen zu beteiligen, trafen sich mit den Projekten des Petersburger Akademikers Kupffer,[29] den er schon in Paris kennengelernt hatte, als jener sich an Aragos Messungen beteiligte, und der in Kazań die erwähnten mit Paris korrespondierenden Beobachtungen ausgeführt hatte. Als Humboldt Anfang November 1829 Moskauer Archive aufsuchte, stellte er zu seiner Überraschung fest, dass kein geringerer als Leibniz sein Vorgänger gewesen war, hatte dieser doch in richtiger Einschätzung der Bedeutung von erdmagnetischen Messungen in dem riesigen russischen Reich für die Erkenntnis der obwaltenden Gesetzmässigkeiten den Zaren Peter I. mit viel diplomatischem Geschick und psychologischem Fingerspitzengefühl zu bewegen versucht, im europäischen wie im asiatischen Teil regelmässig zu bestimmten Zeiten Inklination und Deklination beobachten zu lassen. Mit berechtigter Genugtuung konstatierte Humboldt, dass Leibniz' Wunsch nach 117 Jahren in Erfüllung ging und dass er, Humbolt, es war, der an der Realisierung massgeblichen Anteil hatte;[30] er hätte in Leibniz auch einen Vorläufer in den Bestrebungen erblicken können, aus China geomagnetische Beobachtungsresultate zu erhalten. Humboldt unterstützte auch mit allem Nachdruck Kupffers Vorschläge zur Errichtung eines Physikalischen Zentrallaboratoriums in Petersburg,[31] das dann tatsächlich gebaut wurde und 1848 seine Tätigkeit in vollem Umfange aufnahm.

Einen Meilenstein in der Entwicklung der internationalen Zusammenarbeit zur Erforschung des Erdmagnetismus stellt ein Brief dar, den Humboldt Ende 1831 an Wilhelm Weber nach Göttingen sandte[32] und der mit den Ausschlag dafür gab, dass sich Gauss und Weber nun dem Geomagnetis-

mus zuwandten.   Gewiss, Gauss' Interesse an den geomagnetischen
Phänomenen war alt, aber bei seinem Besuch in Berlin 1828 als Gast
Humboldts wurde das Interesse neu belebt und nach Webers Berufung nach
Göttingen hatte er zusammen mit dem jungen Physiker bereits begonnen,
die Messinstrumente und ihre theoretische Grundlage zu vervollkommnen.
Humboldts Brief diente Gauss aber als Argument in seinem Antrag auf
Bewilligung eines besonderen Gebäudes für magnetische Beobachtungen.
Humboldt hatte bei Abfassung des erwähnten Briefes die Absicht, Göttin-
gen mit in die Erforschung des Erdmagnetismus einzubeziehen und an den
korrespondierenden Beobachtungen zu beteiligen.   Er konnte nicht
vorhersehen, dass Gauss nun bald die führende Rolle in der geomagnetischen Forschung und in der Organisation der Beobachtungen übernehmen
würde, die er selbst bisher innehatte, er ahnte nicht, dass Gauss und
Weber das von ihm benutzte und empfohlene Gambeysche Instrumentarium
nicht benötigen, sondern durch eigene, genauere und zuverlässigere
Instrumente und Methoden ersetzen würden und er konnte noch nicht
absehen, dass die theoretische Durchdringung der anstehenden Fragen von
Gauss ausgehen würde.   Der in Rede stehende Brief markiert eine Zäsur
in der erdmagnetischen Forschung - die Humboldtsche Äre geht ihrem Ende
entgegen, die Gauss-Webersche Äre hebt an.   Aber wir können sagen,
dass letzlich auch diese von Humboldt mit initiiert worden ist.
     Wenig später kam es zur Gründung des sogenannten Göttinger Magnetis-
chen Vereins,33 wir würden heute von einer "Arbeits-gemeinschaft"
sprechen, mit eigenem Publikationsorgan; Gauss führte absolute Massein-
heiten ein, Empiric wurde durch physikalisch-mathematische Analyse
ersetzt, neue Termine wurden eingeführt, neue Methoden angegeben, neue
Messinstrumente entwickelt und hergestellt, darunter transportable, die
von Forschungsreisenden auf Sizilien erprobt wurden, und eine allge-
meinere Theorie des Erdmagnetismus entwickelt.   Die Zahl der Staionen,
die die Gauss-Weberschen Vorschriften und Ausrüstungen anwandten, wuchs,
das Ansehen der Göttinger Vereinigung, die mit dem Namen des bedeutend-
sten lebenden Mathematikers der Welt verknüpft war, stieg rasch an.   Es
ist erklärlich, dass Humboldt dieser Entwicklung zunächst mit etwas
gemischten Gefühlen zusah.   Dass das, was er so lange als seine Domäne
angesehen hatte, nun in andere Hände überging, bereitete ihm doch etwas
Kummer.   Es ist nicht richtig, wenn es in der Literatur meist so
dargestellt wird, als sei der Übergang von der Humboldtschen zur Gaussschen Epoche gewissermassen von Anfang an durch Humboldt gefördert
worden.   Es hat vielmehr eine ganze Zeit gegeben, in der Humboldt dem
Gang der Dinge mit kritischer Skepsis zusah und und seine Prioritäten
reklamierte.34   Es wird dies ohne weiteres verständlich, wenn man sich
vor Augen hält, dass die Gambeyschen Bussolen, die Humboldt benutzte und
empfahl, durch das exakte und wirklich einheitliche Gausssche Magneto-
meter ersetzt wurden, dass die Humboldtschen Beobachtungstermine (8-mal
im Jahr je 44 Stunden) durch Gausssche Termine (6-mal im Jahr je 24
Stunden bei Herabsetzung der Intervalle zwischen den jeweiligen Beobachtungen) abgelöst wurden, dass das Rumboldtsche Prinzip der Anknüpfung
der Termine an natürliche Perioden, wie Solstitien und Äquinoktien,
fallengelassen wurde, dass das Humboldtsche Postulat nach Gleichzeitigkeit der Beobachtungen durch Festlegung auf die Göttinger Zeit
präzisiert wurde, dass die teilnehmenden Beobachter schriftlich oder
persönlich in Göttingen genau instruiert und zur rechnerischen Mittelung
der Beobachtungen Schemata vorgegeben wurden, dass unberücksichtigt
gebliebene Werte wie Torsion   Trägheitsmoment und Schwingungsdauer nun

exakt erfasst wurden,[35] kurz, dass erst durch Gauss und Weber die allgemein gehaltenen Humboldtschen Forderungen nach Einheitlichkeit und
Gleichzeitigkeit der Beobachtungen auf eine wissenschaftlische Grundlage
gestellt und verwirklicht wurden, dass die Humboldtsche Definition der
magnetischen Feldstärke durch die Gausssche Absolutbestimmung hinfällig
wurde und dass Humboldt bei Gauss anfangs irrtümlicherweise eine
Vernachlässigung der Inklination feststellen zu müssen glaubte.   Nimmt
man noch hinzu, dass Humboldt während eines Besuchs in Göttingen im
September 1826 mit Erstaunen bemerkte, dass Gauss offenbar nie zuvor
mit einem Inklinatorium gearbeitet hatte,[36] versteht man seine Empfind-
lichkeit gegenüber der ihn überholenden, von Göttingen ausgehenden
Entwicklung.   Indessen wurden seine Vorbehalte in zunehmendem Masse
abgebaut, als er erkannte, dass es Gauss gelungen war, die absolute
Bestimmung der Intensität mit zuvor nicht erreichbarer Exaktheit
vorzunehmen und entscheidende Fortschritte in der Magnetik zu erzielen.
Indessen haben manche Ressentiments noch einige Jahre nachgewirkt, ohne
dass allerdings das freundschaftliche Verhältnis zwischen Humboldt und
Gauss ernsthaft getrübt worden wäre.

Als sich 1835 auch Kupffer entschloss, der ständig zunehmenden
Erweiterung des Göttinger Vereins Rechnung zu tragen, und seinem früheren Lehrer Gauss erklärte, er wolle sich mit den russischen Stationen
an den Göttinger Terminen beteiligen, zog Humboldt die Konsequenzen und
stellte sich mit einer neuen Initiative noch einmal an die Spitze der
Bewegung.

Bis dahin war nämlich das grosse britische Kolonialreich noch nicht
an den korrespondierenden Messungen beteiligt.   Humboldt nun fasste den
Entschluss, einen Brief an den Präsidenten der Royal Society in London
zu schreiben und ihn zu publizieren, um damit den Anstoss zu weltweiter
Errichtung geomagnetischer Observatorien zu geben.[37]   Dass dies nicht
ohne einige Verstimmungen abging, wird nach dem bereits Gesagten nicht
verwundern, zumal Gauss und Humboldt während der Zeit der Konzipierung
des Briefes von einander nur über einen Mittelsmann (Schumacher) hörten;
ich habe dies ausführlich anhand der Quellen dargestellt[38] und will es
hier nicht widerholen. Für das jetzige Thema ist wichtig, dass der Humboldtsche Appell an die Royal Society vom Februar 1836 Erfolg hatte und
nach einigen Jahren permanente magnetische Stationen auch im britischen
Weltreich zu arbeiten begannen.   Wie Nathan Reingold unlängst ermittelt
hat, scheint Edward Sabine, eine Schlüsselfigur in der erdmagnetischen
Forschung und Organisation in England, den Brief Humboldts an den
Präsidenten der Royal Society angeregt zu haben, indem er sich davon
eine Unterstützung seiner eigenen Vorschläge zur Ausrüstung einer Expedition in die südliche Hemisphäre und zur Errichtung von Messtationen in
den Kolonien versprach.[39]   Humboldt selbst sah dies aber anders: Er
"rühmte" sich, den "magnetischen Drang in England eingeimpft zu haben".[40]
Er war mit Recht stolz auf jenen Brief, denn, wie er sagte, "man musste
sich zwischen vielen Nationaleitelkeiten durchwinden, dem Vorwurfe
ausweichen, als glaube man, nun erst würden die Engländer etwas Magnet-
isches leisten, die Menschen alle nennen, die man schon gebraucht hat
oder notwendig brauchen wird."[41]

Besonders folgenreich für die internationale Zusammenarbeit war ein
kleiner "magnetischer Kongress", der Mitte Oktober 1839 in Göttingen
unter Gauss' Vorsitz stattfand und an dem aus England Edward Sabine, aus
Irland Humphrey Lloyd, aus Russland Kupffer sowie Steinheil aus München
teilnahmen.   Humboldt blieb fern, aber Sabine und Lloyd hatten ihn auf
der Hinreise besucht und seine Vorstellungen erkundet.   Es wurden die

Pläne für die englischen, russischen und deutschen Beobachtungen des
Erdmagnetismus beraten und koordiniert und eine (in der Folge durchge-
führte) Vermehrung, Vergrösserung und Neuausstattung der russischen
Observatorien empfohlen.

Mit seiner Aufforderung an die Royal Society, 1838 durch einen
zweiten Brief an ihren Präsidenten zur Gewährleistung der Publikation
der magnetischen Beobachtungen unterstützt,[42] hatte Humboldt einen
Anstoss gegeben, der so ausserordentlich folgenreich wurde, dass dies
unwiederholbar war, was die Ergebnisse anbetrifft.   Schliesslich waren
an 50 Stationen in der Welt am Beobachtungsnetz beteiligt.   Aber noch
einmal ging von ihm ein bedeutender Impuls aus.

Er, der schon 1800 vor Kuba die Idee von Forschungsschiffen zur
Ermittlung der Meeresströmungen konzipiert hatte, hat wiederholt für
Forschungsreisen auf See Anregungen und Hinweise für vorzunehmende
Untersuchungen gegeben.   Im Oktober 1839, wohl auf Grund der erwähnten
Rücksprache in Berlin mit Sabine, sandte Humboldt einen Brief an den
Ersten Lord der britischen Admiralität mit Instruktionen für die nach
Anregungen der British Association for the Advancement of Science
erfolgende Expedition von Sir James Ross in die Antarktis 1838/43 mit
besonderer Berücksichtigung der Verfolgung der Linien ohne Deklination.[43]
Es war dies nicht die letzte organisatorische Einflussnahme Humboldts
auf dem Gebiet der Erforschung des Geomagnetismus.   Sein Interesse
blieb lebendig und noch manche Anregungen wurden von ihm gegeben, wenn
sie sich auch nicht mehr mit den Ergebnissen seiner bereits genannten
Aktivitäten vergleichen lassen.   Leider sind wir in mancher Hinsicht
dadurch in unserer Urteilsfindung beeinträchtigt, dass Sabine, der Mann,
mit dem er seinen ausgedehntesten Briefwechsel in magneticis geführt
hat, aus unbekannten und nicht erratbaren Gründen letztwillig verfügt
haben soll, dass die Briefe Humboldts an ihn und seine Gattin, die als
Übersetzerin naturwissenschaftlicher Abhandlungen, darunter vor allem
auch Humboldtscher, einen bedeutenden Namen hatte, zu vernichten seien.
Es muss wohl Vorwürfe Humboldts an die Adresse Sabines gegeben haben,
dass letzterer eine Gelegenheit ungenutzt habe vorübergehen lassen,
Humboldts Verdienste um die Lenkung der öffentlichen Aufmerksamkeit auf
die Wichtigkeit und Bedeutung des systematischen Studiums der erdmagnetischen Phänomene gebührend zu würdigen.   Anders ist der wortreiche
Rechtfertigungsversuch Sabines vom 27.10.1851[44] nicht zu verstehen.

Auch Gauss blieb nicht lange an der Spitze der international organi-
sierten geomagnetischen Forschung.   Nach Webers von der Reaktion
erzwungenem vorübergehenden Weggang aus Göttingen erlahmte sein aktives
Interesse am Erdmagnetismus; 1843 erschien der sechste und letzte Band
der "Resultate aus den Beobachtungen des magnetischen Vereins" mit den
Beobachtungen von 1841.

Der bedeutende Astronom Bessel hat 1843 die Entwicklung bis dahin
unter eleganter Vermeidung des Eingehens auf den wie erwähnt nicht
problemlosen Übergang vom Humboldtschen zum Gaussschen Stationsnetz so
charakterisiert: "In diesem Jahrhundert gewinnt der Eifer für den
Magnetismus der Erde neues Leben; Alexander von Humboldt erregt ihn
und steigert ihn durch eigene Erfolge.   Bald wird er kräftig genug,
Untersuchungen hervorzubringen, welche die Vervollständigung der Beobachtung des magnetischen Zustandes der entlegensten Punkte der Erde
zum einzigen oder hauptsächlichen Zweck haben. [....]   Der magnetische
Apparat fängt an, ein hauptsächlicher Teil der Ausrüstung aller
Reisenden zu werden.   In wenigen Jahren liefern sie eine Grundlage für

die Gaussische Theorie.   Es gelingt Humboldt, die Regierungen von
England, Frankreich, Russland, ... für den Magnetismus der Erde zu inter-
essieren und grosse, kostbare Expeditionen sind die Folge davon.   Diese
rüsten sich mit den Gaussischen Apparaten aus und folgen den ihnen von
Humboldt vorgezeichneten Wegen."[45]

Noch mit 88 Jahren machte Humboldt in seinen Instruktionen für die
Wissenschaftler, die an der österreichischen Weltumseglung auf der
"Novara" teilnahmen, auch wieder auf die Beobachtung der "Wanderung der
magnetischen Kurven", besonders der Variation, aufmerksam;[46] die Rückkehr
dieser Expedition sollte er nicht mehr erleben.   So lieferte er noch
einmal einen Beweis dafür, dass es ihm ernst gewesen war, wenn er 1854
konstatiert hatte, er habe nur drei wichtige Originalleistungen voll-
bracht:[47] seine Geographie der Pflanzen, seine in den Schriften der
elitären Société d'Arcueil[48] zuerst publizierte Theorie der isothermen
Linien und drittens seine "Beobachtungen über den Geomagnetismus, welche
die über den ganzen Planeten auf meine Veranlassung verbreiteten magnetischen Stationen zur Folge gehabt haben".

Wenn man über Humboldts übernationales organisatorisches Wirken
spricht, sollten frelich auch seine Bemühungen um die Erlangung zuver-
lässiger meteorologischer Messdaten, vor allem Mitteltemperaturen, die
er gern mit den magnetischen Beobachtungen verbunden sehen wollte, oder
beispielsweise seine Beteiligung an der Verbereitung der russischen
Chronometer-Expedition in der Ostsee von 1833 zur Längenbestimmung[49]
wenigstens erwähnt werden, wirkten doch an dem zuletzt genannten Unter-
nehmen auch Preussen, Dänemark und Schweden durch Errichtung zeitweili-
ger Beobachtungsstationen mit.   Erinnert werden sollte in diesem
Zusammenhang auch an den Einfluss Humboldts auf J.J. Baeyer und seine
Schöpfung, die Internationale Erdmessung,[50] auch wenn diese weltweite
Organisation erst einige Jahre nach Humboldts Tod ihre Tätigkeit aufnahm
und die Geschichte der Bestimmung der Erdgestalt hier sonst unbetrachtet
bleibt.

So ist es nicht nur verständlich, sondern begründet, wenn die beiden
Internationalen Polarjahre 1882/83 und 1932/33, Vorläufer des eingangs
erwähnten Internationalen Geophysikalisehen Jahres, mit ihren magneti-
schen und meteorologischen Beobachtungen an Humboldt und den Gauss-
Weberschen Magnetischen Verein anknüpften; die Verbindung von Humboldts
organisatorischen Initiativen zu internationaler Zusammenarbeit reicht
bis in die Gegenwart.   Dergestalt ist die Anregung Prof. Croslands, des
Initiators dieses der Geschichte internationaler Kooperation auf dem
Gebiet der Naturforschung gowidmeten Symposiums, hier auch Alexander von
Humboldts Wirken zu gedenken, vollauf berechtigt.

Sie ist es umso mehr, als es für Alexander von Humboldt keine ideolo-
gische Koexistenz mit den Ultras seiner Zeit gab, die ihn als einen
"trikoloren Lappen" betrachteten,[51] den man im Falle des Bedarfs bei
revolutionären Erhebungen "entfalten", will sagen, als Beweis der
Konzessionsbereitschaft verzeigen konnte.   Für ihn gab es auf weltan-
schaulischen Gebiet keinen Kompromiss mit den Vertretern der Reaktion,
des Rückschritts; suaviter in mode, fortiter in re hielt er unbeirrbar
an der Überzeugung und Propagierung der Notwendigkeit einer demokrati-
schen Entwicklung fest.   Auf dem Gebiet der Mehrung der Naturerkenntnis
hingegen hielt er eine Kooperation mit Gelehrton abweichender politischer Standpunkte für möglich und praktizierte sie.   Humboldt besass eine
durch keine staatliche Funktion und durch kein akademisches Amt gestützte
Autorität, die einzig und allein auf seiner eigenen wissenschaftlichen

Leistung, vornehmlich als "zweiter, wissenschaftlicher Entdecker
Amerikas", beruhte. Dies Ansehen hat er nicht nur zur Förderung un-
gezählter junger Gelehrter eingesetzt, sondern auch, wie zu zeigen
versucht wurde, auf dem Felde der Organisation internationaler wissenschaftlicher Zusammenarbeit, vor allem bei der Erforschung des Erdmagnetismus, zu der ihn Neigung und Fähigkeiten in gleicher Weise hinzogen.

Anmerkungen

1   Kellner, L[otte], 'Alexander von Humboldt and the history of
       international scientific collaboration.' Scientia, 54 (1960),1-5.
    Kellner, L[otte], 'Alexander von Humboldt and the organisation of
       international collaboration in geophysical research.'
       Contemporary Physics, 1 (1959), 35-48.
    Balmer, Heinz, Beiträge zur Geschichte der Erkenntnis des Erd-
       magnetismus. Aarau 1956.
    Chapman, S[ydney], 'Alexandre von Humboldt et l'étude due
       géomagnétisme.'Ciel et Terre, 65 (1959), 269-284.
    Chapman, Sydney, 'Alexander von Humboldt and geomagnetic science.'
       Archive for the History of Exact Sciences, 2 (1962), Nr.1. 41-51.
    Körber, Hans-Günther, 'Alexander von Humboldts und Carl Friedrich
       Gauss' organisatorisches Wirken auf geomagnetischem Gebiet.'
       Forsch.u.Fortschr, 32 (1958), 1-8.
    Körber, Hans-Günther, 'Aus der Korrespondenz Alexander von Humboldts
       und Carl Friedrich Gauss' mit Teilnehmern an geomagnetischen
       Beobachtungen.' Forsch.u.Fortschr, 33 (1959), 298-303.
    Körber, Hans-Günther, 'Alexander von Humboldts meteorologische und
       geomagnetische Forschungen.' Referentenmaterial d. Ges. zur
       Verbreitung wissenschaftl. Kenntnisse, Sekt.1. (1959), H.1.51-68.
    Schneider, Otto, 'Humboldt y el geomagnetismo.' Ciencia e investi-
       gación, 1 (1945), 543-548.

2   Hellmann, G[ustav], Repertorium der deutschen Meteorologie,
       Leipzig, 1883, 895-902.

3   Arago, D.F.J., Populäre Astronomie, Bd.4, Leipzig, 1859, 34-35.

4   Herrmann, Lieter B, 'Das Astronomentreffen im Jahre 1798 auf dem
       Seeberg bei Gotha.' Archive for History of Exact Sciences, 6
       (1970), Nr.4, 327-344. Insbes. 342.

5   Humboldt, Alexander von Kosmos, Entwurf einer physischen Welt-
       beschreibung, Bd. 1-5. Stuttgart u. Tübingen 1845/62.

6   Humboldt an George von Cotta, 5.5.1839. (Schiller-National-
       museum, Marbach, Cotta-Archiv.)

7   Biermann, Kurt-R., 'Alexander von Humboldts Forschungsprogramm
       von 1812 und dessen Stellung in Humboldts indischen und
       sibirischen Reiseplänen.' Studia z dziejow geografii i karto-
       grafii, (1973), 471-483. Zit. 476.

8   Kosmos (siehe Anm.5), Bd.1, 432.

9   Humboldt, Alexander von, 'Beobachtungen der Intensität magnetischer
       Kräfte und der magnetischen Neigung, angestellt in den Jahren
       1798 bis 1803, [...] in Frankreich, Spanien, den canarischen

Inseln, dem atlantischen Ocean, America und der Südsee.' Annalen der Physik u. Chemie, 15 (1829), 336-355.

10 Humboldt, F[riedrich] A[lexander] von, 'Uber die merkwürdige magnetische Polarität einer Gebirgskuppe von Serpentinstein.' Neues Journal der Physik, 4 (1797), 136-140.

11 Siehe Anm. 5, Bd. 1, 82.

12 Ebd, 197.

13 Ebd, 64.

14 Humboldt, Alexandre de, Examen critique de l'histoire de la géographie du Nouveau Continent et des progrès de l'astronomie nautique aux quinzième et seizième siècles. T.1-5 Paris 1836/39.

15 Siehe Anm. 5, Bd.1, 431-432.

16 Humboldt, Alexandre de, 'Sur les variations du magnétisme terrestre à différentes latitudes; par MM. Humboldt et Biot. Lu par M. Biot à la classe des Sciences mathématiques et physiques de l'Institute national, le 26 frimaire an 13.' Journal de physique 59 (1804), 429-450.

17 Siehe Anm. 5, Bd.4, 64.

18 Humboldt, Alexandre de, 'Observations sur l'intensité et l'inclinaison des forces magnétiques, faites en France, en Suisse, en Italie, et en Allemagne. Lu à l'Institut le 8 septembre 1806.' Mémoires de physique et de chimie, de la Société d'Arcueil, 1 (1807), 1-22, 1 Tafel.

19 Siehe Anm.5, Bd.4, 125-127, 196.

20 Humboldt, Alexander von, Vorwort [für] 'Correspondirende Beobachtungen über die regelmässigen stündlichen Veränderungen und über die Perturbationen der magnetischen Abweichung im mittleren und östlichen Europa, gesammelt und verglichen von H.W. Dove.' Annalen der Physik und Chemie, 19 (1830), 357-361. Insbes 357-358.

21 Humboldt, Alexander von, 'Die vollständigste aller bisherigen Beobachtungen über den Einfluss des Nordlichts auf die Magnetmadel.' Annalen der Physik, 29 (1808), 425-429.

22 Humboldt, Alexander von, Einleitung [zu] Franz Arago's sämmtliche Werke, Bd.1, Leipzig 1854, I-XXII. Insbes, XIII.

23 Siehe Anm.20, 358.

24 Siehe Anm.5, Bd.1, 437.

25 Siehe Anm.20, 359.

26 Humboldt, Alexander von, 'Uber die Mittel, die Ergründung einiger Phänomene des tellurischen Magnetismus zu erleichtern.' Annalen der Physik u. Chemie, 15 (1829), 319-336. Zit. 319-320.

27 Siehe Anm. 20, 359.

28 Humboldt, 'Alexander von, Beobachtungen der Inclination der Magnetnadel, gemacht auf einer Reise nach dem Ural, dem Altai und dem Caspischen Meer.' Annalen der Physik u. Chemie, 18 (1830) 355.

29 Schramm, Hugo, 'Adolf Theodor von Kupffer.' Westermann's Jahrbuch
    der Illustrirten Deutschen Monatshefte, 19 (1866), 502-518, 604-
    620.

30 Humboldt, Alexandre de, 'Remarques servant d'éclaircissements au
    Tableau d'inclinaisons magnétiques, observées en 1829 dans le
    Nord-Est de l'Asie et sur les bords de la mer caspienne.'
    Asie Centrale par A. de Humboldt. T.3. Paris 1843, 440-478.
    Insbes. 469-478. - Siehe auch Anm.5, Bd.4, 203-204.

31 Rykatchew, M[ichajl] A[leksandrovič], Histoire de l'Observatoire
    Physique Central P.1. St.-Pétersbourg 1900.

32 Biermann, Kurt-R., 'Der Brief Alexander von Humboldts an Wilhelm
    Weber Ende 1831 - ein bedeutendes Dokument zur Geschichte der
    Erforschung des Geomagnetismus.' Monstsberichte Dt.Akad.Wiss.
    Berlin, 13 (1971), 234-242.

33 Schering, Ernst, 'Carl Friedrich Gauss und die Erforschung des
    Erdmagnetismus.' Abhandlungen Kgl. Ges. Wiss. Göttingen, 34
    (1887).
    Wiederkehr, Karl Heinrich, 'Aus der Geschichte des Göttinger
    Magnetischen Vereins und seiner Resultate.' Nachrichten Akad.
    Wiss. Göttingen, II. Math.-phys.Kl, (1964), Nr.14.

34 Biermann, Kurt-R, 'Aus der Vorgeschichte der Aufforderung Alexander
    von Humboldts von 1836 an den Präsidenten der Royal Society zur
    Errichtung geomagnetischer Stationen.' Wiss. Zs. Humboldt-Univ.,
    Math.-nat.R, 12 (1963), 209-227.

35 Körber, Hans-Günther, 'Alexander von Humboldts und Carl Friedrich
    Gauss' organisatorisches Wurken auf geomagnetischem Gebiet.'
    Forsch.u.Fortschr, 32 (1958), 1-8.

36 Siehe Anm.34, 211.

37 Humboldt, Alexander von, 'Über die Mittel, den Erdmagnetismus durch
    permanente Anstalten und correspondirende Beobachtungen zu
    erforschen. [Brief an den Präsidenten der Royal Society.]'
    Berlin, April 1836. Astronomische Nachrichten, 13 (1836), 281-
    292.

38 Siehe Anm. 34.

39 Reingold, Nathan, 'Edward Sabine.' C.C. Gillispie, ed. Dict.
    Scient.Biogr., 12 (1975), 49-53. Insbes. 50.

40 Humboldt, Alexander von, Briefe an Christian Carl Josias Freiherr
    von Bunsen, Leipzig 1869. 177.

41 Siehe Anm.34, Zit. 11-12.

42 29.5.1838. The Wellcome Historical Medical Library, London.

43 Humboldt, Alexander von, 'Letter to the Earl of Minto. Berlin,
    12.10.1839.' Royal Society. Report of the Committee of Physics,
    Including Meteorology, London, 1840, 91-95.
    Humboldt, Alexander von, 'Additions fragmentaires aux "Instructions
    for the Scientific Expedition to the Antarctic Regions". Berlin,
    26.10.1839.' Royal Society. Report of the Committee of Physics,
    Including Meteorology, London, 1840. 95-102.

44 (Fragment.) Deutsche Staatsbibliothek Berlin, Handschriftenabt.
    A.v. Humboldt, K.VIII, Nr.13; acc.Darmst.1932.30. Z.Zt. Staats-
    bibl.Pr. Kulturbesitz, Westberlin.

45 Bessel, Friedrich Wilhelm,'Über den Magnetismus der Erde.'Populäre
    Vorlesungen über wissenschaftliche Gegenstände von F.W. Bessel,
    Hamburg, 1848, 326-386. Zit. 385-386. [Zuerst in Jahrbuch für
    1843. Hrsg.v. H.C. Schumacher. Stuttgart u. Tübingen, 1843, 1-56].

46 Humboldt, Alexander von, 'Physikalische und geonostische Erinner-
    ungen,' Berlin, 7.4.1857. [Instruktionen für die 'Novara'.]
    Karl von Scherzer, Reise der Österreichischen Fregatte Novara um
    die Erde in den Jahren 1857, 1858, 1859, Volksausgabe, Bd.1,
    Wien, 1864, Beilage II.

47 Biermann, Kurt-R., 'Streiflichter auf geophysikalische Aktivitäten
    Alexander von Humboldts.' Gerlands Beitr. Geophysik, 80 (1971),
    277-291. Zit. 277-278.

48 Vgl. Crosland, Maurice, The Society of Arcueil. A View of French
    Science at the Time of Napoleon I, London, 1967.

49 Engelmann, Gerhard, 'Die russische Chronometer-Expedition im Jahre
    1833 und ihre Durchführung an der südlichen Ostseeküste,'
    Greifswald-Stralsunder Jahrbuch, 9 (1970), 151-167.

50 Wattenberg, Diedrich, 'Johann Jacob Baeyer.' Vorträge u. Schriften
    Archenhold-Sternwarte Berlin-Treptow, 11 (1962), 11-16.

51 Humboldt, Alexander von, Briefe an Christian Carl Josias Freiherr
    von Bunsen, Leipzig, 1869, 51.

J.A.Cawood

COMMENTS ON PAPERS BY BIERMANN AND CROSLAND

Professor Biermann's position as the director of research on Humboldt's life and work at the University of Berlin gives him unique authority on his subject. Also there is no doubt that Humboldt did make a major contribution to international scientific collaboration in the early nineteenth century.

Professor Biermann, in his paper, attempts to gain insight into Humboldt's role in geomagnetism by breaking the problem down into four questions [Teilfrage].

1. Why did international cooperation in geophysics develop at such an early date?

2. Why did this happen initially at the end of the eighteenth Century and the beginning of the nineteenth?

3. Why was it that Humboldt was attracted to geomagnetism in particular?

and 4. To what extent was he the initiator and organiser of international research and cooperation?

These questions are answered within a Marxist framework which examines the possible economic and political influences on the development of geomagnetism in general and Humboldt's contribution in particular. For Biermann, Humboldt's "progressive" political opinions and his internationalism in the face of national prejudice (particularly British) functioning at a time of rapid and revolutionary change helped lay the basis of worldwide magnetic observation.

It is not necessary to subscribe to the philosophy of dialectical materialism to recognise that geomagnetism by its very nature was directly and indirectly affected by economic and political considerations. Magnetism became an early example of "big" science and by the third decade of the nineteenth century, at least, required large scale financial support to progress beyond the stage of localised or irregular observation. This type of sponsorship, with one or two exceptions,[2] was only obtainable through government funding, and it is thus worthwhile and important to examine "external" factors in an analysis of the growth of geomagnetic research. However, the roots of international collaboration went back into the eighteenth century and after an initial burst of government interest were located in individual exchanges of information. Thus it is not as simple as Biermann suggests to explain early development of international collaboration in geophysics by the expected utility of magnetic observations to "shipping, trade, commerce, and agriculture".[4] As far as terrestrial magnetism is concerned there is no convincing evidence that either governments or merchant marines, in the period under consideration, expected that an investigation of geomagnetism would lead to the formulation of simple laws of variation which would improve navigation. It is true that Humboldt and others[5] used this argument to justify their work or to solicit support for it. However, it was never seen as an easily realisable benefit of research. The difficulties in establishing simple laws of magnetic variation had been recognised in Halley's era and it is significant that Humboldt advocated not the traditional element, declination, but inclination as a possible navigational aid.[6]

If geomagnetic research had no immediate utility then we are still left

with the problem of why international cooperation began so early. I think
that it is possible to answer this without making too rigid a distinction
between internal and external factors. The majority of researchers worked
within some kind of government context (the navy, army, or astronomical
service) which helped the prosecution of magnetic investigations[7] and
probably hoped for practical results in the long term; but they all seem to
have had an interest in the earth's field per se, Humboldt for particular
reasons to be mentioned later. The very nature of geomagnetism meant that
observations had to be carried out on a worldwide scale preferably simultaneously
and at regular intervals and so there was an internal factor at work within
the institutions of magnetic research. Once cooperation progressed from
individual to government sponsored work then political and economic considerations grew in importance. (I should also mention here that Humboldt,
unlike Sabine, Arago, Ross and Hansteen or earlier investigators such as Le
Monnier and Borda was neither a navigational expert nor an astronomer. It
seems to me that he was something of an exception in geomagnetic research and
that his collaborators and contemporaries are much better examples of the
direct influence of practical interests - Sabine having originally been
an artillery officer, Ross a naval officer and Arago an authority on
nautical astronomy for the Bureau des longitudes).

To return to Biermann's second question of why it was in the late
eighteenth century and early nineteenth century that there occurred an
upsurge of activity in terrestrial magnetism and related subjects: we are
told that it was the revolutionary replacement of feudalism by a capitalist
economy, especially in France, that gave power to the bourgeoisie. The
upsurge of scientific research and collaboration according to Biermann was an
expression of that power. I would not like to comment on the processes of
historical change which were then taking place but I am not sure this argument,

using its own criteria, holds for Britain or Germany, not to mention Scandanavia or Russia where magnetic research was also underway. However, as far as France is concerned we can go some way with Biermann in accepting that the changes wrought by the Revolution had important consequences for science and scientific collaboration. I would like to restrict my comments to the narrower and more familiar area of geomagnetism in France.

At the onset of the Revolution in 1789 there was in France an established tradition of magnetic research. The scientific expeditions of La Perouse, Borda, and earlier workers, the contributions of Coulomb, the apparatus and instruments of Lenoir and the facilities of the Paris Observatory made France the leader of geomanetic research. The institutional reforms of the Revolution led to the creation of the Bureau des longitudes which drew together this wealth of experience and provided official support for magnetic observations.[8] It was a member of the Bureau, Borda, who encouraged Humboldt to make magnetic observation an important part of his scientific expeditions. It was another member Arago who became Humboldt's collaborator and friend on his return from America. There is no doubt, as Biermann points out, that Humboldt's interest dates from 1794 when he discovered magnetic rocks in the Fichtel Mountains, four years before he returned to France. But it was in France that his interest was fostered and it was there (from 1804) that he was able to find collaborators (Biot and especially Gay-Lussac) and later to work at the Paris Observatory on regular and systematic magnetic investigations. The Bureau provided support, apparatus, facilities and experience. On his short stay in Germany (1805-8) although Humboldt did excellent work, it was on an ad hoc basis in a difficult situation.[9] I would claim, therefore, that in the early part of the nineteenth century Humboldt's geomagnetic activities were substained by institutions and individuals produced by the French Revolution.

Humboldt, of course, was predisposed to study magnetism but I think that
Professor Biermann might have located this interest more firmly in his overall
view of nature.   Biermann stresses Humboldt's concern for "mean numerical
values" [miltlerer Zahlenwerte] as the "expression of physical laws" [Ausdruck
physicher Gesetze] but, I think, he neglects the context in which they were
sought, Humboldt's "cosmical" approach to the phenomena of nature.   If he had
only been interested in the empirical determination of natural laws then
perhaps Humboldt would have been guilty of Schiller's claim that he sought only
to "measure out" nature.   But although he constantly stressed the need for
empirical evidence he did have a vision of a unified geophysics which was part
of a scientific description of the universe.   After all that is what Kosmos
was all about.   Thus whilst criticising the speculations of Naturphilosophie
and praising inductive method Humboldt had preconceived ideas of his objectives.

The attention which Professor Biermann draws to Humboldt's concern for
the history of physical phenomena I find particularly interesting for the
history of science in this period.   I presume that he is trying to demonstrate
that Humboldt had a kind of "historical materialist" approach to his research.
Whether or not this is correct, and Humboldt does seem to lean towards
materialism, this concern for history is not unique.   I think perhaps that
the historical approach to geophysical phenomena displayed by Humboldt, Sabine,[10]
Arago[11] and others was a continuation of an eighteenth century tradition in
science and that by the 1830s it was completely outmoded as a viable method in
most sciences, even in geomagnetism.   It might be true that long records were
required to identify possible regular, secular variations in the earth's magnetic
elements but some theoretical hypothesis was necessary.   As an example of the
weakness of Humboldt's historical method I would like to cite his failure to
recognise the connection between the sunspot cycle, aurora and the violence of
magnetic storms.   Magnetic storms were Humboldt's pet obsession.   He had

studied them and recorded them from at least 800. Arago and Sabine, who both shared Humboldt's interest in geomagnetism and his approach, had astronomical experience but it was not until 1851 that Sabine established the connection between sunspots and magnetic storms. For all the study of historical records little was achieved. Perhaps a bolder speculation on the nature of aurora might have suggested the relationship much sooner.

The final part of Biermann's paper is devoted to the part played by Humboldt in initiating and organising geomagnetic activity and connects with his predilection for collaboration mentioned earlier in the paper. Here there seems to be disagreement with other historians of science on the value and reason for Humboldt's personal collaboration and, more importantly, on some events in the crucial period of the 1830s.

Firstly Humboldt's need for collaboration was seen by contemporaries and by modern historians as a weakness rather than a strength. He is accused of producing little or no original work[12] and of having personal rather than scientific reasons for seeking out collaborators. [3] Perhaps these criticisms are trivial and show a misunderstanding of Humboldt's achievement, generalisation in an age of developing specialisation. However, it is true that he failed to establish any significant following within the community of professional men of science. His personal reasons for attaching himself to successive collaborators, which have been the subject of recent study, would be of no consequence but Biermann makes Humboldt's desire for collaboration a contributing factor to his organisation of international cooperation and so perhaps should be mentioned here.

There is no doubt that during the first three decades of the nineteenth century Humboldt used his persuasive powers, prestige, and political influence to further the cause of magnetic investigation. By intensive lobbying and his

own hard work and example, he managed to establish magnetic observatories
in or secure magnetic records from an astonishing number of locations.
However, the period of crucial importance is that in which Professor Biermann
claims Humboldt reawakened Gauss' dormant interest in terrestrial magnetism
and then by writing two letters to the Royal Society persuaded the British to
join the German magnetic undertaking. According to most secondary sources
the process was exactly the opposite and suggests that it was Gauss who
rekindled Humboldt's interest when the latter returned to Germany.[14] This
point is perhaps not too important. Both men had a long-standing interest
in geomagnetism and the process may have been one of mutual stimulation.
However, the claim that Humboldt was responsible for stimulating interest in
magnetism in Britain and for Britain participating in the Göttingen Union is
more contentious. According to Biermann, quoting Humboldt, the latter had to
fight his way through many national prejudices to obtain British cooperation.
In contrast Nathan Reingold,[15] Walter Cannon[16] and others suggest that it was
Sabine who persuaded Humboldt to write the first letter in 1836[17] and that the
second letter of 1838[18] was also prompted by Sabine as part of a carefully
orchestrated campaign to ensure the success of the "Magnetic Crusade." I
have not yet examined all the primary sources[19] but from my own reading of the
Society Minutes for the period in question I would side with Reingold. There
is no other way to make sense of what Sabine, Lloyd, et al. were up to. This
is not to detract from Humboldt's achievements but I think that the outburst of
geomagnetic activity in the late 1830s is much more complex and interesting
than would be the case if we attribute it simply to Humboldt's powers of
persuasion.

One final point of criticism. I believe that by 1840 and probably much
earlier Humboldt was out of touch with modern magnetic investigations. As
Professor Biermann points out, the "Magnetic Congress" which took place in

Göttingen in 1839 was attended by Sabine, Lloyd, and Kupffer among others and chaired by Gauss. These were the professional researchers with the necessary institutional support. Humboldt had none of these attributes and stayed away. He belonged to another era. Thus after his departure from France in 1827 his role became increasingly one of provision of inspiration rather than organiser of research. This was not due to Humboldt's weakness but to the changing nature of science.

In conclusion I would like to make a few brief comments on the international activity of the 1830s and 40s which touch on some of the points raised by Professor Crosland. Firstly, there does seem to have been in the 1830s a transition from the first and second types of international activity mentioned by Professor Crosland — that is, correspondence and travel to a new level of collaboration. In geomagnetism this was not so much to international meetings, although one or two did take place, but to some form of on-going coordinated research and in Britain at least there was a transition from personal authority (Sabine) to the collective authority of the British Association for the Advancement of Science — a rational association which particularly after 1840 provided an international stimulus to geomagnetic research.

However, in examining international collaboration, it is important to investigate the exact nature of the proposals emanating from individuals and institutions. Humboldt may have truly been a disinterested interventationalist but the same cannot be said for other men of science or the institutions to which they belonged. In France, Arago displayed a fierce chauvinism in refusing to accept the exact scheme of magnetic observation without which results were not strictly compatible; in Germany Gauss and Weber showed a rational bias and in Britain the proposals of Sabine and the British Association were designed to annexe the activities of other nations whilst declaring a belief in the spirit of international cooperation. In geomagnetism in the mid-nineteenth century, national considerations certainly came first once institutions and governments became involved. Perhaps this was inevitable and a parallel could be drawn here with the political history of the nineteenth century. It may have been necessary for national scientific communities to develop fully their national activities before international cooperation in the modern sense could be established.

References

1. Biermann has a number of criticisms of the British in his paper in particular pp. 2-3. The refusal of permission for Humboldt to mount an expedition in India and p. 8 Sabine's orders to burn Humboldt's letters and his neglect of Humboldt's role in the organisation of international research. In the translation of _Kosmos_ edited by Sabine (London, 1849) there appears a passage in which Sabine replies to criticisms that the British interrupted magnetic observations on Sundays to attend church (preface to 4th edition, vol. i). These criticisms did originate from Humboldt so there may have been some friction between the two.

2. Humboldt's own activities and expeditions were undertaken without state assistance and James Ross in 1829 went on a private expedition financed by Felix Booth, a wealthy distiller. Almost all other geomagnetic surveys were government backed.

3. Humboldt considered Halley's voyages as the first example of government sponsored scientific expeditions. See A. von Humboldt, _Cosmos_, (Eng. Trans, London, 1858) vol. 5, p. 59.

4. See page 1 of Biermann's paper.

5.   See for example Humboldt, Cosmos, vol. 1, p. 171 and Arago, Oeuvres complètes, vol. 4, p. 464.

6.   Humboldt attempted to use the variation of inclination with latitude to improve bad weather navigation.   His ideas are discussed in Encyclopaedi Britannica, vol. 12, (Edinburgh 1815) p. 393.   See also The Philosophica Magazine, vol. 22, p. 307.

7.   It was, for instance, relatively easy to add geomagnetic observations to the objectives of voyages planned primarily for other reasons, if the scientist concerned was already involved in the planning of such undertakings.

8.   According to the act which established it, the official concern of the Bureau des longitudes was geodesy and geophysics (physique du globe). See M.G. Bigourdan "Le Bureau des longitudes. Son Histoire et ses Travaux de l'Origine à ce jour" Annuaire du Bureau des Longitudes (1928) A1-A72.

9.   Humboldt worked with a collection of more or less interested young men of science in the garden of a wealthy brandy distiller.

10. See E. Sabine "Report on the Variation of the Magnetic Intensity observed at different points on the Earths Surface" Seventh Annual Report of the British Association for the Advancement of Science (London, 1838) p. 1.

11. D.F.J. Arago, "Sur le tonnère "Annuaire du Bureau des longitudes" (1828) Appendix.

12. See D. Botting, <u>Humboldt and the Cosmos</u> (London, 1973) and L. Kellner, <u>Alexander von Humboldt</u> (London, 1973) for discussions of Humboldt's contribution to science which are not of the partisan variety.

13. This was Humboldt's preference for male companions. It is discussed in detail in Botting (<u>op. cit.</u>) and de Terva, <u>Humboldt</u> (New York, 1955).

14. This is the opinion of Keller, de Terva, and Botting.

15. See N. Reingold "Edward Sabine" <u>Dictionary of Scientific Biography</u>, vol. 12, p. 50.

16. See W. Cannon "History in Depth: The Early Victorian Period" <u>History of Science</u>, <u>3</u>, (1964) 33, for a description of the "conspiritors" who launched the "Magnetic Crusade" and the means they used to launch it.

17. The letter of 1830 is reproduced in A.C. Becquerel, <u>Traité complet du magnetisme</u>, (Paris, 1846) pp. 435-49.

18. See the printed <u>Minutes of the Council of the Royal Society</u> vol. 1. (1832-46).

19. Primary sources are the Herschel and Sabine correspondence in the Royal Society Library which I am currently studying.

Brigitte Schroeder-Gudehus

TENDANCES DE CENTRALISATION DANS L'ORGANISATION DE LA COOPERATION SCIENTI-FIQUE INTERNATIONALE, 1900-1945

Le nombre d'organisations scientifiques internationales a augmenté de façon si spectaculaire depuis la deuxième guerre mondiale, que leur omniprésence entraîne déjà, ici et là, une tendance à confondre "coopération" et "organisation", à étudier la coopération scientifique internationale tout en se contentant de la considérer dans sa forme organisée. Nous ne sousestimons pas les avantages que présente cette substitution d'un aspect à l'ensemble : les organisations se prêtent plus facilement à l'évaluation quantitative que les rapports de coopération non-formalisés, difficiles à appréhender. Nous ne voudrions pas nier non plus que les organisations et la coopération comme telle soient étroitement liées. La substitution pure et simple risque cependant d'induire en erreur.   Dans la présente communication nous nous bornons à traiter d'<u>organisations</u>, conscient du fait qu'il s'agit là d'un aspect seulement du thème auquel ce symposium est consacré.

   L'étude d'organisations scientifiques internationales se fait généralement sous deux angles :
   1. - Ces organisations sont envisagées, d'abord, sur l'arrière-plan de la croissance du savoir : comme des éléments d'infrastructure de l'activité scientifique; elles facilitent la communication et la collaboration entre chercheurs et contribuent ainsi à intensifier et à accélérer l'accroissement des connaissances.
   Les écrits se situant dans cette perspective - largement descriptifs la plupart - ont tendance à endosser, d'emblée, cette raison d'être des organisations. Ceci est le cas tout particulièrement de la littérature qui s'est développée aux confins de la science et de la politique, dans le sillage des débats portant sur la responsabilité des scientifiques dans la création de l'arme atomique d'abord, puis sur la politique de recherche des Etats de manière générale. Les grandes revues scientifiques, et celles de haute vulgarisation en particulier, en constituent les véhicules les plus importants. Même les études critiques portant sur la performance d'organisations particulières reposent généralement sur le postulat que les organisations scientifiques internationales procèdent directement de la nature spécifique de l'activité scientifique, c'est-à-dire de son caractère intrinsèquement coopératif et universel.
   2. - Il faut attribuer à ce caractère universel et coopératif l'intérêt qu'ont suscité les organisations scientifiques dans d'autres milieux : chez les internationalistes, et plus particulièrement chez certains théoriciens de l'intégration internationale, les fonctionnalistes (*). Ici, l'"internationalité" de la science et la tradition séculaire d'une collaboration par-delà les frontières se présentent comme des illustrations convaincantes à l'appui des thèses selon lesquelles les communautés transnationales de normes

(*) Un avertissement s'impose ici pour éviter une confusion : le terme "fonctionnalisme" dans ce contexte n'a pas la même signification - ni le terme "fonction" d'ailleurs la même précision conceptuelle - que celles que nous leur connaissons en sociologie. Il ne s'agit pas là d'un emprunt de concept, ni d'une dérivation. En fait, les thèses "fonctionnalistes" de l'intégration internationale ont été formulées bien avant que n'apparaissent les écrits de l'école fonctionnaliste en sociologie.

et d'intérêts et les entreprises de coopération fonctionnelle (c'est-à-dire technique, non-politique) constituent une voie privilégiée vers l'intégration internationale et, à plus longue échéance, vers un nouvel ordre mondial.
Aux auteurs proches d'idées fonctionnalistes, la communauté scientifique internationale a constamment servi de référence, de Leonard Woolf qui, en 1916, considérait la science comme un des domaines dans lesquels, déjà, un "gouvernement international" existait (1), jusqu'à certains sociologues contemporains qui fondent leurs espoirs dans l'impact d'associations scientifiques internationales sur le système politique mondial ; cet espoir est basé sur la thèse selon laquelle ces associations contribuent à étendre et à approfondir l'adhésion des scientifiques du monde entier aux normes de la science et, en même temps, à les socialiser dans l'éthique de la coopération transnationale (2). Dans cette vision de la dynamique des relations internationales, l'existence d'organisations tient une place importante comme indicateur du niveau d'intégration, donc comme mesure du progrès déjà réalisé dans l'établissement d'un système de coopération et de paix.
3. - Le but de l'étude faisant l'objet de cette communication est d'explorer un autre aspect des organisations scientifiques internationales que les analyses et interprétations courantes semblent portées à négliger. Il suffit d'examiner de près, en fait, ces organisations, pour s'apercevoir qu'elles sont loin de servir toutes en premier lieu - et quoiqu'en disent leurs statuts - à améliorer directement ou indirectement les conditions de l'avancement coopératif du savoir; elles ne présentent guère non plus, comme le voudraient les fonctionnalistes, des foyers privilégiés pour le développement d'une éthique de la coopération internationale; les engagements éventuels pour la solidarité internationale restent généralement déclaratoires.
En revanche, les organisations scientifiques internationales se révèlent comme des éléments importants de la mécanique socio-professionnelle qui apparaît et s'entregreffe nécessairement avec l'activité scientifique moderne. Ces organisations ne servent pas seulement d'instruments pour défendre les intérêts professionnels des communautés scientifiques vis-à-vis des sociétés qui les entourent; elles jouent également un rôle dans la vie politique interne des communautés scientifiques. En l'absence d'études de cas suffisamment précises et suffisamment nombreuses, nous ne pouvons qu'émettre des hypothèses sur leur fonction de "palier" dans le processus de stratification sociale. Il est certain cependant que les organisations scientifiques internationales possèdent une valeur stratégique quand il s'agit d'avancer la cause de disciplines, d'écoles, d'institutions ou de carrières personnelles. Des observations de ce genre sont des lieux communs entretemps, quand il est question du rôle d'institutions nationales. Mais ce n'est qu'au cours des dernières années que des chercheurs - des politologues surtout - ont commencé à explorer le paysage politique qui s'étend au-delà de la zone un peu brumeuse où règne l'optimisme fonctionnaliste et la discrétion des communautés scientifiques. Que ces études analysent le comportement des organisations en termes de groupe de pression ou qu'elles étendent leur intérêt aux tensions internes des communautés scientifiques dont ces organisations sont à la fois l'instrument et l'enjeu, elles sont pratiquement toutes consacrées à des cas actuels. Les organisations anciennes, celles de la fin du 19e et du début du 20e siècle, sont largement ignorées ou ne sont mentionnées qu'en passant. Elles continuent à paraître comme les témoins d'une époque idyllique où l'activité scientifique n'était pas entravée ni corrompue par la politique; témoins aussi du dévouement désintéressé de leurs membres à l'avancement coopératif du savoir et à l'esprit universaliste de la science.
C'est précisément cette période à laquelle nous nous sommes intéressée; dans une première approche, nous avons limité nos recherches aux organisations-cadre : à l'Association Internationale des Académies, créée au tournant du siècle et disparue entre 1914 et 1918; au Conseil International de Recherches, fondé en 1919 et transformé en 1931 en Conseil International des Unions Scientifiques.

Notre intention n'était pas d'ajouter une autre étude de cas - historique celle-là - à celles qui existent déjà sur des organisations contemporaines, mais de dégager certains schémas d'interaction communs aux relations scientifiques internationales à toutes les fois qu'une organisation-cadre (ou organisation faîtière) essaie de s'établir. L'objectif manifeste de l'Association Internationale des Académies (AIA) et du Conseil International de Recherches (CIR) était de mettre sur pied un mécanisme de coordination centralisée; les deux organisations avaient à faire face à la résistance des "Unions" (*), des associations internationales disciplinaires. Au cours de la période que nous essayons de couvrir - de 1900 jusqu'au lendemain de la 2e guerre mondiale - le schéma d'interaction fut considérablement enrichi par des organisations concurrentes, de sorte qu'un titre tel que "monopole et compétition" ou "tendances monopolistiques et tendances pluralistes" aurait été plus approprié.

Il n'est pas possible dans les limites d'une communcation nécessairement brève, de fournir toute l'information factuelle sur laquelle repose notre perception d'un schéma recourant d'interaction. Nous nous contenterons donc de mentionner quelques aspects <u>structurels</u>. Nous espérons qu'ils sont suffisamment significatifs pour bien illustrer le matériel sur lequel nous basons les principaux éléments de notre conclusion.

L'ASSOCIATION INTERNATIONALE DES ACADEMIES

La seconde moitié du 19e siècle avait vu augmenter de façon spectaculaire le nombre de congrès et d'associations scientifiques internationales. A côté des rapports maintenus et développés entre hommes de science selon une tradition séculaire, se manifestait le besoin de tenir des rencontres plus régulières pour confronter les expériences, pour organiser les disciplines. Il n'y a d'ailleurs rien de surprenant à ce que le processus d'organisation et de professionnalisation caractérisant le développement de certaines disciplines au 19e siècle, débordât ainsi sur le plan international. Certaines de ces nouvelles organisations s'étaient données des tâches précises, telle l'uniformisation de nomenclatures, la coordination d'observations, la compilation de bibliographies, etc.; d'autres étaient plutôt des fédérations de sociétés savantes nationales. Beaucoup de ces associations ne se manifestaient que par l'organisation périodique de congrès. Elles n'avaient alors pas de secrétariat permanent, chaque assemblée décidant de la date et du lieu de la prochaine réunion et confiant à l'association-hôte l'organisation matérielle du congrès. Parfois, les associations se dotaient néanmoins d'un vague organisme intermédiaire, d'autres avaient de véritables secrétariats. Si ce foisonnement d'organisations internationales réjouissait les uns, parce qu'ils y discernaient les signes avant-coureurs d'un nouveau type de relations internationales (5), il déplut à d'autres qui y voyaient au contraire une dispersion regrettable d'efforts et de moyens financiers à laquelle seule une certaine coordination centralisée pourrait porter remède.

En 1899 se constituait alors, dans le domaine des sciences fondamentales et des humanités, la première organisation faîtière de type non-gouvernemental, l'Association Internationale des Académies. Les académies les plus importantes du continent, la Royal Society et la National Academy of Sciences de Washington étaient les membres fondateurs; d'autres Académies adhérèrent par la suite. A la veille de la première guerre mondiale, l'AIA comptait 22 membres. Les structures de l'organisation étaient extrêmement souples, son mandat était très général : encourager la coopération scientifique internationale, faciliter les contacts entre savants de différents pays, prévenir - grâce à des discussions

---

(*) On les appelait alors volontiers des "Unions", créant ainsi un risque regrettable de confusion avec les unions administratives internationales, organismes de droit public.

et une coordination appropriées - des "collisions regrettables", c'est-à-dire des doubles-emplois (6). L'AIA ne survécut pas à la guerre quoique, d'une assemblée à l'autre, les discours officiels et l'écho qu'en reproduisit la presse scientifique eussent exalté la science sans frontières et la collaboration scientifique comme voie de communication entre les peuples, voie qui serait ouverte encore le jour où celle de la diplomatie se trouverait bloquée.

Le bilan des activités de l'AIA était maigre. Sans personnalité juridique, ayant des fonctions uniquement consultatives, elle ne disposait ni de l'autorité ni des fonds nécessaires pour entreprendre elle-même des travaux d'envergure. Et en 1914, les ponts entre les communautés scientifiques nationales devaient céder aussi vite que ceux de la diplomatie. L'acharnement avec lequel on recherchait plus tard dans l'histoire sans éclat de l'AIA des arguments à l'appui de la thèse qu'une organisation scientifique internationale serait plus efficace si l'on réussissait à en exclure des puissances centrales, a obscurci quelques aspects importants de cette première tentative de centralisation.

L'objectif principal de l'Association - auquel les statuts, pudiques, faisaient à peine allusion - était d'ordre pécuniaire ; c'était de découvrir des sources potentielles de financement et d'appuyer qui les solliciterait, en jetant dans la balance tout le prestige de l'Association. La base financière de certains projets présentait ainsi une mosaïque étonnante. Que l'exploitation du prestige de l'Association à de telles fins entrait dans le calcul de ses fondateurs ne fait donc aucun doute. Il y avait cependant, parmi les considérations ayant conduit à sa création, d'autres préoccupations encore dont on faisait état plus discrètement, mais qui n'en étaient pas moins très vives. A première vue, il s'agissait là, encore, du souci d'obtenir de l'argent. Mais dans ce contexte, en recherchant des ressources financières supplémentaires sur le plan national, les Académies soulevaient tout le problème de leur position au sein des communautés scientifiques nationales.

Dans un aide-mémoire de 1892 - donc au début de la période préparatoire de l'AIA - un membre de l'Académie viennoise avait présenté trois arguments en faveur de la collaboration entre Académies : cette collaboration permettrait d'éviter les efforts parallèles; elle permettrait de mettre en commun des ressources humaines et financières; elles permettrait aussi de drainer de manière plus profitable les subventions publiques. Il s'était expliqué sur ce dernier point :

"Une coordination internationale des recherches s'impose, et le fait que les Académies n'aient pas réussi, jusqu'ici, à se mettre d'accord, a déjà entraîné la création d'organisations indépendantes - sous la forme de congrès et d'unions scientifiques, par exemple. Celles-ci ne cessent d'entreprendre des projets de plus en plus nombreux et d'accaparer ainsi une partie toujours plus importante des fonds publics que les gouvernements destineraient peut-être plus volontiers à leurs Académies. N'ont-elles pas été créées spécialement pour assister l'Etat dans tout ce qu'il voudrait voir entrepris pour encourager et promouvoir la recherche scientifique ? Plus l'Etat consacre de moyens financiers à des milieux et à des projets extra-académiques, moins il en restera pour les Académies." (7)

L'Association se révèle donc ici comme un effort des Académies pour raffermir en revendiquant un rôle privilégié sur le plan international , une autorité que la croissance des communautés scientifiques nationales et plus particulièrement le développement des sociétés savantes étaient en train de saper sur la scène interne. Il est nécessaire, en d'autres termes, d'interpréter la création et le fonctionnement de l'AIA non seulement comme une conséquence de l'internationalité intrinsèque des activités scientifiques, mais aussi à la lumière des tensions existant à l'intérieur même des pays au sein des différents milieux scientifiques. Souvent, les sociétés savantes s'étaient créées en opposition aux Académies; elles se développaient, en règle générale, en étroite asso-

ciation avec l'évolution des disciplines, s'adaptant plus facilement aux
exigences de la spécialisation et de l'organisation économique et sociale.
Or, l'essor des congrès et associations internationaux au cours de la seconde
moitié du 19e siècle était dû plus souvent à l'initiative de sociétés savantes
qu'à celle des Académies. En trouvant un moyen de contrôler une partie des
activités scientifiques se déployant en dehors des frontières nationales, les
Académies comptaient recouvrer, du moins sur la scène internationale, une partie de l'autorité qu'elles avaient perdu sur la vie scientifique nationale;
elles voulaient alors établir au moins une coordination centralisée des congrès et des associations et obtenir - solution idéale ! - un droit de regard
sur les décisions des autorités publiques vers lesquelles la communauté
scientifique - donc les sociétés savantes - se tournaient habituellement en
quête d'appuis moraux et matériels (8). tout au long de l'histoire de l'AIA,
les idées semblables à celles que Friedrich von Hartel avait exprimées en
1892 continuaient d'être avancées. En 1899, J. Lister, alors président de la
Royal Society, soulignait les avantages d'un mécanisme

  "qui donnerait l'occasion à des savants consacrés de discuter d'abord et
  d'un point de vue purement scientifique, tout projet de coopération scientifique internationale avant que de tels projets soient soumis à des instances
  gouvernementales pour appui.." (9) George Ellery Hale, tout en se plaignant
amèrement de l'indifférence que la National Academy of Sciences montrait à
l'endroit de l'AIA, voulait voir échouer à celle-ci une "position de commande" (10) (*). Une position centrale - sinon de "commande" - était convoitée
pour l'AIA par d'autres, objectif qui ne se trahit pas seulement dans la correspondance; elle transpire aussi dans les documents officiels. L'Assemblée
générale de l'AIA de 1904 en fit l'objet d'une résolution.[12] Sept ans plus tard,
lors des célébrations du Cinquantenaire de la National Academy of Sciences,
Arthur Schuster, alors secrétaire de la Royal Society, revendiqua à nouveau
pour AIA le statut d'une "autorité centrale" dans le réseau des associations
scientifiques internationales (13).

  Cette discussion n'était manifestement pas confinée aux cercles scientifiques ou à la Royal Society. Dans un de ses ouvrages, un des auteurs désormais
classiques du fonctionnalisme, Leonard S. Woolf, consacra quelques pages non
seulement à la science en tant que domaine particulièrement adapté pour voir
s'y réaliser un "gouvernement international"; il mentionne explicitement
l'AIA et ceci d'une manière qui reflète des informations de première main :

  ".. It is also a sign of the need for and the growth of a new international structure that one resolution of this Association recommends that it
  should be consulted by Governments before they take part in any new International Scientific Association. There can be no question, when one considers the status of the members of this Association, that its request is not
  unreasonable, and if it is complied with, we shall have in this International
  Association a supernational scientific body bearing the same relation to the
  community of States and international society as, for instance, the Royal
  Society bears within the national boundaries to the British Government and
  the British people." (14)

(*) G.E. Hale essayait par ailleurs d'utiliser le rayonnement dont la science
    américaine pourrait selon lui bénéficier grâce à cette organisation, pour
obtenir des avantages tangibles sur le plan national : les moyens financiers
pour ériger un édifice digne d'une Académie nationale pour un jour serait hôte
des Académies du monde. En sondant Elihu Root d'une façon prudente sur les
chances qu'aurait la National Academy d'obtenir des fonds de la Carnegie Corporation, il n'oubliait pas de souligner que l'encouragement et l'extension
de la coopération scientifique internationale représentait un champ d'action
par excellence pour les hommes politiques engagés dans la lutte pour la paix (11)

Il est difficile de se prononcer sur le succès effectif des efforts de coordination centralisée que l'AIA déployait au cours de son existence, aussi longtemps que l'on n'en aura pas systématiquement relevé les traces dans les archives des sociétés savantes. Un certain nombre de projets avait été discuté au sein de l'Association; elle en avait appuyé quelques-uns. Aussi longtemps que l'on ne dispose pas, à titre de comparaison, d'un inventaire des projets qui se sont réalisés sans que l'Association y fût impliquée (manifestement la majorité), toute conclusion que l'on peut en tirer reste évidemment tronquée et provisoire. Il suffit cependant d'examiner les âpres débats qui ont entouré la reprise des efforts d'organisation en 1917-18 pour se rendre compte de la place importante qu'occupait dans ces travaux préparatoires la détermination de régler, une fois pour toutes, le problèmes des "unions" indépendantes.

LE CONSEIL INTERNATIONAL DE RECHERCHES

Le Conseil International de Recherches (CIR) représente l'effort le plus extrême de contrôle et d'hiérarchisation que l'organisation de la collaboration scientifique internationale ait jamais connu. Si la guerre avait provoqué la rupture des rapports entre les milieux scientifiques de l'un et l'autre côté des fronts, elle avait resserré certains réseaux de collaboration entre pays alliés et associés, réseaux dont la National Academy of Sciences avait par ailleurs constitué un des principaux points d'appui. Profitant de cette amorce, et avec une précipitation qui trahissait un sens aigu de la conjoncture, les représentants des Académies des Sciences Inter-alliées préparaient dès l'automne 1918 l'intégration pure et simple des "Unions" dans les structures d'une nouvelle organisation. Il n'était plus question alors, pour l'organisme central, d'influencer la création et les activités des Unions en obtenant un discret droit de regard sur le financement de leurs projets et l'accréditation de leurs délégations au niveau gouvernemental. La nouvelle organisation, le Conseil, constitué par une fédération d'organismes scientifiques représentatifs nationaux, scil. les Académies, se réservait le droit de vérifier les statuts des Unions pour assurer une "symétrie d'organisation" dont le principe était ancré dans les statuts du Conseil. Ces statuts l'habilitaient d'ailleurs à créer des unions de sa propre initiative dans des domaines où cela lui paraissait nécessaire. (15)

Il faut souligner ici les motifs particuliers qui inspiraient la centralisation outrancière que les dirigeants du CIR essayaient de mettre en oeuvre : en fait, certains des fondateurs ne prenaient même pas la peine de déclarer, fût-ce pour la forme, que l'organisation reflétait l'internationalité de la science et répondait aux besoins de l'avancement coopératif du savoir. Bien au contraire, Emile Picard, mathématicien, secrétaire perpétuel de l'Académie des Sciences de Paris, avait souligné à maintes reprises que c'étaient des considérations d'ordre moral qui étaient à la base des préoccupations des hommes de science du continent (16) et qu'il fallait être prêt à sacrifier à ces considérations-là même le progrès des sciences (17). Ces "considérations morales", c'était la mise en oeuvre d'une justice réparatrice, le souci de mettre au ban les scientifiques allemands et autrichiens de la coopération scientifique internationale pour une période allant (selon le degré d'indignation des individus se prononçant sur cette question) de quelques années jusqu'à l'arrivée de nouvelles générations. Les hommes de science des puissances centrales devaient être exclus des congrès et des associations internationales; les services centraux d'observation et de compilation confiés jusqu'en 1914 à des Instituts allemands, leur étaient enlevés.

La mise en oeuvre du "Boycott" impliquait par la nature même de cette mesure l'uniformité du comportement de la communauté scientifique mondiale. C'est par souci d'assurer cette cohésion du front commun que s'explique en grande partie le caractère excessif de la règlementation que les dirigeants du CIR

essayaient d'imposer (*).Une représentativité incontestable était une condition fondamentale de la réussite du CIR. Ses dirigeants - représentants des
Académies des Sciences Inter-alliées - déployaient des efforts considérables
pour obtenir l'adhésion des Académies des pays anciennement neutres, tout en
prenant toutes les précautions afin que ces adhésions ne compromettent pas
la solidité de l'engagement moral que les Académies Inter-alliées avaient
pris à l'endroit des universitaires allemands et autrichiens. Cependant, le
CIR aurait couru de plus graves dangers en ignorant les Académies neutres,
- celles-ci auraient sans doute ranimé, de concert avec les Allemands, l'ancienne Association Internationale des Académies.

Grâce au climat psychologique particulier de l'immédiat après-guerre, le
CIR réussissait à faire accepter son autorité plus facilement que cela n'avait
été le cas de l'AIA : les communautés scientifiques des pays alliés étaient
prêtes à première vue à accepter le principe d'une coordination centralisée;
au cours de la première année, les principales Académies neutres adhéraient
au Conseil. Le refus de reprendre la collaboration avec les Allemands coïnci-
dait avec le courant dominant de l'opinion publique de l'époque. Des difficultés ne tardaient cependant pas à surgir : les dirigeants du CIR, les
membres du Comité exécutif notamment (Arthur Schuster, Emile Picard, Georges
Lecointe, Vito Volterra et G.E. Hale) devenaient la cible de critiques venant
de l'intérieur de leurs communautés scientifiques respectives. La Royal Society,
par exemple, se voyait contester le droit de prendre, pour l'ensemble des
scientifiques britanniques, des engagements aussi décisifs que la limitation
de leurs rapports de collaboration internationale. Une partie de l'opposition
à laquelle les Académies devaient faire face sur le plan national pour des
raisons qui n'avaient aucun rapport avec l'attitude à adopter vis-à-vis des
Allemands trouvaient manifestement un véhicule tout désigné dans la critique
de la politique du CIR. Dès la première Assemblée générale du CIR, les neutres
prenaient position en faveur d'une "vraie internationalité" , c'est-à-dire
en faveur de l'admission des Académies allemandes. En 1925-26, elles imposèrent
l'abolition de la clause d'exclusion. Certaines associations internationales
ne s'affilièrent jamais au CIR; celui-ci avait créé des Unions dont certaines
n'existaient que sur le papier; d'autres commençaient tôt à ignorer les con-
trainte statutaires et invitaient qui bon leur semblait. Elles laissaient par
ailleurs entendre qu'elles étaient prêtes   continuer leur travaux, indépen-
demment du Conseil, s'il le fallait. En 1931, au moment de la révision des
Statuts, elles obtinrent l'autonomie qu'elles avaient réclamée depuis des années :
le CIR devint le Conseil International des Unions Scientifiques.

(*) Cette règlementation ne s'appliquait officiellement qu'aux questions ad-
    ministratives, laissant aux Unions une autonomie totale en matière scien-
tifique; en réalité, l'interdiction d'inviter les Allemands et les Autrichiens
à des réunions ou de les associer à des projets n'était administrative qu'en
apparence. Dans certaines disciplines, les réunions semblaient perdre de leur
intérêt pour les participants du moment où ils ne pouvaient pas y rencontrer
des collègues qui a cette époque se trouvaient engagés dans des recherches de
pointe.
La subordination de l'opportunité du point de vue scientifique aux considérations politiques ne se faisait pas uniquement à l'égard de l'exclusion
des Allemands, d'ailleurs; ainsi, le domaine du magnétisme terrestre fut, en
1919, provisoirement exclu de l'aire de compétence de l'Union géodésique et
géophysique internationale, parce que l'association des Hollandais à ce type
de travaux aurait été inévitable; les rapports entre les milieux scientifiques
français et hollandais étaient cependant si tendus que l'on ne crut pouvoir
courir le risque de heurter les sentiments des délégués français (présents aux
travaux préparatoires du CIR), préférant loger ce domaine ailleurs (en météoro-
logie, par exemple), en attendant que s'établissent, après la signature des
Traités de Paix peut-être, des relations plus amicales (18).

Quoique l'on ne puisse guère parler dans ce contexte de motifs reliés à la sphère socio-professionnelle, il est évident que l'organisation internationale ne servait ici ni l'avancement du savoir, ni la compréhension internationale, et que cette coordination centralisatrice ne pouvait pas ne pas avoir des conséquences sur la distribution du pouvoir dans la cité scientifique, sur la configuration du paysage disciplinaire, sur les carrières personnelles (*).

Au moment où les représentants des Académies neutres attaquèrent pour la première fois la politique d'exclusion du CIR, un danger surgit de manière inattendue.

Au cours des années 1918-20, les promoteurs de CIR avaient montré un souci manifeste de se maintenir près du pouvoir, d'obtenir l'endossement officiel, par les gouvernements de leurs pays, des objectifs de leur organisation. Ils s'étaient alors servis volontiers du prestige des Traités de Paix et de la Société des Nations pour renforcer l'autorité du CIR : ils laissaient entendre que la rupture avec l'Allemagne était imposée à toutes les associations internationales d'avant-guerre en vertu de l'art. 283 du Traité de Versailles portant sur la dénonciation des Conventions (**); ils aimaient également à comparer le CIR à une Société des Nations dans le monde de la science (20). Quand ils s'aperçurent que l'art. 24 du Pacte établissait l'autorité de la Société des Nations sur tout bureau ou commission traitant d'affaires internationales tout en parlant, aussi, d'assistance que le Secrétariat pourrait fournir dans certains cas, l'idée surgit de faire du CIR la "section scientifique" de la Société des Nations, à condition évidemment, comme G.E. Hale faisait remarquer à A. Schuster, que la Société payât les factures sans se mêler des affaires du CIR, "notamment de ses rapports avec les Allemands" (21). Les risques d'ingérence étant finalement plus grands que les avantages hypothétiques, le CIR finit par prendre ses distances.

Or, en 1922, la Société des Nations prit une initiative inquiétante : elle créa la Commission Internationale de Coopération Intellectuelle, organisme censé incarner l'idéal d'une humanité réunie grâce aux progrès de l'intelligence et

(*) Par ailleurs, les promoteurs du CIR n'étaient pas tous animés d'une égale détermination vengeresse; celle-ci se manifestait dans sa forme la plus exclusive et la plus durable dans les académies belge et française. George Ellery Hale avait de l'Organisation scientifique internationale nouvelle une vision qui allait au-delà d'une arme servant à un règlement de comptes avec les Gelehrten allemands. Il voyait dans cette nouvelle organisation une possibilité de mettre en oeuvre, à l'échelle internationale, une idée qui lui était chère et pour laquelle il n'avait cessé de plaider aux Etats-Unis, à savoir la "recherche coopérative", la collaboration entre universités, industries et gouvernement. La guerre en avait démontré l'efficacité; elle semblait avoir démontré également l'importance stratégique de la recherche scientifique, et Hale était bien décidé à consolider la position de cette dernière parmi les grandes préoccupations nationales. L'entérinement de la "recherche coopérative" par une organisation internationale ne pouvait que renforcer, dans l'esprit de Hale, ses chances de réalisation sur le plan national. Il était alors très satifsfait quand la Conférence des Académies des Sciences Inter-alliées de Londres incluait, en octobre 1918, dans son projet d'organisation internationale - la fois la recherche fondamentale, militaire et industrielle (19).

(**) Cette interprétation était erronée.

de la culture. Henri Bergson, le philosophe, en fut nommé président; mais la
Commission comptait parmi ses membres deux savants prestigieux : Albert Einstein et Marie Curie. Se rendant compte des risques que cet organisme faisait
courir pour la position centrale que les promoteurs du CIR s'efforçaient d'assurer à leur organisme, ils réagissaient immédiatement : G.E Hale et R.A. Millikan essayaient par une lettre conjointe de mobiliser Lord Balfour à la dé-
fense des intérêts du CIR.[22] Il s'agissait d'empêcher la Commission de la Société
des Nations d'étendre son aire de compétence au domaine scientifique. Ressortis-
sants d'un pays qui n'allait pas être membre de la Société des Nations, ils
étaient sans doute particulièrement sensibles au danger que courait la position
centrale du CIR. Le Conseil de la Société des Nations essayait de résoudre le
problème en nommant G.E. Hale membre de la Commission.

Malgré les efforts de celle-ci pour ne pas empiéter sur les domaines que le
CIR revendiquait comme étant de sa juridiction exclusive, les relations entre
les deux organismes restèrent marquées d'une distance extrêmement circonspecte.
Quand, en 1937, ils signèrent un protocole d'entente, ce geste ne devait plus
guère intéresser personne.

Les clientèles auxquelles s'adressaient les deux organismes, sans coïncider, se
chevauchaient néanmoins dans une certaine mesure. La position d'une "autorité
centrale" que le CIR avait essayé d'occuper depuis 1919 était potentiellement
menacée dès qu'un autre organe était en mesure d'allouer des valeurs profession-
nelles : un cadre institutionnel pour des projets de collaboration, des fonds,
du prestige. Ce que l'une et l'autre organisation pouvaient offrir aux communautés
scientifiques n'avait rien de spectacuaire; par conséquent, la compétition
qu'elles se livraient était, elle aussi, relativement discrète. Cela allait
changer au cours des années quarante, quand à Londres s'élaboraient les fondements d'une nouvelle organisation inernationale à vocation universelle, dans le
domaine de l'éducation et de la culture.

LE CONSEIL INTERNATIONAL DES UNIONS SCIENTIFIQUES, L'UNECO ET L'UNESCO

Pendant la seconde guerre mondiale, le Conseil International des Unions
Scientifiques (CIUS) et les Unions étaient pratiquement en hibernation.
C'est à Londres, en 1942, au sein de la Conférence des Ministres de l'Education des Pays Alliés, et aux Etats-Unis, à la National Academy of Sciences,
que commença la discussion sur un réorganisation des relations scientifiques
internationales après la guerre. La Commission des relations extérieures du
National Research Council fit envoyer, en mars 1944, un questionnaire - tous
les individus ayant tenu une fonction au sein d'Union ou de Congrèes internationaux dans le domaine scientifique et technique. Le résultat avait cependant une faiblesse, les réponses provenant presque exclusivement des Etats-
Unis, du Royaume-Uni et du Commonwealth. Le rapport constata que la majorité
des correspondants considéraient les organisations scientifiques internationales comme des mécanismes indispensables au bon fonctionnement de la collaboration scientifique, quoique cette majorité se composât, comme souligne le
rapport, principalement d'astronomes et de représentants des sciences de la
terre (23). La question fut débattue lors d'une conférence convoquée par la
section des relations internationales du National Research Council en décembre 1944 et portant sur les problèmes que soulevait le rétablissement
de la coopération scientifique (et universitaire en général) après la guerre.
Dans les procès-verbaux de cette conférence (comme dans le rapport mentionné
plus haut d'ailleurs) on peut relever un grand nombre de commentaires sou-
lignant les aspects négatifs de ces organisations et congrès de l'avant-guerre.
On fait remarquer qu'ils étaient des terrains où se déployaient des rivalités
politiques, où se manifestait une identification souvent excessive avec des
institutions ou des pays, au détriment des activités scientifiques dont les
délégués étaient censés se préoccuper; où s'étalaient morgue et jalousies;

correspondants et participants à la conférence semblaient par ailleurs d'accord pour voir dans les organisations scientifiques internationales des citadelles d'un leadership essentiellement conservateur. Dans l'ensemble cependant, on semblait vouloir continuer plutôt qu'abandonner le CIUS et les Unions. en comparaison aux événements de 1917-18, une réaction marquée se manifestait contre toute velléité d'organisation énergique et précipitée : "solvitur ambulando."

Or, les nouvelles concernant un organisme spécialisé à créer dans le système des Nations Unies allaient mettre fin à cet attentisme. Il convient de rapprocher l'émergence de l'UNESCO de la création de la Commission de Coopération Intellectuelle, bien entendu. Par rapport à 1922, la situation était cependant différente sur plusieurs points. Le système des Nations-Unies s'annonçait plus puissant que la Société des Nations ne l'avait jamais été; une de ses institutions spécialisées comme l'UNESCO devait nécessairement participer à cette puissance : on n'excluait pas, en 1945, que l'UNESCO aurait un budget considérable.

On se trouvait, vers la fin de la guerre, en présence de plusieurs projets d'organisation de la coopération scientifique internationale au lendemain des hostilités, dont les plus importants se situent de façon intéressante par rapport - la tension qui se dessinait, déjà, entre l'organisation traditionnelle, le CIUS, et celle qui serait rattachée à la nouvelle organisation mondiale, la future UNESCO. Un élitisme culturel, englobant la science, avait survécu de l'époque de la "Coopération Intellectuelle" et favorisait une ressurrection de cette organisation dans le cadre des Nations Unies (25). Une telle perspective ne satisfaisait pas les milieux scientifiques proches du CIUS, parce qu'elle ne donnait pas à la science une place suffisamment différenciée et suffisamment prominente. Ces milieux scientifiques étaient d'autant plus décidée d'assurer aux activités scientifiques internationales un cadre approprié et indépendant que, dans sa phase initiale surtout, l'organisation projetée par les Nations Unies semblait voir sa vocation prioritaire dans le domaine de l'éducation . On craignait alors, dans les cercles universitaires, que la nouvelle organisation ne soit dominée par une "mentalité de maîtres secondaires." Une organisation, soit tout à fait indépendante, soit spécialisée et autonome dans le cadre des Nations Unies, semblait rallier les opinions.

Joseph Needham était un des rares scientifiques à avoir élaboré un projet cohérent de coopération scientifique à l'échelle mondiale (26). Ce projet reflétait, à part un fonctionnalisme sommaire, une confiance remarquable dans les moyens que la science et la technique allaient mettre à la disposition des hommes pour le bien commun de l'humanité. Le rôle crucial qui échouait, dans cette vision d'un "International Science Cooperation Service" aux hommes de science était très flatteur et devait renforcer de façon spectaculaire leur statut socio-professionnel. Le projet de J. Needham était cependant loin de susciter l'enthousiasme unanime des milieux scientifiques. L'ambition même du projet déplut à certains proches du CIUS, soucieux d'éviter qu'une trop grande partie des fonds disponibles ne soit engloûtie dans un organisme international et soustraite ainsi à la gestion des communautés scientifiques nationales et leurs organismes représentatifs. Si l'on ne pouvait guère reprocher au projet de J. Needham de véhiculer une mentalité de maîtres d'école, il était évident que ses préoccupations concernant le développement scientifique des régions arriérées du monde n'était pas largements partagées. Les critiques ne se trompaient pas : Le projet du "International Science Cooperation Service" était profondément inspiré par des idées qu'avait suscitées depuis la veille de la guerre, la prise de conscience d'une "fonction sociale de la science". Il était inévitable alors que la discussion sur les mérites comparatifs d'options organisationnelles se doublât de parti-pris politiques précis.

La situation allait évoluer vers un nouveau dualisme. L'UNECO devint l'UNESCO. Joseph Needham, premier directeur de la section des sciences naturelles de

l'UNESCO, marqua alors fortement le programme des premières années (27).
Le CIUS était la première organisation à bénéficier du lien privilégié
que l'UNESCO allait établir avec certaines organisations internationales
non-gouvernementales : il se vit octroyer le stuatut consultatif et l'appui
moral et financier que ce statut implique. Encore, sans se recouvrir, les
clientèles des deux organisations se chevauchent considérablement.

*

Est-il possible, en résumant, de faire ressortir ce schéma d'interaction
que nous avons mentionné au début de ce texte ?

L'argument fondamental avancé pour expliquer et pour justifier l'existence
d'organisations scientifiques internationales était toujours - notamment
dans le cas des organisations faîtières - la nécessité d'avoir un mécanisme
approprié de coordination. Cette coordination permettrait d'éviter une dispersion, inefficace et coûteuse, des efforts; une organisation fournirait
un cadre pour des projets impliquant plus d'une discipline; elle permettrait
d'assurer une collaboration harmonieuse; fédérative, elle représenterait une
dernière expression de l'unité des sciences.

Il est inévitable que de telles organisations développent des tendances
"monopolistiques".: le succès d'une coordination centralisée dépend en grande
partie de la possibilité d'imposer un arbitrage. L'autorité que des organisations scientifiques peuvent exercer dans un domaine donné dépend en somme
du contrôle que ses leaders sont en mesure d'établir en regard de certains
mécanismes du "reward system". Peu de recherches systématiques ont, à notre
connaissance, été entreprises pour élucider la place des activités internationales dans ce "reward system". Il ne peut faire de doute cependant que,
par exemple, l'élection à une position au sein d'une organisation internationale reflète et confère du prestige en ce qui concerne les individus; que
l'endossement, par une organisation internationale, de certaines orientations
inter- ou subdisciplinaires, de projets ou de programmes, constitue généralement un atout sur le plan national.

Le moins ce contrôle sur le "reward system" est partagé, le mieux c'est.
Des organisations scientifiques internationales tendent nécessairement vers
l'institution d'un "monopole". Dans les cas que nous avons évoqués de manière
allusive, cette tendance rencontrait trois types d'obstacles :

1. - une opposition sur le plan national, concernant la légitimité de la repré-
     sentation nationale au sein de l'organisation internationale; cette oppo-
   sition était parfois l'expression d'autres tensions existant à l'intérieur
   de ces communautés;
2. - une résistance de la part des associations disciplinaires, des "Unions";
   les porte-parole de ces organisations - sans rejeter entièrement l'idée
   d'une coordination - mettaient en question la nécessité decréer une "super-orga-
   nisation"à cet effet et faisaient peser sur l'organisation faîtière le soupçon
   d'être parasitaire;
3. - la concurrence d'un nouveau type d'organisation internationale qui,
   s'adressant à la même clientèle, disposait de moyens pour s'ingérer dans
   le fonctionnement du "reward system" et pour saper ainsi la position exclusive
   des organisations faîtières traditionnelles.

Un trait important du schéma d'interaction réside dans le fait que l'"internationalité"de la science s'est révélée comme une lame à double tranchant.
Les nouvelles organisations internationales du XXe siècle, la Société des
Nations et les Nations Unies, ont été fondées toutes les deux sur le principe
de l'universalisme. L'annexion, à leur aire de compétence, de la coopération
scientifique, ne paraît alors que comme une conséquence logique d'une uni-

versalité politiquement pertinente que les scientifiques eux-mêmes avaient
d'ailleurs constamment mis de l'avant. Sous la Société des Nations, l'entreprise coopérative de la science était considérée comme un instrument de
publicité inévaluable. Sous le système des Nations Unies, la coopération
scientifique est considérée comme plus indispensable encore à cause de sa
contribution potentielle au développement économique et social.

L'histoire nous montre que la possiblité d'une concurrence entre ces nouveaux organes spécialisés et les organisations non-gouvernementales traditionnelles était perçue par les communautés scientifiques. Certes, le CIUS
et l'Organisation de Coopération Intellectuelle n'étaient arrivés que tar-
divement à un modus vivendi, tandis que le CIUS avait acquis, dès le début,
le statut consultatif auprès de l'UNESCO. Mais on ne peut s'empêcher
d'avoir l'impression que le CIUS n'a jamais cessé de se considérer comme
la seule organisation légitime représentant la communauté scientifique
internationale (celle-ci n'a par ailleurs guère laissé passer une occasion
pour donner à cette façon de penser la publicité appropriée (28).

Il faut finalement soulever la question des conséquences qu'une situation
de monopole ou de compétition entraîne pour l'avancement du savoir. Il
semble que, pendant un demi-siècle, l'existence d'une organisation faîtière
n'a pas affecté de manière sensible la substance des activités scientifiques.
Leur objectif principal se trouvait dans la sphère politique, dans la
sphère des interactions socio-professionnelles. Jusqu'à quel point cela
est-il le cas quand il s'agit des unions disciplinaires ? Il s'agit là
d'un domaine largement inexploré que l'on ne devait aborder sous ses seuls
a ects contemporains. Il serait notamment intéressant de voir de plus
près les conséquences d'activités internationales sur les carrières des
personnes, des institutions et des disciplines à des époques, ou la solidarité internationale et le maintien de la paix ne figuraient pas encore parmi
les normes incontestées des politiques étrangères des Etats.

NOTES

(1) Leonard WOOLF, International Government. Londres, 1916, p. 317.
(2). W.M. EVAN, "The international professional associations and the interna-
    tional system". Proceedings of the Conference on International SCientific
and Professional Associations on the International System. Philadelphia,
November 1976, p.4.
(5) Voir, par exemple, P. OTLET, Les problèmes internationaux et la guerre.
    Genève, 1916; les compilations plus récentes, publiées dans le cadre
de l'Union des Associations Internationales et inspirées par le même optimisme que l'ouvrage précédent, Les 1978 organisations internationales fondées
depuis le Congrès de Vienne, Bruxelles, 1957.
(6) En ce qui concerne l'histoire de l'AIA, nous nous basons essentiellement
    sur les archives de la Royal Society, de la National Academy of Sciences,
de l'Académie des Sciences de Paris, ainsi que sur les documents et travaux
publiés par l'AIA. Voir aussi W. HIS, "Zur Vorgeschichte des deutschen Kar-
tells und der internationalen Assoziation der Akademien", Berichte über die
Verhandlungen der Kgl. Sächs. Gesellsch. d. Wissenschaften zu Leipzig, Math.-
Phys. Klasse, 54, 1902; R. MEISTER, Geschichte derAkademie der Wissenschaften
in Wien, 1847-1947. Wien, 1947; H. LYONS, The Royal Society, 1660-1940. Londres,
1944.
(7) HIS, loc.cit., p. 14-15.

(8) A part de subsides, il s'agissait généralement d'obtenir de gouvernements
   - dans le cas d'associations internationales - la prise en charge de cô-
tisations annuelles, ou - dans le cas de congrès - l'acheminement par la voie
diplomatique des invitations ou l'octroi d'un statut de délégation nationale
au groupe de scientifiques se rendant à la réunion à l'étranger.

(9) J. LISTER au président de la National Academy of Sciences, le 14 avril
   1899 (NAS Archives - International Association of Academies, 1899-1914).

(10) G.E. HALE à Elihu ROOT, le 29 décembre 1913 (NAS Archives, ibid.).(11) ibid.

(12) A. SCHUSTER, "International Cooperation in Research", in The Semi-Centen-
   nial Anniversary of the National Academy of Sciences, 1863-1913. Washington,
   1913, pp. 19-36.

(13) ibid.

(14) L.S. WOOLF, op.cit., p. 317.

(15) Voir sur l'histoire du Conseil International de Recherches et les re-
   làtions scientifiques pendant l'époque de l'entre-deux-guerres, B.
SCHROEDER-GUDEHUS, Paix, science, politique. La communauté scientifique internationale et la détente en Europe au cours des années vingt.   Montréal, 1978
(à paraître)

(16) E. PICARD à A. SCHUSTER, le 3 octobre 1918 (Royal Society, ARchives,
   Vox 540).

(17) Emile PICARD, Discours et mélanges. 1922, p. 298.

(19) G.E. HALE, "The international organization of research", in R.M. YERKES,
   The New World of Science. Its Development During the War. Freeport, N.Y.,
1920, pp. 405-416; voir aussi correspondance avec sa femme (notamment la
lettre du 30 octobre 1918, in NAS Archives, Hale Microfilm, 69, n. 160-61).

(18) C.E. MENDENHALL à G.E. HALE, le 19 mars 1919 (NAS Archives, NAS-NRC
   Research Information Service, 1919).

(20) International Research Council. Constitutive Assembly held at Brussels,
   July 18th to July 28th, 1919. Reports of Proceedings. Londres, 1919.
p. 63.

(21) G.E. HALE à A. SCHUSTER le 18 novembre 1919 (Royal Society, ARchives,
Box 540).

(22) SdN Registry, 1922, Coop. Int., 13-19588-14297.

(23) W.B. CANNON, R.M. FIELD, "International Relations in Science", Chronica
Botanica, 9, pp. 251-298.

(24) NAS Archives, Foreign Relations, Conference ..., 13 December 1944, transcrip
   of discussion.

(25) Parmi les ouvrages les plus récentes sur la genèse de l'UNESCO, voir
   J.P. SEWELL, Unesco and World Politics. Engaging in International RElations.
Princeton, N.J., 1975. Voir aussi, C.S. ASCHER, Program-Making in UNesco,
1946-1951. Chicago, 1951. Pour un témoignage personnel, voir J.G. CROWTHER,
Fifty Years with Science. Londres, 1970.

(26) J. NEEDHAM, Memorandum on an International Science Cooperation Service.
   1944; The Place of Science and International Scietific Cooperation in
Post-War World Organisation, 1945.

(27) Science and UNESCO. InternationalScientific Cooperation. Tasks and
   Functions. Londres, s.d.

(28) Voir, par exemple, "Unesco's tarnished image", Science, 194. 26 november
   1976, p. 497; "UNESCO cuts its THROAT", Nature, 230, 12 mars 1971, p. 71,

Dr. Nathan Reingold

NATIONAL STYLE IN THE SCIENCES: THE UNITED STATES CASE

With brief casualness, the first volume of John Theodore Merz's A History of European Thought in the Nineteenth Century (1896) limits its treatment to the French, the Germans, and the British among the European nations but surprisingly contains a specific exclusion of the culture of the United States of America:

"...the steady growth and peculiar civilisation of a new and vigorous people on the other side of the Atlantic force me to the twofold confession, that there is a large world of growing importance, of which I have no personal knowledge, and to estimate which I therefore feel unqualified and unprepared; and further that I am equally unable to picture to myself the aspect which the whole of our European culture in its present state may assume to an outside and far-removed observer who is placed in the New World. As this New World grows not only in numbers and national wealth, but also in mental depth, as it becomes more and more intellectualised and spirtualised, so it will no doubt experience the desire of recording its own inner life and culture emphasising the peculiarities which distinguish it as a whole from our civilisation... which has indeed been powerfully influenced by it."[1]

In the Age of Imperialism, Merz saw no need to offer an explanation for omitting the ancient cultures of the Orient and the Near East. Europe ascendant in thought and deed provided the setting of his four volumes. Yet, the United States of America shows up again, a troublesome specter, in still another limiting disclaimer. Here, in the "borderland, where spirit subdues matter, where thought becomes useful..." Merz might have cited materialistic and utilitarian motivations to downgrade the cultural contributions of the civilization of the United States. Instead, his exclusion of the practical in all its realms takes a surprising twist:

"For where are discovery and invention at this moment more at home than in America; where have political theories, the original rights of man, the ideas of liberty, equality, and brotherhood, been more widely put to the test; where have religious beliefs entered into closer contact with the work of the day; or where in our age has the simple rule of early Christianity been more successfully put into practice? An account of the application of thought taken merely from our European experience, where half our endeavour must always be spent in clearing away obstacles, in removing the debris of antiquated institutions, in overcoming prejudices, or battling evils which have grown to uncontrollable magnitude, would give us but a poor notion of the influence of thought over material circumstances, and a very exaggerated one of the inertia of the mechanism of older societies."[2]

Not only was the United States a problem to Merz as a new, different civilization but also because of its success in applying knowledge to attain high ideals of European Civilization. Implicit in Merz is the idea of the uniqueness of the United States of America, an idea attractive to the nationalistically inclined in the North American republic. Historiographically, uniqueness often resulted in neglect by Europeans, as an awkward case for Marxists and an uncomfortable reality for well-intentioned, old-fashioned liberals like Merz. But if one talks of

Western Civilization, not simply European Civilization, the United States of America clearly belongs under that broader rubric. Why then an assertion of alien status rather than a recognition of relationships, perhaps a variation within a general pattern? For that is what the United States represents, not a singular case but a permutation involving elements present in the Old World.

A British-born author of German parentage and a successful industrialist, Merz sometimes reads like a caricature of the cultured Teutonic mandarinate. Educated on the Continent in mathematics, chemistry, and philosophy, Merz well represents a peak in the Western European tradition, the heyday of the German university.[3] Sarton dismisses Merz's History as more philosophy than science;[4] in making this self-revealing judgment, Sarton misses the point. In a broad sense, Merz is typical of much of the history of science to this day. The two disclaimers about the United States are important because they hinge on a key element in his general historiographic position--the presumed disappearance of national styles in science.

National styles existed at the start of the nineteenth century, Merz asserts. Space is consequently given to a trite, superficial survey of the histories of science in Great Britain, France, and Germany.[5] Basically, this is the received account, now being revised by the present generation of a professional history of science. In his view, after the mid-nineteenth century, national differences began to wane:

"In the course of our century Science at least has become international: isolated and secluded centers of thought have become more and more rare. Intercourse, periodicals, and learned societies with their meetings and reports, proclaim to the world the minutest discoveries and the most recent developments. National peculiarities still exist, but are mainly to be sought in those remote and more hidden recesses of thought, where the finer shades, the untranslatable idioms of language suggest, rather than clearly express, a struggling but undefined idea."[6]

Science, the most precise form of Thought, is in the vanguard of this internationalization. What makes this great trend possible, of course, is technology whose fruits in improved means of transportation and communication facilitate the flow of literature, concepts, and savants. Although Merz rates this as the great achievement of Europe in the last century, it was not to be the subject of his history. By adopting the conventional reductionist theory that technology is applied science, a gross over-simplification, Merz can turn his encyclopedic mind to Thought with few qualms.[7]

But underlying this particular conceptual stance is an even more fundamental historiographic position which converts this turn of the century text and its author into exemplifications of still present attitudes once overwhelmingly dominant in the practice of historians of science. On the first page of Merz's work he postulates three levels: "The external events and changes," "the hidden world of desires and motives, of passions and energies," and, finally, "the inner regions of Thought," Merz's subject. The last connects the other two, makes them intelligible, and provides "the only moving principle."[8] Implicitly and explicitly, the history of science was (and is to some today) defined as the analysis of the generation of philosophic constructs (à la Merz) or as the chronicle of the operations, data, and concepts of science (à la Sarton). In both cases there is an idealization removing from serious consideration the very stuff of history--individuals; communities; institutions; and social, economic, and ideological processes. Most important, idealization reduces the history of science to a timeless and placeless recital of concepts and data, often linked trivially to human names functioning as abstractions--bloodless, boneless, and nerveless. At best, the school has yielded superb examples of a species of intellectual history as in the work of Koyre; at worst it produced a kind of higher antiquarianism.

From viewing "Science" as a system of concepts and data rather than as a pattern of human behavior, one can, like Merz, define it as a body of knowledge transcending national boundaries, particularly after the improvement of transportation and communication in the modern era. In this framework, there is little doubt as to the answer to the question: Did particular conditions in the United States in the nineteenth century or later lead to a specifically "American science" different from European science? If by this is meant the persistance, beyond a chronological generation, of scientific concepts and operations (whether originating in the United States or adopted from abroad) after being supplanted elsewhere by later developments, the answer is no. In the Merzian-Sartonian world, there is no other possible response. If one admits the topics excluded by Merz, then the origins, elaborations, diffusions, and differential receptions of scientific advances in the United States of America and elsewhere display degrees of dependence upon particularities of time and place.

A quite different historiography might start with the recognition that history deals with the temporal succession of human environments which include intellectual perceptions and actions arising from these perceptions; and that the historian of science needs awareness of the entire environment to analyze successfully any of its parts. To this one might add two ideas widely present in Western Civilization (perhaps going back to the Greeks) which act as Merzian "moving principles." One is the belief in a value-free science independent of society, probably arising from religion but now secularized. The second is the belief in the fundamental utility of all knowledge. The two ideas obviously have great potential for tensions and clashes, particularly since our proposed historiography presupposes that science and scientists are not sealed from human environments but necessarily immersed in social, political, economic, and ideological realities.

In this perspective Merz's disclaimers about the United States assume a special importance. The foreign quality Merz sensed arose from his inability to place it within his historiographic framework. Only one American scientist is cited by him.[9] Even if Merz had known of one hundred more, he might not have used these examples. Despite his words, the people across the Atlantic were part of Western Civilization, but the texture of their society was alien to his experiences and to his intellectual assumptions.

The United States of America is one instance of the world-wide expansion of European peoples, ideas, and techniques going back to the fifteenth century. That expansion had two aspects: the development of possessions, now nations, largely populated by European peoples and the penetration of areas with indigenous populations with distinctive persisting cultural traditions. The history of science in the latter (Japan, China, the Indian sub-continent, the Islamic world) is essentially one of adaptation of science by non-Western peoples to their own traditions. In the former, we are dealing with the establishment of provincial outposts of European civilization--British, Russian, Iberian,[10] and French. Both the Russian and United States cases are instances in which provincial societies have moved from the margin to the centers of Western Civilization. In the wings are other nations of modern science waiting their turns--Canada,[11] Australia,[12] Brazil. And beyond these former outposts are the heirs of ancient civilizations now tingling with the ferment of modernization.

If there is a national style of science in the United States in the broader sense given previously, it is only a matter of degree, having attributes found also in Western Europe (and in countries like Canada and Australia). And these attributes have changed in time on both sides of the Atlantic, giving rise to discussions of the "Americanization of Europe" and the "Europeanization of America." But matters of degree can count. In the United States of America the early independence, during the Enlightenment, is important. The former physical security from the inanities of European international politics is a factor. The presence of great natural resources is pertinent. But what counts most of all is the population and the social texture of their lives. Diverse in origins, this population from the seventeenth century until today included a minority of European born and trained intellectuals and technicians. Almost by genetic definition, it is an internationalist population open to European influences. (In the century before World War I nearly 10,000 U.S. citizens attended German-language universities.)[13] The structure and dynamics of U.S. society influenced the nature of the American scientific community and its institutional patterns.

What Merz saw as he looked westward around 1900 we may never know. What made the United States of America an alien culture to his eyes? Perhaps we can guess from the mirror image, from what countless Americans saw in Europe around 1900, a reflection of what had sent many of their ancestors across the ocean to a New World. From the West of Ireland to the Urals the mother continent reeked of firm social hierarchies and their associated patterns of deferential behavior. Even today when so much has changed, visitors from North America still sense such survivals. Visitors from Europe, on the other hand, sensed a formless casualness; some, then and now, are put off by the absence of accustomed deferential behavior. Not that the United States of America is a classless society; a structure exists but more blurred, and behavior often lacks the obvious elements of hierarchical deference present in other societies. And that has considerable implications for Science, Thought, in Merz's terms, or high culture in a more recent locution.

In the leading countries of Western Europe there was (and still may be) a rather strong correlation between class and culture; high culture was for the top, often being defined as distinct from or antithetical to the prosaic concerns of daily life. The correlation was much stronger than in the United States. An assured hierarchical structure in Europe hopefully shielded those at higher levels from the broad social and other forces of daily life, the very stuff of history Merz passed by for the "inner regions of thought." In the United States of America few individuals could indulge in such illusions, much as many might so wish. A split of elite from mass, of culture from service, of theory from practice could not survive the realities of a republic at once democratic, eager for material comforts, and ambitious to contribute to the cultural stock of Western Civilization. Despite the strong exertion of some intellectuals, it is a society persisting in somehow blending high culture into the vernacular tradition. Becoming both culture and service, both theory and practice, at once elitist and of the people, science in the United States was (and is) simultaneously value-free and utilitarian. A creative tension resulted, manifested in government policy, educational strategy, and in technological practice.[14]

Even after the post-World War II growth of federal support for research, national science policy was cited by Americans as evidence of United States indifference to basic research, in contrast to a supposed contrary European predisposition. Many Europeans also believed this; it was comforting in the

face of American economic strength to raise the Merzian banner of Thought
over the Old World, to think of themselves acting as Greeks to the Romans
westward over the Atlantic. Recent studies indicate this position is more
a historiographic or ideological artifact than an accurate description.
Writing of Great Britain and France in the last century, W.H. Brock
concludes, that "government patronage was usual only when there were
obvious commercial, military or national benefits."[15] F.R. Pfetsch's work
on Germany notes that "'undirected' science was not greatly emphasised by
the Imperial Government."[16]

   Both conclusions--and the similar one for the United States--rest upon
viewing the entire range of activities related to the pure and applied
sciences, not simply focusing upon peaks of intellectual achievement.
Consider the total 1900 financial expenditures in the United States, Great
Britain, and Imperial Germany. (See Appendix for details.)[17] All three
countries tend to have the same pattern of activities. Deviations exist
but are minor. In particular categories of support, relative and absolute
values differ. In these one finds traces of national modes. In Imperial
marks, in 1900 the United States was spending on the pure and applied
sciences 40,000,000; Britain 19,000,000; the Reich 8,000,000. If one
adds the expenditures of the German States, the last total is 49,000,000.
(Equivalent data for the American state governments are not readily
available). All three countries maintain observatories; do surveys; have
medical programs; send expeditions; engage in international organizations,
support military research, and provide scientific and technical education.

   Patterns emerge from the American figures. There are the strange
organizations, the Smithsonian Institution and the Coast and Geodetic
Survey, whose histories embody the classic American strategies for combining
culture and service, theory and practice, mass and elite. The first was
originally an enclave for research within a vernacular society. By 1900
it was well on its way to becoming a great museum complex with a mandate
for popular diffusion. But its research tradition persisted, not only as
enclave, but as being defined as an integral part of the mission to
popularize.

   A similar broad definition of mission characterized the Coast and
Geodetic Survey--the physics of the earth, not simply an assignment to
produce useful charts. By 1900 its budget was down from that of a quarter
of a century before, reflecting more the vagaries of administrative history
than the coming of a post-classical physics. The U.S. Geological Survey
now had pride of place in Washington. 2,700,000 M far overshadowed
equivalent support in Germany and Great Britain. J.W. Powell and his
successor, C.D. Walcott, stressed doing both the useful and the arcane.
Measuring and understanding the land was a great preoccupation in the
United States of the last century. Using it and the resources of plants
and animals was a great economic activity, one particularly sanctioned by
national ideologies. The more than thirteen million marks for agriculture,
forests, and fisheries are surpassed in the comparative accounts for 1900
only by the expenditures of the German States in support of higher education.
We know from recent publications that these activities were increasingly
directed by trained scientists whose standards were those of their disciplines.
Not surprisingly, even agricultural research was defined to include a necessary
portion of basic research.[18] Under the justification of aiding farmers, the
Department of Agriculture administered an extensive program in meteorology.

   Overwhelmingly, the fields involved were well-established, not the newer
ones at the growing edge of research. It was positional astronomy, not
astrophysics, at the Naval Observatory. That new specialty had an enclave,

the Astrophysical Observatory of the Smithsonian, but was lodged more conspicuously outside the federal government. As D.B. Hermann has shown, astrophysics was flourishing in the United States in the two decades before 1900.[19] It was stratigraphy and paleontology rather than geophysics or geochemistry at the Geological Survey. The entomologist C.W. Stiles at the Agriculture Department spurned the newer cytology and embryology for the older morphology.[20]

Support for the new in Germany and the United Kingdom was largely of two forms: aid to learned societies and aid to higher education. The former was totally absent in the U.S., being part of the private sector. In 1900 the federal government did give 5,000,000 M to the land grant (i.e. state) colleges and universities. As in Germany, most moneys to them came from the state governments. For academic physics, Forman, Heilbron, and Weart estimate expenditures of 2,990,000 M.[21] Part of this came from the state governments. Most probably came from the private sector. Recent research indicates that the private sector in the United States was interested in supporting science both as culture and as utility.[22] As in the case of governmental funds, estimating the extent of the private sector is difficult because most funds were general in nature, not designated specifically for research. Just before the great outpouring of Rockefeller and Carnegie funds, those endowments earmarked for research were estimated at $2,952,642, yielding an annual income of $199,625.[23] This figure is a bare minimum. Recent research also suggests the private sector had a greater role in Western Europe than once believed.[24] In the United States the private sector had a crucial role in the growth of research until the aftermath of World War II. Previously, national science policy largely furthered missions including basic research in older fields while the private sector fostered enclaves of newer theory in higher education and elsewhere. Stiles in Agriculture and the Smithsonian Institution might shun the newer biological fields around 1900; the leading private universities had no such scruples.[25] Until recently, science, as high culture, was like religion in the United States in being a matter for the private sector. Today, national science policy encompasses both science as service and as high culture.

Understandably, Merz has nothing on the educational strategy of the United States but much on the experiences of the three principal European scientific powers. While both France and Great Britain garnered praise, the German university received the laurel. To Merz, it was the great European institution of the nineteenth century. Although we may never know what he thought of education in the United States, we can estimate what his eyes beheld and offer educated guesses at Merz's reactions.

Perhaps he turned in Newcastle-upon-Tyne to the pages of another Victorian, the statistician Michael G. Mulhall. In his 1880 compendium, The Progress of the World... Mulhall rated popular instruction second only to immigration as the great factor in national progress in the United States. What national progress meant to him is quite clear from the following: "Every day that the sun rises upon the American people it sees an addition of Ł500,000 sterling to the accumulated wealth of the republic, which is equal to one-third of the daily accumulation of mankind." As early as 1831, the country was second only to Prussia in the proportion of school children to population. The growth of learned societies by 1880 was on a par with Great Britain, France and Germany.[26] A few years later, Mulhull noted the 364 "university colleges" in 1880 with 8,900 science students.[27] And Mulhall's successor, A.D. Webb, in his later compilation reported 573 "universities, colleges, and technological schools" with 292,80(

students in 1907-1908.[28] These numbers, taken from the reports of the
U.S. Commissioner of Education, were quite startling by contemporary
European standards. Although impressively large, the other educational
data did not deviate markedly from European experience.

In elementary education, for example, Mulhall in 1900 reported an 1896
literacy rate of 83%, well below many countries of Western Europe. An
important trend was obscured in the popular education figures. In 1900
high schools graduated 94,883 students, 6.4% of the 17 year old population.
Mulhall and Webb missed it, but the St. Louis Hegelian in the Commissioner's
office, William T. Harris, wrote: "The increase in the number of high
schools [up 510 in a year to 6,005] and in the number of students enrolled
in them is something phenomenal. It would seem as if the people of all
cities and villages had determined to provide high-school accommodations
for their children." By 1940 the high school graduates were just over 50%
of the 17 year old population.[29] The heritage of Black slavery and the
presence of recent illiterate immigrants from Europe had kept the literacy
figures comparatively low in the last century but only until the elementary
schools did their work. As a leading American educator, Nicholas Murray
Butler, noted proudly in 1900: "The annual expenditure of the United
States for common schools is quite equal to the sum total of the expenditures
of Great Britain, France, and Germany combined on their powerful navies.
It is nearly four-fifths of the total annual expenditures of the armed
camps of France and Germany upon their huge armies."[30] The high schools
would go beyond the elementary rudiments, somehow producing Thomas
Jefferson's enlightened, responsible electorate. Or so hoped a nation
with imprecise beliefs in culture and utility, in service and a value-free
knowledge.

But Mulhall's figures on higher education were both partly misleading
and presciently enlightening. For the late 1880's, for example, he credits
the United States with 360 "universities," comprising 4,240 professors and
60,100 students. Imperial Germany in contrast had only 21 universities
with 1,920 professors and 26,680 students.[31] Given the awe surrounding
the term "university" in many European circles and the careful control
by the state of the use of the designation, the United States numbers
must have astounded. Think how long it took the Technische Hochschulen
in Germany to be allowed to issue degrees. One can imagine European
feelings of disbelief, even contempt, at the United States numbers, perhaps
tinctured with concern it might turn out true after all.

Commissioner Harris knew these numbers better. In 1900 he used _Minerva_
to compile an equivalent table which tacitly contested Mulhall. American
universities now numbered 51; to these he added 48 institutions dealing
with technology and other applied fields. These 99 had 69,831 students.
The 46 German equivalent bodies had 55,813 students. On various comparative
measures, the Imperial Reich did very well indeed; the United States appears
as up to a solid Western European standard, not unique in its devotion to
higher education.[32] But its future rate of growth catches the eye. 382
doctorates in all fields in 1900 would reach 2,299 by 1930 and 3,290 by
1940. Higher education's resources were considerable in 1900. Endowments
were estimated at $166,193,529, yielding $11,995,463 income. Although
private institutions had more students than publicly supported ones (76,735
to 34,177), funds from national, state, and local governments amounted to
$7,448,582. Science was a minority interest; most students took the
classical course.[33] National figures on research in the universities in
1900 are lacking. By 1930, American universities expended over $18,000,000
on "organized research;" by 1940, before the inflow of wartime support,

the total was over $27,000,000.[34] In short, a solid movement had started before the end of the last century.

All those other institutions counted by Mulhall somehow mattered. Yet one wonders why Mulhull only included the 21 German universities and their analogs in other European nations while totalling in every conceivable American body bearing words like "university" and "college." Like the high schools, they reflected the determination of United States localities, regions, states, the national government, and private groups to train a responsible, enlightened electorate. In 1900 slightly more than 4% of the 18 to 21 year old population attended institutions of higher education, a number impressive by contemporary standards; by 1930, over 12%; by 1940, over 15%. There were over 100,000 graduate students in all fields that year.[35] The basis for post-World War II American education was in being. Europe was different; Mulhall probably could not conceive of doing Harris' tabulation for Europe with its combination of "university" and technological schools. But Mulhall took the raw American data on faith.

Merz was aware of analogous trends in Europe. Noting that movements for popular education developed independently of the reform of the German universities following the ideal of Wissenschaft he observed: "Under the latter [i.e. Wissenschaft] were trained the leaders and higher teachers of the nation, as well as the members of the learned professions. The educational influence of this ideal on the more gifted among the student class was the very highest and best; but it hardly reached the multitude of less gifted minds, who always give themselves to bread-studies; and it must necessarily fail yet more when not only the future teachers and leaders, but the masses of the nation, flock into the halls of the universities."[36]

Protestant Europe--Scotland, Switzerland, a few smaller German States-- pioneered in the movement for popular instruction according to Merz. The parallel to the United States is obvious. On both sides of the Atlantic religious concerns became intertwined with civic needs. In the United States the profusion of common schools yielded the proliferation of high schools which helped make possible the efflorescence of higher education. The process originated in the colonial period. By the last century, civic needs persisted while sectarian religious drives abated. Higher education today becomes simply the chronological culmination of popular instruction, much to the distress of some academics looking eastward. By 1900, if not sooner, science had edged out religion as the form of value-free knowledge independent of society. And the knowledge served the utilitarian needs of the nation. The universities provided protected enclaves for science, secular equivalents of monasteries; at the same time the universities defined their utilitarian missions to encompass necessarily the abstractions of Thought à la Merz.

How alien it must have all seemed to him in Newcastle-upon-Tyne. Perhaps Max Weber's view of the American academic scene is how Merz reacted.[37] Later, the configuration outraged Abraham Flexner, a great fan of the German university.[38] But Felix Klein earlier saw something of value, treating engineers like scientists.[39] Merz hailed the German university for the ideal of pure science and its pursuit without perceiving any incongruity with his earlier praise for its giving equal dignity to pure and applied science.[40] In fact, despite the exclusion of engineers from the university world, both assertions were true. The Germans and other Europeans were more like their kinsman overseas than Merz's, Sarton's and Koyré's historiography allows. For example, Pfetsch (like Merz?) rates the German States' support of education as undirected, "zweckfrei" by analogy with pure research.[41] But the universities of Germany existed to provide the staffs of government and industry, to man the free professions, and broadly, to maintain and to further a particular national culture. It is impossible to think of any European university that did not serve some mission or need of its society. The Americans, lacking such impressive terms like Wissenschaft and Kultur, had an open and realistic view of the entire process of education, preventing all but a limited idealization of the university.

As to theory and practice, an older formulation persisted in the United States long after vanishing from many of the best circles in Western Europe. They were seen not as antithetical nor disparate but, even today, as some kind of an identity. All this is most distressing to U.S. scientists committed to the primacy of basic research. The common intellectual strategy of reducing technology to applied science is often interpreted in the United States not as maintaining a proper hierarchy but as a form of leveling upward.

Nor is this surprising. Since the earliest settlements, Americans have persisted in viewing the world as an unfinished object requiring both thought and action. Not only grand thought in the mode Merz learned from his German teachers but also theories for the humble and the prosaic. Scientists in the United States were not simply contemplative but also doers. Science, like the unfinished world, was not wholly an aesthetic object for pleasurable manipulation. To paraphrase a seventeenth-century settler, scientists are mixing their labors with the wilderness.

1. John Theodore Merz, A History of European Thought, 4 vols., 1896-1912. Citations are to the Dover edition of 1965. 1:14-15.
2. 1:78-80.
3. For Merz, see the John Theodore Merz Memorial Number of the Proceedings of the University of Durham Philosophical Society, vol. VI, pt. 4, 1922-1923.
4. George Sarton, A Guide to the History of Science, Waltham, 1952, 119.
5. Part I, Chapts. I-III.
6. 1:19f.
7. Scattered comments on this theme appear at many points; for example 1:305 and 4:260-261. See 1:302-3 for comments on technology.
8. 1:1-5 et passim.
9. The astronomer G.W. Hill appears in a footnote on 3:102. Other references to the United States note F.A. Walker, William James, pragmatism, Josiah Royce, J.B. Stallo. C.S. Peirce is apparently absent.
10. See Marcel Roche, "Early History of Science in Spanish America," Science, 194 (1976) 806-810.
11. T.H. Levere and R.A. Jarrell, eds., A Curious Field-book, Science and Society in Canadian History, Toronto, 1974. Peter Bowler, "The Early Development of Scientific Societies in Canada," in A. Oleson and S. Brown, eds., The Pursuit of Knowledge in the Early American Republic..., Baltimore, 1976. While I sympathize with the first book's assertion of a unique Canadian science, from it and the Bowler article one gets a picture of a society following a pattern common to other former colonies, perhaps a generation or two behind the U.S. The assertion on p. 2 of the former of utilitarianism as the unique character simply will not hold up. Canadians are different from their neighbors to the south but both the English and French speaking ones seem closer in spirit to the United States than do the Europeans.
12. Ann Mozley Moyal, ed., Scientists in Nineteenth Century Australia, A Documentary History, Melbourne, 1976.
13. See Carl Diehl, "Innocents Abroad: American Students in German Universities, 1810-1870," History of Education Quarterly, 16 (1976) 321-341.

14. Much of this and what follows is elaborated in various prior publicati‹ of mine: Science in Nineteenth-Century America, A Documentary History, New York, 1964; "Alexander Dallas Bache: Science and Technology in the America‹ Idiom," Technology and Culture, 11(1970)163-177; "American Indifference to Basic Research: A Reappraisal," in G.H. Daniels, ed., Nineteenth-Century American Science, A Reappraisal, Evanston, 1972, 38-62; "The Professionalization of Science in Nineteenth Century America," in Oleson and Brown, 33-‹ "Reflections on 200 Years of Science in the United States," Nature, 262(197‹ 9-13.

15. G.L.'E. Turner, ed., The Patronage of Science in the Nineteenth Century Leyden, 1976, 177.

16. The quotation is from "Scientific Organisation and Science Policy in Imperial Germany..." Minerva, 8(1970) 557-580. Based on his Zur Entwicklung der Wissenschaftspolitik in Deutschland 1750-1914, Berlin, 1974. See commen below on where Pfetsch finds undirected research in Imperial Germany.

17. In addition to Pfetsch, the numbers that follow are based on R.M. MacL‹ and E.K. Andrews's 1967 mimeographed Selected Science Statistics... for the United Kingdom. The U.S. figures are taken from the U.S. Treasury Departmen official publications on expenditures and appropriations. Details, includin the limitations of the data, are in an Appendix available from the author a‹ the Joseph Henry Papers, Smithsonian Institution, Washington, D.C., 20506. a history of federal involvement in research and development, see A. Hunter Dupree, Science in the Federal Government, Cambridge, USA, 1957.

18. For example, the various articles of Charles Rosenberg such as "The Ada Act: Politics and the Cause of Scientific Research," Agricultural History, 37(1964)1-10. See also Margaret W, Rossiter, The Emergence of Agricultural Science: Justus Liebig and the Americans, 1840-1880, New Haven, 1975.

19. D.B. Herrmann, "Zur Frühentwicklung der Astrophysik in Deutschland und in den USA," NTM-Schriften. Gesch., Naturwiss., Technik, Med, 10(1973)38-44.

20. E.G. Conklin to C.W. Stiles, February 25, 1902; Stiles to Conklin, March 1, 1902, Conklin Papers, Princeton University Library.

21. Historical Studies in the Physical Sciences, 5:6. All 1900 budget figures are converted to marks as done by this pioneering compilation.

22. Howard S. Miller, Dollars for Research, Science and Its Patrons in Nineteenth Century America, Seattle, 1970. See also my forthcoming article on the early years of the Carnegie Institution of Washington, "National Science Policy in a Private Foundation...," to appear in the American Academy of Arts and Sciences' conference on Knowledge in American Society, 1860-1920, forthcoming.

23. Carnegie Institution of Washington, Report of the Executive Committee to the Board of Trustees, Washington, 1902, 247-269.

24. This is one of the conclusions of The Patronage of Science in the Nineteenth Century (fn 15).

25. As witness the careers of T.H. Morgan, E.B. Wilson, E.G. Conklin, H.S. Jennings, and others.

26. Michael G. Mulhall, The Progress of the World, 515-517. Citations are to the 1971 reprint by the Irish University Press of the London 1880 edition.

27. Mulhall, Dictionary of Statistics, 4 ed., London, 1900, 241. Characteristic of the compiler's insouciance with data is that on 232 the total number is given as 360.

28. A.D. Webb, The New Dictionary of Statistics, London, 1911, 228. The 1971 reprint was used.

29. Mulhall, Dictionary, 693. U.S. Bureau of Education, Report, 1899-1900, p. LIV; Historical Statistics of the United States, Washington, 1960, 207. See chapt. 39 of the 1899-1900 report.

30. Bur. of Ed. Rept. 1899-1900, 568.
31. Mulhall, Dictionary, 232.
32. Bur. of Ed. Rept. 1899-1900, 861-864.
33. Ibid, LIII, and Chapt. 25. Historical Statistics, 211f.
34. Historical Statistics, 213.
35. Ibid, 210f.
36. Merz, 1:257-262.
37. The reference is to his essay "Science as Vocation."
38. Abraham Flexner, Universities, American, English, German, New York, 1930.
39. See "Alexander Dallas Bache: Science and Technology in the American Idiom," particularly p. 177.
40. 1:193, 211-212.
41. Zur Entwicklung..., 68.

Grateful acknowledgement is made to the National Endowment for the Humanities and the National Historical Publications and Records Commission for their support in the preparation of both the text and the appendix.

M.G.Yaroshevsky

NATIONAL AND INTERNATIONAL FACTORS IN THE DEVELOPMENT OF SCIENTIFIC
SCHOOLS OF THOUGHT

Science is the creation and possession of all humanity. Its truths,
like the nature of things reflected in them, do not depend on national
and racial distinctions. There is nothing in the structure of human
mind that could impart even a shade of national narrow-mindedness or
national superiority to the process of cognition. The fact that science
belongs to all humanity does not mean that it is a historical fact. Conversely, historism is inherently characteristic of human intellect
which is indefatigably seeking the absolute truth. Although this
process is sometimes uneven from one nation to another, it has a
common direction. Historism which is a pre-condition of the originality
of different cultures and civilisations does not break the single chain
of the development of scientific knowledge; one link of this chain is
inevitably followed by another. This thesis has a direct bearing on
the question of the laws of the development of scientific cognition and
its objective historical logic which is being much discussed nowadays.
The reality of science is not only a "creatively irregular" confrontation
and change of hypotheses, theories, and empirical conclusions; this
apparently "Brownian movement" is characterised by a certain "rhythm"
and regularity.

The attempt of Kuhn, the American historian of physics, to reconstruct
this rhythm has become widely known. His concept is built on the notion
of the paradigm as a common pattern of research. The paradigm was
described in such books as Newton's Opticks, Darwin's Origin of Species,
and others. The cycle "pre-paradigmatic period - paradigm - crisis -
revolution - new paradigm" was represented as an inevitable process,
like the development of a caterpillar into a butterfly. According to
Kuhn, the notion of paradigm is inseparable from the notion of
scientific community, and it is the paradigm that cements the world
of science.

However, Kuhn interprets the scientific community as a socio-psychological
isolate, an autochthonous system which is closed due to its paradigmatic
"riggings," and irrelevant to the processes in the social world at large.
Therefore Kuhn's theory, which at first sight does not recognise any
human ties except for those determined by the common research patterns
and thus asserts boundless internationalism, in fact does not offer any
ideas pertinent to the analysis of the problem of the correlation of
national and international factors in the development of scientific
knowledge. In order to correlate them it is first necessary to distinguish them, and this distinction lies beyond the boundaries of the
closed scientific community as it is presented by Kuhn. In other words,
although Kuhn's concept of paradigm introduces the idea of the common
regular movement of scientific ideas into the historical scientific
thought, thus seemingly buttressing the conclusion about the independence
of this process from the inimitable originality of the national soil
which produced it, it is unable to assert real internationalism. It has

no resources for this. Apparently, these resources should be sought in the theory of society rather than in the theory of science proper. The Marxist theory of society regards science as a sub-system in the system of social and economic relations. The nature of these relations determine both the "ramifications" in the development of scientific thought in various regions and the synthesis of the intellectual accomplishments of different nations which have their distinctive ethnic and cultural features. Before becoming the possession of all humanity, new scientific ideas are conceived by one or several nations.

This most important historical fact is neglected by the "internalist" concept which disregards everything but the immanent logic of cognition. However, if the "external" ethnic, social, or psychological factors are believed to be the final causative agent, there is an immediate danger of ascribing intrinsic intellectual advantages to a nation which leads in scientific progress.

From the point of view of Marxism differences in the level and rates of scientific progress are explained by social and historical circumstances. The approach to the question of the correlation of national and international cooperation of research and the utilisation of its achievements in different regions should be manifold.

Let us examine this problem from the point of view of the inception and activities of the scientific schools of thought. The term "school" is polysemantic and it is not easy to single out its invariant content. Sometimes it is understood as a scientific trend which has emerged in one country and differs in its approach to various problems, concepts, and methods from the practices of scientists in other countries. Such schools are called national and there are distinguished, for example, the German, French, and Russian schools in physiology. It has been pointed out above that the scientific truth is not national in nature, therefore differences between schools are determined by subject-logical factors. The question is, however, why the effect of these factors in different national-cultural media differs so much that a special school emerges and develops in an individual country. Obviously, no answer can be given to this question in terms of the logic of the development of science,[1] because logic is irrelevant to the national differences in the scientific world and its forms and structures contain no prerequisites for different approaches to the development of one and the same problem in different countries.

What then are the extra-logical forces that come into play? These forces lie in the distinctive features of the social and historical development of a country in which a national scientific school is being formed. In other words, we should go beyond the boundaries of the scientific community as such. At the same time if the national school were not an

---

[1] "Logical" in the present context defines any factor pertaining to the logic of the development of science as the dynamics of the inner (substantial) forms of scientific cognition. This line of thought should not be mixed up with other varieties of logical analysis.

integral part of this international community, cemented together by
common norms and values, there would be no reason to consider it
scientific.

The emergence of the national schools reveals most forcefully the "dual"
nature of the object of scientific process in a sense that it is as
much the offspring of the scientific community which knows no national
distinctions as the child of its people with its historic traditions
and current social and philosophical problems.

The intrinsic connection of the national school with world science also
is dual. Such a school cannot emerge in any other way but on the basis
of the achievements, contributed to the common pool of knowledge by
research workers representing different national traditions. Being a
synthesis of these achievements, however, it is growing in significance
commensurately with the extent and stability of its influence on the
development of science beyond the boundaries of the country in which it
has emerged and is being cultivated. On the other hand, all the scientists carrying out research into one and the same field in a given
country constitute a national school with its unique subject-logical
characteristics (which alone give grounds to single it out from the
scientific community as a whole).

The Russian school of physiology, for example, is not identical to the
community of all the physiologists working in Russia which in the
fifties to seventies of the last century boasted such brilliant
research workers as Ovayannikov, Babukhin, Kovalevsky, and Tsion.

It was not they, however, who determined the originality of the national
Russian school which emerged at that period. The recognised founder and
first leader of the Russia school of physiology was Ivan Mikhailovich
Sechenov. His ideas contained the most vivid and characteristic features
of the trend which imparted to the Russian school of physiology a
special nature. This special nature meant not isolation from world
physiology, but the enrichment of the fund of its basic ideas with the
achievements born of the social and scientific climate which was not to
be found in any other country.

The achievements of the German school of physiology which had shaken
off vitalism intersected with the ideas of Claude Bernard, who looked
for the factors of the self-control of the internal medium of an
organism as a whole. Bernard, however, was interested only in the
internal controls of an organism, whereas Sechenov continued to concern
himself with the possibility of applying the determinist principles to
the behaviour of a whole organism in the environment. The German
school thought the environment to be a complex of physical and chemical
stimuli. This approach gave no explanation for the psychically controlled aspect of behaviour which was characterised by the perception
of comprehensive objects, orientation towards them according to internal
pulses, and the development of the new forms of reactions.

Darwin's theory opened basically new prospects for this approach.
Darwin's ideas enabled Sechenov to approach the nervous system as a
system of the different levels of the adaptation of behaviour to the
environment and thus understand the psyche as a special element in

the ascending series of behavioural acts which in the manner of being
committed are reflexes.

In this way the achievements of different national schools and trends
prepared in Sechenov's scientific thought a new synthesis. That was a
creative synthesis rather than an eclectic combination; a synthesis
which, according to Pavlov, constituted "entirely our, Russian contribution".

We have thus characterised the genesis of the Russian school of physiology by contrasting it with the schools and trends in other countries
in terms of conceptual progress, i.e., the battle of ideas. But behind
these ideas was the communication of scientists who proclaimed these
ideas and, since we discuss different national schools, they were
people of different nationalities. Sechenov's research programme
should not only be regarded in one "dimension", the subject-logical
one. It is pervaded with the estimation, evaluation and analysis of
other programmes which Sechenov knew not only from reading, but also
from discussions with their authors.

The logic of the development of science urged the application of the
principle of reflex-signal control to the brain. This logic was
following a winding path, however, finding its way through Sechenov's
numerous contacts with the followers of different, sometimes contra-
dicting theories. In Sechenov's words, the straight way to an
objective becomes clear only after the objective has been achieved.
In the midst of his research he associated with Ludwig, who became his
close friend, but who nevertheless believed that studying the brain
by the methods then available to physiologists was like studying a
clock by shooting at it from a gun. He associated with Helmholtz,
whom he worshipped but whose concept of the "deductions of the eye" he
believed to be isolated because it had no place for the evolutionist
approach. (Hence Sechenov's idea "to reconcile Helmholtz with
Spencer".) He worked with Bernard, in whose laboratory in Paris he
discovered his famous "central inhibition". Although Bernard was the
"godfather" of this discovery and advised Sechenov to publish reports
about it in scientific journals, according to Sechenov's recollections,
his attitude towards it was polite, but indifferent. Thus, Bernard
failed to appreciate the discovery which, according to Sherrington,
became a milestone in the physiology of the nervous system. In this
way the core of the new school was being formed in direct communications
(friendly contacts or scientific confrontation) with the leading
European physiologists of the period. Sechenov both learned from them
and polished his own programme and his own approach which determined the
originality of the Russian school of physiology.

Schools in science are an essential constant factor of its progress.
Contrary to Kuhn's concept, there is no reason to believe that they are
alien to the period of "normal science". This viewpoint is born of the
notion that the scientists of the period of normal science are working
within the limits of a single paradigm as a common pattern of filling
in the "puzzles" devised by Nature.

In the real historical process, however, the splitting up of the community into schools and their constant opposition meant the continuous building up of science through the interaction (confrontation, intersection, synthesis) of the different trends of scientific thought rather than the disintegration of the subject-logical basis of science. Historical experience shows that the development of national schools (which is interpreted by the paradigmatic theory as the symptom of a crisis) results in basically important achievements which pertain not to individual problems, "puzzles" as Kuhn calls them, but to the entire edifice of scientific knowledge, to its fundamentals. Every school contributes its brick to this edifice and consolidates rather than undermines its unity which can only be maintained in science through the dynamic equilibrium of the opposite forces. Taken separately, the German, Russian, French, and British schools of physiology each made its own way (had their own paradigms). The German school exhibited a physical-chemical approach; the British an evolutionist one; the French physiologists regarded the organism as a system; and the Russians studied complex behaviour as a function of signals and controls. Behind these apparent differences was internal harmony which ensured the successful development of the entire scientific "organism". The formula "as many schools so many truths", which concerts with the "anarchic theory of knowledge" proclaimed by American philosopher P. Feierabend, cannot be accepted. The multitude of schools, provided they are truly scientific, reveal certain mechanisms which make it possible to consolidate knowledge valuable as it is, irrespective of the assigned objectives of these schools. The different approaches exhibited by the national schools did not result in the dissolution of physiology as a uniform discipline, but determined its outstanding progress.

The problem of the correlation of national and international factors in the development of science is irrelevant when science is interpreted as an independent system with the inbuilt principles of inception and transformation, and the subject of scientific research as an abstract factor, a human substratum of the "flows of information". In real fact, however, this system is created by people who have been brought up on the soil of a certain national culture. They absorb the riches of this historical culture and its original traditions determined by social consequences. A scientist representing a national school is a link between the national and international factors, and not only in a sense that through his activities the achievements of an individual country become common property. More, his work is a medium through which original and inimitable national culture replenishes world culture. Sechenov's theory of neuropsychic functions was based not only on the logic of the development of the theory of reflexes, from the ideas of Pereira, a Spanish doctor of the 16th century who before Descartes viewed the animal as a machine, to the famous experiments of Sechenov's friend, German biologist Eduard Pfluger, with a decapitated frog.

In certain conditions the national school of science becomes an international one. The analysis of this phenomenon is of special interest not so much from the point of view of the ideas of one scientific school to other, far removed trends of thought, as from the point of view of the emergence of research teams consisting of research workers from different countries. In such teams the relationship of the national and international factors is represented in individuals, in forms which can be immediately perceived. This specific phenomenon provides a "close-up", as it were, of the processes similar to those occurring at the vast "expenses" of the universal history of science over great periods of time. Such a school is an informative model for research into the problems of cooperation between scientists representing different national traditions, and into the effects of this cooperation. Among the national schools of physiology which became international (i.e., trained scientific personnel for many countries) mention should be made of the German school of Karl Ludwig, the Russian school of Ivan Pavlov, and the American school of Walter Cannon. The outstanding contribution of these schools to the development of physiology is well known. That contribution was a result of joint efforts, by the founder of the school and other, mostly younger research workers.

Ludwig did not have such large-scale programmes as Helmholtz, who created new trends in physiology; Ludwig's plan of research envisaged the tackling of more limited tasks within the common physical-chemical approach. The situation seems to be a paradox. Apparently, a research worker should find it easier to cope singlehanded with limited, specific problems, and therefore he should have no motives for engaging in joint research. And when a prospective leader of a scientific school has no motives, the school as a research team cannot be formed. Being a professor, Ludwig did not limit himself to teaching but came out as an organiser of joint research, i.e., of the process of acquiring new know-ledge rather than reviewing the existing information. It is because of the variety and specific nature of his experiments that he needs pupils. He had to have many research workers capable of implementing his ideas which covered a variety of the aspects of physiology. At the same time account should be taken of the qualifications of his assistants: all of them were young scientists who had to master or develop together with Ludwig special techniques before undertaking major research programmes. The variety of the subjects and techniques of physiological research conducted in Ludwig's laboratory, and later on in his institute, is striking indeed.

The basis of Ivan Pavlov's international school was the theory of higher nervous activity. A particular feature of his school was that, apart from training many young scientists who later established centres for studying behaviour reflexes in their home countries, including the United States, Poland and Japan, he sent many of his pupils on scientific missions abroad. These missions served to popularise the innovative theory and its tremendous research potentialities; moreover, they stimulated the emergence and development of trends which determined the future of physiology. In this context mention should be made of the trip of Pavlov's close associate, Rozental, to Great Britain where he developed an experimental installation for the examination of behaviour reflexes at Barcroft's laboratory in Cambridge. According to the plan elaborated by Pavlov, Rozental was working together with budding physiologist G.Walter who therefore may be considered as belonging to Pavlov's school. As a result, Walter developed a basically new idea of the functions of the brain which later on enabled him to produce cyber-netic theories and build models simulating behaviour reflexes.

The cybernetic analysis also was prepared in Walter Cannon's international school. Cannon considered himself to be a "grandson of Ludwig" and believed that every scientist belongs to a "big family" whose members are connected not by blood relationship, but by the community of scientific interests. Speaking at the dinner in honour of the 25th anniversary of his professorate at Harvard, he characterised the genealogy of his physiological "family" in the following way (rephrased): "I am the son of Bowditch who introduced me to physiological research. In his turn Bowditch was the son of Ludwig in whose laboratory in Leipzig he contacted with other young people from many countries. Through my grandfather, Ludwig, I am related to many of his followers, including Italian physiologist Mosso, British pharmacologist Britton, and Russian physiologist Pavlov."

However, each of those scientists was working on his own programme and following his own path. Links between them were no firmer than the ties between physiologists who never learned from Ludwig. Why then does Cannon cite Ludwig's school as an example of scientific genealogy? Most likely, he regarded it as an example of the international community of a big family of research workers. Among its members Cannon mentioned Germans, Englishmen, Americans, Russians, and Italians. He could easily continue this list and name people of other nationalities who also hailed from Ludwig's school.

A comparison of Cannon's school with Ludwig's reveals an important feature which also is caused by specific historical circumstances. Not only the content and direction of research are changing from one period to another. The nature of intercourse and cooperation between scientists carrying out this research is affected by the processes in the big social world. The internationalism of Ludwig's school was determined above all by the need for the development of physiology which at that period was becoming a discipline based on accurate physical and chemical techniques. The Leipzig institute was the world's biggest centre which offered scientists from all countries the most favourable opportunities for mastering these techniques.

The internationalism of Cannon's school was determined not only by the scientific credo of its leader, but also by his political and philosophical predilections. He purposefully sought to make his school a centre for training research personnel for other countries, above all the Latin American ones, in order to promote the progress of science and culture there. This is seen not only from the constant influx of young Latin American scientists to Cannon's laboratory (which could be explained by the proximity of the United States), but above all from Cannon's attitude to his pupils, from Mexico, Chile, Brazil, and Argentina. "Latin America is especially indebted to Cannon", wrote Argentinian physiologists. A student from a Latin American country used to work for many years in Cannon's laboratory, acquired a lot of knowledge and mastered new techniques there, and then returned home to carry out scientific research. It should be mentioned that most of the Latin American countries were dominated by reactionary regimes. Scientists who opposed them were severely persecuted. In such cases Cannon did everything he could to support his pupils.

The analysis of the historical circumstances of the development of international schools is valuable not only for the theory of the scientific development and for studying its laws. It has a direct bearing on the organisation of research work in modern conditions, on the problems of scientists' migration, and on their social responsibility. This responsibility is not limited to the sphere of the production and consumption of scientific ideas, but pertains to the development of national cultures which determine the overall progress of civilisation.

Prof. Masao Watanabe

THE EMERGENCE OF JAPAN IN THE INTERNATIONAL SCIENTIFIC COMMUNITY

A Secluded Nation

As the birth and development of what we now call modern science did not take place in Japan nor in the Far East, though there had been such Oriental arts of medicine and herbs, calender-making, mathematics, etc., the Japanese people had to learn of modern science first from the Western countries where it originated.

Japan's first direct contact with the Westerners occurred in the mid-sixteenth century when a Portuguese ship drifted ashore on a Japanese southern island. This marked the beginning of the introduction of Western religion and learning to Japan, mainly by Portuguese and Spanish missionaries. In the 1630's, however, Christianity was banned and all Europeans were excluded except the Dutch. Even under these restrictions, the Japanese could still absorb some of the Western development through the Dutch visitors.

The publication in 1774 of Kaitai Shinsho (a Japanese translation from the Dutch version of a German treatise on anatomy) was a memorable achievement in the history of the introduction of Western science into Japan. This work so distinctly demonstrated the excellence of Western learning that it was followed by the introduction of Dutch versions of European science with ever-increasing degrees of enthusiasm. The interest of the Japanese was attracted more to the practical aspects, but some important basic knowledge was also partly transmitted, such as Copernican astronomy, Newtonian dynamics, modern human anatomy, Linnean system of taxonomy, and Lavoisier's chemistry. At clan schools, Chinese classics had constituted the core of the curricula, but from about the end of the eighteenth century, Dutch learning came to be added and by the middle of the nineteenth century, one-third of all courses of instruction consisted of scientific disciplines. At the same time, the number of scholars of Western learning in Japan increased remarkably.

Western Impact and Japanese Response

In 1853 the West suddenly appeared in Tokyo Bay, in the form of four "black ships" from America headed by Commodore Matthew Perry, and Japan concluded a peace treaty first with America on Perry's second visit to Japan in the following year. Among the presents Perry carried with him were a Lilliputian steam locomotive and an electric telegraph. Noting the surprise of the Japanese when they saw the presents operating, the Narrative of Perry's Expedition later stated that the presents were a "triumphant revelation, to a partially enlightened people, of the success of science and enterprise."

Japanese swordsmiths were able to copy Western firearms in the sixteenth century, but the situation in the nineteenth century was different. While Japan had been isolated from the world outside, the West had undergone a great transformation, science and technology having developed dramatically. Science was established as a speciality in its own right, and together with the technology connected with it, had clearly demonstrated its worth in such areas as industry, transportation, communication, and the military. In the face of Western civilization impelled outward by

182

the strength of its science and technology, the opening of Japan was inevitable. Now there was no other way but to establish a modern state based on such principles, and these the Japanese had to learn directly from the West, from concept to actual practice.

To meet this urgent need, over ninety students were sent abroad by the Shogunate or by some of the progressive local lords. Besides the Dutch language, English, German, and French were also studied. Naval and army training schools were organized, and an institution devoted to foreign matters was created, and a school of Western medicine was founded by the Shogunate.

The Beginnings of the Meiji Development

With the Meiji Restoration in 1867, traditional Chinese learning was replaced by Western learning, and the Shoheiko as the central institution for the study of Chinese classics was soon closed. Governmental policy was formulated to accord with the statement in the so-called Charter Oath that "Knowledge shall be sought throughout the world, so that the foundations of the Empire may be firmly established." Education was to play a particularly important role in the whole process of Meiji development. The government sent many students to study abroad, primarily to America, Britain, and Germany, and employed scores of foreigners, mostly American, British, German, and French, to help establish modern education and teach Western science and technology.

The pivotal institution for the introduction of Western science was the University of Tokyo, which was established in 1877 as a result of the merger of separate educational institutions originally belonging to the Shogunate. The number of Western teachers there was greatest in the late 1870's, but it decreased rapidly during the following decade, when most of them were replaced by Japanese trained abroad.

A modern educational system was formulated. Technical terms were soon translated into Japanese terminology, and Western learning came to be taught in the Japanese language. Most of the academic societies for basic and applied scientific studies were started between 1877 and 1890. The number of newly founded academic periodicals increased with years (Diagram I), and consequently the number of published articles also increased (Diagram II). In the early stage, however, the contributors were mostly Westerners, but by the late 1880's the majority of them were Japanese (Diagram III). Diagrams I, II and III may be an indication of how, once introduced, scientific activities were rapidly promoted and how the Japanese soon took over what had been initiated by the Westerners.

Mathematics

Although most of the fields of natural science in Meiji Japan were first developed under the leadership of Western teachers, mathematics differed somewhat in this respect. Traditional Japanese mathematics had already been quite highly developed and Western mathematics had also been taught by Dutch officers at the Nagasaki Naval Institute since 1855. During the 1870's two Japanese professors, Dairoku Kikuchi and Rikitaro Fujisawa, returnees from study abroad, rather than Western teachers, were most active in the field of mathematics. The Mathematical Society of Tokyo, established in 1877, was also the first among the academic societies inaugurated in Japan. It consisted of three groups of Japanese mathematicians, namely, scholars of Japanese mathematics, scholars of newly introduced Western mathematics, and naval and army mathematicians. One of Fujisawa's students, Teiji Takagi, after studying for three years in

Germany and receiving stimulus particularly from David Hilbert, gained world-wide recognition for the papers he published on the Arbelian problems in the early part of the twentieth century. Under this tradition, Japan was to produce two recipients of the Fields Medal after World War II.

Physical Sciences

Physics was introduced mainly by American and British professors and by the Japanese who studied in America, Britain, and later in Germany. Since natural calamities such as the earthquake and the typhoon were typical of nature in Japan, seismology and the science of the typhoon became specialities of physical sciences in Japan. John Milne, Thomas Gray, and James Alfred Ewing, these British professors and some others were the initiators of seismology in Japan. They started the investigation of earthquakes, made seismometers, created the Seismological Society of Japan, and published papers in the Transactions of this society. They were followed by Seikei Sekiya, Fusakichi Omori, and Akitsune Imamura. In 1886 at the University of Tokyo, Sekiya was made professor of seismology, the first in the world. As Thomas Corwin Mendenhall, former American physics professor at the University of Tokyo, correctly stated in 1900, "Japan has become within twenty years a vast seismological laboratory in which seismic phenomena are being used as never before.   Indeed, modern seismology had its birth there." However, theoretical aspects of seismology, such as the study of the transmission of seismic waves in relation to the internal structure of the earth, did not develop in Japan until after the 1920's.

As to the science of the typhoon, the theoretical treatment of Jiro Kitao won universal appraisal already in the late 1880's. Kitao had been in Germany for thirteen years and studied under Kirchhoff, Helmholtz, and others.

The study of magnetism was another feature of physical sciences in Japan. J. A. Ewing, the above-mentioned British professor, was the discoverer of "magnetic hysteresis." His study on magnetism already began while he was in Japan and this was succeeded by the Japanese, Hantaro Nagaoka and Kotaro Honda. Both Nagaoka and Honda graduated from the University of Tokyo and afterwards went to Europe to study for several years. Nagaoka became well-known for his proposal of the Saturnian atomic model of 1903. Honda was to be highly successful in his work in physical metallurgy, particularly in producing new kinds of permanent magnet steel.

In astronomy, the most noted early achievement was the discovery by Hisashi Kimura in 1902 of Z-term in relation to the variation of latitude.

In chemistry, we find such early contributions as the discoveries of ephedrine by Nagayoshi Nagai (1885), of adrenalin by Jokichi Takamine (1901), and of vitamin B by Umetaro Suzuki (1910).

Biological Sciences

The first botanists who won world fame at this early stage were Seiichiro Ikeno and Sakugoro Hirase who discovered in 1896 the spermatozoa of the cycad and of the gingko respectively. Equally outstanding was Kametaro Toyama's study on the heredity of silkworms (1906). In this study Toyama showed that Mendel's laws, which were rediscovered only six years before and proved to be applicable to plants, also held good with silkworms.

The opening of Japan coincided with the appearance of Darwin's theory of evolution. In Japan, Darwinism was immediately accepted and exerted a powerful influence on the Japanese mind. The outcome there, however, was quite different from that of the Western world. While the new idea concerning the origin of mankind was most disturbing in the West, this was not the case in Japan. Without any sign of hesitation, the Japanese people welcomed Darwinism as a most up-to-date theory of science, with the only exception of a few Japanese Christians and some of the Westerners in Japan. The theory of "survival of the fittest" was reduced to a simple principle and applied to all sorts of situations. The academic and the political, the conservative and the radical, all used this principle to support their own position or to attack the opponent. That there existed no religious objection to Darwinism in Japan did not imply that the intellectual soil there was ready for the further development of the evolutionary biology. On the contrary, both critical mind and basic studies in biological sciences were yet lacking.

Study of Japan by Western Science Teachers

While employed in Japan, Western science teachers initiated various sorts of scientific investigations. Many of the themes were peculiar to this region, related to, for instance, weather conditions, gravity, earthquakes, animals, plants, minerals, agricultural and marine products, shell-mounds, fossils, anthropological matters, etc. Their works included the chemistry of sake brewing by R. W. Atkinson, a British chemist, research on Japanese pottery, architecture, and ethnography by E. S. Morse, an American naturalist, and the study of Japanese history, politics, foreign relations, fairy tales, myths, religion, Christianity in Japan, and biographical accounts of people active in Japan and Asia by W. E. Griffis, an American science teacher who after his return became a minister. As some of the science teachers continued their study of seismology after they left Japan, so both Morse and Griffis continued their Japanological work and published many papers and books on Japan, among which best known are Griffis' Mikado's Empire (1876) and Morse's Japan Day by Day (1917) and Japanese Homes and Their Surroundings (1886 & 1961). These works contributed a great deal to the introduction of Japan and its culture to the Western world. Thus a few years' sojourn of Western science teachers in Japan also marked the beginning of an extensive cultural exchange between Japan and the West.

Science against a Japanese Background

It is not easy to explain why modern science did not grow in Japan. It is certain, however, that the matter has some relevance to world views. In the West, nature was considered to be a structured entity with logical coherence to it, and man was believed to be the especially privileged creature assuming a position to look down on nature and to regard it as an object of exploitation and of investigation. In Japan, on the contrary, nature was not considered as an object of man's mastery, but nature was his best companion. While the objective treatment of nature produced science and technology in the Western culture, the appreciative approach to nature left in Japan various forms of art and life, but not modern science.

Once Western science was introduced on a full scale, however, the Japanese people were quick enough to absorb it. Several factors may be considered to account for it. At this time in the West, the institutionalization of science was just taking place. A new system of scientific education was being started. Science was becoming a new enterprise and it was not too "big" nor too complicated as yet. A single scientist could cover rather wide fields of research and application and an exchange of comparatively few people could suffice to help transmit Western science into Japan. Japan was not too late to enter the international scientific community.

There had already been the tradition of Dutch learning in Japan which paved the way and enabled the Japanese to make rapid strides in assimilating Western science. Moreover, unlike other Asian countries, Japan was a unified nation, in race, in language, and politically, and the people were extremely receptive. Under the strong leadership of the centralized new government, the modernization of the country and the introduction of Western science and technology were promoted. A national budget was rather bountifully allocated to invite capable science teachers from the West, to send out promising students to study abroad, and to set up a well-equipped national university and other institutions. At the university great emphasis was placed on scientific education and up to about 1920 more than one half of the enrolled students each year at the University of Tokyo belonged to the departments of pure and applied sciences.

These must have been the factors that contributed to the rapid and efficient reception of Western science. Efficient as it was, however, the reception was rather uneven and sometimes superficial, and it was predominantly government-oriented.

Erwin Baelz, a German professor, out of his experience of teaching medicine at the University of Tokyo for twenty-five years——an exceptionally long period of stay as a Western teacher——pointed out in 1901 that the Japanese people were content with receiving the most recent development of science and did not attempt to learn the basic spirit underlying those results. His advice was that the Japanese should aim at becoming "cultivators of the trees of science" and should not be satisfied by being mere receivers of "the fruits of science."

The same sort of criticism had been made in 1878 by another Western teacher, William Wheeler, an American professor of mathematics and civil engineering at Sapporo Agricultural College. He very highly estimated the aptitude for learning of Japanese students on the one hand, but on the other noted their weakness in original thought. He attributed this (1) to the methods of Japanese scholarship which depended heavily on the Chinese classics and focused on memorization of the work of predecessors, and (2) to social constraints which kept the majority under the oppression of the powerful.

Since the modernization of the state was the pressing question, the promotion of national industry and military forces took priority. Scientific activities were commenced primarily at governmental institutions under centralized national authority. Unlike the situation in the Western countries, these activities were not necessarily initiated voluntarily by individuals interested in scientific inquiries, but were promoted as parts of national projects spurred on by immediate national needs. As was rightly pointed out in the above-cited criticisms, science was eagerly pursued but not as a part of Western culture, and creative thinking was not very much encouraged. The Japanese scientists mentioned above as those who attained to the first rank could only produce their results either during the period of study in the West or soon afterwards. Complete transmission of Western science was not an easy task.

The Early Twentieth Century

The situation improved gradually. After World War I particularly, the importance of chemistry and aircraft came to be realized and the weight of the voice of scientists increased. A bill passed the Diet for the first time to start a system of research grants (1918). There were five national universities by this time and the number of professorships in science at the University of Tokyo increased. New governmental and non-governmental research institutes were created. Some of the governmental laboratories devoted more or less to immediate ends were now made affiliated institutions of national universities where research was carried out in an academic atmosphere. The Metal Materials Research Institute, Tōhoku University, was founded in 1919 upon the initiative of Kōtaro Honda, a physicist previously mentioned, and under his leadership this Institute proved to be very productive in the new field of physical metallurgy.

The Institute of Physical and Chemical Research founded in 1917 was in fact the first non-governmental research center of a large scale for the basic study of physical sciences. It was to make important contributions to the development of science in Japan. To take an outstanding example, the Institute sent out Yoshio Nishina to study in Europe. He spent a year under Rutherford in Cambridge and six years at Bohr's Institute in Copenhagen. Having gained a reputation from his famous Klein-Nishina formula, he returned home in 1928 and stimulated young scientists by his introduction of the new quantum mechanics and of the "Copenhagen spirit." Under his direct influence, two Japanese Nobel Laureates in physics were to emerge: Hideki Yukawa and Shinichirō Tomonaga. In 1937 the Institute was equipped with a cyclotron.

The Institute of Physical and Chemical Research could not be operated, however, only on the interest received on its capital investment and it had to secure additional income by selling its patents and also by running subsidiary companies which engaged in manufacturing.

Postwar Japan

Japanese industry had to depend heavily on import not only of the resources but also science and technology, foreign patents and products. The weakness was felt seriously during World War II.

After the War, the defeated nation underwent a drastic change, the scientific aspect of which may be represented by the hapless fate of the glorious Institute of Physical and Chemical Research. The cyclotrons at this Institute, at Kyoto University, and at Osaka University were all dismantled and destroyed by the Occupation Forces. Nishina, the leader of atomic physics at the Institute, declared that his Institute would now deal with biological sciences and fertilizers. The Institute soon started research on the industrial production of penicillin. But, since the Institute had subsidiary companies to finance its research expences, it had to dissolve itself under the occupation policy of liquidation of the big financial combines. It was then reorganized on a far smaller scale into the Scientific Research Institute and Nishina was made its director.

The whole educational system was reformed and remodelled into somewhat of an American form. Many new universities and colleges were created. Graduate courses were expanded, new ones initiated, and now graduate schools rather than undergraduate became the main institutions for the training of young researchers.

As the country revived economically, departments of science and technology were newly set up in many of the non-governmental universities. At the national universities, the affiliated research institutes, both new and old, have come to play a more important part as centers of large-scale research projects in the age of "big science." The necessity of basic research for the industry has become more realized and big undustries have inaugurated their own research institutes.

Thus the number of scientific researchers have increased immensely as may be seen in Diagram IV. International exchange and cooperation also has become more extensive. Consequently, under ever-increasing stimuli Japanese scientists seem to be getting more and more productive both in quality and quantity.

Moreover, there are even signs nowadays that some of the Japanese and East Asian concepts of nature have been contributing to the furtherance of both physical and biological sciences, sometimes by providing them with new perspectives and sometimes by helping them to produce new types of theories, although in most cases the scientists themselves are unaware of these processes.

The Problem

In this respect, Japan appears to have been successful in its effort to catch up with the international scientific community. It seems to me, however, that there remains one important aspect left unattended, and because of this the transmission of science is still incomplete. The Japanese people have not quite learned yet to see science in relation to world views and value systems. They have not cared to regard this newly introduced discipline in relation to its cultural foundations in the West, nor have they tried to correlate it adequately to their own traditional culture. This aspect must now be most seriously taken up, and the entire development of science and technology requires a thorough reexamination.

BIBLIOGRAPHY

Bartholomew, James R., "Why Was There No Scientifice Revolution in Tokugawa Japan?" Japanese Studies in the History of Science, No. 15, 1976, pp.11-125.

Koizumi, Kenkichiro, "The Emergence of Japan's First Physicists: 1868-1900," Historical Studies in the Physical Sciences, Vo. 6, 1975, pp. 3-108.

Nakayama, Shigeru, et al. (ed), Science and Society in Modern Japan, Tokyo, 1974, xxiii+ 337pp.

Tuge, Hideomi, Historical Development of Science and Technology in Japan, Tokyo, 1961, xi+ 200pp.

Watanabe, Masao, "The conception of Nature in Japanese Culture," Science, Vol, 183, No. 4122, 1974, pp. 279-282.

Watanabe, Masao, "American Science Teachers in the Early Meiji Period," Japanese Studies in the History of Science, No. 15, 1976, pp. 127-144.

Yuasa, Mitsutomo, "The Growth of Scientific Communities in Japan," Japanese Studies in the History of Science, No. 9, 1970, pp. 137-158.

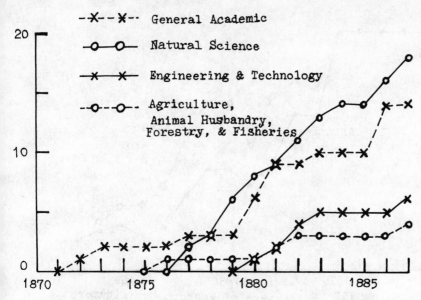

Diagram I Number of Academic Periodicals

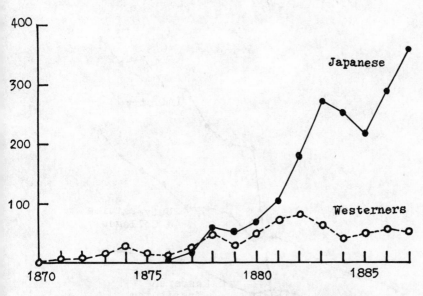

Diagram II Number of Published Articles
on Natural Sciences (excepting Mathematics)

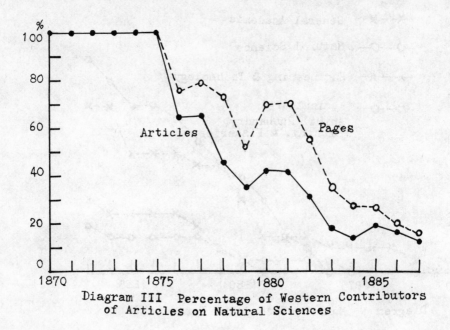

Diagram III Percentage of Western Contributors
of Articles on Natural Sciences

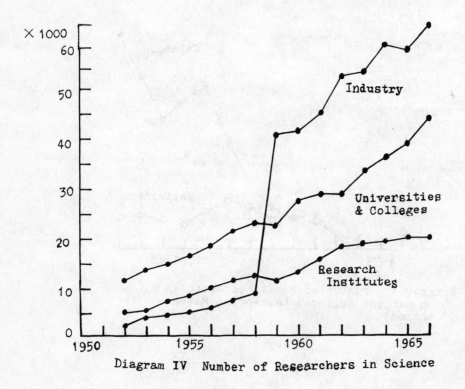

Diagram IV Number of Researchers in Science

Dr D.M. Knight

COMMENTS ON PAPERS BY REINGOLD, YAROSCHEVSKY AND WATANABE

It is curious to find Merz under attack from Dr Reingold, as a purely intellectual historian unaware of technology and social history.   As an engineer living in Newcastle, the greatest coal-port in England and a centre of chemical industry and shipbuilding Merz can hardly be accused of living in an ivory tower.   When an author feels it necessary to demolish a work published eighty years ago, one may suspect that he also has his eyes on an enemy nearer home; and surely another target of Dr Reingold's is the writings of historians of the American Mind, such as Perry Miller and Howard Mumford Jones.   For Reingold, America is an ordinary country, doing much the same as the leading European countries in the support of science, and clearly a part of Western Civilisation. This was not the way it struck various European observers, such as Dickens and Tocqueville, in the nineteenth century; and it is surely a doctrine that would have shocked many of the first generation of independent Americans at the beginning of the nineteenth century.   Science in the Europe of the <u>anciens régimes</u> was an affair of courts, academies, established churches, and patrons; and it was reasonable to hope that when these fetters were struck off, the manly and free heirs of all that was best in the English tradition would be enabled to carry on science in an altogether new and far more efficient manner.   The new republic could be expected to bring forth a generation of Boyles and Newtons.

Miller and Jones describe these hopes and their disappointment; the two American scientists of the revolutionary generation who achieved eminence and are extensively discussed by Merz are Rumford and Wells, who were both (like some of my ancestors) Tories and therefore left the country at independence and came to work in Europe.   If we take Kuhn's view of science as a guide, those in whom Merz was most interested were the revolutionary scientists who gave us a new paradigm - but this kind of 'revolutionary' seems to have been different from the American kind, for during the nineteenth century we do not find Americans providing a new paradigm as Franklin had in the eighteenth century.   The exception is Willard Gibbs; but he proves the rule, because he was a prophet without honour in his own country until his work was taken up by eminent Europeans.   America could provide a translator of Laplace, which was more than could be said of England at the time; the Michelson-Morley experiment was done there, but the paradigms within which it was interpreted were produced by Maxwell, Helmholtz, Fitzgerald, Lorentz and Einstein in Europe; and the evolutionary history of birds and horses was worked out in America, but following Darwin's paradigm.   If one looks, as intellectual historians like Merz and his successors have done, for really and fundamentally original thinking in science, then one will not find much to look at in nineteenth century America.

What Reingold here and elsewhere has shown us is surely that whether or not America was a normal country in the nineteenth century, its science was normal science - and this may have been partly because it was extensively supported by the government.   What is surely most interesting, and comes out in Reingold's edition of Henry's papers and in Sinclair's study of the Franklin Institute in Philadelphia, is that really original science is not a democratic affair; it may not have required a court, but it was not until there was something like an

academy, and until the German model of the University as a centre of
research as well as teaching had been transported across the Atlantic,
that America would begin to measure up to European nations not merely in
expenditure but also in originality. Before that time, Reingold's
recipe of concentrating upon the norm undoubtedly enables us to see the
institutions of American science far more clearly than Merz's kind of
approach does; and it does seem as though careers, utility, and education were more important in American science than was intellectual distinction.

While Reingold's paper is then about normal science and about money,
Professor Yaroshevsky is telling us about more revolutionary innovations
and the new schools of thought that propagate them. His examples from
psychology do perhaps fit the model of different nations with their
different paradigms; and something similar might be made from the
history of nineteenth-century chemistry. One should not forget that
there seem to be other cases where within one country there are competing paradigms, or views of how a science is related to other sciences,
and how it may be expected to advance. Thus in nineteenth-century
England, it seems that Lockyer was at odds with most of his contemporary
astronomers because they were Cambridge men trained in mathematics,
while his more chemical interests led him towards astrophysics. In
contemporary Britain, geophysics is in most universities taught within
physics rather than geology departments; its boundaries and relations
are thus rather differently seen in different institutions:- and we
must know that the history of science looks rather different when viewed
from a history, a philosophy, or a science department.

Be that as it may, even to such micro-schools within one small
country Yaroshevsky's analysis could no doubt be applied; and his
emphasis upon the individual person, with his particular world-view conditioned by his particular background, who founds a school is of great
interest. It seems that skill in building up a school, in securing an
institutional base and an outlet for publications, may be distinct from
the highest abilities in original research. Reingold has reminded us
that scientists have to make a career for themselves, and not just to
think or experiment. A Darwin may need a Huxley to act as his bulldog,
and to incorporate his work into the teaching of an institution so that
a school in Yaroshevsky's sense may arise. Some great and original
scientists have left no school; Davy would be an example from nineteenth-
century chemistry, of a man of more original genius than Liebig, but of
less influence because he did not found a school.

To see the founder of a school as rather like the founder of a family
is also interesting; though there are cases where it seems more appropriate to see him as the founder of an industry turning out graduates,
and generating demand for them. The problem is then what one makes of
this idea. When Galton investigated hereditary genius at the end of
the nineteenth century, he found that eminence ran in families; and to
him this indicated that ability was inherited. We are more cynical,
and assume that all that Galton's statistics show is that the establishment looks after its own - that anybody with well-placed uncles need not
worry about finding a good job. If Yaroshevsky's families really
represent the most talented in a generation being attracted to the most
original teachers, and then going out into the world to propagate his
new views and to develop them, then that is something that everybody
would applaud; but if it means that well-placed and well-known men can

pull wires and get their pupils the best jobs, then things seem perhaps rather less admirable.

Yaroshevsky brings out the dialectical progress of science as the various schools clash and somebody achieves a new synthesis from them. He remarks that this person whoever he is must have a particular cultural background; and for Sechenov he points to the literature of Russia in his day as reflecting interests similar to those that turned out to be important in his psychological work. It is good that one can find such parallels or links; but perhaps it is unsurprising that one can do it in psychology, while it might be harder with chemistry or astronomy which seem relatively arid and value-free disciplines, lending themselves more easily to the sort of internalist and intellectualist treatment that Reingold deplores.

The United States and Russia in the nineteenth century were clearly sharing in a common culture with western Europe, even if they were on the fringes of that culture; but there is no doubt that Japan was beyond the fringe. The European countries had much in common, even though they also differed among themselves; but Japan, with its different religious and political traditions, and its two hundred years of isolation from other countries is a very different case, and it is extraordinary how rapidly and effectively science was transplanted there. Professor Watanabe mentions the ready reception of Darwinism, and it might be that a people suspicious of Christianity and accustomed to the idea of reincarnations did not find repugnant the idea of animal ancestors as Western Christians did; but in his account it seems that the Japanese experience was one of absorbing - at just the right historical moment, when it could still rapidly be done - an alien tradition, under the direction of a government which no doubt wanted to build black ships as powerful as the Americans', and indeed soon did so.

Watanabe reminds us that the soil on which science had to grow was not completely virgin; that through the Dutch, there was already some knowledge of western science when the Americans arrived. One would suspect that Japan had also some advantages not shared by all European countries; a high level of literacy, and also of craft techniques - the Japanese were better at swordmaking, at ceramics, and at colour-printing, for example, than were the westerners. It seems possible that the gap between the old and the new has been made too much of, and that there was in Japan a synthesis of foreign and native traditions, as there was in European countries when new ideas came in - most dramatically when the Renaissance came to north-western Europe, but also all the time as different schools come to prevail in different countries.

When Japanese remark that science does not yet seem to be fully assimilated into their culture, it is possible that they over-estimate its assimilation in other countries. Some British schools (particularly girls' schools) teach very little science and many university students in Arts faculties are quite ignorant of science.

One way in which science was synthesised between cultures in previous centuries which Watanabe does not mention was in natural history; where illustrations were often done by non-Western artists. Kaempfer's History of Japan, compiled in the seventeenth century, was illustrated by Japanese; and Siebold's magnificent Fauna Japonica, published very shortly before Perry's black ships came to Japan, was illustrated partly by eminent westerners like Wolf, and partly by Japanese artists. This was not something unique to Japan. The British Museum (Natural History) has recently published a book of superb illustrations made in Canton for John Reeves, who sent them back to London - some of these are type drawings, being the illustration from which the species was first scientifically described.   In India, General Hardwicke at the beginning of the nineteenth century collected drawings by Indians of animals, later published by Gray of the British Museum in a really splendid book; while Roxburgh at the Calcutta botanical gardens got local artists to draw plants for him, which are now at last being published by the Indian government.   Some of the plates for Jerdon's <u>Illustrations of Indian Ornithology</u> were not only drawn by Indians but lithographed in Madras, where the book was published in 1847.   In South America, Mutis got local Indians to draw plants for him at the end of the eighteenth century - these again are now slowly being published after a wait of 150 years.

What one sees here is surely some kind of international and cross-cultural collaboration, for the illustrations - even when engraved in Europe - do not look quite European because they come from a different aesthetic tradition.   On the other hand, the leader in the enterprise was European; and one is not seeing an equal partnership as one would hope to do today, but something more like the old Indian Army with its European officers.   If one is to get collaborators rather than assistants in the scientific enterprise, no doubt one must get them young; and it is striking that all three of those upon whose papers I have had the honour to comment emphasise the importance of educational institutions.   What perhaps makes this even more striking is that such an emphasis seems so inappropriate if one looks at England before the middle of the nineteenth century; when science was perhaps more assimilated into general culture than it was a hundred years later, without being formally taught.

Symposium 4.   MEDICINE AND INDUSTRIALIZATION IN HISTORY

SYMPOSIUM 4: MEDICINE AND INDUSTRIALIZATION IN HISTORY

Introduction

W.F. Bynum

Symposium 4 became a memorial to one of its invited speakers, Professor George Rosen, who died on his way to the Congress. The morning sessions were devoted to aspects of the late 18th and early 19th century British experience of industrialization. Rosen's paper in the afternoon session widened the geographical and temporal boundaries of the Symposium's theme. In this brief introduction I attempt to summarize the day's discussions.

In the opening paper Dr E.A. Wrigley evaluated English mortality between 1780 and 1870, a period during which English population is known to have increased sharply.(1) Assessment of mortality trends after the introduction in 1837 of civil registration of births and deaths is relatively easy. Before that date one must rely on various contemporary estimates, local parish records, post-1801 census data, or on the backward projection of information collected in the Registrar General's Office under William Farr from 1837. Dr Wrigley chose to use primarily the technique of backward projection (cf. Wrigley's footnote 9), such projection being done on age-related English mortality which William Farr calculated in various English life tables, particularly his third table, based on mortality from 1838-54. Working with these and other data, Wrigley concluded that English mortality was virtually constant from the 1830's to 1870. He then used the relatively accurate data for this later period as the basis of a backward projection (based for instance on the size of the age group 40-49 in 1841 compared with that group, then aged 20-29, in the 1821 census), to estimate mortality trends back to 1781. His sophisticated paper concluded that English mortality decreased slightly during the last two decades of the 18th century (from a crude death rate of 28.9 in 1781 to one of 24.1 in 1806), but that it then remained virtually constant until the 1870's, when it began to fall again. However, as Wrigley noted in his concluding paragraph, the combination of urbanization and the excessive mortality in cities when compared with country towns and rural areas means that 'During the long period 1806-70 when expectation of life nationally was improving only marginally it is quite likely that mortality was falling in most regional or socio-economic subdivisions of the country.'

Both commentators on Wrigley's paper emphasized that, given the validity of the various assumptions which he had introduced, his conclusions readily followed. Professor Peter Mathias noted that Wrigley's discounting out- and in-migration during the period, even if necessary due to a lack of accurate figures, rendered Wrigley's calculations less certain. Mathias particularly stressed the importance of the in-migration of people from Wales, Scotland, and Ireland. These movements clearly affected both the age distribution and the size of the population in England. He further noted that much uncertainty surrounds the accuracy of the 1801 and 1821 censuses which Wrigley used. A shortfall of age returns by almost 13% in 1821 compounds the uncertainties, since it is not necessarily the case that the age distribution of unreported districts and individuals was identical to those reported. The tendency – noted by Wrigley – of young ladies to report their ages in the low 20's imposed additional difficulties, as did the habit of the elderly to exaggerate their ages. These and other difficulties, such as the possibility of large short-run changes in mortality rates, make this exercise in historical demography an exciting, challenging, but treacherous enterprise. As Mathias remarked, causal hypotheses explaining the modern rise in population abound, but there is not even general agreement about the relative importance of fertility, mortality (and migration) in the phenomenon. The data can be moulded to fit a variety of mutually incompatible hypotheses. (2)

Dr Peter Razzell elaborated some of Mathias' remarks. He suggested that mortality and fertility should no longer be seen as separate and distinct causal variables. A number of diseases, including mumps, syphilis, gonorrhea, chickenpox, tuberculosis of the genital tract, enteric fever, and prolonged pyrexias may contribute to male or female infertility. More recently, Phadke and his colleagues working in Bombay have implicated smallpox as the 'most important single etiological factor in India which produces obstructive azoospermia in man'. (3) Diet can also have a significant impact on fertility, as the recent studies of Lee and Devore among the King Bushmen in Africa demonstrate. (4) Thus the relationship between medicine, industrialization and population is undoubtedly much more subtle than many scholars have assumed, since disease and nutritional states can affect both mortality and fertility.

Like Mathias, Razzell questioned the validity of Wrigley's assumption that his data were sufficiently accurate to permit the backward-moving computer programme to calculate English mortality for five year intervals between 1781 and 1821. As Razzell remarked, 'With such large aggregate uncertainties, it would appear that the only sensible strategy is to abandon speculative statistical constructions of national mortality and fertility rates out of census data, and return to much more careful detailed work on local sources'. The accuracy and completeness of local parish records can be cross-matched with census and various civil registration records for the first three decades of the 19th century. Razzell's own work with forty-five parishes from various parts of the country yields the following crude birth and death rates (to be compared with Table 4 in Wrigley's paper):

	Birth Rate	Death Rate
1801 – 10	41.4	30.5
1811 – 20	42.0	28.9
1821 – 30	40.1	25.9
1831 – 40	35.9	22.8

(N.B. Razzell's figures refer to both males and females in England and Wales, whereas Wrigley's refer only to females in England).

These results, arrived at by different methods, show essential agreement on birth-rate, but Razzell's figures suggest a more impressive fall in mortality between 1801 and 1840 than do Wrigley's. The significance of this discrepancy is not entirely clear, given the quality of the data. But as Razzell pointed out, 18th century data are even more ambiguous, and cannot of course be independently estimated by his technique of cross-matching census and parish records.

The potential medical significance of these issues is obvious, since improved medical care has traditionally been imputed in the 18th and 19th century European rise in population. So, too, has industrialization with its resultant increased availability of many manufactured products, such as washable cotton garments. (5) But one conclusion which emerged from the Symposium's first session was that no easy and simple connexions can be drawn between medicine, industrialization, and population increase. Following the work of McKeown and his colleagues, the distinction must be drawn between medical knowledge (which clearly did increase during this period) and therapeutic or prophylatic effectiveness (which, except for smallpox inoculation, probably did not increase significantly). (6) Besides, as both Wrigley and Mathias noted, the beginning of the modern English rise in population can be dated from the middle of the 18th century and hence precedes widespread industrialization.

Nevertheless, it is generally assumed that there was some coherent medical response to the health problems exacerbated by the Industrial Revolution. The Symposium's second session was concerned with this issue. Dr Charles Webster's paper examined 'the crisis of the hospitals during the Industrial Revolution'. He challenged the idea that industrial areas spawned hospitals, noting that the provincial voluntary hospital movement had lost its momentum by the time that industrialization and urbanization were in full swing. Further, as he writes, 'most of the hospitals were founded in small country towns, few in centres of rapid urban growth.' As both Webster and Dr John Pickstone, his commentator, pointed out, Infirmaries and Dispensaries, in industrial as well as non-industrial areas, were primarily demonstrations of philanthropy, not responses to new public health problems. The admissions policy of Infirmaries, their architecture, and their common preoccupation with remunerative lunatics and expensive baths, rather than with fever or accident victims, revealed their predominant orientation towards the interests of the charitable. Even where Infirmaries found themselves surrounded by rapidly growing industrial populations, reformers such as Percival and Ferriar in Manchester and Clark in Newcastle found it difficult or impossible to bend their local institutions into effective instruments for the improvement of public health.

Pickstone's comments, based largely on examples from the Lancashire and Cheshire textile districts, extended the argument from Infirmaries to Dispensaries. Though the central Manchester Dispensary was an extension of the Infirmary brought about by 'enlightened' reform, most of the Dispensaries in the region were traditional institutions; the pattern of Dispensary growth in the North West corresponded far more closely to the pattern of pre-industrial towns than to the distribution of textile workers unable to afford doctors.

Dispensaries appeared earlier in the Cumbrian market towns than in the rather larger textile towns around Manchester and among the latter Dispensaries were first founded in corporate boroughs (Wigan 1798, Preston 1809) or non-corporate towns with substantial gentry interest (Stockport 1792). In such towns Dispensaries reinforced traditional elements, their leadership strongly overlapped with that of the local volunteer troops, and they could draw on the remnants of that "non-denominational" charity which was also evidenced in Sunday Schools. The continuing rapid growth of textile towns elicited little medical charity, because the same processes as produced factory accidents and typhus also undermined the appeal of traditional charities. As social tension increased further, new Dispensary foundations became increasingly obviously factional, until the last Dispensaries to appear in the cotton towns (in Bury 1829 and Rochdale 1831) were little more than adjuncts to the beleaguered parish churches: feeble attempts to project the values of a hierarchical society onto increasingly hostile or indifferent urban populations. In the newer and more radical towns of Oldham and Ashton, there were no medical charities at all until after 1850. Only in the small, older towns and in the very large cities of the North West did Dispensaries reach more than about 5% of the population (annually).

Some Dispensaries were extended to provide fever wards; but this can hardly be called a response to rapid urbanization since in the larger industrial towns outside Manchester there was far less provision than in small towns such as Kendal and Lancaster which had hardly grown at all. Where Dispensaries did aim directly at industrial problems by trying to provide accommodation for accident victims, there was little response from industrialists; accident wards were very few and under-financed. Some Dispensaries, e.g. Rochdale and Bury, excluded all cases of factory accidents from their lists until about 1830; Preston and Wigan Dispensaries barred surgeons from their honorary staffs.

Against this background it is not the failures of medical reformers in industrial areas which need particular explanation but their occasional successes, especially their success in late eight-eenth century Manchester. As Pickstone reminded us, the explanations will have to include both medical politics and urban politics; the concentration of money and professionals in the regional capital, the need of physicians for an arena, and the desire of rising industrialists to control a large democratic charity previously run by Anglican-Tory-Surgeons.

The Symposium's second session thus underscored the failure of medical charity to come to grips with the far-reaching consequences of industrialization. As was pointed out in discussion, however, the negative connotation of the word 'failure' implies that doctors could have done something to alleviate the health problems of the worker, a theme which is also related to the demographic debates raised by the first session. It may well be, however, that demographic and social history part company at this point, for the 'failure' documented by Webster and Pickstone is a social one. As social historians, they must take seriously the 18th century assumption that medicine was a worthwhile activity.

Neither of the first two papers nor discussion following went very far in assessing the applicability of these circumscribed studies - the one on English demography, the other on Manchester and Newcastle Infirmaries - to other patterns of medical response to industrialization. Professor Rosen's paper in the third session, imbued with Rosen's spirit of meliorism, did provide some positive evidence for the American medical response, though even here it should be noted that positive measures did not immediately follow industrialization. Further, as one of Rosen's commentators, Professor Barbara Rosenkrantz, noted, the industrial group which Rosen selected - the garment workers - w.as in many ways atypical.

Following George Rosen's tragic death, his colleague at Yale, Professor A.H. Aaboe, read Rosen's study on 'Urbanization, occupation and disease in the United States; 1870-1920: The Case of New York City'. Rosen had of course noticed the lag between intensive industrialization and effective provision for job-related disorders and injuries. His study focused on the 'needles trade', by far the most important industry in New York City during this period (cf. the Table in Rosen's paper). His paper graphically described living and working conditions in the needles trade, with its high proportion of home employment, emigrant (especially Jewish and Italian) workforce, and sweating. He then traced the growing political consciousness of these workers, and the obvious progress in unionization and in occupational medicine between about 1900 and 1920.

Professor Michael Flinn, Rosen's first commentator, paid tribute to Rosen's immense contribution to our historical knowledge of occupational disease. (Rosen's A History of Miners' Diseases, published in 1943, is still one of the few studies of its kind). Flinn pointed out that industrial pathology and industrial medical services were not welded together on a large scale until late in the 19th century. However, he noted that in Europe a number of pioneering studies and industrial health programmes had been effected during the 18th and early 19th centuries. Among the studies were those of Charles Thackrah on occupational morality in Leeds, (1831); Thomas Bateman on disease in London (1819); and James Mitchell on the actuarial experience of lead miners (1830's). Various early employers also provided health and welfare schemes for their employees, often using salaried surgeons. These enlightened employers included Ambrose Crowley in his iron works; the Hope Family in lead mines which they owned; and the London Lead Company. While all these individuals and industrial concerns were exceptions, the whole area deserves much further study, though it is hampered by the paucity of industrial records from the 18th and early 19th century.

Professor Barbara Rosenkrantz also opened her remarks by recalling how much George Rosen had contributed to the Symposium's theme. She then developed some points arising from Rosen's paper, particularly regarding the general notion of a 'dangerous trade', and the movement for compensation for injuries suffered in the work place. This movement culminated with the passage, in 1912, of federal legislation regarding workmen's compensation. By referring to specific examples of other occupations (street railway employees and metal workers) where activities like those affecting the needles trade might have been expected, she suggested that Rosen's study of the needles trade was a special case of workers with relatively high literacy and political consciousness. Further, the needles trade was labour intensive, highly competitive, and acutely responsive to public pressure when revelations of the evils of 'sweating' and home employment were exposed. Despite the efforts of various social reformers to introduce medical inspections of the work place, and compensatory insurance for industrial injury, both unions and employers generally favoured voluntary agreements rather than federally inforced mandatory legislation. Unions feared the inherent paternalism and the threat to job opportunities. Compensation raised the issue of worker carelessness and gave employers a lever with which to discipline and cow their workers. Medical inspection of workers' health created similar difficulties, since a worker found to be ill might be denied the right to work, and in Gilman Thompson's clinic at Cornell in the 1910's, only a small minority of workers examined were found free from all defects. Within this context, Rosenkrantz pointed out, the establishment in the needles trade of a programme for identifying illness in the workforce and providing health care for workers was a rather singular event. Situations in other occupations differed, as did public, political, union, and employer response.

Rosen's contribution thus provides a superb case study. That we shall have no more from him is an incalculable loss.

FOOTNOTES

1. Convenient introductions to these demographic issues may be found in E.A. Wrigley, ed., An Introduction to English Historical Demography (1966); M. Drake, ed., Population in Industrialization (1969); and E.A. Wrigley, Population and History (1973).

2. The most recent attempt at synthesis is Thomas McKeown, The Modern Rise in Population (1976).

3. A.M. Phadke, N.R. Samant, and S.D. Dewal, 'Smallpox as an etiologic factor in male infertility', Fertility and Sterility, vol. 24, (1973).

4. R.B. Lee and I. De Vore, eds., Kalahari Hunter-Gatherers (1976).

5. See, for instance, A.J. Taylor, ed., The Standard of Living in Britain in the Industrial Revolution (1975).

6. In addition to McKeown (n.2), see Peter Razzell's The Conquest of Smallpox (to be published in November 1977).

Dr E. A. Wrigley

ENGLISH MORTALITY IN THE INDUSTRIAL REVOLUTION PERIOD

Views about mortality levels and trends in England during the industrial revolution have been strikingly diverse. No-one doubts that population rose very sharply between 1780 and 1870 which I shall take to be the industrial revolution period for the purpose of this paper. The population of England and Wales reached 23.7 millions in 1871, almost exactly three times the total of 1780, a much faster rate of growth than in any earlier period of English history of comparable length. Since there was also substantial net outmigration in the latter part of the industrial revolution period, which moderated the growth which would otherwise have occurred, it is evident that, compared with the period before 1780, either fertility must have risen or mortality have fallen substantially (or that both changes may have occurred in lesser degree).

The view that falling mortality accounted for the acceleration in the rate of English population growth either entirely or predominantly has a long pedigree.[1] It has many contemporary advocates, notably Hollingsworth, McKeown and Razzell.[2] Indeed much recent discussion has taken it for granted that mortality fell substantially in the first half of the nineteenth century and has been chiefly concerned with the relative plausibility of the various hypotheses advanced to explain the fall. The view that all or most of the increased impetus of population growth to be attributed to a rise in fertility has had fewer advocates though it includes some respected names.[3] In this paper I shall present evidence which demonstrates that there was very little fall in mortality between 1806 and 1871, though in the preceding twenty years the fall was more substantial. Since the rate of population grew considerably in the early decades of the industrial revolution, such a conclusion implies, of course, that there was a rise in fertility in this period, but I shall touch on this related issue only cursorily.

It is convenient to work backwards from the end of the period to the beginning, dealing first with the period after the inception of civil registration in 1837 when the data are good, before turning to the more contentious, earlier period.

During the whole of the period from 1837 to 1871 the dominant figure within the Registrar-General's office in marshalling evidence about mortality was William Farr.[4] His third English life table, based on the mortality experience of England and Wales during the period 1838-54, provides the first authoritative, detailed information about mortality in nineteenth century England (the first and second life tables, based on information for 1841 and 1838-44 respectively, do not differ significantly from the third).[5] The third life table gives an expectation of life at birth (sexes combined) of 40.87 years.[6] Farr was aware of the very substantial underrecording of births in the early years of civil registration and made allowance both for this and for the marked misreporting of age in the young adult female age groups (especially 20-4) in nineteenth century censuses. He appears to have made no attempt to correct the marked tendency to overstate age among the elderly at census time, which caused all age groups above 70 to be inflated above their true size. The inflation grows with age and above 90 the recorded totals are almost twice the true figure.[7] Since, however, the inaccuracies are confined to the later age groups which contribute relatively little to the cumulative total of years lived in a life table (T_{80} where expectation of life at birth is c.40 years is only about one hundredth of T_0), the third life table figure for expectation of life may be taken as highly accurate.

The next fully tabulated English life table to be published covers the decade 1871–80 and yields a figure for expectation of life at birth of 42.95 years. This represents an increase of just over two years compared with the third life table and suggests that between 1854 and 1870 there was an appreciable, if modest, improvement in mortality. The improvement appears, however, to have been confined almost entirely to the 1870s themselves, since William Farr published information in the Supplement to the 35th Annual Report (8) which can easily be converted into a life table covering the years 1855–71. As part of his work for this decennial report covering the years 1861–70 he had done the pre-liminary work necessary for a new full set of life table calculations but concluded that they would differ so little from those of the third life table that it would be otiose to complete the exercise.(9) He did, however, publish a summary mortality table to show why he had not thought it necessary to proceed further. In the table he gave age-specific mortality rates for five year age groups to age 15 and thereafter for ten year age groups both for the seventeen years of the third life table (1838–54) and for this period combined with a further seventeen year period (1838–71). It is therefore possible with only a very small margin of possible error to calculate a life table for 1855–71.(10) This in turn yields an estimate of expectation of life at birth of 41.33 years, a figure less than half a year larger than that of the third life table. Mortality over the whole period from the start of civil registration to 1870 was therefore virtually unchanging at a level which gives an expectation of life at birth very close to 41 years.

Given the absence of change in the period 1838–70 it is perhaps not surprising that it should prove to be the case that there was also very little change in mortality levels in the next two decades moving backwards in time. The period 1821–41 is a convenient time period to consider next since there is age data in the censuses at the beginning and end of the period The censuses of 1821 and 1841 were indeed the first two English censuses to contain such information.

The 1821 census divides the population into five year age groups to age 20 and into ten year age groups at greater ages. The 1841 census has five year divisions throughout. It is therefore always possible to 'match' age groups in the two censuses; 0–4 in 1821 with 20–4 in 1841, or 20–9 in 1821 with 40–9 in 1841 and so on. Because of this, it is also possible to test whether it is true that the mortality experience reflected in the third life table also describes accurately the mortality of the 1821–41 period. For example the third life table suggests that the ratio between the population 20–9 in 1821 to the population 40–9 in 1841 should be 1.2608: 1.0000 (that is $_{10}L_{20}$ stands to $_{10}L_{40}$ as 1.2608:1.000 in the life table). If at this and other ages the recorded population in 1821 agrees well with the 'predicted' total derived from the 1841 matching age group inflated by the appropriate ratio, then it may be assumed that mortality had not changed during the intervening period. Table 1 shows the results of such an exercise.

In considering table 1 several preliminary points must be borne in mind. First the population totals refer to England only, and not to England and Wales. The material used in this paper is drawn from a larger study covering English population history throughout the parish register

period (1538–1837), and since the quality and coverage of Welsh
parochial registration is less good then the English, England alone was
studied. For the same reason England, here and throughout, excludes
Monmouth unless otherwise indicated. Second, age information was
incomplete in both censuses. In 1841 the shortfall was trivial: 44870

Table 1

	A 1841 pop. (M+F)	B Inflation ratio		C 'Predicted' 1821 pop.	D 1821 census population	E C/Dx100
20–4	1451968	1.2654	0–4	1837320	1668330	110.13
25–9	1201010	1.1628	5–9	1396534	1459809	95.67
30–4	1093697	1.1788	10–4	1289250	1243430	103.69
35–9	828495	1.2122	15–9	1004302	1109385	90.53
40–9	1431626	1.2608	20–9	1804994	1767171	102.14
50–9	998041	1.3632	30–9	1360529	1324283	102.74
60–9	664999	1.6576	40–9	1102302	1047886	105.19
70–9	288151	2.7022	50–9	778642	737739	105.54
80–9	59511	7.9285	60–9	471833	506430	93.17

persons were returned as age unknown, or only 0.3 per cent of the
English total. In 1821, on the other hand, 1,430,596 ages were unknown,
or 12.8 per cent of the population. If there were reason to think that
the omissions were heavily concentrated in particular age groups, the
1821 age data would be of little value. Omissions, however, took the
form of absence of age data about whole communities rather than groups
within them, so that it seems justifiable to assume that the age
structure of those omitted was not greatly different from those
included.(11) Accordingly, I have increased the size of all 1821 age
groups on the assumption that the age distribution of the whole population
was the same as that of the population for whom ages were given. The 1841
data were treated similarly. Third, the totals in the 1841 population in
the four age groups above 50 have been adjusted to offset the misreporting of ages(12) (the uncorrected totals for the four age groups
50–9 to 80–9 are 955695, 648561, 318568 and 85147, compared with 998041,
664999, 288151 and 59511 in table 1). Finally, above the age of 55 the
life table death rates of the 1891–1900 English life table have been
substituted for those of the third life table. It is demonstrable that
the rates in the third life table are too low because age reporting
tended to inflate numbers in the higher age groups. Death rates late in
life changed little during the nineteenth century and age reporting had
greatly improved before its end.
   There are too many uncertainties in the reporting of ages in 1821 and
1841 for too much weight to be placed on the individual ratios
representing the relationship between 'matched' age groups. For example,
the age reporting in 1841 showed a marked tendency for totals to be
disproportionately large in the first quinquennium of each decade of
life. With a suitable adjustment to offset this influence(13), the
first four ratios of column E become 103.29, 99.74, 97.29 and 97.90, a
much more even result than the 'untreated' set. Age misreporting in
1821, however, is less easy to detect or correct but probably explains
some part of the remaining discrepancies. It is particularly notable
that the cumulative totals of the 'predicted' and recorded population

aged 5-69 in 1821 are 9208386 and 9196133 respectively (or 100.13:100.00). This is a notably close agreement. The much larger discrepancy in the 'predicted' and recorded population 0-4 may confidently be set down to the very marked 'heaping' of female ages 20-4 which is visible in all nineteenth century censuses. For males only the 'predicted' and recorded totals 0-4 are 837829 and 838752 respectively (99.89:100.00).

Table 1 would therefore represent strong reason to believe that mortality between 1821 and 1841 was approximately at the level of the third English life table if it were safe to assume that the English population in this period was effectively 'closed', or alternatively that net migration was very slight whatever might be true of gross migration totals. The first of these two alternatives clearly did not hold true of early nineteenth century England. The second may be quite close to the truth, bearing in mind the movements of Welsh, Scots and Irish into England offsetting in part at least the emigration from England to North America and Australia. If, as would probably be the safest assumption, there was net outmigration from England but on a modest scale it might be expected to cause the 'predicted' totals to fall short of the recorded since the 1841 population would have been depleted by emigration which would cause the 'predicted' total to be less than it would otherwise have been. The age groups under 30 in 1821, which would be those most affected by emigration over the next twenty years, are those in which the 'predicted' totals tend to be smaller than the 1821 census totals, though by modest percentages, while for those over 30, amongst whom there would have been relatively little migration, the opposite is found. While therefore the net effect of migration was probably too slight to affect the 'matching' operation significantly, the pattern in the individual matched age groups suggests that the third life table is more likely to have overstated rather than to have under-stated the mortality level of the years 1821-41.

In order to assess the significance of the totals in table 1 it is important to consider how different the result would have been if different assumptions had been made about mortality between 1821 and 1841, and therefore a different set of inflation ratios had been used. If, for example, the absolute L_xs in another life table had been quite different but then relative sizes had been similar to those of the third life table (that is that the ratios of, say, $_{10}L_{20}/_{10}L_{40}$ and other matched age groups were almost the same), the apparent implication of table 1 might be misleading. It would not constitute a reason for supposing that mortality changed little over the period. Such ratios do, however, differ significantly according to the level of expectation of life at birth which they reflect. For example, if expectation of life at birth over the period 1821-41 had been 38.39(14) rather than 40.87 years (the third life table figure modified to reflect the changed q_xs above age 55 described above), the 'predicted' population 5-69 would have been 9510548, a markedly worse agreement with the recorded total than that found when using the third life table (the ratio of 9510548 to the recorded total 5-69 is as 103.42 to 100.00 which may be compared with the figure of 100.13 quoted above when using the third life table mortality rates). Yet some of those who have written about mortality during this period have suggested that mortality was falling so rapidly in the early nineteenth century that an expectation of life even lower than 38 years would have been an appropriate choice for a backward projection of the type described.

If the evidence suggests strongly little change in mortality during

1821-41, what of the preceding twenty years ? Once again the census
returns provide more support for the view that mortality was changing
little than for any other conclusion. Since there are no age data in
censuses before 1821, no matching of linked age groups in successive
censuses is possible. It is, however, possible to estimate the size of
the population 0-79 in 1811 and 0-69 in 1801 by backward projection from
the population aged 10-89 and 20-89 respectively in 1821. In order to
compare a 'predicted' with a recorded population in 1811 and 1801, the
'predicted' populations 0-79 and 0-69 have been inflated in table 2 by
the ratios in which the total population in 1821 stood to the population
0-79 and 0-69 in 1821. This introduces a further element of uncertainty
into the exercise but it gives an intelligible point of departure for
discussion. The exercise has been carried out on the female population
only since the male totals in both 1801 and 1811 were so seriously
affected by the absence of men in the armed services. The full working

Table 2 Female population

A	B	C	D	E
1821 pop. 10-89	1811 pop. 0-79 predicted	B inflated to give 1811 total pop.	Census pop. 1811	C/Dx100
4182144	4933596	4968673	4931924	100.75
1821 pop. 20-89	1801 pop. 0-69 predicted	B inflated to give 1801 total pop.	Census pop. 1801	
3004251	4199029	4328215	4320090	100.19

is not shown in table 2 since the totals are built up from individual
ten-year age categories in 1821 exactly as in table 1 but the absence of
age data in the earlier censuses makes it pointless to reproduce the
operation in detail.

It will be seen that the agreement between the 'predicted' and the
census populations is remarkably close. Is it suspiciously close ?
There are two strong grounds for suspicion. The first lies in the
distortion likely to arise from the overstatement of the aged in 1821.
Since the highest age-group (80-9) is greatly inflated in backward
projection, any serious overstatement will increase the 'predicted'
population at earlier dates. If the individuals concerned had been
placed in their true age group they would have been subject to a lower
inflation factor. It can be shown that the numbers in the highest age
groups were overstated in 1821, just as in 1841, but the effect of the
overstatement can be estimated by excluding the 80-9 age group from the
back projection, basing the operation therefore on the age groups 10-79
and 20-79 for 1811 and 1801 respectively, and inflating the resultant
totals by the ratios in which total population in 1821 stood to population
0-69 and 0-59. When this is done the figures in the final column of
table 2 fall to 100.16 for 1811 and 98.54 for 1801. This test suggests,
therefore that the third life table mortality remains appropriate for
the decade 1811-20 but there is less certainty about the decade 1801-10
when the figure of 98.54 suggests that mortality was at a higher level

since the backward projection based on the third life table fails to produce a sufficiently large inflation in numbers. Such a conclusion would be premature, however, since the age structure of the population in 1821 is unlikely to have been the same as in 1801 because population was growing exceptionally quickly in the first two decades of the century and it is therefore to be expected that the ratio of the total population to the population 0-59 would be lower in 1821 than in 1801. The use of model life tables to explore the issue suggests that the assumption of a continuance of mortality at the level of the third life table is not incompatible with the difference between the 'predicted' and enumerated population in 1801. In any case any change in mortality would be minor on this evidence.

The second ground for suspicion in relation to the figures of table 2 lies in the possibility that enumeration became increasingly complete with the passage of time. Krause, for example, suggested that the fraction of the population missed in the 1801 census was 5 per cent of the total falling to 3 per cent in 1811 and 1 per cent in 1821.(15) Clearly to the extent to which these assumptions are justified the inference that mortality changed little in the first twenty years of the century would be undermined (and, indeed, a comparable argument might also undermine the conclusion of unchanging mortality 1821-41). If Krause was right alternative calculations of mortality levels might be made, but I have not done this because I doubt whether his assumption is correct. To settle the matter would require a substantial research effort, but seems feasible. Preliminary study shows a substantial number of instances in which the published totals in the 1801 census are clearly far too high, and they appear to occur more frequently than contrary cases.(16)

We have now seen that the first seventy years of the nineteenth century saw little change in mortality levels. What of the earliest part of the industrial revolution period, the last two decades of the eighteenth century ?

There were no censuses before 1801 so that the problem must be approached differently. Space prohibits a full description of the techniques used, but a brief summary will make clear the sequence of operations involved in providing estimates for 1781-1801. If the argument and evidence presented so far can be regarded as conclusive in establishing the applicability of the third life table to the whole period 1801-41, it follows that the number of births occurring in each decade can be calculated without difficulty (if, for example, $_{10}L_{10}$ were 684427, and the census population 10-19 in 1841 were 3135516, then the births 1821-30 might be taken to be 4581227 (3135516 $\div$ 0.684427). Since the total increase of population between censuses is known, it is also simple to calculate total deaths over intercensal periods. Both the estimation of births and deaths is dependent on the unimportance of net migration over the period in relation to the other totals involved. To minimise any disturbance from this and, in the earlier decades, to avoid the difficulties occasioned by the large numbers of men abroad in the army or navy, all calculations were based on the female population only, with subsequent inflations of the numbers to represent total births and deaths rather than female events only.

The operations described in the last paragraph by providing totals of births and deaths, also enable the shortfall between events occurring and events registered to be calculated. It has proved possible to make independent estimates of the scale of the contribution to the total shortfall made by several well-known sources of underregistration - for

example, prevalence of Nonconformity, and the effect of the delay
between birth and baptism (which in the case of an early death before
baptism affects the completeness of both birth and death coverage in
Anglican registers). Since the size of the total shortfall is known,
and the scale of some of its major components is also known, the extent
of the residual sources of underregistration can be calculated.

Table 3, for example, shows the size of some of the elements in the
shortfall between deaths and burials over the period 1781–1830. Before
1801 the estimation of the total of deaths (and hence of the size of
residual underregistration) is less straightforward than for the period
in which regular censuses were taken. It is beyond the scope of this
article to describe the method used but it may be noted that the
proportion of deaths escaping registration for reasons other than the
delay between birth and baptism (the burial of such children was seldom
recorded)(17) had fallen to very low levels by the 1780s. If the scale
of most sources of underregistration can be estimated, and the relationship between these in turn and any residual underregistration is also
capable of being estimated, it is possible to derive totals of births
and deaths throughout the parish register period as well as after 1801.(18)

Table 3

	A	B	C		D
1781–90	89.9	1.1	9.0	(6.0)	100.0
1791–1800	86.6	1.3	12.1	(6.7)	100.0
1801–10	79.9	1.4	18.7	(7.4)	100.0
1811–20	75.4	1.5	23.1	(8.1)	100.0
1821–30	82.4	1.8	15.8	(9.1)	100.0

A: registered Anglican burials
B: registered Nonconformist burials
C: deaths escaping registration (in brackets the totals
   whose burials went unregistered because death occurred
   before baptism)
D: 'true' deaths

Recent advances in the use of aggregative totals of births and deaths
allows a fuller use to be made of these data than had at one time
seemed possible. A backward projection program has been written which,
moving backwards in time from a date when the size and age structure
of the population is known, generates <u>inter alia</u> estimates of population
size and structure and of expectation of life at birth. The program is
too complicated to describe here (19) but can be shown to produce very
acceptable results when available data allow it to be rigorously tested,
as with the excellent Swedish demographic data of the eighteenth and
nineteenth centuries. Since the 1841 English census is the first to
provide reasonably comprehensive information about age, the program was
run from that date backwards over the three preceding centuries for
which the parish register aggregative tabulations held at the Cambridge
Group permit estimations to be made. This provides an alternative method
of calculating mortality between 1801 and 1841 and allows the series to
be carried back to 1781 and beyond.

In table 4 the results of the exercise are set out in the form of
estimates of expectation of life at birth, together with the crude birth,
death and natural increase rates over the same period. All the expectation

of life figures refer to current rather than cohort calculations, and
are therefore sensitive to short term fluctuations in totals of deaths.
Each figure refers to the five year period centring on the date shown.

Table 4

All figures relate to the period of 5 years centring on the years shown

	A	B	C	D
	e_o	CBR	CDR	CRNI
1781	35.0	36.3	28.9	7.4
1786	36.4	37.7	27.0	10.7
1791	37.4	39.0	26.2	12.8
1796	36.9	38.0	26.5	11.5
1801	35.6	37.0	26.8	10.2
1806	39.5	40.6	24.1	16.5
1811	39.1	39.7	24.4	15.3
1816	39.3	40.4	24.1	16.3
1821	40.7	38.9	22.6	16.3
1826	41.3	37.3	22.1	15.2
1831	40.7	36.0	22.1	13.9
1836	40.4	35.3	22.4	12.9
1841	41.1	36.5	21.8	14.7

A: expectation of life at birth (sexes combined). Note
   1838-54 (third life table) 40.9; 1855-71, 41.3

B: births per 1,000 population

C: deaths per 1,000 population

D: natural increase per 1,000 population

As might be expected in view of the earlier argument, expectation of
life at birth changed little during 1806-41 but deteriorated very sharply
in the 1801 quinquennium, reflecting the very high totals of deaths in
the years 1800-3. In the preceding twenty years, covering the 1780s and
1790s mortality was in general at a higher level than after 1800 and
expectation of life at birth was in the range 35-37 years, compared with
the range 39-41 years which prevailed with little change from the 1806
quinquennium until 1870.
     Space does not permit an adequate discussion of the evidence about
mortality 1780-1870. As a result I have confined myself to a brief
presentation of findings, but it is important to emphasise in
conclusion the probable significance of compositional change in the
English population during the industrial revolution period. During the
long period 1806-70 when expectation of life nationally was improving
only marginally it is quite likely that mortality was falling in most
regional or socio-economic subdivisions of the country. Indeed, each
such subdivision taken separately – big cities, country towns,
industrial areas, the countryside – may have seen substantial improvement,
but the much higher rates of growth of population in areas with high
death rates tended to conceal the improvement when all areas were
combined to form the national aggregate.

Notes

1. The view is to be found among contemporaries and was repeated at intervals during the nineteenth century. It was very vigorously argued by Griffith half a century ago, and has been the predominant view since that time: G. T. Griffith, Population problems of the age of Malthus (Cambridge, 1926), esp. ch.2.

2. T. H. Hollingsworth, Historical Demography (London and Southampton, 1969), pp.339-53; T. McKeown, The modern rise of population (London, 1976), esp. ch.2; P. E. Razzell, 'The evaluation of baptism as a form of registration through cross-matching census and parish register data: a study in methodology', Population Studies, xxvi (1972), pp.121-46.

3. Notably H. J. Habakkuk, 'English population in the eighteenth century', Economic History Review, 2nd ser., vi (1953), pp.117-33; and J. T. Krause, 'Changes in English fertility and mortality, 1781-1850', Economic History Review, 2nd ser., xi (1958), pp.52-70.

4. William Farr served in the Registrar-General's office over a period of 41 years from 1839 to 1880.

5. For the first English life table see Fifth Annual Report of the Registrar General, pp.342-67; for the second life table see Twelfth Annual Report, appendix, pp.i-lxv and 1-152 (males only).

6. This figure differs very slightly from Farr's own figure (40.86 years) because Farr's figure is based on the accumulation of years lived by single years, whereas my figure was derived from accumulations over five year intervals and reflects also modifications to mortality rates late in life described later in this paper. Full details of the third life table may be found in W. Farr, English life table, Table of lifetimes, annuities and premiums (London, 1864).

7. I expect to present evidence for this assertion in a forthcoming book covering English population history 1540-1840.

8. Supplement to Thirty-Fifth Annual Report of the Registrar General, p.xxvi.

9. As Farr put it, 'The English Life Table is constructed on ... the deaths registered ... in the years 1838-54; since those dates two more censuses having been taken, and the deaths registered and abstracted at the several ages down to 1872, I thought that it might be desirable to construct a new Life Table on the basis of the more extended experience. But the mortality at the several ages having been calculated for each of the 34 years 1838-71, the mean of the rates was found to agree so closely with the mean rates on which the Life Table was based that a new construction became unnecessary. The law of mortality had fluctuated from year to year, but had as yet remained constant; so that the persistence of the force of death as it affects different ages is beyond doubt.'

10. For example, the m_{35-44} figure for males 1838-54 was 1.28 per 100; that for 1838-71 was 1.30. I have assumed that the rate for 1855-71 was twice the 1838-71 rate minus the 1838-54 rate (2x1.30 - 1.28 = 1.32).

11. 14374 enumeration returns were received for England. Only 394 contained no answer to the question concerning age, but these included many large towns, such as Leeds, Manchester and Birmingham. Most country summaries also contain a note to the effect that a 'small', 'very small', or 'remarkably small' proportion of the returns of ages were deficient, or redundant, or incorrect in the numbers of males and females. In Westmorland's case, the note remarks that the 'ages as returned are rather redundant than deficient' since 51,374 ages were returned for 51,359 people.

12. See note 7.

13. See note 7.

14. The derivation of the life table which yields this figure will be described in the book already referred to in earlier notes.

15. J. T. Krause, 'Changes in English fertility and mortality, 1781-1850', Economic History Review, 2nd ser., xi (1958), p.60.

16. As an example of the type of evidence of overcounting which may be found in the 1801 census, consider the information given in 1801 and 1811 for Chittlehampton in Devon where at the earlier date a population of 3003 is recorded divided among 281 families, whereas ten years later there were only 1527 people in the parish in 327 families. There can be little doubt that the family count is accurate but that the population total was substantially overstated. Similar cases are not uncommon.

17. See E. A. Wrigley, 'Births and baptisms: the use of Anglican baptism registers as a source of information about the numbers of births in England before the beginning of civil registration', forthcoming in Population Studies, xxxi, 2(1977).

18. Once more pressure of space prevents a full discussion of this very complex issue. I believe, however, that it can be shown that the remaining uncertainties are modest in scale, and unlikely to involve any substantial margin of uncertainty round the estimates given below for the period before 1801.

19. The program differs from that developed by R. Lee both in that it moves backwards in time from a population whose size and age structure is known rather than moving forwards from an 'artificial' population whose size and age structure is based upon informed 'guesstimates', and in that the population is not assumed to be closed (net migration estimates are derived by the program). Nevertheless Lee's pioneering work supplied much of the

intellectual impetus for attempts to make better use of aggregate
data. R. Lee, 'Estimating series of vital rates and age structures
from baptisms and burials: a new technique with applications to
pre-industrial England', <u>Population Studies</u>, xxviii (1974),
pp.495-512.

Dr. Charles Webster

THE CRISIS OF THE HOSPITALS DURING THE INDUSTRIAL REVOLUTION

Infirmaries were one of the most characteristic expressions of christian philanthropy of the eighteenth century. The first English provincial infirmary was founded in 1736. Thereafter infirmaries became one of the main avenues for provincial philanthropic activity. Between 1736 and 1779 no fewer than twenty-three infirmaries were established in English provincial towns. In 1771 Aikin claimed with satisfaction that "general Infirmaries have been established at almost every considerable town upon the most liberal and extensive principles". Dispensaries, sister institutions to the infirmaries, were founded on an equally extensive scale in the latter half of the century. One or other of these institutions stood in most of the major population centres and country towns as an object of civic pride, a conspicuous symbol of the charitable impulses of the rich, and as a spur to the gratitude and submission of the poor.

The hospital movement has attracted relatively little detailed historical comment. Until recently the hospitals were simply regarded as a reflection of an urge within the medical philanthropic movement to "bring such knowledge as it had to the services of the mass of the people". We were then warned that this urge might have done more harm than good. Recently the pendulum has swung back again to refute the notion that hospitals were "gateways to death". There is once again an inclination to regard hospitals as centres of medical enlightenment, bringing benefits not only to their inmates, but also acting as agencies for the improvement of health standards of the population at large within their catchment areas.

This present study is a preliminary examination of medical charities in the rapidly expanding northern industrial towns of the later eighteenth century. It is designed to draw attention to the importance of institutional frameworks in determining the scope of medical innovation. It can be by no means assumed that infirmaries were created, expanded and adapted in accordance with the needs of local communities, or even that the advocates of modest medical improvements could rely on support from the wealthier classes who traditionally supported medical charities.

Any idea of a close relationship between the voluntary hospital movement and the industrial revolution is dispelled by a glance at the chronology. A rapid spate of hospital foundations occurred in the years immediately after the establishment of the first provincial hospital at Winchester in 1736. In the 1740s seven hospitals were established; in the next two decades nine; in the next three decades ten. Thus momentum was gradually lost in the later years of the century, precisely at the time when demand was increasing most sharply. After the initial burst of hospital creation, further progress depended on the vagaries of charitable initiative, and was hence highly unsystematic. Most of the hospitals were founded in small county towns, few in centres of rapid urban growth. Yorkshire had only one hospital until 1782, when Hull Infirmary was established. Sheffield was without a hospital until 1797, by which time the population of the town itself was over 30,000. The movement to establish small infirmaries at Derby and in the Potteries began only after the turn of the century. The Infirmary at Newcastle-upon-Tyne, founded in 1751, served the counties of Durham and Northumberland and that founded at Nottingham in 1782 served both Nottinghamshire and Derbyshire. The Manchester and Birmingham Infirmaries had even more populous catchment areas.

Infirmaries which were found adequate when gracing small cathedral towns proved to be so small, and so widely scattered that their facilities were swamped by embarrassing hordes of sick poor of the industrial towns. The inadequacy of infirmary provision was to a certain extent compensated for by the establishment of dispensaries which, although not offering beds, had the inestimable advantage of providing care for large numbers of patients suffering from all categories of ailment. Before the end of the century dispensaries were founded in towns already having an infirmary like Birmingham, Liverpool, York, and Newcastle-upon-Tyne, as well as in other towns such as Wakefield, Doncaster, Lancaster, Stockport, Wigan, Carlisle, Kendal and Whitehaven.

In general the benefactors of the provincial infirmaries embarked on expansion of in-patient facilities with caution, ever reluctant to modify rules inherited from the Reformation which forbad the admission of children, pregnant women, the consumptive, the dying, the epileptic and those suffering from venereal or infectious disease. By this means, and by the exclusion of servants, soldiers, and usually those outside some particular catchment area, as well as by rules requiring nomination by a subscriber, some hospitals could keep down in-patient numbers to manageable proportions. The limited evidence relating to hospital admissions suggests that voluntary hospitals of the eighteenth century differed little from the London hospitals of the sixteenth century in the spectrum of ailments treated. Admissions were dominated by abscesses, tumours, ulcers, burns, and a variety of skin complaints. The only indicators of changing social conditions were occasional admissions for burns or fractures occasioned by industrial accidents.

Even with a restricted admissions policy the trustees of those hospitals in growth centres found that they were not able to meet local demand. Even at Newcastle, one of the larger infirmaries at its foundation in 1751, there were reports from 1767 onwards that patients were being turned away; but the number of beds was not increased until after the turn of the century. Usually the hospitals were expanded on a piecemeal basis by the addition of new wards. The Liverpool Infirmary at its opening in 1752 had 54 beds. After a series of additions it had increased to 130 by 1800. The Nottingham Infirmary supplied 44 beds at its inception in 1782. Ten more were added in 1784, and a further ten in 1787 (the "Derbyshire ward"). In 1771 the Leeds Infirmary opened with only 27 beds. After expanding in three stages it housed in 1802 128 beds, but like other hospitals was not allowed to operate at full capacity. By the end of the century most of the infirmaries in industrial towns were short of funds; some were in desperate difficulties.

The infirmaries of the later eighteenth century could claim to assist the recovery of the sick poor by offering "suitable diet, cleanliness and medicine" lacking in their houses. Usually by facilitating convalescence, sometimes by efficient minor surgery, occasionally by well-attested medical treatment, each hospital brought genuine relief to a modest number of In-patients. But as critics like Aikin, Howard and Percival increasingly appreciated, even this function would not continue to be exercised efficiently if the hospitals were not modernised.

Like their French counterparts they might become dismal prisons "where
the sick are shut up from the rest of mankind to perish by mutual
contagion". The boards of trustees, conscious of the deleterious effect
on subscriptions of adverse publicity, improved cleanliness and ventilation in their wards, but it is a significant reflection upon their
susceptibilities that their energies tended to be siphoned off into
developments which had least relevance to the growing health hazards of
industrial towns. They were deeply suspicious of plans for fever,
venereal or lying-in wards, slow even to expand in-patient wards, but
they were positively enthusiastic about lunacy and balneotherapy. Small
baths were installed in most of the infirmaries from the outset. More
elaborate baths were constructed at Manchester in 1779, offering a choice
of temperature, as well as vapour and Buxton water baths. At Nottingham
the Second Annual Report proclaimed the existence of two sets of medi-
cated baths, and gardens were established partly to provide herbs for
these baths. Large and sophisticated baths became a central feature of
the new Derby Infirmary at the turn of the century.

Manchester was the first provincial infirmary to come forward with
plans for a 'lunatic hospital'. This was established only a decade after
the building of the Infirmary. At Leicester also the asylum was opened
a decade after the Infirmary. In both cases the costs were unexpectedly
high, the asylums coming to rival the parent infirmaries in cost of
building and maintenance. In Nottingham an appeal for an asylum was
initiated in 1788, but funds accumulated so slowly that it was not opened
until 1812. At Liverpool the Lunatic Hospital extension was opened in
1790 at a cost of £6,000.

Asylums and baths, like musical festivals, organised in aid of hospital
funds, and, indeed, the elaborate democratic ritual of the infirmaries,
cemented the involvement of the wealthier classes in institutions with
which they might otherwise have had only remote contact. The dependence
of the asylums on patients' fees was so great that it was difficult to
ensure that poor lunatics could gain admission. Even then asylums
tended to be a financial liability to the trustees. Baths were less for
the hygiene of poor In-patients than for public "convenience and elegance"
and profit. The more elaborate baths could give industrial towns a taste
of the elegance of the fashionable spa resorts. Thus baths were worth
extending and became the object of great expense and the technical
ingenuity of local engineers. Almost in the same category as balneotherapy were electrotherapy and pneumatic medicine. These again catered
more for middle-class curiosity than for easing the lot of the sick poor.
Typically the Nottingham trustees ordered "Electrical machines to be paid
for and that Any Person who may come to the Hospital to be Electrified
shall give six pence each time ... but the Poor to be electrified gratis".

Notwithstanding their diligence on certain matters the trustees could
be accused of betraying the intentions of the founders, by taking in
patients who could not benefit by treatment, then dismissing them as
'relieved' when in fact they were incurable. It was asserted that
infirmaries had degenerated into almshouses rather than evolving into
centres for the effective relief of the sick. Critics gradually realised
that changing social conditions were rendering infirmaries obsolete.
The chief, and limiting, design of County Infirmaries had only been 'to
administer relief to those who suffer by accidents, and require the
assistance of surgery; or to those who labour under such distempers as
will admit of the removal of patients from a considerable distance
without injury '. Infirmaries were thus useful to the agricultural

labourer whose leg had been fractured by the kick of a horse, but they were of little value to industrial labourers whose subsistence was regularly threatened by an attack of an infectious disease. Their diseases, it was pointed out, were sudden and violent hence unsuited to weekly admission arrangements. They could not be moved safely any distance to an infirmary. It was also impracticable to separate the sick from their family situation. Infirmary rules excluded the admission of the predominant diseases of parents and children alike. Finally it was argued that even those cases admitted by the infirmaries were severely exacerbated by the failure of the institutions to deal with the pre-disposing factor of contagious disease.

It can be concluded that in general infirmaries made only minimal concessions to the crisis of health of the industrial revolution. Their trustees were wedded to a traditional concept of philanthropy which insulated their infirmaries from the problems of venereal disease, maternal and infant mortality and contagious diseases, malnutrition and industrial diseases, which increasingly dictated the pattern of mortality and morbidity. If these problems were faced at all, it was only by those few infirmaries having a more vigorous out-patient policy, or, more usually, by independent, but underfinanced dispensaries. Such an arrange-ment was increasingly unpalatable and offensive to a small but active body of medical men and laymen in some of the industrial towns. To these men of the enlightenment, the correspondents of Condorcet, the founders of Literary and Philosophical Societies, the advocates of a new spirit of experimental enquiry into medicine and science, the maintenance of the traditional boundaries of philanthropy was as untenable as it was irrational. They believed that it was essential for the evolution of a well-ordered industrial society for "the useful classes" to be restored to health, as it was for machinery to be repaired, or factories planned for maximum efficiency. Notwithstanding their care to stress a concern to benefit all social classes, a desire to contribute to the maintenance of social stability, their scientific public health proposals were greeted with suspicion by many members of the medical establishment, as well as by those classes whose financial backing was essential for the expansion of medical charities. Nevertheless their experiments in public health were pursued with the vigour and tenacity which characterised their work on social statistics or chemistry.

The manner in which attempts were made to modernise the infirmaries consistently with the dominant scientific ethos of the late eighteenth century can be appropriately illustrated by reference to developments at two of the major infirmaries, Manchester and Newcastle. These institutions were established within a year of each other; the Manchester Infirmary was by almost every comparative standard the most receptive of any to innovation; while by the same standards Newcastle proved to be among the most resistant to change. The Manchester Infirmary regularly expanded its In-patient capacity and it was early to develop its Lunatic Hospital and Public Baths. Its Quarterly and Weekly Board Minutes indicate an active spirit of democratic participation among the subscribers, and there were strong ties between leading lay trustees and the medical staff. One indication of the lively state of the infirmary is the almost continuous debate about the emendation of its Rules, often with the expansion of scope of its work in mind.

In 1781 the Rules were modified to allow those patients who could not "be admitted into the House according to the Rules, and are Unable to attend as Out-Patients to be visited in their homes by the medical staff

of the infirmary". This minor adjustment led to a fundamental
reorientation in the work of the Infirmary, by directing physicians into
the homes of the poor to treat patients suffering from serious illnesses.
There was no better way for the medical men to familiarise themselves
with the magnitude of the problem of disease in industrial societies,
and for them to learn that curative medicine alone offered no hope of
containing this problem. Investigations into epidemics in the slums and
factories of Manchester made Thomas Percival and John Ferriar leading
authorities on contagious diseases and it cemented their commitment to
public health reform. There were soon more 'Home-patients' than In-
patients. In the 1790s the Home-patient numbers were of the same order
as the Out-patient.

The staff of the Infirmary and subscribers became divided over policy
towards Home-patients. Eventually three physicians specifically for Home-
Patients were appointed. Immediately a second contentious issue emerged,
when William Simmons offered to join the Infirmary staff to take responsibility for delivering pregnant women in their homes. This extension of
the charity was vigorously opposed by the Surgeons and Physicians to the
Infirmary, led by Charles White, one of Britain's leading obstetricians.
In order to pre-empt this plan, White and his colleagues immediately set
about founding a separate in-patient lying-in charity under their own
control.

Yet a further issue was raised at this time when it was suggested that
an extension of the Infirmary should be built for improving facilities
for Home and Out-patients and for wards to house fever patients.
Opponents of fever wards quickly appreciated that a modest scheme to house
in-patients "seized with a Fever or any other infectious Disorder during
their stay" could be readily expropriated for the isolation of patients
contracting fevers outside the infirmary.

There followed a brief but dramatic confrontation between the expansionists and their critics. The former were in firm control of the democratic
mechanisms of the Infirmary and they won convincingly. They immediately
set up a committee to frame plans for modernising the charity. Rather
than face humiliation all six Honorary Physicians and Surgeons resigned
and relinquished their connection with the Infirmary. The only senior
figure to remain was Thomas Percival the distinguished and sole
Physician Extraordinary, and natural leader of the expansionist party.

The way was now open for experiment. A new staff of physicians and
surgeons was recruited and their number increased from six to twelve.
Drs. Ferriar, Bew and Darby moved up from their junior positions as
physicians to Home-patients to the full rank of Honorary Physician.
William Simmons came into the hospital as its senior surgeon. The Rules
were extensively revised. Henceforth the entire medical staff was
obliged to participate in Home-patient work. Manchester was divided into
six districts, each being in charge of a medical officer on a six-month
rota basis. Plans were made for a library and museum, so equipping the
enlarged institution to proceed with plans for becoming a teaching
hospital, at which it was intended to develop a novel and economical
system of medical training. Very quickly the new committee agreed plans
for an extension to the Infirmary, which was designed to provide new
surgical and fever wards, accommodation for resident staff, a hall for
Out-patients, receiving rooms, and a new Dispensary. The social outlook
of the expansionist party was reflected by Ferriar's To the Committee of
Police in Manchester (1791). It was at this time also that Percival
drafted his famous Medical Ethics.

After one year under the "New System" it was claimed that "not only a strict attendance on the Patients is secured but every disorder that the human Frame is subject to, whether of the Body or Mind finds compleat assistance at these Charities". The new rules were circulated to hospitals elsewhere to publicise the charity which operated to the "universal benefit to our fellow Creatures".

Expansion was not accomplished smoothly. External critics conducted a vigorous campaign against the new system, alleging among other things abuse of funds, improvidence, breaking of rules, and endangering of the health of In-patients, indeed of the population at large. There was a polarisation between expansionists the leaders of whom were predominantly radical and Unitarian, and their critics, who were tory and Anglican. The conflict between these factions was deepened by their strong partisan loyalties over issues connected with the French Revolution. Even in the face of active defamatory publicity however the charity attracted donations to support expansion. The Dispensary was opened in 1792, and the West Wing in 1793. Inoculation for small-pox had been practised to a limited extent since 1784; in 1792 unanimous agreement was given for a new intensive inoculation campaign. In 1793 patients contracting fever <u>inside</u> the hospital were being accommodated in the fever wards, despite the fact that the plan to use the new wards for this purpose was at one stage rescinded. Finally in 1795 Simmons' scheme for delivering pregnant Home-patients was put into operation. But the accomplishment of these notable objectives left the advocates of the New System with their greatest problem still unsolved. The charity had not conceded the principle of setting aside wards for the isolation of patients contracting fevers <u>outside</u> the hospital. John Haygarth of Chester believed that the two wards set aside for this purpose at Chester did "ten times more real good in the prevention of misery, than all the other parts of the Infirmary". Epidemiological investigations conducted by Ferriar, Percival, and colleagues in the Manchester area suggested that prompt isolation could suppress the spread of contagion.

The isolation principle was not new. It had long been practised on the continent, and sporadically in England during plague epidemics. The Manchester theorists were also repeating previous practice when they proposed more general public health measures, such as cleansing, fumigation and whitewashing of infected houses, to counteract the spread of contagion. But they went beyond Haygarth or the continental boards of health in the degree to which they evolved a comprehensive social welfare programme aimed at improving the general standard of health within the working population. Their repertoire of proposals embraced: improvement of building standards for working-class dwellings; measures including public housing to counteract overcrowding and cellar dwelling, the registration and inspection of lodgings; smoke abatement and the control of noxious trades; improvement of street-cleaning, paving, refuse collection and sanitation in poor districts; the establishment of public cold and warm baths; and the superintendence of markets, and control of the quality of retailed food. At a still more general level they produced proposals for the regulation of the conditions and hours of work in factories, particularly with respect to child labour. A mechanism of formal education was regarded as the most effective vehicle for preventing the further erosion of family life and for introducing the poor to the basic principles of health and domestic economy. Although these special welfare proposals were evolved by the infirmary expansionist party, the vehicle by which they articulated these ideas was a sister charitable body, formed in 1796, which was appropriately entitled The Board of Health. The unveiling of this separate charity dealing specifically with problems relating to contagious diseases, effectively protected the New System at the Infirmary from further criticism, and created an institution which could legitimately embark on agitation for wide-ranging preventative measures.

The Proceedings of the Board of Health for 1796 contain vigorous and imaginative discussions of almost every aspect of public health. But in the face of sustained opposition from a party styled the "Board of Protection", Percival and his colleagues were forced to fall back into a defence of their central objective, the creation of a hospital for the reception of patients contracting contagious diseases within the community. In order to allay popular fears about such an institution, it was disarmingly designated the "House of Recovery".

The House of Recovery was opened in May 1796. It was organically linked with the Infirmary by situation and a largely common staff and trustees, but formal independence was preserved, and it was perhaps essential for their mutual well-being.

In each of the first four years the House of Recovery admitted about 350 patients; in 1800 the figure increased to 739, in 1802 it reached 1031. This level of activity did not occur again until 1825. After the initial burst of enthusiasm the ambitious public health interests of the Board of Health dwindled, until it was concerned with only the more obvious and traditional adjuncts of isolation, such as fumigation, cleansing, and relief of infected households. In 1804 it was painfully admitted that "the evils against which we thought to provide a remedy, increase and multiply upon us daily, whilst our embryo projects to oppose them either never came to birth, or to such a feeble and imperfect one, as leaves them incapable of grappling with their gigantic and growing antagonist. Although the charity continued to be known as the Board of Health, after 1805 it did little other than administer the House of Recovery. Gradually as the objectives of the expansionists became more curtailed and as it appeared that the House of Recovery had succeeded in containing contagious diseases, opposition softened. The House of Recovery became accepted as a useful and conventional charity.

The Infirmary at Newcastle was founded virtually simultaneously with that at Manchester. They were organised along similar lines and each was at the centre of a large and ever more populous catchment area. Initially Newcastle was better endowed, and it was established more quickly, but then it remained virtually unchanged until the turn of the century. Its social function remained that of a traditional infirmary; no concession was made either to middle-class taste for baths and lunatic hospitals, or to pressure from below for the treatment of contagious diseases. The Infirmary was old-fashioned and it had a poor mortality record. In the absence of impetus for diversification or improvement, its economic fortunes flagged until it was in severe financial difficulties during the 1790s - precisely the moment of the most rapid expansion of the Manchester medical charities.

In Newcastle the chief steps for extending the scope of medical charity were taken outside the Infirmary. A "Hospital for Lunaticks" for the same catchment area as the Infirmary, was initiated in 1765 and opened in 1767. This was soon taken over by John Hall, one of the Honorary Physicians to the Infirmary and run as a private mad-house. Poor lunatics were admitted by virtue of an annual subscription from the Newcastle Corporation.

The main initiative for establishing an institution to deal with the diseases of the "useful classes" was undertaken by John Clark, soon after he settled in Newcastle after a varied career which had involved successful service with the East India Company. Clark came to realise that his experience with tropical fevers and ship diseases had been an ideal apprenticeship for work among the poor of Newcastle. As a newcomer Clark had no association with the Infirmary. Following the success of the recently established Aldersgate Dispensary in London he and his associates founded a similar dispensary in Newcastle in 1777. Although initially opposed by the staff of the Infirmary, the Dispensary succeeded in establishing a place among the charities of Newcastle. While operating at a more modest level, the Dispensary fulfilled similar functions to the Out-patient and Home-patient departments of the Manchester Infirmary. The Dispensary far outstripped the Infirmary in the number of patients treated. The statistics for the first two years of operation of the Dispensary illustrate the accuracy of Clark's contention that the major categories of patient would have been ineligible for treatment at the Infirmary. Out of 1,346 patients treated, 725 were suffering from fevers or other infectious diseases; 446 could not admit delay in treatment; 193 could only be treated in their houses. From the outset the Dispensary offered a free smallpox inoculation service; in 1779 it initiated a plan for the 'general inoculation of the poor', but despite Clark's estimate that this would eliminate virtually a thousand cases per annum in Newcastle, the plan was not brought into effect until 1786, partly through lack of funds, partly because of the apathy of the poor. From 1791 onwards the Dispensary distributed printed "Rules for preserving health" and "Rules for preventing fevers" to the poor, introducing on a self-help basis the kind of sanitary measures later advocated by the Manchester Board of Health.

Clark and his allies at the Dispensary gained in confidence and authority. They established the Philosophical and Medical Society (1786) and they were a major influence in the first committee of the Literary and Philosophical Society (1793). The members of Clark's group were not so closely identified with the radical and unitarian interest as their colleagues in Manchester, but it is noticeable that in the debate over the extension of the Infirmary the leading advocate of Clark's proposals was openly accused of having been a Jacobin, at a time when his critic "conceived every man was called upon to stand forth against such principles".

Not surprisingly the expansionists in Newcastle looked towards the Dispensary rather than the Infirmary as the basis for their operations. In papers delivered to the Medical Society in 1793 John Ramsay urged the advantages of a Fever House (styled by Clark a "House of Reception"), while Clark himself outlined a plan for a "Society for Preserving the Health of the Poor". It is not difficult to see in these proposals the analogues of the Manchester House of Recovery and Board of Health. The relationship is confirmed by the general similarity of the public health proposals in Manchester and Newcastle. This unity of outlook is partly explained by independent operation on the basis of common philosophical and medical premises, and partly by the existence of many points of indirect contact between the two groups. In the course of the 1790s these links were consolidated, with the result that the Newcastle Papers of 1802 and the Manchester Proceedings of 1805 utilised a largely similar body of testimony in support of their House of Recovery schemes.

The Newcastle Dispensary Committee agreed to the expansionist plans, but funds were inadequate even for a modest House of Reception. But a second opportunity for action occurred when Clark and Ramsay were firmly established as physicians to the Infirmary. They worked for the modernisation of the Infirmary along Manchester lines. There was now a conscious attempt to replicate the Manchester Board of Health, and to ensure that a House of Recovery would be integral to any expanded infirmary. Here Clark and his allies miscalculated. They secured general support for a modest extension of the Infirmary, and for essential improvements in the wards, but they encountered bitter opposition to their allegedly disingenuous scheme to appropriate this extension for use as a House of Recovery, without making the purpose evident to the subscribers. So confident was Clark in the efficacy of isolation in controlling the spread of contagious fevers that he was willing to situate fever wards within the Infirmary complex itself. In this respect Clark believed that the Newcastle Infirmary could rival Manchester, and become a model for modern hospital planning. All that emerged in the short term was an unedifying debate which, despite the supporting testimonies amassed in the Papers issued in 1802, ended with defeat for Clark's party. These bitter and uncompromising exchanges had much in common with those recorded in the Manchester Proceedings. But in the case of Newcastle the conservative party controlled the situation. The position was only saved by the intervention of the Grand Visitor to the Infirmary, the Bishop of Durham, Shute Barrington, who, fortuitously, was one of the most active members of the Society for Bettering the Condition of the Poor. Among its armoury of philanthropic proposals, the Society had taken up the cause of the Fever Hospitals, after witnessing the successes of the Manchester House of Recovery. A compromise emerged; the charity accepted an obligation to treat fever patients, but this was accomplished at a House of Recovery situated outside the town wall. The extension to the Infirmary was used to accommodate the traditional classes of In-patient. Thus the Infirmary was extended, its rules slightly modified, a House of Recovery established, but as at Manchester these events took place at the cost of sacrificing larger plans for a Board of Health exercising an active influence in the control of public health.

The pattern of events at Manchester and Newcastle cleared the way for the calm acceptance of the House of Recovery into the fabric of charitable organisations at many other towns. Within a few years Houses of Recovery were founded at Stockport, Liverpool, Leeds, the Potteries, London (London Fever Hospital), Waterford, Cork, and Dublin. But the coverage was not systematic. In other towns like Leicester, Nottingham, Sheffield and Birmingham, with an equally serious health problem, fever wards found insufficient support, with the result that the issue was not reopened until twenty-five years later.

The Manchester and Newcastle case histories indicate the seriousness with which small groups within the medical and lay intelligentsia reacted to the changing pattern of disease within industrial populations. Significantly this reaction was coherent and sustained in towns in which there were established two of the most effective and permanent Literary and Philosophical Societies. The spirit of investigation into questions relating to demography and health and the programmes initiated in the sphere of public health were consistent with the ideology of the founders of the Literary and Philosophical Societies. The associates of the embryonic Boards of Health had much in common with members of the Société Royale de Médecine in France. Ramsay spoke of the movement in which he was participating as representative of the "æra of the cultivation of all physical science, by simple experiment and observation". Among other things the proponents of "experimental physic" brought a new level of precision to vital statistics, enabling them to monitor the dramatic population growth and changing pattern of disease of industrial towns.

It was demonstrated in quantitative terms that the very institutions founded to relieve the sick were constitutionally ill-adapted to deal with the major problems of disease in industrial societies. The exponents of "Experimental Physick" were confident that they had both come to under- stand the nature of contagion, and evolved and tested methods for the control of contagious diseases. But what seem to the modern observer to be obviously beneficial improvements were likely to be regarded by contemp- oraries as dangerous social experiments.

The eighteenth-century hospital foundations were therefore not readily adaptable into health centres responsive to the prevalent medical needs of their communities. They were traditional institutions designed to serve a specific and limited function, and there was in general little pressure for this role to be changed. The records of the hospitals situated in areas of industrialisation contain few reflections of the escalation of ill-health, or the changing pattern of disease in their neighbourhoods. The Manchester and Newcastle case histories demonstrate the strength of opposition to the modernisation of infirmaries, where this was attempted. The eighteenth-century movement to establish separate dispensaries and fever hospitals was a response to the unchanging character of the infirmaries. It represented a limited but positive expansion of the concept of philanthropy. It was these later medical charities which reflected the medical needs of industrialised towns, but they were few in number and severely underfinanced compared with the more prestigious infirmaries.

The maximisation of the potential of hospitals, fever hospitals and dispensaries could have been important for the health of industrial communities. But as the advocates of Boards of Health clearly appreciated, this network of charities was likely to remain impotent unless its work was supported by a body of public health measures which the philanthropic spirit of their age proved unwilling to countenance.

Dr George Rosen

URBANIZATION, OCCUPATION AND DISEASE IN THE UNITED STATES; 1870-1920.
THE CASE OF NEW YORK CITY.

In October, 1884, the American Public Health Association held its
twelfth annual meeting in St. Louis. On that occasion George H. Rohé,
professor of hygiene at the College of Physicians and Surgeons in
Baltimore, presented a paper on the hygiene of occupations in which he
stressed the importance of the subject and deplored the fact that very
little had been done in the United States to investigate and to elucidate
the nature of occupational hazards and their consequences. As evidence
supporting this assertion, he noted that for the twelve years since
1872 he had been able to find only 24 papers in the "American medical
or sanitary literature in which the hygiene of occupations is considered."[1] Yet there was little apparent change in this situation over
the next few decades. Indeed, as late as 1910, Henry W. Farnam,
president of the American Association for Labor Legislation, struck a
similar though somewhat more optimistic note in opening the First
National Conference on Industrial Diseases in Chicago. "Our country
is just beginning to appreciate the importance of industrial hygiene",
he said, and went on to describe the situation of those concerned with
the health of workers as being "like that of a watchman on a high
tower. He does not know exactly how the attack is to be made but he
knows enough to justify him in giving the alarm and in advising that
scouts be sent out to ascertain more precisely the strength and
position of the foe."[2]
A number of contemporary events provide evidence that Farnam's
cautious optimism had a firm basis in reality. The first American text
on industrial health, a pioneer work prepared by George M. Kober, had
appeared in 1908, and by 1910 there was further evidence of growing
attention to occupational health. That year also saw not only the con-
vening of the first National Conference on Industrial Diseases, but also
the creation of the United States Bureau of Mines, the establishment by
W. Gilman Thompson of the first clinic for occupational diseases at the
Cornell Medical College in New York, the publication by John B. Andrews
of his studies on phosphorus poisoning in the American Match industry,
the appearance of Alice Hamilton's report on industrial lead poisoning,
and the issuance by the United States Bureau of Labor of a list of
industrial poisons. The same year the Joint Board of Sanitary Control
of the Cloak, Suit and Skirt Industry of Greater New York began the
study and control of health conditions in clothing factories and shops.
The U.S. Public Health Service also began in 1910 to take cognizance
of dangers to health in the working environment and of the need to prevent them, so that four years later a Division of Industrial Hygiene
and Sanitation, headed by J.W. Schereschewsky was set up. That year,
1914, Schereschewsky studied the health of garment workers in New York
City, revealing an excessive prevalence of tuberculosis.[3]
Clearly, by 1910, a number of socially conscious Americans were
aware that problems of occupational health were not receiving adequate
attention, and efforts had been started to improve the situation. By

224

the second decade of the 20th century, the movement to better health conditions in industry was in full swing. Emerging developments during this period were, however, the result of a cumulative movement extending back over several decades and influenced in some degree by European experience. This movement can be traced to the latter decades of the 19th century, but the economic and social world from which it sprang and in which it was rooted was produced by three intertwining developments in American society during the last quarter of the 19th century and the early decades of the 20th century-industrial expansion, urban growth and a mounting wave of immigration. From 1860 to 1910 the urban portion of the population rose from 19 to 45 percent of the total, due in large measure to the immigrants who poured into the cities and industrial towns where workers were in demand. The majority of the immigrants who came after 1880 had left the backward, wretched circumstances of countryside and hamlet in southern and eastern Europe to seek a better life in the New World. Though the newcomers dispersed to various parts of the United States, a large proportion settled in the eastern half of the country, mostly in urban centers. By 1890, the foreign-born inhabitants of Chicago almost equaled the entire population of the city in 1880. In 1890, one-quarter of the people of Philadelphia, and a third of those living in Boston were foreign-born. But the cities of New York and Brooklyn became the greatest center of immigrant settlement. When Greater New York was formed in 1898, four out of every five residents within its boundaries were foreigners or of foreign parentage.[4]

The concentration of immigrants in New York and other large cities was due chiefly to the economic opportunities for unskilled and semi-skilled labor that existed there. Indeed, the location of the newcomers in urban centers throughout the country coincided very closely with the availability of jobs they could fill. In this respect, the city of New York occupied a very favorable position. The decades following the Civil War saw New York become a great industrial as well as a commercial and financial metropolis. The beginnings of this growth were already evident in the years preceding the war. Ship building and sugar refining had developed into large-scale industries in the 1820's, and the manufacture of ready-made clothing became increasingly prominent during the following thirty years. Just before the outbreak of the Civil War, the product of the latter industry was valued at about $20,000,000.[5] Though its Southern market was lost during the war, the demand for uniforms to clothe the Union Army of over a million men enabled the industry to expand and grow.[6] By the 1870's, the shores of Manhattan Island were lined with iron works, brass foundries, engine works, ship yards, slaughter houses, as well as fat rendering and soap making plants. Chemical and paint factories were beginning to crowd the Brooklyn shore of the East River. Food processing, baking, brewing, cigar making, construction, and printing were still other components of the thriving economy of New York.[7] Industrial growth continued unabated and at a rapid rate in the subsequent decades. In the 1880's alone the number of plants in the city doubled, and the work force increased by 49 percent. By 1914, almost 10 percent of the manufactured goods produced in the United States came from New York City. [8]

The nature of this thriving manufacturing complex, a major factor in the city's economy, was a consequence of interacting natural and social factors. Two basic factors shaping this industrial development have been the natural and physical peculiarities of the vicinity, particularly the shape and location of Manhattan Island, and the convergence of transportation routes upon lower Manhattan. As a result New York became a major commercial and financial center with ample credit facilities, wholesale markets, and expanding opportunities for international trade. But there were also negative factors involved. Concentration of finance, commerce and industry in a relatively small area of the city, the lower part of Manhattan, led to intensive land use as expressed by high land values and rentals, as well as expensive construction in erecting lofty buildings. To these factors must be added high taxes and insurance rates as well as excessive transportation charges.[9]

Nevertheless, manufacturing establishments occupied these exceedingly costly sites and goods were produced. But only certain industries could operate and grow under these conditions, namely those with a small profit margin, but where the cost of production could be kept at a still lower level by employment of cheap labor and intensive utilization of building space. For the most part these industries involved the manufacture and distribution of a wide range of consumer goods requiring specialized skills or capable of being organized in production patterns able to accomodate semi-skilled or unskilled workers. Industrial expansion went hand in hand with an ever growing labor force, supplied chiefly by immigration. For many decades the largest proportion of the immigrants to the United States (67% in 1870-79 to 79.1% in 1890-99) entered through the port of New York, providing an almost inexhaustible supply of workers for New York's rapidly growing manufacturing enterprises.[10] The increase in various industries from 1860 to 1910, as well as in the number of people employed in them is shown in the following table.[11]

Selected Industries in New York City and Workers Employed in them, 1860-1910

Industry	No. of Establishments		No. of Employees	
	1860	1906	1860	1910
Clothing, Millinery and Laundry	594	10,189	30,158	235,565
Metals, Machines and Conveyances	918	3,557	15,897	99,867
Printing and Paper Goods	425	2,193	10,050	71,290
Food, Liquors and Tobacco	610	3,987	6,784	65,081
Wood Manufacturers	563	1,693	8,895	37,587

Clearly evident is that over a span of fifty years, clothing manufacture had emerged as the city's major industry. By the first decade of the 20th century about 40 percent of the factories in New York City and an almost equal proportion of the industrial workers were engaged in the business of garment making.[12] Viewed in terms of these developments between 1870 and 1920, New York City provides a useful case for studying the relations between industrial technology and organization, transit facilities, living conditions, particularly housing, and the health problems of workers.

When the health administration of the city was assumed in 1866 by a

Board of Health empowered to act within the Metropolitan Sanitary District of New York State, its responsibilities included the inspection of tenement houses and factories and the regulation of occupations detrimental to health.[13] In practice, this activity was one of the tasks performed by a staff of sanitary inspectors, who with few exceptions were physicians. Despite changes in the organization and geographic scope of the health agency, the sanitary inspectors went about their business, and their reports throw light on the working conditions and health problems of several occupational groups.

In 1873, Allan McLane Hamilton, physician and assistant health inspector reported on manufacturing establishments where lead and arsenic were used, among them type foundries, lead smelting, and shot and lead pipe plants. Factories where wall paper was printed were also visited.[14] Hamilton's inspection covered not only possible deleterious effects of lead and arsenic, but also such matters as dusty atmospheres and ventilation, hours of work, and the habits of workers. In the course of these investigations 1500 men, women and children were interviewed, but "not a single case of true lead-disease" was found. This apparently favorable situation was, however, not entirely consistent with other findings. Ventilation and sanitary conditions were poor. In most type foundries and printing plants, the windows were tightly shut and there was very little circulation of air. Though cases of paralysis due to lead poisoning were seemingly infrequent, lead colic was more common. Moreover, Hamilton described the appearance of the printers as "bad", and noted that in the inspected factories "the men seem to be generally anemic, and the children brought up in these places are weak and puny... The susceptibility of these persons, apprenticed at an early age, to lead diseases, is marked." Both in printing plants and factories producing wall paper, the air was full of dust containing lead and arsenic which entered the nose, ears, mouth and eyes, irritating and ulcerating these parts. As a result, conjunctivitis, rhinitis, epistaxis and bronchitis often developed. A few establishments used machines to trap dust but this was not popular. Hamilton also observed that the chance of being poisoned would diminish considerably if the workers washed their hands to remove metallic dust before eating lunch, something they rarely did. Finally, he reported that arsenic poisoning among printers of wall paper was declining because arsenic compounds used to produce green papers were being supplanted by other chemicals that were rarely injurious.

The following year Hamilton discussed the occupational pathology of another group of workers, the car drivers on the various horse railways. In 1866, New York had 16 separate horse railway lines with 800 cars and almost 8,000 horses. [15] The men who drove the cars worked long hours, rarely less than twelve, sometimes as many as fifteen to seventeen, with scarcely an hour for meals and relaxation. As a result, said one driver, describing how the men ate their meals, "we bolts it down and takes it in our hands and eats it on the cars." At work the driver had to stand, and a new employee soon found his legs beginning to swell, the skin becoming tense and shiny, and that it was impossible to wear the same size boots or shoes as before. The worker also experienced a sensation of tension as well as pain in his legs which persisted until his "day off", when a stay in bed relieved these complaints. A large number of drivers developed varicose veins, which in many cases burst and bled extensively. Varicose ulcers were not uncommon, and in winter these men were also subject to chilblains and frost bite. Finally Hamilton turned to the effect that the constant jolting of the moving cars might have on the drivers. Though he does not refer to John Erichsen's concept of "railway spine" (1866), his comments are clearly consistent with this syndrome and its alleged pathogenesis.[16] Since the workers stand on their heels, Hamilton says, "the perpetual jolting of the cars then is transmitted directly through the bones of the leg and thigh to the vertebrae, and acts upon the spinal cord. A constant jarring of this organ therefore occurs, and in many of the drivers a congestion of the cord or meninges, followed by a heavy pain and sense of weight in the legs, symptomatizes various nervous diseases." Considering the exposure to which these workers are subjected and the unreasonable amount of work required of them, Hamilton concludes, "morbid pathological changes are fairly invited. Degeneration of tissue, through mechanical changes, disturbance of digestion, secretion, and excretion, are the evils that come to the car-driver." To ameliorate this situation he proposed the provision of a comfortable seat for the driver and the regulation of working hours, changes which would advance the physical comfort and health of the driver. Nevertheless, Hamilton was not too sanguine about the acceptance of these improvements. "It may be impossible to aim at the root of the system," he observed at the end of his report, "for it is useless to attempt to work a change, sanitary or otherwise, that will touch the pockets of the railway director".[17]

The third group on which the sanitary inspectors reported were workers employed in the manufacture of cigars. As a consequence of the Civil War, the making of cigars moved northward, centering in New York, where both domestic and imported tobaccos were easily available. A very large number of cigar-makers were German, but immigrants of other nationalities were prominently represented. In the early years of the twentieth century there were 15,000 workers in the city who made one-quarter of the cigars produced in the United States. Of these workers, 3000 were East European Jews.[18] In 1872, Roger S. Tracy reported the results of a study of tobacco and cigar manufacturing establishments. This was followed in 1874 by a report on the manufacture of cigars in tenement houses prepared by Nathaniel B. Emerson together with Tracy.[19]

Among the workers in the factories, there were large numbers of female employees. According to Tracy, in the first report, working with tobacco was apparently not harmful, except for very young girls whose growth appeared to be stunted. The second report presented the findings of a survey of 214 families engaged in making cigars at home. The group included all those residing in 17 large tenements, as well as in a few other houses, and comprised 796 persons, included infants and children. The room in which these families worked was generally also their kitchen, dining room and living room. As a rule, the families were small, so that relatively few persons worked together in the same room. The average number of persons to a room was almost four.

Emerson and Tracy compared working conditions in tenement houses with those found in three factories employing a total of 325 people. Lighting was satisfactory for workers in the shops as well as for those employed at home, though the latter had to work at the windows. Though both factories and tenements could be ventilated, windows and doors were usually kept closed during working hours to conserve warmth; as a result the air was poor for both groups. Home workers had one advantage in that they could have a hot, cooked midday meal, while factory employees generally ate cold sausage or cheese with bread, or some dish which they warmed up on a stove. Some patronized a neighboring restaurant, or if they lived close by went home to eat. Tenement work rooms were cleaner than the factories, with less litter on the floors, and less of an odor in the air. Factory employees went to work at about 7 AM and stopped at 6 PM. The working day for home workers began and ended when they pleased, but was influenced considerably by the number of family members who worked and the price they received for their work. On the average, a single factory worker earned higher wages than an individual home worker because the former made cigars from choice quality tobacco, while the latter worked with tobacco of less quality for which a lower price per thousand cigars was paid. Nevertheless, home workers could earn more because it was possible for various family members to lend a hand toward swelling the family income. These might include housewives, mothers with nursing children, the aged, young girls and school children when at home. In this way, by working at home, widows with dependent children could support themselves and keep the family together, thus dealing with a contemporary social problem of considerable importance.[20]

The last question considered by Emerson and Tracy is the effect of tobacco on the health of workers making cigars. Their survey led them to conclude that tobacco leaves do not act as fomites, and that cigar making in tenement houses did not spread contagious diseases. Nevertheless, they expressed the opinion that the manufacture of cigars did have some ill effects, manifested particularly in the relative infertility of tobacco workers. Five years later, in 1879, in his chapter on the hygiene of occupations prepared for A.H. Buck's Treatise on Hygiene and Public Health, Tracy expanded this observation. "It has appeared to me, from observation made in cigar factories and in dispensary practice", he wrote, "that sexual development is decidedly retarded in young girls who enter the factories before sexual evolution has begun, and in an investigation made by Dr. N.B. Emerson and myself on the condition of the cigar-makers who worked at their trade at home in crowded tenements, were very much surprised at the smallness of the families. In the 124 families of which one of us took notes, there were only 136 children, or an average of 1.09 to each married couple, and in the 201 families visited by the other there were only 329 children, making an average of 1.63 to each married couple. When we consider the swarms of children that usually grow up in tenement-houses, in the families of the laboring and artisan classes, the paucity of offspring in a particular class becomes significant." However, "the cause of this lack of fecundity among cigar-makers was not investigated."[21] Tracy suggested, however, that this observation was the result of frequent miscarriages, basing his opinion on a study by Kostial, an Austrian investigator, that abortions were frequent among women employed in tobacco factories due to the death of the fetus, and on the finding of nicotine in the amniotic fluid by Ruef, a French physician.[22] Finally, Tracy concluded that no special preventive measures were necessary in tobacco factories, except to forbid "the employment of children under puberty, and possibly of all females."

These reports are significant for the specific information provided on the health problems of certain urban workers related to environmental conditions in their places of employment, whether in factories or at

home. Equally important, however, is the light they cast on the prevailing forms of industrial production. Three distinct types of work
places are described in the reports of the inspecting physicians: the
home workshop, the factory and the service workplace.

As exemplified by the cigar makers, the home workshop consisted of
one or more rooms where members of the family living there produced
articles for outside employers. This system was widespread throughout
the period under consideration and did not decline until after World War
I. In 1913, the New York State Factory Investigating Commission reported that homework had been found involved in some process of manufacture
in numerous industries, but that there were nor reliable data on the
number of domestic workshops nor of the total number of persons engaged
in homework. One of the investigators for the Commission noted that,
"At the present time it is impossible to estimate with any degree of
accuracy the number of people engaged in homework in New York City. If
the figures obtained form any basis for computation, the number of
homeworkers must run into the hundreds of thousands".[23] Several years
earlier, in 1909, a similar view had been expressed in a Federal report
on the condition of wage earning women and children. Referring to
certain operations in the production of clothing, the report stated that
"Nowhere are there accurate statistics to indicate the extend of home
finishing or other homework. It is resorted to more extensively in New
York and more proportionately in Chicago than elsewhere. There are solid
blocks in New York where by actual count more than three-fourths of the
apartment contain home finishers".[24] These general estimates are borne
out by the available, though limited numerical information. According
to the Factory Investigating Commission, in 1913, New York State permitted the manufacture of 41 articles in tenements, and 66 other articles
were being made in tenement workplaces even though not legally permitted
There were more than 15,000 licensed tenements in New York State in
which every apartment could be used as a domestic workplace. At the
same time, an investigation of 18 embroidery firms in New York City revealed an average of 103 home workers per firm.[25] Mary Van Kleeck, who
headed a Committee on Women's Work established by the Russell Sage
Foundation in 1910, made a series of investigations of women's labor in
various trades, among them artificial flower making. Among her findings,
published in 1913, she reported that there were some 7000 domestic
workers in the artificial flower and feather trade, and that they out-
numbered the workers employed in establishments outside the home.[26]
The same year, Annie S. Daniel, a physician at the New York Infirmary
for Women and Children, reported on 357 families whom she attended during
the eleven months ending September 1, 1912. Of these families, 176 were
engaged in some kind of tenement house manufacturing. The articles produced by these workers included "Literally every article of personal
wear from the hat to the shoes; all kinds and grades of embroidery, including....church vestments and altar cloths, passemanterie, buttons,
flowers, feathers, hair switches, cigars, boxes of all descriptions,
furs and fur trimmings - articles allowed by the labor bureau, and some
they never thought of; anything that can be carried by the worker."[27]
These observations coincide with the findings of the New York State
Factory Investigation Commission which listed six industrial groups in
which homework was prominent: (1) clothing, millinery, including artificial flowers, feathers and the like, (2) fur and leather goods, (3)
textiles, (4) food and tobacco, (5) paper boxes and other paper products,

and (6) several miscellaneous industries.[28]

Clearly, throughout the latter 19th century and the early decades of the 20th century, production in the home workshop existed side by side with factory production, chiefly in industries making consumer goods and employing cheaply paid workers. Though mainly a phenomenon of large cities both in the United States and in European countries, homework could also be found in smaller communities and rural areas.[29]

Industrial production outside the home was carried on in factories and workshops, varying in size from a large complex of several buildings to part of a floor in a converted tenement, as well as in the number of workers employed. A New York business directory, published in 1884, provides a number of illustrative examples. Warner Brothers, corset manufacturers, a firm established in 1874 by two brothers who were physicians, had a factory in Bridgeport, Connecticut, occupying over 2 acres and employing 1200 workers, four-fifths of them women.[30] At the other end of the spectrum was the firm of C.G. and F. Neumann, book-binders, who occupied two large floors in a business building and em-ployed between ten and fifteen workers on the average.[31] Comparable in size was the firm of F.G. Otto and Sons, makers of surgical and dental instruments, who employed about fifty skilled workers in their factory located at Jersey City Heights.[32] On the other hand, the New Home Sewing Machine Company had about 800 workers in their extensive plant comprising a series of buildings.[33] Between these two extremes were a large number of firms whose factories and shops employed from 100 to 200 workers. The Dale Tile Manufacturing Company, for example, had over 100 men in its employ. Stern and Co., manufacturer of shirts and overalls, occupied the first story and basement of a building on lower Broadway and provided employment for 200 workers. The fur manufacturing trade comprised firms of similar size. C. C. Shayne, for example, occupied premises consisting of a large salesroom on the street level and a basement devoted to the manufacture of fur garments. Another firm in the same trade, Ph. Weinberg and Co., had a factory with more than 150 workers. The premises occupied by Stern Bros. and Falk, manufacturers of boy's and children's clothing, comprised four floors where 135 workers produced the garments sold in the United States and Canada.[34] The production of wearing apparel for women and misses was an important division of the clothing industry and had several branches of which one of the most prominent during the middle and later 19th century was the manufacture of skirts, hoopskirts, bustles and other goods of this nature. In 1859, Douglas and Sherwood's Hoop Skirt Factory comprised three buildings, employed 800 women, and turned out 3000 hoop skirts daily. At the same time, Thomson's Crown-Skirt Factory provided employment for 1000 women who produced from three to four thousand skirts daily.[35] In 1884, established firms producing goods of this type were Rosenstock and Cohn, and Huntington Brothers and Company. The former manufactured hoop skirts in an extensive plant with a large work-force, mostly women and girls. The latter produced suits and cloaks, felt and flannel shirts, and women's underwear in a plant consisting of three attached buildings, four stories high, with an average work force numbering between 250 and 300.[36]

Only superficial attention can be given to the service workplace, but it should at least be noted that thousands were employed in service occupations as sales personnel, bookkeepers, stenographers, typists, telephone and telegraph operators, draymen, streetcleaners, plumbers, peddlers and transit workers, to name but a few. Furthermore, to judge by the demands of the motormen, conductors and workers in the electrical power plants who struck the Brooklyn street car lines in 1895, working conditions had hardly improved since Hamilton's report on horse railway drivers in 1874. The electrification of the street railways in the late 1880's and 1890's made it possible to expand the boundaries of the city of Brooklyn, but the transit workers were still demanding a ten-hour working day and that motormen and conductors be allowed to leave their cars long enough to eat lunch.[37] An examination of working conditions in various service occupations and their relation to health conditions among those employed in them is undoubtedly desirable, but cannot be pursued here due to the wide range of occupations and the multiplicity of circumstances within which they were practised.

The industry in which the factors previously presented can best be seen in thier interrelationships is the manufacture of clothing, New York City's major industry for almost a hundred years. In 1846, Elias Howe patented the sewing machine and four years later Issac Singer began to market an improved model, thus opening the way for mass production of clothing, hats, shoes, harness and other articles. About this same time the cloak, the mantilla and the hoop-skirt came into vogue in women's wear, but the latter garment was difficult to make at home. Wholesale and retail dealers in women's apparel who had previously imported European products now decided to make their own. Factories sprang up, some of considerable size. Wests, Bradley and Carey's Hoop Skirt Works on West 20th Street employed as many as 1600 workers, about 30% of the employees in the industry, thus exceeding in size the firms previously mentioned. This development continued into the 1880's.[38] Before 1880 the labor force consisted mainly of Irish and German immigrant women. Not all of them worked in factories; often they worked at home. Even though men largely replaced women in cutting out garments following the development of the Gotham cutting knife in 1876 and the introduction of the slotted table, the labor force in 1880 was still only 12 percent male.[39] Thus, it is hardly surprising that the earliest medical investigations of the garment industry dealt with the influence of the sewing machine on the health of women.

On November 21, 1860, Augustus K. Gardner, professor of clinical midwifery and women's diseases at the New York Medical College, addressed the New York Academy of Medicine on the sewing machine from a hygiene viewpoint.[40] Gardner's interest in this problem was not just a matter of chance. He was an active member of the Academy's committee on public health, involved with studies of milk, dispensaries and other problems.[41]

According to Gardner, the principal conditions blamed on the sewing machine were the so-called "female diseases" and spinal complaints. Based upon his experience as a gynecologist, however, he denied ever having seen a patient "who gained her living by working a sewing machine who was affected with leucorrhoea, 'falling of the womb', 'ulcerations of the womb', or spinal difficulty-who ever had an abortion while using it, or who in any way could trace any injury from it."[42] To check his own observation, Gardner investigated conditions in the factories, making inquiries among employers, women workers and physicians. One employer claimed that absenteeism due to illness was rare among the girls he employed. Nonetheless, he admitted that "when they first come, after a day's work, [they] are obliged to ride home from fatigue", but went on to add that the girls soon became accustomed to the job. Fatigue was also a major complaint among workers in other garment factories. After questioning a number of girls employed at the Seligman & Co. factory, Gardner reported that "when working on heavy goods for ten hours, they did feel fatigued, but they never suffered from any special diseases, never heard of any spinal difficulties, neuralgia, amenorhoea, or leuccorhoeas". As a result, after investigating the problem for six months, Gardner concluded that "the sewing maching is a blessing to mankind, and especially to the female, and that without an appreciable drawback".[43] Not only was factory work more remunerative than home work, but it was also healthier to walk to and from the factory than to spend the whole day without interruption in a small apartment "containing bed, cooking stove, children, work bench etc."[44]

Following Gardner's report the health of women employed in clothing factories was studied by a number of European and American investigators. The most extensive investigation, published in 1870, was made by Decaisne, a French physician, who examined 661 women.[45] He found that when young women first began to work on sewing machines, they suffered greatly from muscular pains and fatigue in the legs and often in the lower back. A large number had uterine disorders, but for the most part these conditions antedated their employment. Many were anemic and dyspeptic, but no more proportionately than occurred among other seamstresses. Decaisne concluded that symptoms and complaints were due chiefly to pre-existing medical conditions, to the nutritional and living conditions of the workers, and to unaccustomed muscular exertion. As the young women became accustomed to their work, the latter factor generally became insignificant. A.H. Nichols, an American, who prepared an exhaustive report on the same subject which appeared in the Report of the Massachusetts Board of Health for 1872, arrived at essentially the same conclusions.[46]

By the end of the decade, in 1879, Roger S. Tracy, the New York sanitary inspector who had studied tobacco workers, summed up these and other investigations. There was general agreement, he wrote, that moderate work on the sewing machine for three to four hours a day was rather beneficial to health than otherwise. Only in large workshops where young women worked at the treadle for eight to ten hours were harmful consequences apparent. The chief cause of their complaints was fatigue, but significant contributing factors were their way of life, nutritional habits, forms of recreation and the like. To prevent ill health among the workers, Tracy proposed that sewing machines should be operated by some other motor power than the muscles of women, and where practicable steam or water power should be used. "In almost all large establishments in New York", he said, "the machines are worked by steam, and the only harmful conditions attached to the occupation are the confinement, the cramped position of the thorax, and the strain upon the eyes."[47] Nevertheless, the mechanization of the factory recommended by Tracy was not accomplished until after the turn of the century with the electrification of the sewing machine and the introduction of the steam-pressing iron to replace irons heated by gas or coal. Instead, an unanticipated historic event initiated a profound change in the organization of the garment industry, creating new working conditions and leading to increasing attention to the health problems of its workers.

This event was the migration of hundreds of thousands of Jews from

Eastern Europe to America. Between 1881 and 1905 some 850,000 Jewish immigrants arrived in the United States. Although this great wave of migration began with the flight from progroms they followed the assassination of Czar Alexander II in 1881, its continuation and magnitude were a consequence of profound demographic, economic and social developments. A high rate of natural increase, due largely to a lower infant mortality than that of the general population, produced a population surplus that could not be absorbed in the overcrowded crafts by which most Jews earned a meager livelihood. This situation was intensified by the expanding industrialization of Eastern Europe, especially in Poland, where the textile industry in such cities as Lodz and Bialystok led the way. As crafts declined economically, the portion of Jewish factory workers increased, rising to almost 28% of the factory hands in Poland and to more than 21% of those in the Pale of Settlement by the last decade of the 19th century. For most Jews the industrial transformation and its associated economic crisis only accentuated the harsh poverty which dominated their existence and spurred them to migrate, above all to the United States.[48] Within this world of mounting wretchedness, intensified political reaction and religious persecution finally made life intolerable, and led approximately one-third of the East European Jews, within some thirty years, to flee their homelands for the haven and the hope of a human existence to be found in the land of Columbus.

The Jews were part of the surplus population which eastern and southern Europe exported to America at the end of the 19th century, and which played a vital role in accelerating the process of industrialization after it had been slowed down by the depression of 1873-1879. The Jewish immigrants found a place in the needle trades, into which they were directed by a number of interacting causes. The incoming shiploads were arriving in New York at a time when the women's apparel industry was expanding phenomenally, not only in volume but also in variety of products. Stemming from its origins in the manufacture of cloaks and skirts, the single largest section of the industry produced coats and suits, still known in the jargon of the trade as "cloaks". Though women's dresses were still made at home, frequently by a neighborhood dressmaker, new branches of the garment industry spring up. In 1891, the first shop making shirtwaists was established in New York, and over the following nine years, 471 others came into existence. The manufacture of women's underwear was still another new development. In fact, by 1905, a definition of the needle industry included "the clothing trade, and the manufacture of cloaks, waists, wrappers, skirts, shirts, overalls and underwear."[49] This was the industry that made the shirtwaists and probably also the undergarments worn by the Gibson Girls, who for several decades epitomized the fashionable ideal of young womanhood. A parallel growth was evident in fur manufacture which developed alongside the women's apparel industry.[50]

As the needle trades exploded, more workers were required and the large scale immigration after 1880 provided this essential element. Tailoring had long been a Jewish occupation in Europe, at least as far back as the Middle Ages, due in part perhaps to a preference among Jews for tailors who would observe the biblical prohibition against mixing linen and wool. [51] At the beginning of the 18th century, Bernardino Ramazzini included a chapter on the diseases of Jews in his treatise De morbis artificum. "Nearly all Jews", he wrote, "especially the lower classes to which most of them belong are employed in work at which they

must sit or stand. They are mostly given to sewing and furbishing up
old clothes. ...As for the men: All day long in their booths they sit
patching clothes or they stand on the look out for customers to whom
they can sell their old patchwork...."[52] Tailoring remained an established craft among Jews so that some of the immigrants from Eastern
Europe brought with them skills and experience, and naturally entered
the garment industry. Immigrants who lacked these skills could find
employment with their co-religionists under conditions more or less
consistent with their religious and cultural requirements, since little
skill was required for most of the available jobs which were numerous
in an expanding industry. Indeed, it was not uncommon for employers
or their agents to meet incoming ships and seek out immigrants with some
experience as tailors, whom they hired on the spot.[53]

In his novel, The Rise of David Levinsky, Abraham Cahan, himself an
immigrant and editor for many years of the Jewish Daily Forward, graphically portrays such an episode. Shortly after Levinsky and his ship-
board companion, Gitelson, pass through the immigration station at
Castle Garden in New York, they are accosted by an apparently affluent
middle-aged man who addresses Gitelson in Yiddish. "'You're a tailor,
aren't you?' he questioned him. 'My steerage companion nodded. 'I am
a ladies' tailor, but I have worked on men's clothing, too,' he added.
"A ladies' tailor?' the well-dressed stranger echoed with ill-
concealed delight. 'Very well; come along. I have work for you'....As
I learned subsequently the man who accosted us on State Street was a
cloak contractor, and his presence in the neighborhood of Castle Garden
was anything but a matter of chance. He came there quite often, in
fact, his purpose being to angle for cheap labor among newly arrived
immigrants."[54]

The contractor and the worker whom he hired represent pivotal elements
in the major changes that followed the entrance of Jewish immigrants
into the clothing industry, changes which were to have a decided impact
on the health and welfare of the workers. A supply of skilled workers
as well as a large pool of cheap labor were the basic assets which the
immigration after 1880 gave New York City's garment industry. The
pressing needs of the immigrants as well as the economic and technical
conditions of the industry led to the organization of the labor force in
terms of the contracting system. The beginning of the system may be
seen in industrial homework of the type described by Emerson and Tracy
in 1874 in their investigation of cigar making in tenements. Under this
arrangement, the manufacturers maintained control of production, pur-
chasing the goods, cutting them into garments at their own establishments, and then having the garments put together by individuals or
families at home. The majority of these operatives were Irish and
German immigrants, many of them women who worked within a so-called
"family system", in which the head of a family, usually an experienced
tailor, was joined by his wife and older children in producing the
finished garment.[55]

In the 1870's manufacture in the home had declined as factories sprang
up, but this trend was reversed by the early 1880's, giving way to a
system of "inside" and "outside" shops. The "inside" shops were owned
by the large manufacturers and merchants who controlled the industry,
and in some of them the entire manufacturing process took place under
the same roof. In others, only the cutting of the garments and exam-
ing of the finished product was done, while the garments were sent to

"outside" shops to be made up. The operators of the "outside" shops were contractors who agreed to assemble precut materials into garments for the manufacturers at so much per garment. As middlemen, the contractors made their money by keeping their production costs lower than the receipts from the manufacturers. Several factors combined to make this possible and to produce the characteristic features of the contracting system, exhibited most starkly in the sweatshop. The means of production were comparatively simple and cheap, building space could be used intensively, and there was an abundant supply of cheap labor. The sewing machine was not expensive and could be rented by the month or purchased on installments. Because the machines were small and portable, a number of them could easily be installed in tenement house rooms. Furthermore, the energy to operate them was provided by the worker's foot power, a situation still prevalent in the early years of this century. Nor did the worker have to be highly skilled. The necessary routine tasks required of an operator or presser could be mastered in a month or two.

The contractors performed a dual function. On the one hand they facilitated and accelerated the growth of the garment industry by making it possible for the manufacturer to take advantage most economically of the rapidly expanding market. By letting out his work on contract, the manufacturer saved the expense of running his factory and of having to manage a large labor force. The responsibility for finding, training, organizing and supervising the workers, and of adjusting this labor force to the fluctuations of season and style were shifted to the contractor who thus became "an organizer of labor."[57] His function in the needle trades was comparable to that of the Italian padrone who arranged the recruiting and employment of Italian laborers for railroad work. Both were controllers of immigrant labor who provided the hands required by the burgeoning industrial economy of the United States.

Basic to the contracting system was cheap labor, but since the contractor operated on a narrow profit margin, he endeavored to enlarge it as much as possible by shifting to the worker wherever he could various costs of production. Workers were required to buy or rent their machines, to supply thread and needle, or to pay for the privilege of entering the shop and learning the job. Often the machine was rented from the contractor, in which case the operator had to pay for repairs. Fines were levied for various reasons such as damaging a garment or coming to late to work.[58]

These conditions were intensified still further as more and more workers crowded into the garment industry and the number of contract shops mushroomed. Competition among contractors was intensified as manufacturers played off one contractor against another. Survival depended on keeping wages as low as possible, maintaining an excessively long working day, and rationalizing the production process through a minute division of labor. Furthermore, wherever it was profitable to do so, the contractor could farm out work to be done by cheap home labor.

The shops were generally located in tenement houses, and in many instances the same quarters served as both home and work shop. A graphic, vivid portrayal of these shops was provided in 1890 by Jacob A. Riis, based on his own investigations. "Take the Second Avenue Elevated Railroad at Chatham Square," he wrote, and ride up half a mile through the sweater's district. Every open window of the big tenements, that stand like a continuous brick wall on both sides of the way, gives you a

236

glimpse of one of these shops as the train speeds by. Men and women bending over their machine, or ironing clothes at the window, half naked."[59] The scene within the shops corroborated Riis' impression from the train. Albert Edward's novel Comrade Yetta, published in 1913, recounts the development of a young Jewish woman from shop worker to trade union organizer and socialist journalist. Seeking a job in a shop making vests,"Yetta followed her [cousin] docilely up two flights of stairs into a long workroom which had been made by knocking the partitions out of a tenement house flat. It was a gloomy place, for the side windows were faced by a dingy brick wall three feet away....The tracks of the elevated were on a level with the floor, and every few minutes the light which might have been expected from this quarter was cut off by the rush of a train. Artificial illumination was needed all the year round."[60] Yetta's first grim impression of the shop was powerfully reinforced as she moved from her initial job as a "learner" to that of an operator at one of the twelve noisy machines. "She saw the broken door to the shamefully filthy toilet, saw the closed, unwashed windows, which meant vitiated tuberculosis-laden air, saw the backs of the women bent into unhealthy attitudes, saw the strained look in their eyes."[61]

Worse conditions prevailed where the tenement dwelling served both domestic and industrial purposes. Riis describes such a shop, visited in 1890. "A sweater, this, in a small way. Five men and a woman, two young girls, not fifteen, and a boy who says unasked that he is fifteen, and lies in saying it, are at the machines sewing knickerbockers, "knee-pants" in the Ludlow Street dialect. The floor is littered ankle-deep with half-sewn garments. In the alcove, on a couch of many dozens of 'pants' ready for the finisher, a bare-legged baby with pinched face is asleep. A fence of piled-up clothing keeps him from rolling off on the floor. The faces, hands, and arms to the elbows of everyone in the room are black with the color of the cloth on which they are working.... There are ten machines in the room; six are hired at two dollars a month." The rent for "the two shabby, smoke-begrimed rooms, one somewhat larger than ordinary," was twenty dollars a month. Work began at day break and continued to nine in the evening, with the contractor and his wife working alongside the others. The premises served also as the living quarters for the contractor's family, which included four young children, none over six, and for two boarders who helped to pay the rent.[62]

As Riis pointed out, in these tenements each floor had at least two, and sometimes three or four such shops, making cloaks, shirtwaists, boy's jackets, neckties, suspenders and other articles of apparel. And in the same rooms people also carried on the necessary activities of daily life, preparing meals, performing required religious rituals, dealing with illness and other problems. Even under the best of circumstances it would have been difficult to maintain minimum levels of cleanliness and hygiene. But conditions on the Lower East Side of the city were hardly of the best. Around the turn of the century, the district was the most densely congested area in New York. Arnold Bennett's observation on Rivington Street, that "the architecture seemed to sweat humanity at every window and door", may serve as an appropriate metaphor for the appalling congestion characteristic of the quarter.[63] Furthermore, the tenement apartments which housed this mass of humanity and where many earned a living had few facilities for personal cleanliness. With few exceptions, toilets and running hot water were not available in each apartment. Water closets were most frequently situated in the hallways between apartments and were used by two to four families. In tenements built before 1881 toilets were located in back yards. As investigation of 578 buildings made for the Tenement House Commission of 1900, covering tenements in various parts of the city revealed that 20 percent had toilets within the apartments, and that these buildings were "generally a good grade of modern tenement."[64] Bath tubs in apartments were found only in five buildings. This situation is borne out by a study of Russian Jewish families reported in 1902. Only eight percent of these families had baths, and often even they had no hot water. [65] Small wonder that grime, dirt and unsanitary conditions were associated with the contractor system and the sweatshop.

Though these conditions were most heavily concentrated and their effects most keenly felt and publicized in New York City, the system that produced them also developed with similar consequences in Chicago, Philadelphia and Boston. Outside of New York, the worst sweat shops in the apparel industry were found in Chicago. As Florence Kelley, first Chief Inspector of Factories for Illinois, reported in 1895, in the square mile immediately surrounding Hull House, "All the tenement houses selected either for shops or home finishers are of the worst and most crowded description. The staircases are narrow, and are used in common by tenants and garment workers, so that infectious diseases breaking out among the swarming children can scarcely fail to be communicated to garments anywhere under the same roof, because the utmost laxity prevails in the matter of isolation. The unsanitary condition of many of these tenement houses, and the ignorance and abject poverty of the tenants, insure the maximum probability of disease...."[66]

Shocking conditions in sweat shops were also reported in Boston by Harry Linenthal, a physician on the out-patient staff of the Massachusetts General Hospital, who in 1907 was appointed inspector of factories and tenement work shops where men's clothing was made.[67] "In most of the shops," he noted the following year, "the unsanitary conditions of the buildings are made worse by the slovenly, uncleanly habits of the contractors as well as of their employees. Dust, rags and other refuse are allowed to accumulate in the corners of the shops and under the tables. The habit of spitting on the floor is evident in almost every shop. To appreciate fully the significance of this disgusting and dangerous habit one need but visit the shop in operation. Various parts of the garment are made by different operators. When the operator has finished the part allotted to him he throws it on the floor. The garment is then picked up by the foreman, who passes it to the next operator. The dried as well as the moist sputum may thus be freely circulated among the operators by the contaminated clothing. When the shop is swept, as happens on rare occasions, the sweeping is usually dry and helps to stir the dust. Is it surprising that people continually exposed to such conditions become tuberculous?" [68]

Similar conditions were prevalent in Philadelphia as well as in other American cities. [69] One might also point to the sweatshop belt of London as portrayed by various government committees as well as by social investigators such as Charles Booth and Beatrice Potter (later Mrs. Webb)[70] There was general agreement, however, that three main features characterized the contract system and the sweatshop, extremely low wages, excessively long hours and insanitary conditions, and that these circumstances were detrimental both to the workers and to the community at large. An editorial on sweated industries and the health of the masses

in the New York Medical Record of March 7, 1908, made these points
bluntly. "Very low wages are productive of ill health and of physical
degeneration amongst the workers themselves, that is to say, wages
which are not sufficient to procure for the worker and those dependent
upon him a proper amount of nourishing food and a fairly comfortable
home. Sweated labor is, therefore, to a certain extent, a menace to the
health of the entire community, for those who perform this description
of labor, as a rule, work under unsanitary conditions and receive quite
inadequate wages. Further, home work is the worst paid of the sweated
industries, and is usually accomplished under circumstances which render
it not only dangerous to the health of the workers themselves but a
distinct source of danger to the public at large." [71]

Concern for the health of the workers in the various branches of the
needle trades grew out of a widening awareness that congestion and unsanitary working conditions harbored the potential danger of spreading
communicable diseases. This was, after all, the period, largely between
1876 and 1900, when the microbial causes of numerous human and animal
diseases were revealed and their mode of action began to be understood.
This growing awareness fueled fears that outbreaks of disease might be
propagated by the workers or through the articles which they made. Nor
was this attitude unfounded. Florence Kelley's statement in 1895 that
"diphtheria, scarlet fever, smallpox, typhoid, scabies and worse forms
of skin diseases have been found in alarming proximity to garments of
excellent quality in process of manufacture for leading firms," was
based on personal experience during the Chicago smallpox epidemic of
1893. Judge Andrew Bruce, who had been attorney for the Illinois factory inspectors, later recalled that at the time of the epidemic "both
she and Julia Latherop were risking their lives in the sweatshop
district of Chicago and were fearlessly entering the rooms and tenements
of the west side and not merely alleviating the sufferings of the sick
but preventing the sending abroad of the infected garments to further
contaminate the community." This action later induced the Chicago
health department to destroy "thousands of dollars' worth of clothing." [72]

The Report of the New York State Tenement House Commission of 1900
noted with alarm the serious menace to public health in tenement house
manufacture. Since the work was generally carried on in the living
quarters of the family operating the shop, where young and old mingled
closely with the workers, there was felt to be a great danger of contagion when a member of the family became ill. A member of the Commission
found garments on the floor of a shop amidst dirt and rubbish, and
garments were stacked on a bed with some being used as pillows for sick
children. In one place, garments were stored in a room where a sick
man, apparently in an advanced stage of tuberculosis, was lodged. [73]

Further evidence of concern with the relation of poor environmental
conditions to ill health among tenement workers and the wider implications of this nexus was offered by a number of other investigators.
Owing to the continuing growth of the garment industry and the consequent
increasing demand for workers, large numbers of Italian immigrants,
chiefly women, joined the Jews in the needle trades in the 1890's. By
1900, Italian workers made up about 15 percent of the industry's labor
force in New York. [74] Their health experience was in a number of
respects much worse than the of the Jewish workers, since the Italians
were much more susceptible to pulmonary tuberculosis and other respiratory disorders. [75] In 1908, Antonio Stella, a physician studying the

effects of urban congestion on Italian women and children commented that "for those whose constitution is sapped in the labor of sweatshop or other exhausting trades, urbanization is only responsible in a very limited sense for their physical breakdown." However, he added that many of the young women between the ages of twelve and twenty-five coming from rural areas were healthy when they arrived in America. "But the pernicious conditions of their new atmosphere, the unhealthy and crowded habitations, the long hours passed in the factories, the thousand privations imposed upon them by poverty, combined with the intense labor that, from a desire to make money, they push to utter exhaustion, without being sustained by food proportionate to the demands of the excessive waste and rapid growth of this period, all this explains in the clearest possible manner why so many young lives become an easy prey to tuberculosis in America."[76]

A similar point was made by Linenthal for Boston where the work or "finishing pants" was carried on almost entirely by Italian women living in the most congested area of the city. The work was done in overcrowded, dark and poorly ventilated rooms, in bedrooms with windows opening into narrow, uncleaned alleys or air shafts, or in rooms with no windows at all. Not only did these women work excessively long hours, but other members of the family including children were drafted into service. Linenthal was convinced that the resistance to disease of these workers was lowered by the very long hours, the unsanitary conditions in their homes and their inadequate nutrition, so that they fell easy prey to various illnesses, especially contagious diseases, of which tuberculosis was one of the most prominent. There was also the danger in tenement workrooms, he observed, of exposing garments "to the contagious matter of certain skin diseases which are quite common among the tenement dwellers."[77]

Concern for the health of the workers in the needle trades focused preeminently on the prevalence of tuberculosis among them. This was an aspect of the anti-tuberculosis movement which developed in the United States around the turn of the century. Rober Koch's discovery in 1882 of the tubercle bacillus as the specific etiologic agent in tuberculosis had provided a means of opening up the complex of interdependent causative factors, involving the social allies of the bacillus, poverty and low living standards, poor housing, overcrowded dwellings, unsanitary working conditions, ignorance and poor hygienic practices. Beginning in 1889 the New York City Health Department had developed a series of measures designed to prevent the disease by informing the public about its nature, spread and means of control, and by interrupting its transmission from one person to another.[78] Although the campaign against tuberculosis was initiated by professionals, large community forces were soon mobilized. The strategy employed to fight the disease may be characterized as socio-medical, and was linked to the larger movement for social betterment.

Awareness of the connection between occupation and tuberculosis spread to different groups, creating alarm and leading to a variety of activities. As Lillian Wald of the Henry Street Settlement commented, "tuberculosis is the disease most to be dreaded....We see so much of it we call it the tailor's disease."[79] Though the mortality from tuberculosis was considerably lower among Jews than among other immigrant groups or native-born Americans, there were still enough victims to arouse concern in the Jewish community.[80] Consciousness of tuberculosis as a serious problem was not limited to New York. Theodore B. Sachs, a Chicago physician, after investigating tuberculosis among Jews in Chicago during 1902-1903, concluded that the disease was very prevalent among the Jewish poor, and that unsanitary conditions of home and factory were chiefly responsible for its widespread occurence. A considerable number of the patients studies were employed in the garment industry as operators, cutters and finishers. Of these a majority worked in the sweat shops of the Jewish district. Sachs emphasized that the alleged immunity of the Jews to tuberculosis had been greatly overestimated and that their immunity to certain communicable diseases varied in degree according to the economic and hygienic conditions under which they lived. Thus, tuberculosis was comparatively rare among well-to-do Jews, but much more common among their poor co-religionists. [81]

Tuberculosis in the sweat shop provided a rallying cry and a focus for action to improve conditions. In his novel, Comrade Yetta, Albert Edwards describes a meeting of the "Skirt Finisher's Union" at which a socialist doctor links the prevention of tuberculosis to the establishment of socialism, and exhorts his auditors to adopt his political creed for this reason. "Tuberculosis! How many of you are there in this audience who haven't lost a relative from lungs?" he asked. "As I sat here a moment ago I heard at least a dozen tubercular coughs. It's preventable-it's curable. There's no reason why anyone should have it-less still that any of you should die of it - if Capitalistic Greed didn't force you to live in rotten tenements, to work long hours in worse shops." [82]

Though radical ideologies and utopian visions had their attractions for many garment workers, measures taken by the worker's organizations, by government and professionals to improve working conditions and to prevent ill health were essentially pragmatic. Efforts to curb home work on garments in tenements through legislation began in the 1890's and over the subsequent two decades sweat shops of the worst type had to a considerable extent been stamped out. Prevented from carrying on production in the home, the contractor shops of the garment industry moved into multi-storied loft buildings where conditions in many cases were not much better. At the end of the first decade of this century, work in most shops producing garments, and especially in the cloak and suit trade in New York, was still being carried on under poor sanitary conditions. Though tenement home work had been curbed, it had not yet been totally eliminated and still represented an important sanitary and health problem. Between 1900 and 1910, it became clear that efforts to cope with these problems through State regulation and inspection alone were unsatisfactory, and that other means would have to be developed for this purpose, with the result that actions were undertaken from several sides.

In 1900, cloakmakers from New York, Newark, Philadelphia and Baltimore met on Manhattan's East Side and united to form the International Ladies' Garment Worker's Union (I.L.G.W.U.), a national organization chartered by the American Federation of Labor. Initially small and powerless, the I.L.G.W.U. made little progress until 1909 when its Local 25, the shirtwaist maker's union, called a general strike of the industry. Twenty thousand shirtwaist makers, for the most part young women, about two-thirds of them Jewish, the others Italian, went on strike and closed the shops. Some four months later, in February, 1910, the strike was settled with some improvements in working conditions, and

Local 25 had expanded from about 100 members to 10,000. This "Uprising of the Twenty Thousand" was followed in July by the "Great Revolt" of the cloakmakers, who stayed out for three months, when the strike was ended by a Protocol of Peace proposed by Louis D. Brandeis, the future justice of the Supreme Court.[83] Article Fifteen of this document provided for a Joint Board of Sanitary Control with power to establish standards of sanitation for shops in the industry, which both manufacturers and workers agreed to maintain. The idea of standardizing sanitary conditions developed from two directions. A desire to counter unfavorable publicity on unsanitary conditions in their shops and factories led the manufacturers to propose ways of improving the situation. For the union leaders, setting agreed-upon sanitary standards seemed an effective means for achieving the abolition of home work, a long desired aim.[84]

During the first two years of its existence, beginning in October, 1910, the Joint Board laid the foundations of its work, and on the basis of an investigation to determine specific sanitary defects established 28 standards adapted to the needs of the industry. A major role in these activities was taken by George M. Price (1864-1942), a physician who had immigrated to the United States in 1882, and who had long been concerned with the health problems of workers and tenement dwellers. During 1911-1913, Price was chairman of the executive committee of the Joint Board, serving at the same time as Director of Investigation of the New York Factory Investigating Commission. The situation in the needle trades, opened up to community scrutiny by the strikers of 1909 and 1910, was brought even more dramatically to public attention in New York by the disastrous Triangle Shirtwaist Factory fire in the spring of 1911, which caused the death of 145 workers, mostly young women. Public opinion was strongly aroused and the Commission was appointed to study safety and health conditions.[85]

As director of investigations for the Commission, Price arranged for the medical examination of workers in a number of industries, among them 800 cloakmakers, of whom 30 percent were women.[86] Apparently, the only precedent in the United States for this undertaking was the examination of 200 printers by James Alexander Miller, a prominent New York tuberculosis specialist, on which he had reported in 1908 at the International Congress on Tuberculosis in Washington, D.C.[87] Medical examination of workers was not new, but it had previously been limited to those who came to a hospital clinic as patients or who requested medical attention at home.

In 1910, W. Gilman Thompson, professor of internal medicine at Cornell University Medical School in New York, established the first occupational disease clinic at that institution. The clinic was discontinued in 1916 when Thompson resigned from the school.[88] During this period, he was able to study a wide range of industries and their workers, among them those employed in the needle trades. In 1912, discussing occupational diseases in New York City, Thompson observed that "A large proportion of these employees are Hebrews who are prone to neuroses of various kinds and in these employments many acquire tuberculosis as well. They have long hours of work, often in dusty, ill ventilated, overcrowded rooms. They frequently use gas stoves or charcoal ovens for heating flatirons, and often work under the pressure of 'speeding up', to finish tasks under contract, and to comport with changes in the seasons. They also often work in strained positions over machines which require operation by hand or foot....If a patient enters my clinic with a cough and says he is a 'presser' or 'machine operator', he is almost certain to have pulmonary tuberculosis with chronic bronchitis. The leaky gas stoves and gas flatirons contribute chronic illuminating gas poisoning, with symptoms of marked anemia, headache, and digestive disorders." [89]

As a visiting tenement physician for the New York Infirmary for Women and Children, and as physician in charge of its Out Practice Service, Annie S. Daniel (1858-1944) had examined many patients engaged in tenement work.[90] She was especially concerned with the effect of the work on the physicial condition of children compelled to work in unsanitary, crowded sweatshops for long hours.[91] "Much is being said and written upon occupational diseases", she stated in 1912, "and some attention must be paid to the physical condition of these youngest workers. At their age all parts of their bodies are constantly changing, growing and developing; the work is performed in rooms built for living purposes, not for manufacturing. All the industries require close attention, giving rise to well-defined eye strain, always a serious condition. The constant bending over the work prevents the development of the chest wall and hinders the expansion of the lungs; the spinal column can not be maintained in the proper position." Based on her observations extending over more than two decades, Dr. Daniel concluded that "The pay these tenement folk receive is extremely dear; the price they pay is the stunted bodies and dwarfed minds of their little children." [92]

As he viewed the problem of occupational diseases, Price saw a large group of pathological conditions which owed their origin to an occupation or to some of its elements, but quite indirectly so that it was difficult to trace their incidence to an occupational source. In studying these diseases, he maintained, it was not enough to know the processes and working conditions involved in the particular trade. It was also necessary to determine the comparative mortality and morbidity rates of workers in different occupations. But since mortality statistics suffered from a number of inadequacies, Price decided that "The best standard of the healthfulness or dangers of an industry would be a direct study of the physical condition of its employees and especially of those who have worked in it sufficiently long to be affected by it".[93] The examinations conducted by Price in 1911-1912 were made by experienced physicians, among them a female doctor, who examined the women. Aware that this project was only a preliminary effort to achieve the aim he sought, Price urged further activity along this line. "Our ideal", he said, "should be a compulsory preliminary examination of all workers entering each trade, supplemented by periodical examination during their employment in the trade. Only by such authoritative examinations will it be possible to determine the effects of an industry upon the health of the workers in it and to determine effective measures for the prevention of occupational disease." [94]

Nevertheless, the examinations revealed that of the 800 workers studies, 10.1 percent suffered from acute and chronic bronchitis, 1.6 percent from acute pulmonary tuberculosis, 4.5 percent had cardiac troubles, and 6.4 percent eye diseases. In fact, the general impression which the workers made upon Price was that they were not a very healthy group of men and women, many exhibiting some degree of anemia as well as postural defects. These characteristics appeared most prominently among the pressers, of whom a large number suffered from spinal curvature, and various manifestations of chronic carbon monoxide poisoning such as headaches, anorexia and anemia. [95] On the basis of the findings disclosed by his investigation, Price began to urge the New York locals of the I.L.G.W.U. to create tuberculosis benefit funds for the relief of their sick members, a policy which was increasingly adopted during the next few years. One local even adopted a policy of not admitting new members without a physical examination, and applicants found to have symptoms of tuberculosis were rejected. [96] This incipient health and welfare program received a further impetus as a result of an investigation to determine the effect of the industry on the health of its workers undertaken at the request of the Joint Board of Sanitary Control and carried out in 1914 by J.W. Schereschewsky, a physician of the U.S. Public Health Service.

In 1910, the Public Health Service had begun to take note of dangers to health in the working environment, and in 1914 a Division of Industrial Hygiene headed by Schereschewsky was set up. The objective of this unit was to obtain information which would make it possible to maintain "the physical efficiency of the worker and the length of the period of his economic productivity" by making industry safe in terms of health. As a large part of the loss due to ill health is preventable, Schereschewsky argued, "society is not fostering foolish fads nor indulging in vague humanitarianism by displaying active interest" in the health of its workers. "It is only natural, therefore, that the question of the physical examination of workers (or medical supervision, as I prefer to term it) should have attracted increasing attention in recent years."[97] The linkage of ill health with economic losses in industry reflects ideas current at the time on the need for prevention of disease and promotion of health so as to conserve the human resources of the nation and thereby enhance productivity. [98]

In the course of the investigation conducted by Schereschewsky, over 3000 workers (2139 men and 971 women) were examined. [99] A medical and social history was taken and the workers received a physical examination lasting on an average 40 to 50 minutes, and including determination of blood pressure, pulse rate, hemoglobin, albumen and sugar in the urine (in males), and in 259 cases tests for syphilus. Workers with defects or diseases requiring immediate attention were directed, through the co-operation of W. Gilman Thompson, to the out-patient department of the Cornell Medical School.

Tuberculosis was found to be the most prevalent disease, but it was three times more prevalent among men than among women. Furthermore, it was most prevalent among finishers, the poorest paid group of garment workers. Apart from tuberculosis, the most common defects and diseases were defective vision, faulty posture, chronic catarrhal conditions of the nose and throat, defective teeth and pyorrhea, digestive troubles and nervous affections. Chronic bronchitis and pulmonary emphysema were found principally among older men (tailors, pressers, finishers). Only about 2 percent of those examined were free from any defects. While no diseases peculiar to garment workers were found, it appeared that low wages and irregularity of employment due to seasonal activity were important factors in causing certain conditions such as tuberculosis. This tentative conclusion was supported by a subsequent study of the socio economic data collected during the investigation. B.S. Warren and Edgar Sydenstricker analyzed this material, focusing on the relation of econ-

omic status to health. They concluded that "the greatest number of
poorly nourished, anemic, tuberculous workers in an extremely seasonal
industry were in that group composed of the lowest paid and the least
regularly employed."[100]

The facts revealed by the Public Health Service study were brought to
the attention of the union locals by Price. With their cooperation, he
established a medical clinic as part of the Joint Board of Sanitary Control, and then in May, 1917, a dental clinic was organized for the workers in the industry. In 1919, however, the New York locals which were
supporting these clinics decided to separate them from the Board, by
forming a Union Health Center Association under whose aegis they would
be operated. A building was acquired at 131 East 17th Street in Manhattan where the expanding activities would be located and George M.
Price became the director. This pioneer facility remained unique for
some thirty years until other unions began to establish health centers
for their members in the 1950's. The Union Health Center in New York
has continued to serve the members of its supporting unions uninterrupt-
edly up to the present.

By 1920, obvious progress in dealing with problems of occupational
health in the New York garment industry had been made through the efforts, over a period of years, of organized workers, employers, social
reformers, physicians and government officials. Although this examination of New York City and its major industry, the needle trades, over a
period of fifty years in relation to urban growth, industrialization and
occupational health is only a case study, certain of its aspects have
broader implications. There is a need to study other industries as
facets of urban growth and economic development, both in the United States
and in other countries, and to investigate the occurrence, recognition
and handling of occupational health problems as integral elements within
this process. In New York City, for example, painters, plumbers, sheet-
metal workers, workers in brass, typesetters, electrotypers and hatters
are only a few of the workers who might be studied from this view point.
The relation of this aspect to the growth of medical knowledge and its
application as forms of social action also remains to be investigated.
Clearly, a knowledge of technology and industrial processes is required
to understand the peculiar hazards to which workers may be exposed. Here
the history of medicine meets the history of technology and one should
consider the possibility of fruitful collaboration on specific questions.
There are, of course, numerous other aspects to which attention might be
drawn, such as the role of trade unions, or the significance of political
and social movements such as Progressivism in the United States around
the turn of the century. Enough has been presented here, however, to
indicate that this investigation has only dealt with a very small part
of a very broad and largely unexplored territory.

References

1. George H. Rohé: The Hygiene of Occupations, Public Health, Papers and Reports, American Public Health Association, Vol. X, 1885, Pp. 165-173 (p. 165)

2. Henry W. Farnam: Introductory Address, First National Conference on Industrial Diseases, Chicago, June 1, 1910, New York, American Association for Labor Legislation, 1910, Pp. 5-6. The Association had been founded in 1906 and existed until 1942. Farnam was professor of economics at Yale University from 1880 to 1918, and a founder of the Association as well as its president during its first four years. He had a major interest in social legislation.

3. George M. Kober: Industrial and Personal Hygiene. A Report of the Committee on Social Betterment, Washington, President's Homes Commission, 1908; George Rosen: From Frontier Surgeon to Industrial Hygienist. The Strange Career of George M. Kober, Amer. J. Public Health 65: 638-643, 1975; J.W. Schereschewsky: The Health of Garment Workers (Public Health Bulletin No. 71, Studies in Vocational Diseases), Washington, D.C., G.P.O., 1915, Pp. 13-103

4. U.S. Bureau of the Census: Historical Statistics of the United States Colonial Times to 1957, Washington, D.C., G.P.O., 1960; Arthur M. Schlesinger: The Rise of the City 1878-1898, New York, Macmillan, 1933, Pp. 64-65, 72-75

5. R.G. Albion: Rise of New York Port 1815-1960, New York, 1939; Oscar Handlin: The Newcomers, Cambridge, Mass., Harvard Univ. Press, 1959 Pp. 9-10

6. Emerson D. Fite: Social and Industrial Conditions in the North during the Civil War, New York, Macmillan, 1910, Pp. 88-90; Max Hall (ed.): Made in New York, Cambridge, Mass., Harvard Univ. Press, 1959, p.47

7. Edgar M. Hoover and Raymond Vernon: Anatomy of a Metropolis, Cambridge Mass., Harvard University Press, 1962, Pp. 36; Seymour J. Mandelbaum: Boss Tweed's New York, New York, John Wiley & Sons, 1965, Pp. 34-35; Moses Riselim: The Promised City, New York's Jews 1870-1914, Cambridge, Mass., Harvard Univ. Press. 1962, Pp. 6-7

8. Rischin, op.cit., p. 7

9. Edward E. Pratt: Industrial Causes of Congestion of Population in New York City (Studies in History, Economics and Public Law No. 109), New York, Columbia University, 1911, Pp. 19-25

10. Ibid., p. 38

11. Ibid., p. 40. The 1860 data are from the U.S. Census of 1860; the 1906, 1910 data are from the Report of the New York State Department of Factory Inspection, 1910

12. Rischin, op. cit., p. 67 sets the proportion somewhat higher (47% of the factories and 46% of the industrial labor force) for the production of apparel. Nevertheless, the basic trend remains.

13. Manual of the Metropolitan Board of Health...,June, 1869, Pp. 4-5,71 73; Manual of the Board of Health of the Health Department of the City of New York, 1874, New York, D. Appleton & Co., 1874, Pp. 6, 40-41, 56

14. Allan McLane Hamilton: Report on Manufacturing Establishments, Where Lead and Arsenic are used, Report of the Board of Health, New York, 1872-73, New York, 1873, Pp. 312-314

15. Idem: The Liability of Car-Drivers to Certain Forms of Disease, Report of the Board of Health, New York, 1873-74, New York, 1874, Pp. 444-

445; Allan Nevins: The Emergence of Modern America 1865-1878, New York, Macmillan, 1927, Pp. 81-82

16. John E. Erichsen: On railway and other injuries of the nervous system, London, 1866; E. Fischer-Homberger: Die Traumatische Neuronse. Vom somatischen zum sozialen Leiden, Bern, Verlag Hans Huber, 1975, Pp. 16-22

17. Hamilton, op. cit., p. 445

18. W.N. Baer: The Economic Development of the Cigar Industry in the United States, Lancaster, Pa., 1933, Pp. 41,92; Lucy W. Killough: The Tobacco Products Industry in New York and its Environs, New York, 1924, p. 33; Samuel Gompers: Seventy Years of Life and Labor, 2 Vols. New York, 1925, Vol. 1., p. 34; Rischin, op. cit., p. 68

19. Roger S. Tracy: Report on Tobacco and Cigar Manufactories, New York Board of Health, Annual Report, 1872-1873, Pp. 308-310; Nathaniel B. Emerson and Roger S. Tracy: Report on the Manufacture of Cigars in Tenement-Houses, New York Board of Health, Annual Report, 1874-75, Pp. 649-656

20. The Sanitary and Moral Condition of New York City, The Catholic World 7: 553-566, 1868, (p. 563); New York Association for Improving the Condition of the Poor, Twenty-Sixth Annual Report, 1869, Pp. 57-58

21. Albert H. Buck (ed.): A Treatise on Hygiene and Public Health, 2 Vols. New York, William Wood and Co., 1879, Vol. II, Pp. 42-43

22. T. Kostial: Statistisch-medicinische Studien über die Sanitäts-Verhältwisse der weiblichen Bevölkerung der k.k. Cigarrenfabrik in Iglau, Wochenblatt der k.k. Gesellschaft der Ärzte in Wien, Vol.VIII. 313; 333; 341; 349; 357; 365; 373; 1868; M. Ruef: De l'influence de la fabrication du tabac sur la santé des ouvriers, Gazette medicale de Strasbourg, ----, 1845

23. New York, Second Report of the Factory Investigating Commission, 1913, 2 Vols., Albany, J.B. Lyon Co., 1913, Vol. II, p. 677

24. United States Bureau of Labor, Report on Condition of Women and Child Wage Earners in the United States, 19 Vols., Washington, D.C., G.P.O. 1910-1913, Vol. II, Pp. 677-678

25. Second Report of Factory Investigating Commission Vol. II, Pp. 677-78

26. Mary Van Kleeck: Artifical Flower Makers, New York, Russell Sage Foundation, 1913, p. 90

27. Annie S. Daniels: The Causes, Evils and Remedy for Tenement-House Manufacturing, Transactions, 15th International Congress on Hygiene and Demography, Vol. 3, Washington, D.C., G.P.O., 1913, Pp. 1011-1014 (see p. 1011)

28. Second Report of Factory Investigating Commission, Vol. I, p. 92

29. George M. Price: The Modern Factory, Safety, Sanitation and Welfare, New York, John Wiley and Sons, 1914, Pp. 43,45

30. Richard Edwards (ed.): New York's Great Industries, Exchange and Commercial Review, embracing also Historical and Descriptive Sketch of the City, its Leading Merchants and Manufacturers....,New York and Chicago, Historical Publishing Co., 1884, p. 114

31. Ibid., p. 116

32. Ibid., p. 137

33. Ibid., p. 161

34. Ibid., Pp. 142, 245, 300, 240, 180

35. Douglas and Sherwood's Hoop Skirt Factory, Harper's Weekly, Jan. 29, 1859

36. Edwards, op. cit., (footnote 30), Pp. 131, 213

37. The War of the Trolleys, Harper's Weekly, 1895, p. 79
38. Cloak and Suit, Fortune, June 30, Pp. 92-100
39. Ibid., p. 97
40. A.K. Gardner: Hygiene of the Sewing Machine, American Medical Times Dec. 15, 1860, Pp. 420-421
41. John Duffy: A History of Public Health in New York City 1625-1866, New York, Russell Sage Foundation, 1968, Pp. 429-431
42. Gardner, op. cit., p. 421
43. Ibid., Pp. 435-436
44. Ibid., p. 421
45. E. Decaisne: La machine à coudre et la santé des ouvriers, Annales d'hygiène publique 34: 105, 1870
46. A.H. Nichols: Sewing Machines, Mass. State Board of Health, Report, 1872, Pp. 180-221
47. Tracy, op. cit., (footnote --), Pp. 62-63
48. H.H. Ben-Sasson (ed.): A History of the Jewish People, Cambridge, Mass., Harvard University Press, 1976, Pp. 859-69; Salo Baron: The Russian Jew under Tsars and Soviets, New York, Columbia University Press, 1964, Pp. 65, 113-114; Ezra Mendelsohn: The Class Struggle in the Pale, Cambridge,(Mass.),University Press, 1970, Pp. 1-5; Rischin, op. cit., Pp. 24-31; Louis Levine: The Women's Garment Workers, New York, B.W. Huebsch, 1924, Pp. 12-13; Hall, op. cit., Pp. 48-49
49. Charles Bernheimer (ed.): The Russian Jew in the United States, Phila., John C. Winston, 1905, p. 122
50. A.L. Belden: The Fur Trade in America and some of the Men who Made and Maintained It, New York, 1917
51. Salo W. Baron: A Social and Religious History of the Jews, 2nd ed., revised and enlarged, New York, Columbia University Press, 1967, Vol. XII, Pp. 45, 56-57; Abraham A. Neuman: The Jews in Spain, 2 Vols., Philadelphia, Jewish Publication Society, 1942, Vol. I, Pp. 167, 183, 187; Georg Caro: Sozial-und Wirtschaftsgeschichte der Juden in Mittelalter und in der Neuzeit, 2 Vols., Leipzig, Gustav Fock, 1920, Vol. II, Pp. 108, 269
52. Bernardino Ramazzini: Diseases of Workers, translated from the Latin text De morbis artificum of 1713 by Wilmer Cave Wright, New York, Hafner Publishing Co., 1964, Pp. 287, 289
53. Levine, op. cit., p. 15
54. Abraham Cahan: The Rise of David Leviwsky, (1917), New York, Harper & Row-Harper Torch Books, 1960, Pp. 90-91; see also C. Bezalel Sherman: The Jew Within American Society, Detroit, 1960, p. 164
55. Burton J. Hendrick: The Jewish Invasion of America, McClure's, March, 1913, Pp. 125-165 (see p. 128)
56. Levine, op. cit., Pp. 15-16
57. The Phrase is taken from an analysis of the industry prepared by John R. Commons for the U.S. Industrial Commission (1900-1902) cited by Aaron Antonovsky: The Early Jewish Labor Movement in the United States, (Author) Yivo Institute for Jewish Research, New York, 1961, Pp. 163-164
58. E. Tcherikower (ed.): Geshikhte fun der yidisher arbeter-bavegung in di Fareynikte Shtatn (History of the Jewish Labor Movement in the United States), 2 Vols., New York, Yiddish Scientific Institute-Yivo, 1943-45, Vol. II, Pp. 348-349
59. Jacob A. Riis: How The Other Half Lives. Studies Among the Tenements

of New York (1890), New York, Dover Publications, 1971, p. 100

60. Albert E. Edwards: _Comrade Yetta_, New York, Macmillan, 1913, p. 13
61. _Ibid._, p. 108
62. Riis, _op. cit._, Pp. 100-102
63. Arnold Bennett: _Your United States_, New York, Doran Publishing, 1912, p. 187
64. Albert L. Webster: Tenement House Sanitation, in Robert W. De Forest and Lawrence Veiller (eds.): _The Tenement House Problem_, 2 Vols., New York, Macmillan, 1903, Pp. 303-327 (p. 313)
65. M. Fishberg: Health and Sanitation of the Immigrant Jewish Population of New York, _Menorah_, August-September-1902, Pp. 4, 14 cited by Rischin, _op. cit._, p. 87
66. Florence Kelley: The Sweating System, In _Hull-House Maps and Papers_....by Residents of Hull-House, New York, Thomas Y. Crowell, 1895, Pp. 27-45 (Pp. 33-34). See also Chicago's Housing Conditions, _The Charities Review_ 10: 292-297, 1900-1901 (Pp. 294-295); Josephine Goldmark: _Impatient Crusader_, _Florence Kelley's Life Story_, Urbana, Ill., University of Illinois Press, 1953, Pp. 33-35
67. On June 19, 1907, the governor of Massachusetts approved an act establishing inspection of factories and workshops in order to control tuberculosis. The function of the inspectors was to determine the prevalence of tuberculosis and other diseases dangerous to public health, the health of minors employed in factories, and the sanitation of factories, slaughterhouses, public buildings and tenements in which clothing was manufactured. See William C. Hanson: The Work of the State Inspectors of Health, in Edwin A. Locke (ed.): _Tuberculosis in Massachusetts_, Boston, Wright & Potter, 1908, Pp.18-27; Ida M. Cannon: _On the Social Frontier of Medicine. Pioneering in Medical Social Work_, Cambridge, Mass., Harvard University Press, 1952, Pp. 175-176, 179
68. H. Linenthal: Sanitation of Clothing Factories and Tenement-House Workrooms, in Locke, _op. cit._, Pp. 28-36 (p. 29)
69. Mabel Hay Barrows Mussey: Holding the Mirror up to Industry, _Charities_ 17: 591-598, 1906-07 (Pp. 592-593)
70. E. Royston Pike (ed.): "BusyTimes." _Human Documents of the Age of the Forsytes_, New York, Praeger, 1970, Pp. 203-251; Charles Booth: _Life and Labor of the People in London_, 17 Vols., London, Macmillan, 1902-1903, Vol. IV, Pp. 328-347
71. Sweated Industries and the Health of the Masses, _Medical Record_, March 7, 1908, p. 398
72. Kelley, _op. cit._, p. 34; Jane Addams: _My Friend, Julia Lathrop_, New York, Macmillan, 1935, p. 118
73. DeForest and Veiller, _op. cit._, Vol. I, p. 53
74. Hall, _op. cit._, p. 51
75. Lilian Brandt: The Social Aspects of Tuberculosis Based on a Study of Statistics, _A Handbook on the Prevention of Tuberculosis_...., New York, Charity Organization Society, 1903, p. 55; Herman M. Biggs: _Preventive Medicine in the City of New York_, New York, Health Department, Oct. 1897, p. 23
76. Antonio Stella: The Effects of Urban Congestion on Italian Women and Children, _Medical Record_ 73: 722-732, 1908, (p. 730)
77. Linenthal, _op. cit._, Pp. 31-33
78. Biggs, _op. cit._, Pp. 11-15
79. Antonovsky, _op. cit._, p. 171

80. Maurice Fishberg: The Relative Infrequency of Tuberculosis among Jews, American Medicine 2: 695-699, 1901; Ibid.: Tuberculosis Among the Jews, International Congress on Tuberculosis, Washington D.C., September 30, 1908

81. Theodore B. Sachs: Tuberculosis in the Jewish District of Chicago, J.A.M.A. 43: 390-395, 1904

82. Edwards, op. cit., p. 77

83. Irving Howe: World of Our Fathers, New York and London, Harcourt Brace Jovanovich, 1976, Pp. 296-302

84. Levine, op. cit., Pp. 466-467. Among the members of the Joint Board as representatives of the public were Lillian D. Wald and Henry Moskowitz, a social worker and reformer, who headed the Madison Street Settlement

85. Leon Stein: The Triangle Fire, Philadelphia, (1st ed.) Lippincott Press, 1962

86. George M. Price: Occupational Diseases and the Physical Examination of Workers, Transactions, 15th International Congress on Hygiene and Demography, Washington, D.C., G.P.O., 1913, Vol. 3, Pp.844-849

87. Van Ingen, op. cit., p. 477

88. William Gilman Thompson, M.D., Med. J. and Record, Nov. 16, 1927, p. 632; William Gilman Thompson, J.A.M.A. Dec. 3, 1927, p. 1982; George M. Kober and Emery R. Hayhurst (eds.): Industrial Health, Philadelphia, P. Blakistorn's Son, 1924, p. 112

89. W. Gilman Thompson: Occupational Diseases in New York, Transactions, 15th International Congress on Hygiene and Demography, Washington, D.C., G.P.O., 1913, Vol. 3, Pp. 829-835 (see p. 835)

90. Roy Lubove: Annie Sturges Daniel, Notable American Women 1607-1950, ed. Edward T. James, Cambridge, Mass., Harvard University Press, 1971, Vol. I, Pp. 429-431

91. Annie S. Daniel: The Wreck of the Home. How Wearing Apparel is Fashioned in the Tenements, Charities 14: 624-629, 1905

92. Annie S. Daniel: The Causes, Evils, and Remedy for Tenement-House Manufacturing, Transactions, 15th International Congress on Hygiene and Demography, Washington, D.C., G.P.O., 1913, Vol. 3, Pp. 1011-1014

93. Price, op. cit., (footnote 86), p. 845

94. Ibid., p. 849

95. George M. Price: A General Survey of the Sanitary Conditions of the Shops in the Cloak Industry, First Annual Report of the Joint Board of Sanitary Control in the Cloak, Suit and Skirt Industry of Greater New York, New York, 1911, Pp. 35-95 (Pp. 40-41)

96. Levine, op. cit., p. 476

97. J.W. Schereschewsky: Physical Examination of Workers, Public Health Reports, Nov. 20, 1914, Pp. 3107-3113

98. George Rosen: Preventive Medicine in the United States 1900-1975. Trends and Interpretations, New York, Science History Publications 1975, Pp. 14-16

99. J.W. Schereschewsky: The Health of Garment Workers; J.W. Schereschewsky and D.H. Tuck: The Hygienic Conditions of Illumination in Workshops of the Women's Garment Industry (Studies in Vocational Diseases, Public Health Bulletin No. 71), Washington, Government Printing Office, 1915

100. B.S. Warren and Edgar Sydenstricker: Health of Garment Workers in Relation to Their Economic Status, Public Health Reports, May 26, 1916, Pp. 1298-1305 (p. 1305).

Symposium 5.   PHYSICS AND METAPHYSICS IN THE SCIENTIFIC REVOLUTION

SYMPOSIUM 5

'PHYSICS AND METAPHYSICS IN THE SCIENTIFIC REVOLUTION'

Introduction : A.R.Hall

Between sixty and one hundred persons at various times listened to the
three principal speakers and the invited commentators, of whom quite a
number participated in the discussion periods.

Dr W L Wisan (Oneonta, New York), initiating the consideration of the
paper by Professor Stillman Drake, showed herself largely in agreement
with his 'anti-metaphysical' view of Galileo's scientific activity.
'Metaphysical', she pointed out, is in any case a dubious word, often
better replaced by more precise epithets such as 'theological' or
'cosmological'. She agreed with the speaker that Galileo 'did not
attempt to build his science of motion on a foundation of metaphysical
premises' and this agreement she found common among students of Galileo's
writings, not least A E Burtt. Far from denying any such view, it had
rather been Burtt's object to discover what sort of mathematical metaphysics had emerged from Galileo's <u>experimental</u> discovery that Nature
must have a fundamentally mathematical character. 'Burtt does not make
Galileo deduce science from metaphysics, but, on the contrary, makes him
get a metaphysics from his science'. However, the further contention by
Alexandre Koyre that Galileo was an <u>a priorist</u> philosopher and a
Platonist foundered on the new manuscript evidence brought to light by
Drake and others, and the speaker expressed her own scepticism of 'a
genuinely Platonic theory of knowledge' in Galileo, and of his vision
of the material world as reflecting an ultimate mathematical reality.

This comment, prefaced by an acknowledgement of the rich and complex
intellectual background of which Galileo was a master, led Dr Wisan into
a discussion of Galileo's treatment of causation, where she begged to
differ from Professor Drake. Notably, she believed that at all points
in his career Galileo's attitudes to the <u>causes</u> of physical phenomena on
the one hand, and to mathematical analysis of them on the other, remained
pretty consistent. Even in the <u>Two New Sciences</u> Galileo is still
interested in <u>cause</u> (of cohesion, for example). The speaker referred to
a passage in <u>which</u> Galileo suggests that knowledge of mathematical
properties may help us 'to philosophize better about other and more
controversial qualities of natural substance'. Thus there was for him a
level of physical thinking deeper than the mathematical, which might
indeed lead the philosopher into theology, as when he concerns himself
with the problem of the creation of the universe. Dr Wisan drew attention
to the evidence of Galileo's interest in the 'Plato Problem' (of finding
the origin of the planetary speeds in their fall from a single point
towards the sun) at many times in his life. It would be wrong to suppose
such a hypothesis trivial in Galileo's eyes, and to ignore his desire to
attain a broader vision and a deeper understanding.

The audience appeared to be deeply impressed by the learning of both
speakers, as well as conscious of the sincerity of Professor Drake's
search for the authentic Galileo in his manuscripts, and of Dr Wisan's
concern for Galileo's humanity. There was only a brief discussion,

managed by Dr Harry Woolf (Director of the Institute for Advanced
Studies at Princeton) with characteristic urbanity, which largely turned
on the definition of Galileo's epistemological notions. Certainly no
one present rose to dispute Professor Drake's interpretation of Galileo's
private notes.

M. Costabel's paper on Descartes was then presented, with references
both to recent French work on the frequency of Descartes' use of words
and the re-edition of his correspondence. Consideration of this paper
was deferred until after the lunch break, when under the presidency of
Professor M S Asimov (Institute of the History of Science and Technology,
Moscow) who had taken the place of Dr Woolf, Dr A C Crombie (Oxford) spoke
first. Dr Crombie suffered under the disadvantage of not having been
able to study in advance the full text of M. Costabel's lecture, with
which however he showed no disinclination to disagree forcibly, and
therefore (while commenting en passant on the main speaker's views) he
organised his comments around an analysis of the methodology of Descartes,
noting its descent from the postulational method of Greek mathematics
which, when applied to physics, leads to the construction of a mental
world that cannot be otherwise. Here the 'clear and distinct ideas' of
Descartes fill a role analogous to that of mathematical axioms (and so
was created the method of reasoning more geometrico). Dr Crombie also
pointed out, however, that Descartes was no less interesting for his use
of scientific models, particularly of course the self-regulating
mechanical model (as in his physiology) - though here he had again been
at least once anticipated by Aristotle, who was also a master of the
postulatory method.

In the course of discussion, Professor E W Tielsch made the interesting
claim that Descartes had been vastly overrated at the expense of
Gassendi; in her view, the revival of the atomist theory had been the
outstanding feature of seventeenth-century physics, despite efforts to
ban the doctrines of Epicurus.

The final paper on Newton by Professor Paolo Casini (Bologna) was followed
by two invited comments; in the first of these Professor Alan E Shapiro
(University of Minnesota) showed less distrust than Dr Wisan of the very
word 'metaphysics'. He applauded Professor Casini's vindication of the
traditional contention that the bases of Newton's natural philosophy
were mathematical rather than metaphysical in the traditional sense, and
his tracing of Newton's ideas of method to Henry More and Gassendi, the
critics of Descartes' derivation of physics from metaphysics. For Newton,
as Professor Casini said, metaphysics and knowledge of final causes were
posterior to natural philosophy. Dr Shapiro drew attention to Newton's
criticism in his Optical Lectures of the Aristotelian endeavour to discover the essence of things, while failing to produce an adequate account
of the actual operations of nature. The terms of Peripatetic philosophic
discussion were, Newton found, largely irrelevant to an investigation of
refraction and the improvement of lenses; and it was his success in actual
investigation that makes Newton's methodological concepts significant.
Here Dr Shapiro quoted Newton's early statement of his ambition (never
completely realised) to make the science of colours mathematical, and
his injunction: 'I therefore urge geometers to investigate nature more
thoroughly, and those devoted to natural science to learn geometry first.'

Turning to the Laws of Motion, he agreed with Professor Casini's distinction between their metaphysical origin in Descartes, and their origin with Newton in the mathematisation of experience, but considered that Newton no less than Descartes held them to proceed from God's fiat. As for Newton's use of the word 'law' in the modern scientific sense, confirmed by experiment and mathematical reasoning, Dr Shapiro gave examples from all periods of Newton's life. However, commenting on Professor Casini's distinction between a 'strong' metaphysics (as a basis of science) which Newton eschewed, and a 'weak' metaphysics which really served only to justify traditional notions of theology and philosophy, he found this inadequate to account for the 'paradox' of Newton's double life as revolutionary scientist and conservative scholar. Dr Shapiro inclined to believe that the 'paradox' was an optical illusion created by our inadequate understanding of the totality of Newton's intellectual work in the context of its time.

Dr Zev Bechler (Haifa University) in his contribution argued that the metaphysical aspects of Newton's thoughts were more weighty and important than previous speakers had allowed, and that their content and structure was more complex than Professor Casini had taken them to be. His approach had in general been too positivistic.

Professor Henri Laboucheix (University of Paris-Sorbonne) returned to the point made by Professor Shapiro in concluding, and maintained that for all the variety of Newton's activity and the complexity of his thought, this thought was essentially consistent. He did not believe that Newton had said different things in different places.

Professor Casini replied elegantly and courteously to the discussion in several languages, whereupon Professor Asimov concluded the afternoon session from the Chair with the observation that there always has been and will be a close relation between physics and metaphysics; but this is with the metaphysics of materialism, whether we are concerned with the seventeenth century or the twentieth.

GALILEO'S PROCEDURES, AND METAPHYSICS

In this symposium we shall be discussing physics and metaphysics in the
scientific revolution, particularly during the 17th century, taking as
focal points the cases of Galileo, Descartes, and Newton. In the case
of Galileo, with which I have been entrusted, two distinct periods can
be discerned in which the relation of physics to metaphysics as
reflected in his writings was quite different, the transition coming in
the years 1602-1604. For convenience, these may be distinguished as
his youthful and his mature views, though "youthful" may be not quite
the right term for a man up to the age of forty. Our principal
interest, I take it, will lie in Galileo's mature view of the relation
between physics and metaphysics. Since he did not begin to publish
books until after 1604, and since it was through his published works
that he may be presumed to have had his main influence on the scientific
revolution of the 17th century, it seems appropriate to concentrate on
his mature concept. Nevertheless it will be good to begin with some
remarks about Galileo's youthful view of the place of metaphysics with
respect to physics, inculcated in him by his professors of philosophy
at the University of Pisa.

For Aristotle, physics and metaphysics were quite separate topics, to
be studied in that order. The prefix "meta" implies this, but the
word "metaphysics" was not Aristotle's; it was the invention of his
editors, designed to indicate the proper order of study of the two
separate disciplines. Aristotle's own terminology may seem to reverse
that of his later editors, for what they called "metaphysics" he had
called "first philosophy", sometimes referring to physics as "second
philosophy"; nevertheless his Physica had been completed before his
Metaphysica was written. By the adjective "first", Aristotle meant
"in order of excellence", not "in order of study"; he regarded physics
as a necessary preparation for the study of metaphysics and therefore
inferior to it. It follows that in Aristotle's view, one might master
physics and then never go on to metaphysics. As we shall see, that
was to be also Galileo's mature position.

During the centuries of university instruction up to Galileo's time,
however, a subtle change had taken place in philosophy so that the
relation between physics and metaphysics had become something of a
chicken-or-egg paradox. It is easy for us to understand the historical
situation in 1600 because of the foundational implications of the word
"metaphysics" today that have given rise in our century to a number of
other disciplines similarly named, as metalanguage and metamathematics,
for instance. Sophisticated professors of Aristotelian philosophy by
1600 had come to regard metaphysics as the egg from which good physics
should be hatched, whence any ideas in physics that conflicted with
Peripatetic metaphysics were by then pretty much taboo, at least in
university circles. What was inculcated in the youthful Galileo at
Pisa therefore was not the original thought of Aristotle, in which
physics was in principle free to pursue its investigations of nature

unhampered by metaphysical restrictions. This was necessary for
Aristotle in order that his higher metaphysical explorations might stand
on a solid and independent base. In the later Peripatetic tradition,
however, it would be considered pointless to pursue any physical
investigations that might conflict with Aristotelian metaphysics and
must therefore be worthless and fallacious.

Thus it came about that Galileo's Pisan De motu, his youthful treatise
on motion completed in 1591, was mainly chicken-physics hatched from
the metaphysical egg of the Aristotelian notion of cause. Very
different was Galileo's Two New Sciences, half a century later, in
which we have an egg-physics called by him the "elements of a new
science." Causal investigations for their own sake were therein
relegated to an indefinite future - and, if I am not mistaken, to a
separate discipline which the mature Galileo was willing to leave for
the most part to others, to be pursued at such a time as what he called
his elements of a new science of motion had been sufficiently developed,
by minds more able than his own, to prepare them for higher and more
profound speculations.

You will find what I have just said set forth in Galileo's own words
in his Two New Sciences, in passages that are now frequently ignored
or treated only as mock-humility on his part. The same view had been
nevertheless expressed in Galileo's letters and private papers for
many years before that book was printed.

In De motu, Galileo had persuaded himself by causal reasoning that
acceleration in natural motion is merely a brief temporary event at
the very beginning of motion from rest, whence it could be ignored when
seeking ratios of speeds in descent. Also, taking heaviness to be the
cause of downward motion, he reasoned that the more weight the more
motion, and hence the more speed. In this way Galileo early deduced
ratios for speeds along different inclined planes that were not borne
out by actual test, as he duly noted and accounted for by what he
called "material impediments". (3) I mention this part of the chicken-
physics of Galileo's youth, hatched from the causal egg, because it is
important to the later events which resulted in his turning to egg-
physics in 1604 and after.

Current disputes among historians and philosophers of science concerning Galileo's thought have arisen primarily, it seems to me, from
the reading of his mature metaphysical speculations in The Assayer,
the Dialogue, and Two New Sciences as if such ideas were necessary
foundations of his original discoveries and conclusions in science,
though historically those had been arrived at earlier and perhaps
without deep metaphysical reflections. Many modern disputes might be
resolved, and some may turn out to have been idle, when Galileo's
speculations are read not as foundations but as the mature reflections
of a man who had already completed his scientific investigations. Such
a reading seems to me especially indicated in the light of recently
published documents bearing on his discovery of the law of free fall,(3)
whence I shall have no more to say about disputes over Galileo's
Platonism and will be concerned primarily with the newly emerging, or
rather re-emerging, picture of Galileo as a scientist.

As to method, this is what Galileo said in his <u>Dialogue</u>, addressing Simplicio as representative of academic philosophers of his time:

> What you refer to is the method Aristotle uses in
> writing his doctrine, but I do not believe it to be
> that with which he investigated it. Rather, I think
> it certain that he first obtained it by means of the
> senses, experiences, and observations, to assure himself
> as much as possible of his conclusions. Afterward he
> sought means to make those demonstrable. That is what
> is done for the most part in the demonstrative sciences.
> ... The certainty of a conclusion assists not a little
> in the discovery of its proof - meaning always in the
> demonstrative sciences.

I have omitted a penultimate sentence giving Galileo's reason for this practice, to return to that later. It is evident here that Galileo did not reject, but on the contrary approved the procedure he ascribed to Aristotle, saying (in the present tense) that it <u>is</u> followed for the most part in such sciences as Galileo himself cultivated. In this procedure, two different orders are involved, one in original investigation and the other in presentation of conclusions reached, for an evident reason. This is that in searching for means to demonstrate to others something already made certain by observation and experience, the definitions and postulates required for a rigorous demonstration became clear, often for the first time, to the investigator himself. He then presents those first, in order that the reasoning offered shall be rigorous and compelling. Only afterward is it necessary for him to adduce experimental evidence, the purpose of which in presentation to others is to show that what has been demonstrated applies to the real world, and not just to a world on paper.

Now, just as Simplicio had looked no further than to the procedure adopted by Aristotle in presenting his conclusions - since that was all that Simplicio had to go on from Aristotle's extant writings - so it is that many historians have looked only to Galileo's published demonstration of the law of free fall in their attempts to determine how he had discovered it in the first place; for, like Aristotle, Galileo did not publish an account of his original investigations. In that way, many have made the same mistake as Simplicio, supposing that Galileo arrived at the law in the same way he finally presented it; that is, by pure thought. Yet that is not only psychologically implausible, but there is also documentary evidence against it in 1604, as we shall see. Besides, there are indications even in the passage just cited that Galileo was speaking from experience as an investigator when he supposed Aristotle to have proceeded in ways similar to those that he himself had found useful in arriving at acceptable demonstrations. In a similar way 19th-century historians of science, who were usually also scientists, presumed Galileo to have worked in much the same way they did, making sure by actual test that some rule was correct before seeking means to demonstrate the fact. In our own century historians are more severe; they demand to be shown documentary evidence that Galileo did proceed in that way, and did not simply continue in the egg-metaphysics chicken-physics style of his philosophical mentors and of his own Pisan <u>De motu</u>. As long as documentary evidence of his experimental work was lacking, they preferred to assume Galileo to have discovered the law of free fall *a priori*, as for instance by using the mean-speed postulate created in the 14th century.

Yet it had long been known that in October 1604 Galileo was already certain of the times-squared law, and that he had discussed it explicitly with Fra Paolo Sarpi at a time when he was still unable to offer in its support more than a very puzzling attempted demonstration, although the mean-speed postulate would have put into his hands immediately a very compelling proof. The basis for Galileo's confidence about the law in 1604 was also puzzling, since that basis could not reasonably have been some other and more rational *a priori* proof, while to attribute his discovery to careful experiment seemed anachronistic. The one elaborate experiment he later described in Two New Sciences could not very well date back to 1604, since that particular experiment would hardly have been devised by anyone before he knew what kind of law he was going to test. Indeed, I can show you that Galileo reached the year 1604 without any belief that a mathematical law of acceleration in our sense even existed, or that if one did exist it could be subjected to careful experiment.

Some years ago I described one possible way in which Galileo could have discovered the odd-number rule for distances in successive equal times as the by-product of a very simple experiment designed for a different purpose. (6) This did remove the objection of anachronism against an experimental basis for his confidence in 1604, but it lacked the documentary support required by historians. Subsequently examination of Galileo's manuscript notes on motion and the tedious, though rewarding, task of arranging them in order of composition cleared up for me the grounds of Galileo's confidence in late 1604, removed for me the puzzle about his idiosyncratic demonstration for Sarpi, and disclosed that the conceptual problems encountered by Galileo in connection with this discovery and its eventual proof were not those ordinarily supposed, but others, traceable to Aristotelian and arithmetical traditions against the notion of mathematically continuous change.

It was late in 1602 that Galileo hit upon his first correct result concerning motions along different inclined planes, a result that I think should be called Galileo's Theorem. (7) This states that descent along any chord to the lowest point of a vertical circle takes the same time regardless of the length and slope of the chord. The manner in which Galileo arrived at this result is evident from an unpublished diagram on the verso of folio 121 in volume 72 of the Galilean manuscripts at Florence, a diagram which he left unlettered but tabbed for future reference. Ignoring acceleration, and assuming the incorrect rule previously set forth in De motu which had made speeds along planes of the same height inverse to the lengths of the planes, any Euclidean geometer could promptly have reached Galileo's Theorem by simple inspection of that diagram. (8) A letter written by Galileo to his patron Guidobaldo del Monte in October 1602 announced proof of this theorem and revealed much else about Galileo's investigations of motion at that time, including the fact that he had as yet no expectation that close agreement with mathematical deductions could be found in actual experiments of motion on inclined planes.

Galileo's Theorem, which is easily capable of approximate verification
by simple test, led him at once to another result, communicated in the
same letter, which in turn reinterested him in the question of acceleration and gave him reason to question the causal postulate (that weight
was the cause of speed) which in De motu had resulted in his deducing
ratios for speeds along inclined planes that were not borne out by test.
To correct his earlier conclusions he now needed some measure of speeds
capable of actual application, rather than just the conventional
assignment of "degrees of speed" represented by integers that had
sufficed in De motu. It was probably in 1604, after a serious illness
in 1603, that Galileo turned his attention to this problem of actually
measuring speeds. This marked the beginning of his abandonment of
causal postulates in favour of postulates involving only measurable
physical entities such as weight, distance, and time. It was at this
stage that procedures became more important than metaphysics in advancing Galileo's science of natural motion, including experimental as
well as mathematical procedures.

It is on f. 107v of the volume previously mentioned that we find
documentary evidence of a careful experiment that accounts for Galileo's
confidence in the times-squared law late in 1604. A column of figures
there agrees very closely, but not quite exactly, with the distances
traversed by a body under natural acceleration during eight equal time
intervals from rest. Notes beside them show that the figures were
obtained by measurement with a ruler divided into 60 equal parts. These
measurements of distances were used by Galileo as measures of successive
speeds, probably of a ball rolling down a grooved plane inclined 60
units in 2000, or about 1.7°, the times being a little over half a
second. After a preliminary guess that successive speeds were as the
alternate odd numbers, neatly cancelled on this page, Galileo found them
to be as the successive odd numbers, and he drew in the margin the
Archimedean representation of an arithmetic progression, and then
showed this as contained in a triangle. Finally he noted the first
eight square numbers in line with the experimental data, from which
it is seen by inspection that those, multiplied by the first datum,
agree almost exactly with the experimental measurements.

It is evident that this experiment was not conducted to test a law
already known, and it is equally evident that soon after the experiment,
originally designed simply to get a measure of successive speeds,
Galileo was in possession of the law of free fall. I shall not attempt
to discuss the other related notes of 1604 and the proof written out
for Sarpi, having done that in papers published in journals of history
of science, unlike the paper in which I offered a full reconstruction
of the experiment behind f. 107v. (9) What seems to me important here
is that the very close agreement of the data recorded on f. 107v with
the times-squared calculations must have surprised Galileo as much as
it delighted him. So far as I know, the only physical law then known
that agreed with careful measurement to three and four significant
places had been the law of the lever. Galileo had believed up to this
time that material impediments prevented close agreement between
absolute mathematical deductions about motion and actual trials. To

find that natural motion lent itself to careful measurement no less precisely than the law of the lever opened for the first time a prospect that the science of motion could rival the science of statics in mathematical certainty of results. It is to this realisation that I attribute Galileo's having preserved f. 107v all his life, long after it could have been of further use to him, for he probably remembered it as the beginning of his new science of motion and could never bear to throw it away.

Now, if we test the two rival hypotheses about Galileo's discovery of the law of free fall, the currently popular notion that he was assisted by the medieval mean-speed postulate finds no support at all among his working papers or his published books. There is no instance known to me in which Galileo ever used mean-speed reasoning. On the other hand, f. 107v does contain data from an experiment of the kind that 19th-century historians of science assumed Galileo to have made, an experiment involving careful measurements. You may say that too much faith should not be placed in a hypothesis supported by only a single document. I can only reply by asking how much faith should be put in the popular hypothesis which is supported by no known document from Galileo's hand. The discovery of Galileo's Theorem from correct reasoning based on two false assumptions; detection of their falsity from further consequences of this correct theorem; replacement of inverse proportionality of speeds to lengths of planes of equal height by direct proportionality of times to those lengths – those were all consequences of his procedures, not of a change in his metaphysics. Galileo did not derive the law of free fall by adopting a different metaphysics; still less, by clinging to Aristotle's. Rather, it was by neglecting metaphysical restrictions on his actual procedures that he was put on the track of a kind of investigation new to him and repugnant to philosophers of his day, Aristotelian and Platonist alike. I do not deny that before he embarked on that road, he worked in the old way from causal and other metaphysical assumptions, nor can I deny that his first correct result about motions on inclined planes came through reasoning from metaphysical assumptions. But I do not think it helpful to portray his abandonment of metaphysical restrictions as itself a metaphysical decision.

Next, because it is appropriate topically, if not chronologically, I wish to speak of two passages from Galileo's last book that have long seemed incompatible with one another. Both relate to the causal concept, that central source of metaphysical postulates in physics. The first comes when Sagredo has just attempted to find the cause of natural acceleration by adopting reasoning that Galileo had used in De motu half a century before and had long since discarded. Salviati now replies to the effect that causal inquiries are not necessary in physics. (10) What Galileo did curtly dismiss were causal fantasies indulged in by philosophers without their having established a basis for them by careful measurement. Galileo did not thereby outlaw the concept of cause; he merely relegated its discussion to a domain beyond physics. That some cause did operate in natural acceleration was evident to him from the close agreement between mathematical deductions and experimental tests; but whereas Aristotle had assumed all causes of natural motion to reside in the nature of being and in the essences

of existing things, which then became the subjects of his first
philosophy or metaphysics, Galileo was content to stop short of that
subject, for practical reasons that I shall presently endeavour to make
clear. The second passage, seemingly incompatible with this, comes
when Sagredo has admired the power of Galileo's demonstration of a
certain fact long known empirically to artillerymen, because the demonstration led on to additional facts not previously known. Salviati now
replies:

> You say well. The knowledge of one single effect
> acquired through its causes opens the mind to the
> understanding and certainty of other effects without
> need of recourse to experiments. (11)

Galileo's prescription for "knowledge of one single effect acquired
through its causes" had already been given above; it consisted in
mathematical demonstration confirmed by actual test. One need not be
able to name and describe causes in order to know that causes are
operating. The "certainty of other effects" needs no recourse to
further experiments, so long as these effects follow mathematically
from some previous demonstration duly tested and confirmed; but there is
no getting round the necessity of an experimental verification somewhere back along the line. As early as De motu Galileo had remarked
that causes are not given by experience, an Aristotelian position that
he never forsook. (12) It is appropriate next to set forth Galileo's
conception of the proper foundation of science, and to describe the
limitations he placed on science, explaining also Galileo's motives
for imposing unprecedented boundaries on scientific inquiry. His
position on these matters was formed mainly between 1604 and 1615,
particularly during 1613 while he was writing replies to his philosopher-
opponents on the matter of bodies in water. His mature views appeared
explicitly in the Letter to Christina, circulated in 1615. The purpose
of that document was to insure, if possible, freedom of scientific
inquiry from theological interference. The only available strategy
was so to restrict the domain of scientific inquiry that theologians
would hesitate to invade it, and Galileo's previous battles with
invading philosophers had taught him how best to proceed. In his
Letter to Christina he staked out for science those areas, and only
those areas, in which no appeal was necessary beyond "sensate experience
and necessary demonstrations". (13) That phrase, repeated in
Galileo's later books and incorporated in the Latin title of his Letter
to Christina as printed in 1636, described the foundations of his
science and thereby severely restricted its scope.

If science were to restrict its field in this way, then he was sure
that no conflict could arise between scientific propositions and
scriptural passages. He was sure of this because God had both created
Nature and inspired the Bible. That was the burden of his Letter to
Christina, conceived I think not metaphysically but strategically, as
a practical matter of best procedure in trying circumstances. I
believe that Galileo personally was willing to circumscribe scientific
inquiry drastically as the price of freedom to proceed without interference within this narrow domain. The fact that hardly any other
thinker before Laplace was willing again to isolate science from

theology does not prove that Galileo was unwilling to do so; it merely shows that his revolutionary proposal was abortive at this time.

As time went on, Galileo's abortive scientific revolution went still further, and by 1623 at the latest it was his intention to separate science also from philosophy, though without thereby disparaging philosophical opinion any more than he disparaged theological opinion. Galileo implied in the Dialogue that both science and philosophy would benefit from separation, even though some individual philosophers might suffer. (14) Science stood to benefit by gaining freedom from philosophical restrictions such as the Aristotelian dichotomy between elemental and celestial matter, or the Platonic one between the real world and the real real world, that lay behind it. Philosophy stood to benefit by gaining more, and more dependable, physical knowledge on which to build its metaphysical conclusions. Galileo was notoriously of a sanguine disposition, and probably at the outset he had expected philosophers to welcome his 1615 program and even to recognise it as enlightened Aristotelianism. Instead, however, they universally opposed Galileo, so that in his last three books nearly every reference to philosophy or philosophers became ironical, or even sarcastic. Nevertheless he continued to hold out an olive branch to philosophers, and toward the end of his life he described himself as a true Aristotelian to the professor of philosophy at the University of Padua. (15) It is not surprising that such professors, traditional rulers of physics, rejected any reduction of their territory by a mathematician. Their grandiose promises continued to "excite the natural curiosity of mankind and keep men wandering in a labyrinth", as Galileo put it in The Assayer of 1623. (16)

Galileo recognised that independent instances in which close agreement could be found between sensate experience and necessary demonstrations were unlikely to be numerous, whence the domain of science would remain very narrow as compared with that of all human knowledge; yet in another way he saw science as boundless in extent, being capable of indefinite expansion and improvement by minds more intelligent than his. He said this expressly of telescopic astronomy, (17) of his new science of natural motion, and again, in a parable, of acoustics. (18) In his old age he remarked that good luck is an ingredient in the discovery of close agreement between mathematical deductions and physical events, again speaking from his own experience. He was also conscious of the speculative character of every extension of a rule beyond the range of phenomena within which it had been verified by measurement, explaining that things necessarily neglected in order to formulate any general rule might need to be taken into account in a given application of it, or in any extension of its domain. Such things show further that, as Professor Randall said, Galileo was not a philosopher, but a scientist; for good luck in discovery, or principles limited in application, play little or no part in philosophy. It is true that when applying for his position at the Tuscan court in 1610, Galileo insisted on having the title of "philosopher" as well as that of mathematician, but when there he demonstrated his right to that title not by debating metaphysical points, but by defeating professors of philosophy in a dispute over the physics of bodies in water. In all of Galileo's extensive correspondence there are very few letters discussing any traditional philosophical issue, in sharp contrast with the surviving letters Of Kepler, Mersenne, or Descartes. The few philosophical letters written by Galileo were composed under pressure of external circumstances, as in 1615, or at the request of a Medici prince, as in 1640. Much more frequently Galileo's letters dealt with purely scientific questions such as lunar mountains, sunspots, motion, hydraulics, mechanics, or strength of materials; and it is to Galileo's conception of proper physics, rather than to some obsessive metaphysical principle such as perfection of the circle, that we may reasonably attribute his so-called failure to state the inertial law.

With regard to the metaphysical speculations that are found in Galileo's books, it seems to me that the most significant thing to note is that he did not use them as the basis for any physical demonstrations. Here I have in mind such things as Galileo's account of primary and secondary qualities in The Assayer, (19) his reflections on circular motion as the only truly uniform motion in the Dialogue, (20) and his suggestion in Two New Sciences that the ultimate constitution of matter may consist of infinite aggregates of filled and unfilled mathematical points. (21) These, and others like them, present points of much interest with respect to later physical science, but none appear to have been regarded by Galileo as having scientific status. Apart from such speculations, Galileo's writings contain also certain explicit principles, assumed without question, such as the statement that nature acts always in the simplest way; (22) also some implicit assumptions, such as the uniformity of nature in the operation of physical laws. These have the character of axioms, and as Aristotle remarked every science has axioms on which its demonstrations must depend, since everything cannot be proved. (23) The discussion of axioms was, for Aristotle, a topic for metaphysics and not for the science using them; and since Galileo refrained from debates concerning the meaning and applicability of axioms, their relevance to our symposium seems to me highly dubious, and I shall introduce none.

A third class of propositions that may be deemed metaphysical in character consists neither of speculations, like Galileo's on the ultimate constitution of matter, nor of assumptions like the simplicity of nature. As an example of this other class I shall cite Galileo's statement that in order to reach any specified speed, starting from rest, a body must pass through all possible lesser speeds. (24) This certainly cannot be shown by any sensate experience, even today with equipment vastly superior to Galileo's; nor yet can it be mathematically demonstrated, since mathematics as such is loftily unconcerned with heavy bodies. Hence Galileo, who discussed this proposition several times and at some length, because it was hard for his contemporaries to accept and was ridiculed even by Descartes, (25) offered a justification for it. His justification was embedded in language itself and is sometimes called the negative principle of sufficient reason; if any intermediate speed is granted, why not all possible intermediate speeds? This resembles the way in which the postulate behind the lever law is justified; no reason for disturbance of equilibrium can be given for equal weights suspended at equal distances. Such propositions may be properly classed as metaphysical, but the assumption of the uniformity and simplicity of language was no more a source of opposition to Galileo's science than was the assumption of the simplicity and uniformity of nature.

Perhaps something from Galileo's writings will be brought up in our discussions that will change my present view, that his science was built as nearly as possible without deliberate metaphysical foundations by neglecting causes and essences in favour of procedures in observation and in measurement. Galileo had reason to mistrust the universals with which philosophers concerned themselves, and preferred to adopt postulates of limited applicability. That is why his physics dealt only with heavy bodies, as the late Professor Koyré liked to stress, (26) and not only that, but heavy bodies near the earth's surface. Whether or not that was a fault is a matter of opinion, but it was certainly not a source of scientific error in the ordinary sense of those words. Even today a man might restrict his researches to terrestrial physics and not be thereby exposed to the charge of being a poor scientist. Galileo's view in 1613 was that physics was not itself to remain a kind of philosophising, since it was to be newly founded not on philosophical opinion but on sensate experience and necessary demonstrations. This was to remain his view until 1638, when he had Simplicio opine that some great mystery of nature might be contained in Galileo's science of naturally accelerated motion.

Such, then, in Galileo's view, was the proper place of science with respect to philosophy and theology. It was not the view of Descartes, nor yet that of Newton, whose conceptions were to dominate the development of science after the death of Galileo.

Notes

1) Opere VIII, 190; TNS, p. 147

2) Opere I, p. 302; On Motion, p. 69

3) S. Drake, "Galileo's Discovery of the Law of Free Fall,"
Scientific American, May 1973, pp. 84-92; "Galileo's Work on
Free Fall in 1604," Physis 16:4 (1974), pp. 309-322; "The Role
of Music in Galileo's Experiments," Scientific American, June,
1975, pp. 98-104.

4) J. H. Randall, Jr., The Career of Philosophy, vol. 1 (New York,
1962), p. 339

5) Opere VII, 75-76; Dialogue, p. 51

6) S. Drake, "Galileo's 1604 Fragment on Falling Bodies," British
Jrnl. for Hist. of Science 4:4 (1969), p. 349

7) Opere X, 97-100

8) S. Drake, "Mathematics and Discovery in Galileo's Physics,"
Historia Mathematica 1 (1974), p. 136, where I used an inverted
diagram, having not then discovered f. 172v as Galileo's own
diagram.

9) See last paper cited in note 3, above

10) Opere VIII, 202; TNS, pp. 158-159

11) Opere VIII, 296; TNS, p. 245

12) Opere I, 263; On Motion, p. 27

13) Opere V, 319; Discoveries, p. 186

14) Opere VII, 62; Dialogue, pp. 37-38

15) Opere XVIII, 248-249

16) Opere VI, 237; Discoveries, p. 240

17) Opere VII, 91-2; Dialogue, p. 67

18) Opere VI, 280-281; Discoveries, pp. 256-258

19) Opere VI, 347-352; Discoveries, pp. 274-278

20) Opere VII, 44,53,57; Dialogue, pp. 20, 28, 32

21) Opere VIII, 96; TNS, p. 57

22) Opere VIII, 197; TNS, pp. 153-154

23) Metaphysica, 996b, 1006a

24) Opere VIII, 199-201; TNS, pp. 155-157

25) Correspondance du P. Marin Mersenne, ed. C. De Waard, vol. 8

    (Paris, 1963), p. 114

26) A. Koyré, Études Galiléennes, III (Paris, 1939), p. 231 (=III-81)

Abbreviated titles used in notes

Opere   Le Opere di Galileo Galilei, Edizione Nazionale (Florence,

             1929-1939), 20 volumes

TNS   Two New Sciences, tr. S. Drake (Madison, 1974)

On Motion   De motu, tr. I. E. Drabkin, Galileo on Motion and

             On Mechanics (Madison, 1960)

Dialogue   Dialogue Concerning the Two Chief World Systems, tr.

             S. Drake (Berkeley, 1967)

Discoveries   S. Drake, Discoveries and Opinions of Galileo

              (New York, 1957)

P. Costabel

PHYSIQUE ET MÉTAPHYSIQUE CHEZ DESCARTES

Le propos du Symposium qui nous réunit me paraît provenir d'un sentiment de plus en plus fort depuis quelques années chez un grand nombre d'entre nous, sentiment selon lequel une histoire de la raison est encore à faire, et à faire sur d'autres bases que celles adoptées jusqu'ici. Certes ce n'est pas d'aujourd'hui que les historiens des sciences sont avertis des dangers de la projection rétrospective et que leur méfiance est éveillée sur les lumières que l'on trouve trop aisément dans le passé éclairé par le présent. Mais quelques mises en accusation tapageuses de la rationalité classique ont amené progressivement à l'ordre du jour la nécessité de travaux capables de rendre compte de la complexité des démarches à travers lesquelles s'est effectué en réalité le progrès de la connaissance scientifique. Les recherches se multiplient dans ce sens. Et cependant la question de savoir si une objectivité complète dans la restitution du passé est concevable et possible reste posée, et ne saurait guère être résolue, me semble-t-il, au niveau de la méthodologie générale. J'apprécie pour ma part que le présent débat nous convie à porter le souci de cette question dans l'étude précise d'un thème particulier, celui des relations entre Physique et Métaphysique chez de grands promoteurs au XVIIe siècle. L'espoir d'avancer dans la solution d'un problème aussi fondamental que celui de la possibilité d'une histoire objective passe par la qualité des réponses susceptibles d'être formulées dans un cas comme celui du substrat métaphysique dont trois siècles nous séparent.

Si j'ai quelque titre à traiter du cas de Descartes, je n'en suis pas plus à l'aise car ma participation à l'Equipe qui a repris depuis cinq ans la publication d'un bulletin annuel de bibliographie critique des études cartésiennes, m'assure que le thème de Descartes métaphysicien est à la fois d'une grande banalité et affecté de bruits de fond puissants et confus. Le philosophe des idées claires et distinctes connaît cette infortune suprême, suscite périodiquement des contradictions passionnées et prend sa part des mises en accusation tapageuses auxquelles je viens de faire allusion. Témoin le petit livre récemment publié en France sous le titre *Descartes inutile et incertain*, emprunté à un mot de Pascal. On condamne souvent Descartes pour un rationalisme excessif et on ironise volontiers sur sa Physique sans expériences, élaborée de manière artificielle sur des fondements "métaphysiques". Renouveler le sujet est difficile, dans la mesure même où ces critiques s'exercent à partir de données textuelles que tout le monde connaît, mais il vaut la peine de s'y essayer, ne serait-ce que pour s'interroger sur l'application du qualificatif ne serait-ce que pour s'interroger sur l'application du qualificatif même de métaphysique. De nos jours ce qualificatif est en général appliqué à tout ce qui paraît déduit d'une vision religieuse du monde, à tout ce qui résulte d'une cosmologie informée par la théologie, et à ce titre il intervient d'ordinaire dans les controverses pour disqualifier telle ou telle théorie, telle ou telle œuvre sur le plan scientifique. C'est un qualificatif négatif. Mais il arrive qu'il soit employé dans un sens que suggère la préposition grecque de *méta* : ce qui est au-delà de nos prises sur les choses, au-delà de nos observations et de nos expériences. Et à cet égard une référence s'impose, considérable en raison de la personnalité éminente dont elle émane et particulièrement adéquate à notre sujet. Il convient de lui consacrer quelque attention.

C'est dans son ouvrage sur *La formation de l'esprit scientifique* (1938) que Gaston Bachelard a traité d'un "exemple d'obstacle verbal : l'éponge — extension abusive des images familières" et évoqué le cas de Descartes à partir du texte clé que constitue le paragraphe 7 de la IIᵉ Partie des *Principes*.

"Je ne sais pourquoi, déclare ce texte, lorsqu'on a voulu expliquer comment un corps est raréfié, on a mieux aimé dire que c'était par augmentation de sa quantité que de se servir de l'exemple de l'éponge", l'éponge dont l'augmentation de volume lorsqu'elle est gorgée d'eau est naturellement rapportée au remplissage et à l'extension de pores ou d'intervalles entre les parties de l'éponge.

Et Descartes, ou plus exactement son traducteur du latin en français, ajoute : "Nous ne devons pas faire difficulté de croire que la raréfaction ne se fasse ainsi que je dis." Même si cette phrase interpole par rapport au texte latin, Descartes ne l'a pas censurée et on peut souscrire sans réserve au commentaire de Gaston Bachelard lorsqu'il voit chez Descartes une confiance dans la clarté de l'image de l'éponge "qui nous montre comment une matière particulière peut se remplir d'une autre matière".

Les expressions du texte cartésien original sont conformes à cette confiance. "Bien que nous n'apercevions par aucun de nos sens le corps qui remplit les pores d'un corps raréfié", et parce qu'il n'y a "aucune raison qui nous oblige à croire que nous devions apercevoir par nos sens *tous* les corps qui sont autour de nous", la spongiosité est une "explication très aisée" de la raréfaction et celle-ci "ne peut être conçue autrement". C'est-à-dire que pour Descartes la manière dont la spongiosité s'imagine, conjugue deux qualités : celle de fournir une *explication simple* et celle de respecter la *conception* d'une physique du plein.

Lorsque Gaston Bachelard conclut : "la métaphysique de l'espace est chez Descartes la métaphysique de l'éponge", il est évident qu'il exprime sous une forme frappante et suggestive ce qui est au fond de l'affaire, à savoir que l'exclusion du vide est chez Descartes aussi fondamentale pour la conception de l'espace que pour l'élaboration et la fiabilité du modèle de l'éponge. Comme cette exclusion du vide est nécessaire pour des raisons proprement métaphysiques — nous entendrons Descartes à ce sujet tout à l'heure —, on peut suivre l'entraînement du discours tout en restant, avec Gaston Bachelard, parfaitement maître de cet entraînement. On peut parler d'"une métaphysique de l'éponge". Mais le succès de cette formule auprès de beaucoup de lecteurs va-t-il sans ambiguïté ? Je ne le crois pas.

La déclaration du texte cartésien, selon laquelle "il n'y a pas de raison qui nous oblige à croire que nous devons apercevoir par nos sens tous les corps qui sont autour de nous", autorise l'application du modèle de l'éponge au-delà de la perception sensible et il y a fort à parier que c'est pour cela que le qualificatif de métaphysique paraît au lecteur moderne singulièrement approprié. De l'adjectif métaphysique, utilisé à la limite de la permissivité lexicologique provenant de la préposition *méta*, on n'a pas aujourd'hui de difficulté à passer au substantif. Il y a là un phénomène d'usage dont il existe bien d'autres exemples — et dans la langue même de la science — et qui est toujours caractéristique d'un glissement dans le sens des mots parce que les adjectifs se prêtent davantage, à travers leur emploi marginal, à une sorte d'osmose continue. L'attrait que la "métaphysique de l'éponge" a eu et a encore sur le lecteur de Bachelard, peut donc fort bien résulter d'une connivence secrète avec une évolution et un certain laxisme auxquels nous participons tous lorsque nous prononçons le mot de métaphysique. Un mot dont la signification originelle est pour nous trop lointaine. La formule de Gaston Bachelard oblige en tout cas à se poser la question de savoir si ce qui nous plaît en elle ne déroge pas à la pensée de Descartes.

La question en fait immédiatement surgir d'autres. Car la pensée de Descartes est-elle celle que nous livrent les écrits publiés de son vivant, par lui ou avec son approbation, formelle ou tacite ? Est-elle plus largement celle que rendent accessible les publications successives faites après sa mort et que jusqu'à nos jours restituent des lettres et des manuscrits ? Elle est certainement tout cela à la fois, mais les textes qui permettent de l'atteindre imposent eux-mêmes de distinguer chez Descartes l'écrivain publié et l'homme.

Sans doute est-ce là une distinction qu'il faut toujours faire à propos de tout promoteur d'un mouvement d'idées de quelque importance. Il est cependant davantage de rigueur dans le cas de Descartes en raison de la conjoncture dans laquelle il a eu à œuvrer, en raison des précautions dont il a dû user et des difficultés qu'il a rencontrées pour se faire entendre. Difficultés d'ailleurs — il faut le reconnaître — dont il n'a pas vigoureusement cherché la solution, car dans plusieurs domaines il a pratiqué vers la fin de sa vie un curieux détachement par rapport à la diffusion de son bien propre. Et la tentation est grande de relier ce détachement à l'aveu qui figure dès le début des *Cogitationes privatae* (A-T, X, 213), dans les notes manuscrites de jeunesse, datées de janvier 1619, qui ne nous sont connues que par la transcription que Foucher de Careil en fit à Hanovre, sur les papiers de Leibniz, au siècle dernier. "Larvatus prodeo", "Je m'avance masqué", écrivait Descartes à son propre usage. L'interprétation simpliste qui consiste à mettre dans la conscience de l'auteur un mélange de prudence et de duplicité, est aujourd'hui dépassée. L'aveu ne peut pas se détacher d'un contexte qui évoque ce que le masque d'un acteur de théâtre cache au spectateur, à savoir la rougeur du visage, c'est-à-dire l'effet d'émotions diverses. En souhaitant devenir acteur sur le théâtre du monde, Descartes a su très tôt que cela ne va pas sans aliéner une grande part de sa personnalité profonde. Ce qui est beaucoup plus complexe qu'une dissimulation pure et simple. Et Descartes a souffert plus tard, à maintes reprises, d'être jugé par beaucoup sur les apparences de son personnage.

Nous avons aujourd'hui un avantage sur les contemporains de Descartes, celui de disposer de documents, lettres et manuscrits, qui nous permettent de mieux comprendre l'homme, distinct de l'écrivain. Mais cet avantage n'est réel que dans la mesure où nous acceptons l'idée que cette distinction est susceptible de nous apprendre quelque chose. Notre disponibilité d'accueil est à cet égard une condition fondamentale. Favorable du point de vue psychologique, mais aussi, et surtout peut-être, sur le plan méthodologique. Car si nous considérons, comme nous devons le faire pour notre propos actuel, les déclarations de Descartes en faveur des fondements métaphysiques de sa physique, nous les trouvons aussi bien dans ses écrits publiés que dans ses lettres et nous pouvons avoir l'impression qu'il est inutile de s'embarrasser d'une distinction qui ne semble pas avoir une grande importance dans le cas envisagé. Face à l'apparence d'une forte cohésion doctrinale chez Descartes, il y a pourtant intérêt à ne pas mettre la méthode historique sous le boisseau et c'est essentiellement ce que je voudrais montrer.

Mais j'ai besoin, dès l'abord, de vous proposer quelques précisions de langage. Quand je parlerai de manuscrits cartésiens il ne s'agira pas de documents encore inédits — documents de plus en plus rares —, il s'agira de tout ce que l'auteur a, de son vivant, laissé soustrait à la diffusion publique, même si — comme c'est le cas des lettres — il y a eu une certaine diffusion dans un milieu restreint. D'autre part je voudrais, à titre expérimental, user d'un adjectif permettant de traduire la nuance que "cartésius" exprime en latin par rapport à "cartesianus". Et conservant "cartésien" pour ce qui est lu chez Descartes écrivain et relève de la cohésion doctrinale évoquée à l'instant, j'appellerai "péronien" ce qui correspond à une expression plus personnelle de Descartes, sieur du Péron, gentilhomme poitevin.

Tout ceci étant admis, il est temps d'entrer dans le sujet proprement dit.

L'image de l'arbre qui a pour tronc la physique et dont les racines sont la métaphysique est bien connue. Descartes la donne dans la préface à la IIe Partie des *Principia*, mais ce qu'elle traduit animait déjà les *Meditationes* trois ans plus tôt. L'aveu de l'auteur s'accompagnait seulement à cette époque du désir que l'on soit discret à cet égard. La conception des fondements métaphysiques de la physique est donc à la fois cartésienne et péronienne et c'est de 1641 à 1644 que Descartes est passé à son endroit d'une certaine réserve à l'affirmation publique. Tel est le fait qui s'impose à notre attention.

Il correspond évidemment à une démarche péronienne antérieure, probablement d'assez longue durée. La correspondance de Descartes, de 1630 à 1640, donne de cette démarche quelques témoignages positifs. A Mersenne, qui est alors le confident privilégié, Descartes n'hésite pas à déclarer que la Métaphysique est une science dont les vérités sont plus certaines que celles de la Géométrie, et sur lesquelles il espère bâtir solidement sa Physique. Mais il ajoute que bien peu d'esprits sont susceptibles d'entendre ce langage, c'est-à-dire qu'il a conscience de concevoir le statut de la Métaphysique sur un plan de rationalité qui est de nature à surprendre ses contemporains. Et qui nécessite une maïeutique, prudente et progressive. Tandis que le péronien cherche ainsi sa voie à partir d'un optimisme foncier concernant la métaphysique, le cartésien n'en laisse rien paraître. Le *Discours de la Méthode*, en 1637, situe l'origine de la démarche de l'auteur dans la découverte de l'excellence des mathématiques et dans son étonnement que l'on n'ait pas jusque là bâti sur elles quelque chose de plus relevé. Les *Essais* de la méthode, dont deux sur trois intéressent la Physique, ne contiennent pas un mot qui soit en désaccord avec un itinéraire intellectuel inspiré par le modèle de la logique mathématique. Il y a donc manifestement, au moment de la première publication de Descartes, une différence entre ce qu'il livre au lecteur et le projet profond dont il parle seulement à quelques intimes, de manière plus ou moins explicite. Le premier problème qu'il y a lieu de résoudre est celui de savoir si un jugement objectif est possible et si la différence, positive, engageait ou non la sincérité du philosophe.

Essayer de répondre exige de remonter à la décennie 1619-1629 au cours de laquelle Descartes dispose de toute la vigueur physique et intellectuelle du jeune homme qui franchit l'âge adulte, et en use de diverses manières. Deux manuscrits importants encadrent cette période : le *Compendium musicae* et les *Regulae*. L'Equipe Descartes les a tous les deux soumis à l'analyse automatique que permettent les moyens très modernes du Laboratoire du Pr Delatte à Liège et c'est du résultat de cette analyse que je tire, rapidement, quelques éléments utiles pour notre propos. Le *Compendium musicae*, rédigé dans la foulée de la rencontre avec Isaac Beeckman en 1618, relève de la seule passion de son jeune auteur à faire œuvre scientifique en introduisant une rationalité dans un domaine jusque là réservé à l'art. C'est un véritable Essai de la Méthode, mais naturellement de la Méthode avant la lettre. Si celle-ci n'est pas encore acquise explicitement en 1629, lorsque Descartes travaille encore à la rédaction des *Regulae* après plusieurs années de contacts avec Mersenne et avec son groupe, c'est bien cependant de la Méthode qu'il s'agit tout au long de ce manuscrit consacré aux "règles de la direction de l'esprit". Manuscrit inachevé, d'ailleurs, mais dont les titres des "règles" en attente d'un contenu témoignent de l'inspiration par la logique mathématique. Il est bien clair que de 1619 à 1629 ce sont les sciences, leur profonde unité et l'instrument privilégié que constituent les mathématiques, qui occupent le champ de conscience de Descartes. Les publications de 1637 sont homogènes à cette situation et prouvent que l'auteur n'a pas eu, à quelques années de distance, à se renier par rapport à un point de vue précédemment adopté et par rapport à son efficacité.

C'est là une constatation d'ensemble. Mais on peut donner sur la situation péronienne de 1629 quelques détails utiles. Et d'abord, l'auteur des *Regulae* manifeste dans son vocabulaire et par la structure même de maints raisonnements son attachement à la formation aristotélicienne reçue au collège jésuite de La Flèche. Jean-Luc Marion a parfaitement montré cela dans son ouvrage au titre suggestif, et je dois me contenter ici de renvoyer à cette étude très rigoureuse qui renouvelle considérablement la connaissance de Descartes. L'ontologie aristotélicienne a incontestablement conservé aux yeux du jeune philosophe un très grand attrait dans le même temps où il explorait sur le modèle mathématique la structure de la rationalité. Mais l'être des choses dont le même jeune philosophe avait à traiter en matière de science physique lui est apparu comme pouvant rester sous un certain voile, la rationalité ayant prise sur un contenant aux contours suffisamment précis sans que le contenu soit pour autant objet de connaissance complète. Et c'est pourquoi M. Marion qualifie la philosophie péronienne d'ontologie grise. Je crois l'adjectif adéquat, dans son acception commune, pour exprimer la nuance qui est ici nécessaire. S'il est vrai, comme le montre Marion, que Descartes se démarque essentiellement du discours aristotélicien par le flou qu'il laisse possible sous la rigueur logique du langage, c'est bien d'une grisaille qu'il s'agit, d'un affaiblissement de la teinte à contraste maximal que le nominalisme scolastique conçoit pour le produit de ses définitions.

Je crois même l'image en accord avec ce que les *Regulae* révèlent de la position de leur auteur par rapport au phénomène physique de la couleur. Car si cet auteur est informé de diverses tentatives théoriques basées sur l'existence de "couleurs primitives", il maintient sa faveur, du point de vue de la méthode et pour l'étude de la coloration des surfaces, à la notion aristotélicienne de mélange de blanc et de noir. Le noir étant pour lui ce qui caractérise les lignes de force le long desquelles la pression de la lumière est maximale et ces lignes de force se disposant suivant des figures géométriques aptes à réaliser le pavage du plan. L'analogie avec les notions simples et absolues auxquelles l'analyse permet de remonter et dont les diverses compositions s'imposent à la considération du physicien n'est pas factice. Et les règles pour la direction de l'esprit du Descartes de 1629 renvoient bien à une ontologie structurée comme la vision d'une surface éclairée, avec alternance plus ou moins dense de force et d'absence, de noir et de blanc, c'est-à-dire à une ontologie que nous ne pouvons dire que grise.

Mais c'est là une appréciation personnelle sur laquelle je ne saurais m'attarder. Il importe plutôt de relever ici quelques éléments significatifs. On sait que les *Regulae* situent à l'origine de la démarche de l'esprit un acte appelé *intuitio*, qu'il semblerait à première vue opportun de comprendre comme étant une sorte de pénétration immédiate. M. Marion a montré, à l'aide des correspondances de contexte que fournit aisément l'analyse automatique, que la traduction française d'*intuitio* dans la langue de Descartes est le mot *regard*. Le mot *regard* tel qu'il est en français, proche de son sens concret et dépourvu des dérivations morales qu'il a acquises en anglais. Bien loin d'être une vision pénétrante, l'*intuitio* péronienne est donc une saisie préalable d'ensemble, globale. Ce qui ne signifie pas, certes, qu'elle soit grossière, car le regard de l'observateur averti discerne des détails là où le vulgaire ne remarque rien, mais il s'agit bien avec le "regard" de ce qui est livré à l'attention de l'esprit avant toute activité d'analyse. Voilà qui corrige déjà beaucoup l'image que l'on propose trop souvent de Descartes, et il serait aisé d'apporter à l'appui de nombreuses références qui montrent combien le Descartes physicien a procédé par réflexion sur des données d'observation soigneusement rassemblées.

Les *Regulae* témoignent cependant à un autre point de vue, non moins important. Le point de vue de la *mathesis universalis*, c'est-à-dire d'une méthode inspirée par le modèle mathématique, mais d'application universelle. Descartes a donné à l'expression, qu'il trouvait devant lui, une dimension nouvelle. Celle de l'"ordre" que la correspondance entre les diverses catégories d'objets mathématiques révèle comme fondement de la démarche de l'esprit lorsque ayant épuisé les données du "regard" il s'agit de constituer un processus de compréhension. Ce processus, qui va par degrés et exige en chacun un choix raisonnable dans l'énumération exhaustive des possibles, remonte une chaîne ordonnée jusqu'aux principes premiers pour redescendre ensuite en prolongeant l'organisation des moyens et permettre ainsi la découverte. Il y a là beaucoup plus que l'illustration des vertus de la déduction et des chaînes logiques.

D'ailleurs, c'est une critique sévère de la littérature mathématique de son temps qui inspire l'auteur des *Regulae* et l'opposition qu'il manifeste à l'algèbre est liée au fait que les ouvrages qui se réclament du terme sont encombrés d'une nomenclature surabondante, remplis de recettes particulières, incapables de dégager la suffisance de quelques définitions précises et de quelques structures de relations formelles. Le modèle que l'auteur des *Regulae* cherche dans les mathématiques, et dont il pense que l'"Analyse" des Anciens contenait une ébauche voilée, est donc en réalité le fruit d'une réforme profonde. Réforme qui ramène catégoriquement la technique opératoire au rang de moyen et met au premier plan une logique relationnelle. Réforme placée sous le signe de l'économie, dans tous les sens du terme : le minimum de matériel opératoire et conceptuel pour le maximum d'efficacité, à travers des règles de composition itérables et réversibles.

Sans doute cette réforme impose-t-elle au langage mathématique des restrictions, sur lesquelles ce n'est pas ici le lieu de s'arrêter et dont Descartes est parfaitement conscient. Mais elle donne à ce langage un caractère de généralité propice à des applications dans d'autres domaines, en particulier celui des phénomènes physiques, et un rôle que la *Géométrie* de 1637 précisera avec discrétion, mais netteté. A savoir que l'algèbre péronienne est un code qui permet de "chiffrer" la géométrie et d'en transposer les problèmes dans une langue artificielle dont il s'agit ensuite de décoder les produits.

Il n'y a pas de doute sur le fait que les publications de 1637 présentent ainsi l'émergence d'une nouveauté radicale, mûrie durant une dizaine d'années et concernant la nature même du langage de la science.

Leibniz, qui fut celui qui, à la fin du siècle, comprit le mieux le projet, ne cessa de répéter qu'il n'allait pas aussi loin que son promoteur le croyait et s'appliqua à montrer comment il convenait de le dépasser. Mais dans les limites où Descartes se plaçait — exclusion du maniement de l'infini, exclusion de la considération du temps, etc. — le projet correspondait à la volonté de mettre à l'épreuve l'économie de moyens, aussi longtemps que possible, et il avait une consistance propre suffisante pour servir la réflexion philosophique en matière de connaissance certaine.

Sans doute — et c'est ici qu'il convient de revenir à ce que nous avons déjà relevé plus haut à propos de la correspondance avec Mersenne à partir de 1630 — Descartes a-t-il été amené parallèlement et progressivement à réfléchir à la "certitude" sur un tout autre registre. Celui des "vérités éternelles", celui de l'"existence", registre qui permet de situer comment l'appel à l'ontologie est inévitable. Car l'existence du moi pensant n'aboutit qu'à une solitude si rien d'extérieur à lui ne peut être assuré. Et il est bien vrai que Descartes entrevoit la nécessité d'une science de l'être, supérieure à la "Géométrie" et dont la clé de voûte est Dieu , ce Dieu qui "est" et qui garantit, par ce qu'il met en nous, la possibilité même d'une connaissance objective. Mais il est très clair que cette métaphysique n'interfère pas avec la science des phénomènes naturels pour contester sa légitimité et qu'elle vise essentiellement à conforter cette légitimité. En un mot, la métaphysique est passée dans les préoccupations profondes de Descartes afin de justifier une science déjà en voie d'élaboration de manière autonome.

Au cours de la période que nous évoquons en ce moment, Descartes a d'ailleurs fait une fois à Mersenne une déclaration bien surprenante. A savoir qu'il était en mesure de démontrer que la couleur de l'hostie consacrée dans l'Eucharistie est nécessairement blanche ! Il n'y est jamais revenu par la suite et il nous a laissé le soin de comprendre — ce que je crois pour ma part possible. Mais peu importe la démonstration. L'important est le fait que la physique péronienne possédait dans l'esprit de son promoteur une consistance suffisante pour qu'on puisse lui demander d'intervenir en semblable affaire et de fournir une vérité complémentaire de la spéculation théologique.

En définitive, je ne crois pas que le Descartes d'avant les *Méditations* et les *Principes* ait caché son jeu. Il n'a pas commencé par être métaphysicien, il l'est devenu, afin d'assurer davantage les bases d'une connaissance dont, par ailleurs, il s'était attaché à élaborer des moyens cohérents.

Et il est devenu métaphysicien au point de s'engager tout entier dans une œuvre plus philosophique que scientifique. Ce qui lui a valu beaucoup d'ennuis. Sa correspondance avec Chanut en 1646 évoque la leçon qu'il tirait de son aventure.

"MM^{rs} les Régents, écrivait-il le 1er novembre 1646, sont si animés contre moi à cause des innocents Principes de Physique qu'ils ont vu ... que si je traitais après cela de la Morale, ils ne me laisseraient aucun repos... Le mieux que je puisse faire désormais est de m'abstenir de faire des livres et de n'étudier plus que pour m'instruire." C'est-à-dire que le sieur du Péron persistait dans la conscience que ses Principes de la Philosophie naturelle n'entretenaient avec la Métaphysique que des rapports "innocents" et qu'il y avait pour lui une déception amère à constater les procès d'intention auxquels on le soumettait sans cesse. Il n'avait probablement qu'à s'en prendre à lui-même puisqu'il avait pris la plume pour faire des livres et pour montrer que sa Physique s'insérait dans un vaste système philosophique. D'où son aversion maintenant déclarée pour ce genre de littérature. Mais il avait certainement aussi réfléchi davantage, à travers une expérience littéraire dans laquelle il avait donné le change aux autres et un peu à lui-même, et il convient de relire ici ce qu'il disait au même Chanut quelques mois plus tôt, en mars 1646.

"Une seule observation que je fis de la neige hexagone en l'année 1635 a été cause du Traité que j'en ai fait [des Météores]. *Si toutes les expériences dont j'ai besoin pour le reste de ma Physique* me pouvaient ainsi tomber des nues et qu'il ne me fallût que des yeux pour les connaître, je me promettrais de l'achever en peu de temps. Mais pour ce qu'il faut aussi des mains pour les faire et que je n'en ai point qui y soient propres, je perds entièrement l'envie d'y travailler davantage."

Et il ajoutait : "Ce qui n'empêche pas néanmoins que je ne cherche toujours quelque chose ... afin d'en pouvoir conférer en particulier avec mes amis pour lesquels je ne sçaurais avoir rien de caché."

Ce texte admirable se passe de commentaire. Jusqu'en 1637, ce que Descartes ne cachait pas à ses amis, c'était sa préoccupation de fondements métaphysiques, mais la situation s'est renversée. Face à l'incompréhension des philosophes et des théologiens, Descartes mesure en 1646 le temps qu'il a perdu et, confiant à un ami les raisons pour lesquelles il n'espère plus terminer sa Physique, il parle d'expériences nécessaires et non plus de métaphysique.

Est-ce à dire que le péronien meurtri par les polémiques philosophiques en soit venu à estimer moins le problème des fondements de la connaissance ? Je ne le crois absolument pas, Leibniz nous a conservé le contenu de remarques que Descartes avait écrites à propos de ses *Principes* et j'ai donné de ce texte (original latin A-T, XI, 654-657) une traduction française dans l'Appendice de la deuxième édition du nouveau tome (A-T, IX$_2$, p. 361-362). C'est sur ce document que je me fonde.

Une étude d'ensemble me convainc que ce document non daté est contemporain des débats avec Regius, le disciple infidèle, donc qu'il se situe lui aussi en 1646 ou 1647. Je suis évidemment obligé ici de me contenter d'en faire quelques extraits significatifs.

"C'est une grande preuve de vérité de ne pouvoir pas ne pas être conçu, et de fausseté de n'être pas concevable, comme le vide, l'indivisible, le monde fini, etc." Telle est la première phrase qui manifeste d'emblée combien Descartes demeurait alors convaincu de l'importance du moi pensant. Mais il ajoutait : "Le premier cas *implique* l'être, l'autre le non-être *(Haec enim implicant esse, alia non esse)* et je n'ai pas besoin de souligner combien l'emploi du verbe "impliquer" introduit dans le fondement métaphysique une nuance très importante. Il ne s'agit plus de "démonstration" contraignante.

Et Descartes dit d'ailleurs un peu plus loin : "De tout ce qui inclut contradiction absolument on *peut* dire que cela ne peut pas être, bien que cependant on ne doive pas nier que Dieu puisse lui donner d'être, en changeant bien entendu les lois de la nature. Ce que nous ne devons jamais supposer qu'il ait fait, à moins qu'il ne l'ait révélé lui-même. Ainsi pour le cas d'un monde infiniment éternel, des atomes, du vide, etc." On voit que Descartes module ses principes fondamentaux de physique, tels la divisibilité de la matière à l'infini et l'exclusion du vide : l'indivisibilité et le vide ne sont pas concevables parce qu'ils incluent contradiction, mais à condition de le révéler Dieu aurait pu leur donner d'être et il ne l'a heureusement pas fait.

Forte est la *présomption (conjectura)* pour affirmer quelque chose, dit encore Descartes, quand, celle-ci posée, nous concevons Dieu plus grand et le monde plus parfait". Il n'y a ainsi aucun doute. La métaphysique, à laquelle Descartes s'adresse après les luttes qu'il a eues à soutenir, est bien toujours une science de l'être, mais elle n'a plus le caractère absolu qu'il avait d'abord imaginé. Elle se nourrit d'"implications", de "possibilités" et de "présomptions".

Permettez-moi encore une citation. Vers la fin du texte Descartes évoque le fait qu'il est peu raisonnable de considérer le monde comme fini, et il ajoute : "A nous, qui tenons pour l'infini, n'incombe pas la charge de résoudre les contradictions que l'on a coutume de proposer à son sujet, mais *nous sommes délivrés de toutes les difficultés par cet aveu très simple et très vrai* de ce que notre entendement n'est pas infini et qu'il est par conséquent incapable de comprendre telles quelles les choses qui concernent l'infini." Je suis personnellement, et en raison même de mon appartenance à une famille religieuse qui a eu avec Descartes une affinité notable, particulièrement sensible à l'accent de profonde modestie intellectuelle que traduit "cet aveu très simple et très vrai". Mais je crois que je ne serai pas seul ici dans ce sentiment.

Il y aurait évidemment beaucoup d'autres choses à dire, notamment en reprenant méthodiquement l'examen des divers *Essais* de la Physique de Descartes, pour montrer comment le terme d'Essai a toujours correspondu chez ce savant philosophe à une conscience lucide en même temps qu'à l'espoir d'ouvrir des "chemins". Mais le temps fait défaut et il faut finir. J'ai simplement relevé dans le péronien d'avant et d'après les prises de position publiques (1641-1644) quelques traits susceptibles de corriger la notion de la Métaphysique servante de la Physique qu'on croit trop aisément devoir attribuer à Descartes, notion à la fois rigide et immuable.

Puissiez-vous au moins, en réfléchissant chacun sur ces traits, reconstituer à votre usage un "regard" dont je souhaite, en terminant, qu'il soit pleinement anglais.

Prof. Paolo Casini

NEWTON, LES LOIS DE LA NATURE ET LE "GRAND OCÉAN DE LA VÉRITE'"

Dans quelques études récentes sur Newton on a souligné la tendance à la synthèse d'une pensée axée sur différentes perspectives intellectuelles. Il ne s'agit pas seulement de la synthèse scientifique proprement dite et de ses hors-d'oeuvre métaphysiques ou théologiques, mais aussi de ses études de critique biblique, de chronologie, d'alchimie, et du thème de la prisca philosophia. Même à défaut d'une méthode unitaire explicite, un fil continu aurait mené Newton dans le labyrinthe. Ses voies d'approche, quoique variées, conduiraient à un but unique: "Viewed in the light of the religious dedication of all his work, a écrit M. Manuel, Newton's secret researches were a form of daily prayer and communication with God. They are thus not wholly separate from his scientific discoveries. History sacred and profane was part of one divine order and the world physical and the world historical were not essentially different in nature" (I).

Les philosophes systématiques contemporains qui rêvèrent, comme Spinoza, Malebranche ou Leibniz, de réduire à l'unité le monde de la nature et celui de l'esprit, fondèrent leur rêve sur quelques notions-clé, comme l'ens ou la substantia, héritées de l'ancienne science première, la métaphysique. On a souvent parlé de la métaphysique de Newton. On se demande si chez lui la lecture parallèle des deux livres, la Nature et l'Ecriture, présuppose-t-elle à son tour, ou renvoie-t-elle à un savoir métaphysique voué à sonder "le grand océan de la vérité".

Newton n'a rien écrit sous le titre de 'métaphysique'. Il emploie parfois le mot, mais pour proscrire la chose du domaine de la philosophie expérimentale. Dans un passage célèbre du Scholium generale il frappe d'ostracisme au même titre les hypothèses et la métaphysique: "Hypotheses seu metaphysicas seu mechanicas seu qualitatum occultarum fugio..." Toutefois cette phrase est un peu outrée. On a démontré qu'elle n'exprime guère une attitude cohérente, car en fait Newton a formulé plusieurs hypothèses tout au long de sa carrière. De même, quelques propositions 'métaphysiques' subreptices soutiennent-elles les fondements et les voûtes du grand bâtiment. Son adepte mécréant Voltaire ne s'y trompa pas. La première partie de l'ouvrage où il mit "à la portée de tout le monde" les éléments de la philosophie de Newton s'intitule tout simplement "Métaphysique". Titre polémique, sans doute, dirigé contre Leibniz et les cartésiens. Titre de seconde main, comme les arguments que Voltaire emprunte de Clarke; mais non arbitraire, car il met en évidence un curieux paradoxe épistémologique. Malgré ses anathèmes contre les hypothèses métaphysiques, Newton croyait en effet que l'espace est Sensorium Divinitatis et que Causae finales in philosophia locum habent. A prix de la cohérence la métaphysique semble-t-elle donc tolérée, à peine déguisée, au sein de la physique newtonienne.

A cette contradiction apparente Newton et ses disciples les plus proches auraient pu objecter que l'espace et le temps absolus, l'attraction, la finalité de l'univers n'étaient nullement des essences métaphysiques, mais des évidences tirées des structures objectives du monde physique. Par sa procédure d'analyse et de synthèse, Newton n'avait-il pas renversé la méthode coutumière des auteurs de systèmes qui, depuis Aristote jusqu'à Descartes, obligeaient la recherche empirique à accréditer les hypothèses arbitraires préfabriquées par le philosophe dans son cabinet ?

Sans doute, chez Newton toute démarche conceptuelle débute de l'expérience et vise à la généralisation mathématique de ses données. Cependant

comme il arrive souvent dans l'histoire des idées, une formule trop simple
ne rend pas compte d'une situation compliquée. Malgré la simplicité et la
rigueur 'finale' des "Regulae philosophandi" (dont la plus importante, la
IIIe, ne figure que dans la deuxième édition des Principia), Newton est
loin d'avoir cultivé une conception positive et empiriste de la nature
comme le prétendaient les historiens du siècle passé. Depuis sa jeunesse,
il avait pratiqué des compromis avec ces "hypothèses métaphysiques" qu'il
écarte avec dédain en 1713. Au fur et à mesure qu'on a tiré de l'oubli
le Waste book, le De gravitatione et aequipondio fluidorum, les ébauches
des Queries et des "Classical scholia", on a compris jusqu'à quel point
il était allé dans ses compromis dont les traces, effacées avec soin,
demeurent presque illisibles dans les textes imprimés. De plus, les iné=
dits ont révélé l'ampleur de ses lectures: Galilée, Charleton, Gassendi,
Descartes, Henry More. Aussi peut-on réduire la question 'Newton et la
métaphysique' à l'influence que ces auteurs ont exercée sur lui. Croit-on
avoir tout dit là-dessus, que la discussion se ranime à cause des diffé=
rents points de vue où les interprètes se placent.

Car le paradoxe subsiste. Il paraît même plus criant si l'on réflé=
chit sans partialité sur la querelle toujours renaissante parmi les
Newtonian scholars de langue anglaise concernant les deux visages tradi=
tionnellement juxtaposées de Newton, mathématicien et savant positif d'une
part, et d'autre part 'magicien', comme disait Lord Keynes, ou adepte de
l'hermétisme et de l'alchimie, tel que nous viennet de le révéler les
travaux de MM Mc Guire, Rattansi, Westfall. Faut-il se résigner au cli=
ché du génie déchiré par sa névrose, en proie aux contradictions issues
d'un complexe d'Oedipe latent ? On peut s'y résigner, pourvu qu'on renonce
aussi à expliquer ces contradictions à la lumière d'une analyse synchroni=
que correcte.

La mot "métaphysique" est ambigu. Lorsqu'on parle de la "métaphysique
de Newton" il faut s'entendre sur le mot avant que sur la chose. D'abord
il faut préciser en quel sens Newton lui-même a-t-il employé ce mot, qui
avait alors un sens fort, une signification technique très précise et
très différente du sens faible, post-kantien, qu'on lui prête couramment.
Car au sens faible, moderne du mot, deux propositions contradictoires
comme 'Newton est un métaphysicien' / 'Newton rejette la métaphysique'
paraissent également soutenables. Si, au contrarire, on se place dans une
perspective synchronique, il faut rejeter avant tout en dehors de la mé=
taphysique (au sens fort du mot) les recherches que Newton mena dans la
no man's land de l'alchimie, et même ses études bibliques et théologiques,
ses références à la prisca philosophia, à l'hermétisme et à la flûte du
dieu Pan.

Loin de nier qu'il y ait là quelques éléments métaphysiques dont il
faut tenir compte, tâchons de préciser quelle était l'acception technique
du mot surtout chez les auteurs que le jeune Newton lisait. Après Descarte-s
et après ses critiques – Gassendi, Hobbes, Henry More – la métaphysique
ne s'identifiait plus avec les rêves théosophiques à la manière de Robert
Fludd ou des Rosacruciens, ni même avec les débris des philosophies natu=
relles de la Renaissance. Un grand effort de rationalisation et de géomé=
trisation avait marqué la nouvelle métaphysique cartésienne, qui avait
l'ambition de regagner son ancien rôle de science première. Descartes
prétendait en faire le fondement a priori et la contre-partie spirituelle
de la philosophie mécanique. Il ne géométrisa le domaine de la res extensa
qu'en s'appuyant, contre son malin génie, sur la véracité divine, source de
ses nombreuses hypothèses arbitraires.

Dans ses travaux, Robert Lenoble a marqué très nettement le clivage profond séparant les philosophies mécanistes de la nature des dernières efflorescences de l'hermétisme et de la magie. Avec la diffusion du car= tésianisme ces vestiges arcaïques furent repoussés en marge de la commu= nauté des savants. Même en Angleterre, à Cambridge, où l'ancienne tradition platonisante véhiculant l'esprit de nature, l'âme du monde et les natures plastiques survécut plus longtemps, elle fut assujéttie à un processus de rationalisation. A cet égard Henry More est une figure-clé. Au cours de sa polémique avec Descartes, tout en lui opposant sa méthode de métaphysicien, il apprit à traduire dans le langage philosophique de son adversaire les notions propres de l'ancienne tradition théosophique : l'espace, le temps, l'esprit de nature. Cependant les deux systèmes s'opposaient l'un à l'autre de façon nette. Citons deux textes concernant la hiérarchie du savoir et la procédure de la recherche. Selon Descartes un homme qui désire s'in= struire doit commencer par l'étude des mathématiques;

> puis,lorsqu'il s'est acquis quelque abitude à trouver la vérité en ces questions, il doit commencer tout de bon à s'appliquer à la vraie phi= losophie, dont la première partie est la métaphysique, qui contient les principes de la connaissance... La seconde est la phyisique... Ainsi toute la science est comme un arbre, dont les racines sont la métaphysique, le tronc est la physique, et les branches qui sortent de ce tronc toutes les autres sciences(2).

De son côté, Henry More refuse cette subordination de la physique et renverse cet ordre. C'est la métaphysique qu'on doit placer après la phy= sique, μετά τὰ φϑσικά, dont elle est le fruit:

> Metaphysica autem haec ars dicitur... quae post physicam doceride= bet,ut quae est nobilissimus philosophiae fructus. Nam ex accuratiori naturae sive mundi corporei cognitione in satis claram Dei caete rarum= que rerum incorporearum cognitione emergimus (3).

Loin d'opérer un réarrangement simplement formel, More entend détruire la méthode déductive de Descartes. Il débarrasse le terrain des questions concernant l'être, la substance, les attributs, qui sont du ressort de la logique. Héritier de la tradition nominaliste anglaise, il circonscrit le domaine propre de la métaphysique aux choses: "Sola relinquitur substantia corporea quae legitimum metaphysicae objectum esse possit". L'auteur de l'Enchiridion metaphysicum s'excuse d'avoir fourré dans son ouvrage un tel tas d'expériences de physique tirées des registres de la Royal Society. Mais pourqui parle-t-il encore de métaphysique ? C'est que "rerum in= corporearum existentia e corporeis mundi phaenomenis erat demonstranda", et qu'il faut commencer par détruire le "dogme pervers" de la distinction des deux substances. La démonstration du vacuum dépend en effet des expé= riences de physique, auxquelles More juxtapose sa théosophie.

A quelque détail près,ce même point de vue se retrouve chez l'autre grand adversaire de Descartes, l'atomiste Gassendi. Suivant la tradition épicurienne, Gassendi exclue la métaphysique de son Syntagma philosophicum: à son avis la science de l'être n'est autre chose que la recherche de natura rerum, dont les objets sont motus, mutationes; qualitates, ortus, interitus" etc. Bref, à l'ontologie aristotélicienne et à la métaphysique cartésienne Gassendi substitue des catégories physiques destinées à se remplir d'un contenu d'expérience (4).

Comme on l'a maintes fois remarqué, Newton se trouvait à la croisée de ces chemins. Vu le silence relatif desses textes, les interprètes ne tombent pas d'accord sur ce qu'il doit à chacun de ces auteurs. Il est vrai que dans

ses Quaestiones quaedam philosophicae il s'attache à quelques aspects de
la physique cartésienne sans en discuter les prémisses métaphysiques. Rien
n'est d'ailleurs plus étranger à son esprit que la naïveté et l'archaïsme
des arguments soi-disant physiques par lesquels H. More s'efforçait d'éta=
blir la réalité du vide, la cause de la gravité et ses rêveries mystiques.
Cependant, on affirme d'habitude que Newton a emprunté de More les concepts
de l'espace et du temps absolus, "émanations" de Dieu. On admet de même,
en se bornant à ces concepts, que Newton a dû méditer sur la section "De
loco et tempore, seu spatio et duratione" du Syntagma de Gassendi.

Enfin, la découverte de l'essai De gravitatione et aequipondio fluido=
rum a fourni la pièce justificative manquante de la réaction anticartésienne
du jeune Newton. On y constate une réfutation très articulée de la mécani=
que cartésienne dans son ensemble. Malheureusement, si Newton cite les textes
de Descartes qu'il réfute, il ne mentionne même pas les démarches critiques
analogues à la sienne de More ou de Gassendi. L'influence de ces auteurs
est admise e silentio. More, Gassendi, peut-être aussi Barrow ou Wallis :
quelles que soient ses sources immédiates, le jeune Newton a affirmé la
réalité métaphysique du temps et de l'espace par des arguments géométriques,
une thèse assez banale à cette époque, comme l'a remarqué A. Koyré.

Cet essay incomplet est donc une pièce maîtresse qui éclaire la for=
mation philosophique du jeune Newton. Une pièce, ajoute-t-on, méta=
physique, la seule que Newton nous ait léguée et qui nous révèle la
métaphysique latente des Principia mathematica. Mais en disant ceci on
emploie le mot ambigu dans sa signification faible. En fait, dans cet essai
Newton partage entièrement l'attitude phénoméniste et sceptique des advers=
aires de Descartes. En rejetant la géométrisation à outrance de la matière
et la conception relativiste du mouvement, il déclasse en même temps la
métaphysique à l'instar de Gassendi et de More. Il en conteste les catégo=
ries, le langage et même la possibilité. Quant à Gassendi, Newton ne se
borne pas à emprunter de la doxographie du Syntagma les notions détachées
de l'espace, du temps et de la matière. Il partage aussi sa démarche no=
minaliste, caractéristique de l'épistémologie atomiste, sa critique de
toute connaissance per causas, sa conception de la science limitée au do=
maine des perceptions. Autrement dit, Newton rejette en bloc la métaphy=
sique en tant que procédure déductive et systématique, quitte à affirmer
l'existence des entités 'trans-physiques' de l'espace et du temps.

L'attitude de Newton résulte de sa terminologie. Dans cet essai juvénil
il n'emploie que deux fois le mot "metaphysica", dans un sens négatif.
Il résulte du contexte, en premier lieu, que la "metaphysica constitutio"
des choses nous est tout à fait inconnue; en deuxième lieu, la structure
corpusculaire et discontinue de la matière qu'il oppose ici à la res exten=
sa de Descartes implique, dit-il, "praecipuas metaphysices veritates"(5).
Autrement dit, Newton prend de contre-pied l'ordre déductif cartésien.
Ce n'est pas la métaphysique de l'Etre qui prouve l'existence des corps;
au contraire, ce sont les corps sensibles, la durée et l'espace qui ren=
voyent à l'Etre suprême, sans que nous puissions pénétrer l'essence de
n'importe quoi. La terminologie du jeune Newton est précise. L'espace n'est
ni substance ni accident, mais Dei effectus emanativus(6). Les mots de l'é=
cole sont d'ailleurs suspects. Dans un long développement critique dirigé
contre le dualisme cartésien, Newton remarque qu'il dépend d'un équivoque
linguistique et conceptuel. En appelant également substantia l'essence in=
connue des créatures aussi bien que celle du créateur, par un préjugé en=
fantin et scholastique, les philosophes tombent en hallucinations. Ils
prêchent l'athéisme, ou séparent de Dieu les choses qui en dépendent (7).

A première vue,cette discussion paraît un peu sophistique. On a cru pou=
voir l'interpéter comme une négation immatérialiste des corps physiques;
mais à tort, car Newton se borne à nier la notion de substance. Il combat
le verbiage métaphysique par ses propres armes. Tout en parlant substance et
attributs, il affirme la réalité indépendante du vide où bougent les corps
graves, impénétrables, douées de masse. Exactement comme chez Gassendi, nous
trouvons ici une méta-critique des concepts de l'ontologie cartésienne.
Newton réduit les catégories de substance et attribut à un rôle purement
métaphorique, il les adapte à la conception atomiste. En affirmant que
nous n'apercevons que les qualités sensibles des corps dont nous ignorons
le support, il est très proche de la critique de l'idée de substance que
Locke développait à son insu, en ces mêmes années, dans l'ébauche de l'Essay

Loin de jeter les fondements d'une métaphysique, dans le De gravita=
tione Newton les a sapés.. Vers 1670 il avait déjà éclairci à soi-même
les principes de son épistémologie, où se fonde tout son travail des an=
nées suivantes. Témoignage précieux, ce texte prouve que pour batir le
cadre conceptuel des Principia et pour axiomatiser la science du mouve=
ment, Newton dut tout d'abord se débarrasser du jargon scolastique et écar=e
ter les préjugés métaphysiques qui entravaient la mécanique cartésienne.
Quarante ans après, la IIIe "Regula philosophandi" et le Scholium genera=
le expliqueront au lecteur l'a,b,c, du phénoménisme de 1670.

Plusieurs pages du De gravitatione sont consacrées à la réfutation de
la relativité du mouvement. Malgré tout, cette discussion retient des no=
tions cartésiennes, p.ex. l'équivalence des deux états de mouvements et
de repos, comme l'a remarqué Alexandre Koyré (8). Mais il n'a pas vu avec
la mêe clarté que Newton y dégage la dynamique de toute arrière-pensée
théologique. Descartes identifiait tout cort la cause première et la con=
stance du mouvement avec la volonté de Dieu qui, dit-il,"conserve mainte=
nant en l'univers, par son concours ordinaire, autant de mouvement et de
repos qu'il y en a mis en le créant"(9). C'est de l'immutabilité divine
qu'il déduit "certaines règles" ou "lois de la nature", parmi lesquelles
le principe d'inertie. La préhistoire de ce principe est bien connue. M
Herivel a montré que la première formulation de la loi d'inertie chez
Newton, aux Axiomes 1,2 et 100 du Waste book (1665 ?) est un calque pres=
que littéral des textes de Descartes: "Descartes deduced both parts of the
principle from the nature of God,and his working in the universe; the first
as a special case of a more general principle mirroring the immutability
of God, the second from the immutability of God and the simplicity of the
operation by which God conserved motion. Both these peculiar features
figure in Newton's presentation of the principle" (10).

La remarque est exacte, si l'on précise qu'en soudant les deux parties
du principe cartésien, Newton en laisse tomber l'arrière-plan métaphysique.
Dans tous les essais publiés par Herivel la transcription des énoncés car=
tésiens est parfaitement neutre,a-métaphysique. Car, malgré la garantie
divine et tout l'échafaudage de son système, Descartes s'était lourdement
trompé dans la formulation de ses "lois"; et pour cause, puisqu'il s'agis=
sait - citons le mot de Koyré - de "lois pour la nature" tirées des illu=
sions du métaphysicien se mettant à la place de Dieu. Pour réfléchir en
liberté sur le mouvement comme phénomène, même un croyant tel que Newton
devait renoncer à s'élever aussi en haut. Il y renonça dans ses manuscrits
des années '60, où il n'est jamais question de la Cause Première du mouve=
ment. Procédant par tentatives et erreurs, Newton tâchait d'établir un
cadre axiomatique nouveau où l'on pût mathématiser les mouvements réels
des corps. Aussi le principe d'inertie se présente-t-il à son tour comme
un cas-limite conceptuel, destiné à quantifier tout mouvement composé.

Il n'est pas question de suivre ici les développements techniques de la dynamique newtonienne, que d'éminents spécialistes ont examiné en dé= tail. Bornons-nous à remarquer que le principe d'inertie se rattache à une conception passive de la matière, à la _vis insita_ de la définition IIIe des _Principia mathematica_. Chez Newton, la tendance à demeurer en son état n'est qu'un aspect des propriétés de la matière: s'il en avait fait un dogme, comment aurait-il pu concilier cette propriété purement p a s s i v e avec les autres propriétés actives ? La matière chez Des= cartes était toute passive sans exceptions; elle nétait pas susceptible d'action à distance. Newton, depuis ses premiers calculs sur l'orbite de la Lune, avait été amené à réintroduire en astronomie l'ancienne notion d'attraction que Descartes avait bannie. Avant d'afficher une attitude agnostique (la _mathematical way_ des _Principia_) il formula plusieurs hypo= thèses sur la nature de l'attraction. Or, aucune métaphysique dogmatique de type cartésien n'aurait pu tolérer de vues aussi contradictoires sur les propriétés de la matière, tour à tour inerte ou _actuosa_. En ceci Newton s'inspirait, encore une fois, des hypothèses de l'école atomiste; peut-être de Gassendi, qui avait laissé libre jeu aux propriétés de la matière, quitte à y juxtaposer l'action créatrice de Dieu.

Les obstacles que Newton dut surmonter pour fixer ses notions de la force, de la matière, de l'attraction - notions dont il ne fut jamais satisfait - se reflètent dans les hésitations de sa terminologie. A défaut d'un lexique adéquat qui permette un repérage rapide dans ses écrits de différentes époques, essayons de faire quelques remarques concernant l'u= sage du mot "loi".

Dan le _Waste book_, renonçant à l'action constante de Dieu, Newton re= nonce aussi à l'expression cartésienne _lex naturae_: expression assez inu= sitée à cette époque, que Descartes emprunta du langage de la morale et de la jurisprudence et employa le premier de façon systématique. Le terme _Lex motus_ (ou _naturae_) disparaît des manuscrits newtoniens pour n'y réap= paraître qu'à la veille de la composition des _Principia_, dans les ébauches _De motu_, où cette expression désigne de façon définitive les axiomes fon= damentaux de la dynamique inertiale. Jusqu'à ce moment-là (1684) Newton a remanié sans cesse leur nombre, leur contenu et leur titre - "axiomata", "definitiones","hypotheses" - pour les réduire enfin à trois (11).

On peut considérer ces oscillations comme insignifiantes; on peut y dé= celer, au contraire, un aspect de la méta-critique de la mécanique carté= sienne, un indice du refus de ses implications métaphysiques: car le mot "loi" évoquait chez Descartes le rôle du Dieu législateur. Mais si cela est vrai, comment expliquer le retour tardif de Newton à la terminologie cartésienne ? S'agit-il aussi d'un retour à la métaphysique ? Evidemment non. La double diction des _Principia_ , "Axiomata seu leges motus" prouve un certain dégré d'indifférence nominaliste chez Newton. _Cet_ usage du mot est ici fonctionnel, non pas substantiel. Ces 'lois du mouvement' ne sont pas les décrets éternels de la divinité, mais des catégories de notre expe= rience. Newton le suggère aussi par un renvoi très remarquable à Galilée et à sa loi de la chute des graves (12); malgré son peu de précision his= torique, cette attribution à Galilée des deux premières lois du mouvement souligne le sens phénoméniste et anti-cartésien que Newton leur prête.

Que penser alors de l'autre expression "leges naturae ", dont le champ sémantique est beaucoup plus étendu et qui revient dans les lettres à Bentley, dans les "Classical scholia" et dans la dernière _Query_ de l'_Op= tique_ ? C'est ici que Newton introduit une distinction nette entre l'iner= tie de la matière, qui donne "les lois passives du mouvement", et les prin= cipes actifs, d'où résultent "les lois générales de la nature"(13).

Mais nous ignorons ces lois et leurs causes, dépendant immédiatement
-Newton n'a aucun doute là-dessus -du Dieu législateur. Il s'agit évidem=
ment des causes physiques de l'attraction, des interactions chimiques et
de l'essaïm des forces élémentaires que M etMme Hall ont appelées "first
order mechanism" (I4) et qui échappent à la mathematical way. Remarquons
que Newton n'appelle jamais loi la formule de la gravitation universelle,
qui à nos yeux est sa loi par excellence.

Newton, en d'autre termes, refuse de bâtir une métaphysique factice
des leges naturae. La science se borne à l'usage des leges motus - du
moins pour le moment. Toute référence à une science 'autre' est stérile,
même si la raison humaine est tentée sans cesse de plonger dans "le
grand océan de la vérité", où agissent les forces infra-dynamiques qu'a=
nime la volonté immédiate de Dieu au moyen de l'espace ou de l'éther.
Aucun raisonnement a priori ne saurait dàcouvrir ces lois, ni même l'exis=
tence de Dieu. La science première n'est donc autre chose que la science
mathématique du mouvement ou des phénomènes, où se fondent à leur tour
les inférences théologiques: "Even arguments for a Deity if not taken
from Phaenomena are slippery and serve only for ostentation... Metaphysi=
cal arguments are intricate and understood by few. The argument which all
men are capable of understanding and by which the beliefⁱöⁱ a Deity has
hitherto subsisted in the world is taken from Phaenomena. We see the ef=
fects of a Deity in the creation and thence gather the Cause and therefore
the proof of a Deity and what are his properties belong to experimental
Philosophy" (I5).

Newton fit table rase de l'ancienne ontologie. Il coupa à sa racine
l'arbre cartésien du savoir. Au XVIIIe siècle la réceptien de la physique
newtonienne bouleversa l'encyclopédie traditionnelle des sciences. La mé=
thode expérimentale ne se rendit pas seulement autonome; elle remplaça
la métaphysique en tant que clé de toute recherche. Ce caput mortuum de
théologie finaliste que Newton avait réadmis à la fin des Principes et
de l'Optique apparut comme un acte de foi édifiant mais indigne, hélas,
de l'immense effort critique que présupposait sa synthèse. D'Alembert,
Hume, Lambert, Kant et les autres jugèrent naïve et fade cette "métaphy=
sique de Newton" dont Voltaire était devenu le héraut. Elle dépassait les
bornes de la raison expérimentale, que Newton lui-même avait défini in=
franchissables. Les épistémologues repartirent à zéro, rapportant toute
proposition métaphysique, newtonienne ou non, aux critères restrictifs
fixés par les "Regulae philosophandi".

La célèbre feuille volant concernant la modestie de la connaissance
humaine face au "grand océan de la vérité" a une saveur toute pascalienne.
Newton y exprime une attitude ambivalente. Tout en écartant la vaine scien=
ce métaphysique, il n'y renonce pas en principe à l'idée d'un savoir exhau=
stif et total, idée sans laquelle on ne saurait pas même parler de "lois
de la nature". Pour suppléer au défaut de cet idéal-limite, Newton eut re=
cours aux vénérables ruines de l'alchimie, de l'hermétisme, de la prisca
philosophia, de la chronologie, de l'exégèse biblique, dont il tira le
parti que l'on sait. En définitive, il aurait pu dire avec le poète de
The Waste Land : "These fragments I have shored against my ruins". Mais
aucun miracle du Dieu mathématicien et géomètre n'aurait su assurer la
unité d'une synthèse qui, aux yeux de la postérité, marque la ruine défi=
nitive de la métaphysique et ne contient que les éléments fragmentaires
mais solides d'une science in fieri.

NOTES

(I) F.E.MANUEL, A Portrait of Isaac Newton, Cambridge Mass. 1968,p.363.
(2) DESCARTES, Principes de la philosophie, "Lettre de l'auteur à celui qui a traduit le livre", in Oeuvres et lettres, Ed.Pléiade,pp.565-66.
(3) H.MORE, Enchiridion metaphysicum, Ch.I, in Opera, London 1679,I,141.
(4) P. GASSENDI, Syntagma Philosophicum, in Opera, Lyon, 1658, I,233ab, 134. Voir: O.R.BLOCH, La philosophie de Gassendi, La Haye,1971, p.172.
(5) Unpublished Scientific Papers of I. Newton, éd. Hall & Hall, Cambridge 1962; pp. 105, 109.
(6) Ibidem, p. 99. L'expression vient de More.
(7) Ibidem, pp. 109-III.
(8) A.KOYRE', Newtonian Studies, London, 1965, pp. 70 sqq.
(9) DESCARTES, Principes de la philosophie, Lib. II, par. 39.
(10)J.HERIVEL, The Background to Newton's Principia, Oxford,1965,p.50.
(II)Ibidem, pp. 257 sqq, surtout 299 et 307.
(12)Principia Mathematica, Lib.I, def.VI, scholium.
(13)"It seems to me farther that these Particles have not only a vis iner= tiae, accompanied with such passive Laws of Motion that naturally result from that Force, but also that they are moved by certain active Principles ...These Principles I consider... as General Laws of Nature, by which the things themselves are form'd; their truth appearing to us by Phaenomena, though their causes be not yet discover'd"; Opticks (I730), Query 3I.
(14) A.Rupert HALL and M. BOAS HALL, "Newton and the Theory of Matter", in The Annus Mirabilis of Sir Isaac Newton, 1666-1966, edited by R.Palter, The M.I.T.Press, London, 1970; p. 60.
(15) University Library Cambridge, Add Ms 3970.9 f. 619. Ebauche de la Query 23 (3I), citée par R.S.WESTFALL, Force in Newton's Physics, London-New York, 1971, pp. 417-18.

Symposium 6.   ASPECTS OF THE HISTORY OF THERMODYNAMICS:
              THEORY AND PRACTICE

Introduction: Cardwell

Few, if any, of the major branches of science have received less attention from historians than has thermodynamics. This seems strange when we consider that the subject presents a number of fascinating problems, some of them profound, and that its ramifications cover important aspects of physics, chemistry, life sciences, cosmology, meteorology and several branches of engineering. It is even more curious when we recall that there can be few major difficulties attending the study of the history of thermodynamics. The subject was emphatically the creation of the nineteenth century. Although hints and suggestions that were later to bear fruit can be discerned in the works of eighteenth-century scientists and engineers, such as Watt and Lavoisier and Laplace, there is no doubt that the science itself was created, <u>ab initio</u>, in the first half of the following century and by a quite remarkably small group of men of genius. Not only is the material for the history of thermodynamics abundantly available, but the historian does not have to face the enormous problem of trying to think himself into attitudes of mind that are radically different from our own. The Medieval world view was compounded of strands of mysticism, magic, spirituality, superstition and materialism that the historian of science interested in the period must unravel and assess in each individual case if his studies are to be fruitful. But nineteenth-century ideas were much more akin to our own.

We know now, thanks to recent historical research, that the caloric theory and the kinetic theory of gases were not mutually exclusive; in fact a number of intermediate positions could be held that comprised elements of both. We could say, for instance, that heat is the kinetic energy of the caloric; in which case - it is an idle but entertaining speculation - we may retain the caloric theory while rejecting the axiom of the conservation of heat! Although Mach and Duhem were authoritative sceptics of the view that the kinetic theory was an essential factor in the rise of thermodynamics, historically the facts are against them. For example, it is certain that Joule depended on his atomic-kinetic theory to provide a rational basis for his dynamical theory of heat. Embarrassing though it may be those little billiard-ball atoms seem to have been essential for the construction of most of nineteenth-century physics and chemistry. Positivists tend to be wise after the event.

A symposium on the history of thermodynamics must necessarily include some discussion of Sadi Carnot, the father of science. The papers by Professors Hoyer and Lervig require no elaboration; they are models of clarity and they deal with key problems in the development of Carnot's thought. The subsequent discussion about Carnot's possible anticipation of the second law of thermodynamics was, I believe, essentially a discussion of a semantic problem: how exact should one require the statement of a scientific law to be before it can be accepted as final and authoritative? Certainly Carnot's brilliant insight that a cold body is just as essential as a hot body for the generation of motive power is of the <u>essence</u> of the second law; but the energy concept is missing and surely this concept is required for the <u>comprehensive</u>

statement of the law.

The twenty-five years that elapsed between the publication of Carnot's book and the appearance of Kelvin's seminal paper - "An Account of Carnot's Theory" - in which he coined the word "thermodynamics", raise, by implication, intriguing problems for the historian which I shall shortly be discussing elsewhere! (1) It is regrettable but unavoidable that this Symposium did not include a paper devoted to Kelvin. In fact it was doubly regrettable, for a justification for including a Symposium on Thermodynamics in the proceedings of the Congress was to acknowledge the contribution made to the subject by members of the neighbouring Universities of Edinburgh and Glasgow. Maxwell and Rankine both studied at Edinburgh, and Kelvin spent virtually the whole of his long and immensely creative adult life at Glasgow University: nowadays less than an hour from Edinburgh by road or rail.

Besides mathematical skill and a first-class scientific mind Kelvin possessed two other attributes that, taken together, make him a nearly unique figure in the history of science. He possessed great power of persuasion; even as a young man people listened when he spoke. And, secondly, he had a remarkable flair for spotting important ideas that his contemporaries had overlooked; an extraordinary aptitude for nosing out the intellectual growth points in an age that was, by any standards, remarkably creative. So, it was Kelvin who realised the importance of some of Faraday's confused speculations about the electromagnetic field, who thereafter sorted out the wheat from the chaff and who put Maxwell on the right track (2); it was Kelvin who, at the British Association meeting in 1847, rescued Joule's work from oblivion even though he thought that Joule's ideas ran counter to what he believed at the time to be the basic axiom on which the whole science of heat rested. And finally, it was Kelvin who in his paper of 1849 started the Carnot-Clapeyron theory on its path to acceptance and ultimate triumph. But Kelvin was, in the full sense of the word, a magnanimous man; in every case he gives the credit to the original thinker: to Faraday, to Carnot and Clapeyron, to Joule. Such magnanimity has not always been apparent in the history of science.

If the absence of Kelvin was (almost) a case of the play of _Hamlet_ without the Prince, amends were made, to Edinburgh at least, by Professor Daub's paper, dealing with Clausius and Rankine and by Professor Channell's extended comment, emphasising the importance of the conceptual framework from which Clausius' entropy concept emerged. Of particular interest is the point that Clausius was led to generalise his theory of the equivalence of transformations into the entropy concept in order to refute Rankine's suggestion that in another part of the Universe energy may be concentrating and not dissipating, thereby universalising a theory that had been narrowly applicable to a special region of physics and to the theory of the steam engine. In doing so he significantly advanced Carnot's original insight. The latter had seen clearly enough that his principles, derived from steam-engine practice and the physics of heat of his time, were of cosmic significance. Whether or not this particular part of the _Réflexions_ made any impression on Clausius must remain conjecture - if, that is, he ever read that book. (3) But whatever the influence upon him, the concepts of thermodynamics that he and Kelvin enunciated were immediately to be applied in entirely new fields during the succeeding twenty-five years: to chemistry and then the life sciences as well as to physics, cosmology and engineering.

From what has been said it might be supposed that it was disproportionate to include two papers devoted to Willard Gibbs. This may possibly be so; but it was certainly not unjust. The papers contributed by Professors Hornix and Klein not only carry forward the development of thermodynamics from the foundations established by Clausius and Kelvin but also constitute a proper acknowledgement of the achievements of the elusive Gibbs, who of all the great figures in the history of science has received perhaps the least recognition from the historians. Two evident reasons account for this neglect. In the first place Gibbs' work is admittedly recondite - he is no "soft option" for the historian - and in the second place he was a modest and retiring figure; not the sort of man to call attention loudly to his own achievements.

At a time like the present when scientists of all nations are thoroughly and uniformly professional it is fascinating to observe the wide variety of types who contributed to the dramatic advance of science in the nine-teenth century. The history of thermodynamics is no exception here. Was it purely coincidental that the great initiators, as we may call them, Carnot, Joule, Mayer and Gibbs, were all retiring men; devotees of science, who gave their time and money to the advancement of science and who made little or no material gains thereby? Their personalities, transcending differences of language, background and nationality, seem to have a great deal in common. On the other hand their theories and discoveries would surely have been ignored had it not been for such different - more worldly? - men as Clausius, Kelvin, and Helmholtz. We may summarise the matter by saying that it takes all types of scientists to make science. Perhaps this is something worth remembering at a time when the pressures for achieving uniformity are so strong.

Finally, with Professor Hiebert's admirable paper the story is brought up to the present century with the work of Walther Nernst and his third law of thermodynamics, or the heat theorem as it is sometimes called. Subsequent developments could well form a suitable theme for a later Congress. But there is still much to learn about the period with which we **have** been concerned. Indeed it would have been an unsatisfactory Symposium if it had not given rise to more questions than it settled. So, apart from the problem I mentioned above, I would like to suggest two problems worthy of investigation. The first concerns the role played by Rankine in the establishment of thermodynamics as an engi-neering discipline. Although thermodynamics had one of its main roots, perhaps its tap-root, in steam-engine technology there is no doubt that its development in the mid-nineteenth century was very largely due to physicists. Rankine continued the engineering tradition and may be seen as the key figure in this respect in the English speaking world. The problem then becomes: which of his immediate disciples were the most influential in establishing engineering thermodynamics as an essential intellectual tool for the practising engineer? Was it perhaps Macfarlane Gray - another Edinburgh man, by the way - who played the key role? And one may also wonder who were the influential engineers who carried out the same task in France, Germany, Russia. Hirn, Zeuner,

Linde, are obvious names; but which others were influential?

The second question, also a parochial one, is simply: why is it that British scientists and engineers played so small, indeed one might say negligible, part in the development of thermodynamics after Maxwell's time? The possible answers to this question would no doubt take us out of the special realm of the history of thermodynamics and into the general field of social history. But then this is true of all fundamental questions in the history of science.

NOTES

1. In my forthcoming Dickinson Memorial lecture to the Newcomen Society. Two of these problems are concerned with the work and personality of Emile Clapeyron, who wrote a famous and widely read paper in 1834 putting Carnot's ideas into analytic form and relating Carnot's cycle to the familiar indicator diagram. This was translated from the French and published in Richard Taylor's Scientific memoirs selected from the transactions of foreign academics of science and learned societies and from foreign journals 1 (London, 1837), p.347. One wonders why Clapeyron - or for that matter other contemporary engineers and physicists - failed to develop thermodynamics further.

2. I am much indebted to my friend and colleague, Mr J. O. Marsh, for pointing this out to me.

3. In his first paper on thermodynamics, published in 1850, Clausius states that he knows of Carnot's theory only through reading Clapeyron and Kelvin. At no subsequent time, as far as I know, did he say that he had read the Réflexions.

P.Lervig

SADI CARNOT AND NICOLAS CLEMENT

In 1970, Robert Fox gave [6] a very interesting discussion of sources
for Sadi Carnot's work, in particular Clément's theoretical calculations
on the steam engine. The paper to be presented here can be seen as a
parallel to that of Fox, treating single points in much more detail but
giving strong support to the picture suggested by Fox. In addition, I
try to throw light upon some passages in Carnot's text that seem to
show a direct influence from Clément's considerations.

In August 1819, Clément and Desormes presented to the French Academy of
Sciences a memoir with the title: "General Theory of the Mechanical
Power of Steam". The memoir was never published and the manuscript has
disappeared; an abstract, consisting of 4 pages [1] is all that is left.
Sadi Carnot referred to the manuscript in a footnote in his "Reflections
on the Motive Power of Heat" (1824), explaining that he had been allowed
to see it by its author. It is of course not completely unthinkable
that this manuscript could reappear; if so, it might give insight into
the status of the theory of heat in a circle where Sadi Carnot presumably
moved at a time before he wrote "Reflections" and invented his cycle.

Several other manuscripts have, however, rather unexpectedly been brought
to light and have turned out to convey new important information concerning
the very same questions. I refer in particular to the two sets of notes,
taken by participants in Clément's course on Industrial Chemistry at the
Conservatoire des Arts et Métiers; one set by the mathematician L.B.
Francoeur, from 1823-24, another set by a certain J.M. Baudot, from
1824-28. They give a convincing and clear picture of Clément's* theory
of the steam engine. The following considerations will be based upon
these two sets of notes, in particular the Francoeur notes, which, as
will be seen later, are in certain respects particularly informative.

The two sets of notes differ much in style. Baudot's notes seem to be
almost verbatim with many repetitions. Francoeur, however, was an
experienced mathematician, his notes are short, concise and neat, containing only essentials, easy to read and understand. In consequence
of the different nature of the two sets of notes they are useful for
different purposes. Quotations from the Baudot notes can be assumed to
represent a pretty accurate reproduction of Clément's words concerning,
for instance, the nature of heat, as in a statement like the following
(Baudot notes, course 1824-25 p. 65):

---

* and Desormes'. It is difficult and in the present connection hardly
  of interest to distinguish too clearly between the contributions of
  the two men. Carnot seems primarily to have known Clément, and the
  latter seems to be responsible for the investigations leading to
  Clément's law. (See Clément's biographical note, mentioned in (1);
  a similar impression is obtained from Francoeur's notes, page 47.)

With respect to heat, physicists distinguish between two things, its cause and its effect; they give to the cause the name of caloric, and to the effect that of heat (chaleur).

We here recognise Lavoisier's formulation (Traité Elémentaire de Chimie 1789, Chap. 1). Somewhat later, Clement mentions the erroneous belief of William Congreve, known for his rockets, that the heating power of coal can be increased by intermixing it with limestone, concluding (p. 71):

> It is impossible to create heat, it can only be obtained from a combustible substance, the heating value of which is fixed; one may extract a larger or smaller part of it, but it cannot be increased.

On the other hand, we are also told (p. 65):

> Heat has several sources; it may be produced by chemical or mechanical action, by combination or by friction. The latter means may sometimes be very efficient, as some experiments by Rumford have shown. He succeeded in bringing to boil a very large quantity of water, contained in a cavity together with a cannon which was being bored. But experiments of this sort are merely curiosities; the only economical way of producing heat is by combustion.

The conflict between the views expressed in the last two quotations is precisely the one that is brought into daylight in Carnot's posthumous notes. Serious speculations upon this problem, however, are never found in Clément's writings, his theory is - like Carnot's in Réflexions - based entirely upon the idea of heat as a material or immaterial substance, the amount of which is always conserved.*

On the so-called Clément's law or Watt's law

Clément's theory of the steam engine is based upon a certain fundamental law which Clément claims to have proved by experiments. The law is stated in the following words (Francoeur notes, p. 46):

> A certain quantity of water reduced to steam contains always the same quantity of heat, whatever the temperature and the pressure, and one kilogram steam requires 650 calories ** ***

The question of the heat content of steam was already raised by Desormes and Clément in 1811 in connection with a paper they wrote on an experi-

---

* Although this theory is today known to be incorrect since it contradicts the energy principle, I shall everywhere in the following take it for granted. I shall in particular without comment use the expressions correct/incorrect in the sense: correct or incorrect from the point of view of the caloric theory.

(over)

ment by Leslie who had placed a bowl of water and a bowl of sulphuric acid together in the recipient of an air pump. When the air was pumped out, the water was soon covered with ice even when the initial temperature was somewhat above 0°. From known values of the latent heats, Desormes and Clément could calculate that 8.5 grammes of water at 0° would be transformed into 7.5 grammes of ice, when the remaining 1 gramme of water had been evaporated. The number was based upon the assumption that the heat of evaporation could be taken to be the same at 0° as the one known to be valid at 100°; whether this was true or not was unknown. [10]

Clément's law was first formulated by Clément and Desormes in their no-longer extant memoir of 1819; the law was explained in the corresponding abstract from the same year [1] where it was told that the law had been found by experiments. The experiments in question are not described in the abstract but can easily be reconstructed from Clément's lecture notes.

Three types of experiments were described in support of the law. The first type was Leslie's above-mentioned experiment; it was here, however, stated that 9.66 grammes of water would give 8.66 grammes of ice and 1 gramme of steam. By the second type of experiment explained in Francoeur's notes, 8.66 kilogrammes of ice produced, when mixed with steam in a calorimeter, 9.66 kilogrammes of water at 0°C. Since, according to Clément, 75 calories were needed to melt one kilogramme of ice, one kilogramme of steam must contain 75x8.66=650 calories. By the third type of experiment, 400 kilogrammes of water at 0° were mixed in a barrel with 20 kilogrammes of steam entering through a tube from the boiler of a "12 horsepower steam engine" in such a way that all the steam was condensed. The amount of steam could then be measured by the increase in the amount of water in the barrel. It was noted that the water had been heated 30.95°, and it was then concluded that 20 kilogrammes of steam contained 420 x 30.95 = 12999 calories. 1 kilogramme of steam then contains 12999/20 = 650 calories.

Of these three types of experiments, the first two are in practice hardly suited for temperatures much higher than 0°, whereas the third type can be used from 100° upwards. The experiments of the third type are the only ones which can be called quantitative.

Since detailed descriptions of the experiments have not survived, it is difficult to say how carefully they were performed. It is, of course, possible to reconstruct from modern data the results which should have been obtained from ideal measurements. It is evident from Clément's

---

** A calorie is defined as the amount of heat needed to raise 1 kilogramme of water 1° centigrade.

*** The formulation "whatever the temperature and pressure" seems slightly confusing - the steam should, of course, be saturated, as is also stressed by Clément in his lecture the following year. (Baudot notes, 24. January 1925 p. 37)

results that he must have ascribed the highest degree of accuracy to the experiments of the last type. Indeed, one may doubt whether Clément actually ever performed the first two types of experiments. Despretz, however, must have done something of the kind [2]

Clément's results were convincing only because he confined himself to a relatively narrow temperature region, between $100°$ and $152°$C. Despretz, who in 1820 [2] supported Clément, later [3] expanded that interval and found deviations. From the way the experiment was conducted, we can conclude that the 650 calories which Clément ascribed to 1 kilogramme of water should be compared to a standard state, which is liquid water at $0°$.

Clément saw his law as an important new example of the law of constant proportions which, from about 1800, was very much in the centre of attention in French chemistry (see, for instance, Partington: A Short History of Chemistry (1960) p. 153 ff).

In Clément's theory of the steam engine his law entered, as will be seen, in two different ways. The first of these ways was in the form of a simple conversion factor. The calculation gave "mechanical power" per kilogramme of steam; the engineer would prefer to know it per kilogramme of coal; here, Clément's law was useful. It stated that 650 calories were needed in order to produce 1 kilogramme of steam; since 1 kilogramme of coal "of good quality" when burned was known to produce 7050 calories, it could then give 7050/650 = 10.96 or practically 11 kilogrammes of steam. It followed that

> one must multiply the power produced by 1 kilogramme of steam with 11 in order to obtain the one produced by 1 kilogramme of coal

(Francoeur notes, p. 48).

This statement had, of course, mostly the character of a practical rule, not depending much upon the accuracy of Clément's experiments.

The second application of Clément's law was of a more fundamental nature. It was the following. Steam would, of course, like any other gas be cooled by an adiabatic expansion. From Clément's law it could, however, be concluded that a quantity of saturated steam would, by such a process, remain saturated. The steam pressure curve was an adiabatic curve.*

From this theorem Clément could, as we shall see, calculate the power gained per kilogramme of steam. It is of a certain interest to note that in the form of this theorem Clément's law can be ascribed a meaning not only in the caloric theory but in current thermodynamics as well. When considered as a modern result it is, however, not valid or rather, it is only valid as an approximation. Saturated steam becomes by adiabatic expansion, oversaturated and condenses. The approximation is, however,

---

* It must be stressed that Clément never formulated this theorem which is meant as a concentrated representation of his theory.

tolerable.

Clément's theory of the steam engine

I shall now present Clément's calculations of the "mechanical power" in modern terms, work - of the idealised steam engine. The calculations are fully described only in the Francoeur notes from 1823-24, they could not be reconstructed on the basis of the Baudot notes (from 1824-28); this makes Francoeur's manuscript a particularly precious source.

Clément's calculations are given in the form of a table and some examples of its use. The table was later sold in an enlarged version in connection with Clément's lectures in 1826, a single copy of this later version exists at Musée Carnavalet and was published by J. Payen in 1971 (see (1)). I shall base my discussion on the earlier Francoeur version of the table (Fig. 1) since this version is more self-explanatory and is the one that is accompanied by the examples.

The column p and v give the steam pressure and the corresponding volume of 1 kilogramme of steam, the last mentioned quality is uncritically calculated by means of the ideal gas equation.

$$pv \cdot 1.375 = 760 \cdot 1.7(1 + 0.00375t) \quad ,$$

where t is the temperature in degrees centigrade. In the third column, production, the mechanical power (work) by the evaporation of 1 kilogramme of steam is given by

$$production = \frac{10.4}{760} \, p \cdot v = 0.01368 \, p \cdot v \quad ,$$

because pressure is in millimeters of Mercury whereas mechanical power is in Dynamies, one Dynamie being defined as 1 cubic meter of water raised 1 meter (same unit as used - without name - in Réflexions (see, for instance, [5], p. 84) but not in Carnot's manuscript "Recherche d'une formule" (see (1)). In the fourth column, expansion, the mechanical power gained by a step of the subsequent expansion is calculated. It is easy to reconstruct the procedure (see note (3)); it is the numbers

$$expansion = \frac{10.4}{760} \, p_{av} \Delta v$$

that are listed, Δv is the volume increase and p_{av} is - as one can easily control - some mean value of the pressure in the corresponding interval. In the fifth and last column, détente, these contributions are summed from above such as to give the total mechanical power gained by the expansion stroke

$$détente = \frac{10.4}{760} \sum p_{av} \Delta v \cong \frac{10.4}{760} \int p dv \quad .$$

The procedure is illustrated in Fig. 2.

Mechanical power of one kilogramme of steam

D	p	v	production	expansion	Détente
	mm	m cubes	Dynamies	Dynamies	Dynamies
0°	5.06	185	12.89	3.95	"
5	6.95	137	13.13	3.85	3.95
10	9.48	103	13.37	3.83	7.80
15	12.84	77.30	13.62	3.80	11.63
20	17.32	58.20	13.86	3.87	15.43
25	23.10	44.30	14.10	3.75	19.20
30	30.64	34.10	14.34	3.75	22.95
35	40.40	26.30	14.59	3.75	26.70
40	53.00	20.40	14.83	3.64	30.45
45	68.75	16.00	15.07	3.56	34.09
50	88.74	12.60	15.31	3.66	37.65
55	113.70	9.94	15.55	3.51	41.31
60	144.60	7.96	15.79	3.50	44.82
65	182.70	6.38	16.03	3.40	48.32
70	229.10	5.17	16.27	3.35	51.72
75	285.10	4.22	16.52	3.24	55.07
80	352.10	3.47	16.76	3.29	58.31
85	431.10	2.87	17.00	3.11	61.60
90	525.30	2.39	17.24	3.14	64.71
95	634.30	2.00	17.48	2.97	67.85
100	760	1.70	17.72	2.80	70.82
105	904	1.45	17.96	2.56	73.62
110	1055	1.256	18.20	2.82	76.18
115	1247	1.077	18.44	2.52	79.00
120	1452	0.936	18.68	2.62	81.52
125	1690	0.816	18.92	2.58	84.14
130	1961	0.712	18.17*	2.72	86.72
135	2286	0.618	19.41	2.55	89.44
140	2637	0.544	19.65	2.68	91.99
145	3056	0.475	19.89	2.44	94.67
150	3494	0.420	20.13	2.45	97.11
155	3991	0.372	20.37	2.48	99.56
160	4554	0.330	20.61	9.84	102.04
182	7600	0.262	21.68	38.62	111.88
215.5	26600	0.637	23.80		150.50

Fig. 1   Clément's table (Francoeur's notes, page 53)

* this is evidently an error for 19.17

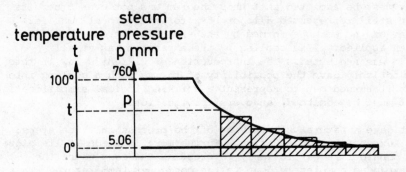

Fig. 2 In the table Fig. 1, each number of the column "Expansion" corresponds to one of the hatched rectangles; the number in the column "Détente" corresponding to the temperature t^{o} and the pressure p is the entire hatched area.

Clément twice exemplifies this procedure for the case of a steam engine working without a feed pump, though he does not refer explicitly to this elementary model. He then gives a further four examples for various existing types of steam engines, and his results reveal that the difference between actual performances of existing steam engines at the time and those that could be obtained theoretically from the steam engine principle. They showed the relatively good efficiency of the huge Cornwall engines with their 141 Dynamies per kilogramme of coal as compared to the ideal figure, 181 Dynamies. The small engines were, as should be expected from their inevitably greater losses, poorer with their 68-70 Dynamies. The improvement to 300 Dynamies that could in principle be obtained from the same type of engine by incorporating a not unrealistic expansion stroke of about 50% ("2nd system Maudsley") was only poorly reflected in the performances of 93 Dynamies for small engines. Large engines of these types did not exist. The same feature, the same enormous distance between theory and engineering practice, was seen in the two last types with their higher pressure. For the Woolf engine one would theoretically get 407 and in practice 190 Dynamies. In the last system, the Oliver Evans engine where there was no condensation in the proper sense since the steam was after use blown out into the atmosphere - the theoretical performance was 344, the real one only 68 Dynamies. Although it would be clear that losses in steam engines were unavoidable, and even if some of Clément's assumptions might turn out to be only approximately correct, the message conveyed by these numbers was strong; in presenting them, Clément must have hoped to provoke inventers to try to bridge this gulf between theory and practice.

I want to finish this discussion of Clément's calculations by making a point. What was, in a nutshell, the essential progress that Clément brought about in connection with steam engine theory? My answer may seem paradoxical: I claim that Clément's essential contribution was his application of Clément's law that the steam pressure curve was an adiabatic curve.

Although it must be admitted that this theorem was not even true, its application still represented a major step forward. In all earlier calculations of the work performed by the expansion stroke, Boyle's law had been applied. If a cooling by the expansion was at all expected, it was neglected. The introduction of Clément's law in this connection not only gave the possibility of taking such a cooling into account but it turned out to represent a striking logical simplification. Perhaps, it must be admitted, an oversimplification.

However, it gave a natural subdivision of the process in three steps: production, expansion, condensation. Furthermore, it made it lie close at hand to consider as particularly interesting the case where the steam was expanded completely down to the condenser temperature. This gave the process the vitally important property that it could be performed forwards as well as backwards.

Intermission

The above is the material I want to present. Anyone who knows Sadi Carnot and Réflexions will recognise its style. That Carnot knew this material in some detail is easily corroborated, in particular by the manuscript published by Gabbey and Herivel (see (1)), where precisely the same case was treated with boiler temperature, condenser temperature and adiabatic expansion to an intermediate temperature, cautiously accepting Clément's law.

My intention is in the remaining part of this paper to present some speculations as to how Clément's calculations and the ideas behind them were utilised by Carnot as a point of departure in the construction of his theory, and to point out that certain, usually neglected parts of the introductory sections of Réflexions throw light on these questions. Which was the path followed by Carnot? This is what I am going to discuss.

Réflexions is not a perfect creation. This fact presumably reflects Carnot's doubts concerning the whole basis of his theory. Such an incomplete creation, however, will often exhibit remnants of the very creation process, the artist's finger prints in the clay, traces that are normally removed in a finished work.

Carnot's "Réflexions sur la puissance motrice du feu" (1824) took its point of departure in the consideration that the heat in a working steam engine was seen to move from hot to cold - from the fireplace through boiler, cylinder and condenser to the cooling water. It was as if one employed the tendency of heat to move towards a colder place in order to gain work. This feature was, however, characteristic of heat engines in general. If heat phenomena were to be exploited to gain motion, it was necessary to utilise the ability of bodies to change form or volume when heated or cooled. In order to give rise to motion the body in question should be heated, that is to say be brought into contact with some hot body so that it could change its form and produce the motion. Subsequently, in order to be returned to its original state so the process could be repeated, the body had to be cooled again, that is to say be brought into contact with some cold body. As the result, motive power was gained and heat had been moved from a hot to a cold body.

As his first example of such a process where heat was moved and motive power gained, Carnot now - in the very beginning of Réflexions, before the formulation of the proper Carnot cycle - described a process performed with a certain amount of water. The process was the following one (D 10)* :

One imagined two bodies, A at a certain high, B at a certain low temperature.

(1) The water was evaporated at the temperature of A and in contact with A.

(2) Having been received in a space capable of expansion, such as a cylinder with piston, the steam was expanded and its temperature would consequently fall; it was assumed that this expansion could be continued until the steam had reached the temperature of B.**

(3) The steam was condensed in contact with B.

This example has always been a puzzle to readers of Carnot. Why did Carnot take his point of departure in this unsatisfactory example, when a few pages later he was actually able to present the logically satisfactory Carnot cycle? The finding of the text from Clément's lectures answers this question. Carnot's audience, the readers he had in mind, must more or less have been Clément and his audience. It was then natural to begin with something they knew. It is in connection with this example that Carnot in a long and often quoted footnote discusses the question of the impossibility of a perpetuum mobile.

Although this example is only provisional and is later to be replaced by the correct proof, I would like to point out, what seems not to have been noticed before, that Carnot's argument would prove a related theorem, not one that would seem particularly interesting to Carnot but one that would on the other hand give us a feeling of the path that led to Carnot's cycle. The theorem proved would be this:

> Among all processes whereby a given amount of steam is taken from a container with saturated steam at a given high and given to a container with saturated steam at a given low temperature, one that can be reversed will give maximum of motive power.

---

* Page numbers marked with D refer to the Dover edition, Ref. (5).

** We see that Clément's law was assumed here although it is not explicitly stated at this place.

Such a theorem would then define a temperature scale, not a universal one but a _water scale_ (4); any evaporable liquid would define its own individual scale, the change of which was to be defined as the motive power to be gained from moving a unit of the liquid in question from one temperature to another. The theorem would mean that a maximum of motive power would be gained by completing the expansive stroke the whole way down to the condenser temperature.

Carnot, however, was not satisfied with this conclusion, and sought to improve upon his demonstration. In a series of footnotes (p. D14), he shows in effect that the motive power gained by Clément's process is not maximum. He constructs another process capable of yielding more motive power, which will be effective whenever contact occurs between bodies of different temperatures. The cycle which now bears his name may be seen as an ingenious trick to avoid such a contact, thus its discovery could well have been made via the line of reasoning outlined above. The _perpetuum mobile_ would, in that case, have been introduced subsequently as an aid to the necessary mathematical proof of maximum. The essential simplification introduced in Carnot's cycle is that the _gas_ passes through a complete cycle but the _heat_ does not since the latter is initially in the hot and ultimately in the cold reservoir.

Notes

(1)   The manuscripts in question are primarily the following ones:
Two sets of notes from Clément's course on Industrial Chemistry
at Conservatoire des Arts et Métiers:

a)   One set taken by L.B. Francoeur (1823-24). The main content
of these notes is described in Fox [6] (1970).

The notes were found by Professor E. Mendoza and are now, as
explained in Fox: The caloric theory of gases (Clarendon 1971),
at the library of the Ecole Supérieure des Beaux-Arts as
MS.407.

b)   One set taken by J.M. Baudot (1824-28). A description of
this set has been given by J. Payen (Arch.int.Hist.Sci. 21
(1968) 15-32). They are at the library of Conservatoire des
Arts et Métiers in Paris.

In connection with the course, a table corresponding to the
disappeared memoir from 1819 was published by J. Payen together
with an interesting autobiographical note by Clément. (Revue
d'Hist. des Sciences XXIV (1971) p. 45-60) The table is
essentially the same as the one given in the Francoeur notes,
given here p. 10, but differs in makeup. The following manuscript by Sadi Carnot is also of interest:

"Recherche d'une formule ..." published by W. A. Gabbey and
J. W. Herivel (Revue d'Hist. des Sciences 19 (1966) 151-166).

(2)   The conclusion from the calorimetric experiments to the adia-
bacity of the steam pressure curve is, of course, directly
based upon the caloric theory of heat. Carnot's not too clear,
critical remarks of that theorem in the manuscript published
by Gabbey and Herivel "Recherche d'une formule ..." (where he,
by the way, directly uses the theorem (for the reference, see
(1)), is presumably to be understood in that light.

(3)   The study of the Francoeur notes (1823) gives the strong impression that Clément did not have too clear a picture of all
aspects of his own calculations but had probably been guided
by someone.

It is in this connection worth mentioning that none of the
two models used by Clément as basis for his calculations gave
any understanding of why to subtract the condensation term, and
the question is never explained or even mentioned in his text.
This is particularly striking, since it is actually the first
place where this feat is encountered.

The calculation of the mechanical power (work) in the expansion
as $\int p\,dv$ was claimed in Francoeur's text to be done on the basis
of the first of Clément's models, the bubble model (see below)
although this model is particularly poorly suited for an
understanding of this. The model has been described in the

literature ([6], [7], [8]).

4.   The water scale $A(t)$ mentioned in the text would be defined in such a way that

$$A(t_1) - A(t_2)$$

is the work gained when 1 kilogramme of steam is transformed reversibly from a container with saturated steam at t_1^o to one at t_2^o, and can under assumption of Clément's law (and, of course, Carnot's theory) easily be expressed by the absolute temperature θ in Carnot's theory.

References

[1]   Mémoire sur la Théorie des machines à feu; par MM. Desormes et Clément. (Extrait) (Bull. des Sc. par la Soc. Philomatique de Paris, 1819, p. 115-118).

[2]   Despretz, César Mansuete (Bull. des Sc. par la Soc. Philomatique de Paris, 1820, p. 1-4) (Clément's law is here experimentally confirmed for a series of substances.)

[3]   Despretz, César Mansuete: Traité élémentaire de physique (4 ed, 1837) no. 185.

[4]   Sadi Carnot, Biographie et Manuscrit, ed. Picard, Paris 1927.

[5]   Sadi Carnot, Reflections on the Motive Power of Fire, (Ed. Mendoza, Dover 1960).

[6]   R. Fox: Watt's expansive Principle in the Work of Sadi Carnot and Nicolas Clément. (Notes and Records of the Roy.Soc. of London, 24, (1970) p. 233).

[7]   E. Mendoza: Sadi Carnot and the Cagniard engine (Isis, 54 (1963) p. 262).

[8]   D.S.L. Cardwell: From Watt to Clausius (Heinemann 1971).

[9]   M.J. Klein: Closing the Carnot Cycle (Sadi Carnot et l'essor de la thermodynamique. Ed. Taton, Paris 1976, p. 213).

[10]   MM. Desormes et Clément: Sur le nouveau procédé de la congélation de M. Leslie (Ann.Chim. 78 (1811) p. 183-202).

E.N.Hiebert

CHEMICAL THERMODYNAMICS AND THE THIRD LAW: 1884-1914

In a lecture on "The calculation of chemical equilibria from thermal data," delivered at the Academy of Sciences in Göttingen in 1905, Walther Nernst (1864-1941) demonstrated that thermal data alone would suffice to predict the position of chemical equilibrium and hence the degree of spontaneity for chemical reactions. Nernst explored the practical and theoretical implications of his heat theorem for a decade. After World War I he turned his attention to other problems connected with the application of physics to chemistry. For contributions to chemical thermodynamics, his crowning life's work, he received the Nobel Prize for chemistry in 1920.

The third law of thermodynamics, as Nernst preferred to call his theorem, was the subject of vigorous and fertile controversies that continued to be explored well into the 20th century. Originally the merits of the Nernst heat theorem were based chiefly upon its practical value for computing chemical equilibria. The feasibility of evaluating entropy constants for gases from quantum calculations led, in 1912, to renewed interest in the theoretical significance of Nernst's theorem. During the 1920's quantum statistical developments gave rise to polemics concerning the general validity of Nernst's theorem for solids. By the last 1920's, when Nernst no longer was actively engaged in thermodynamic investigations, several special formulations of the heat theorem, and notably that of Franz Simon, led to the acceptance of Nernst's fundamental idea in a more refined form as a fundamental law of thermodynamics. In spite of the fact that the reformulated heat theorem was on far firmer theoretical footing by the 1930's than in 1905, Nernst did not accept the riders that had been annexed to the original theorem. Others took over where he left off; they did so without his wholehearted sanction.

In this paper, Nernst's formulation of the third law of thermodynamics, and its reception, is examined against the background of the state of physics and physical chemistry over the thirty-year period that stretches from Le Châtelier's study of chemical equilibrium in 1884 through the end of Nernst's active involvement in third law deliberations in 1914. In order fully to comprehend the approach that Nernst took to chemical thermodynamics, it is important to show in what form the problem that he tackled had been posed by other investigators at the end of the 19th century. Gibbs, Helmholtz, and Boltzmann had provided theoretically sound models for incorporating second law considerations into the study of chemical reactivity. Le Châtelier and Haber, recognising the theoretical inadequacy of the Thomsen-Berthelot principle, had suggested that the indeterminate integration constant in the integrated form of the Gibbs-Helmholtz equation probably was a determinate function of certain physical properties of the constituents of a chemical reaction.

The third law of thermodynamics, which had its origin in chemistry, was conceived by Nernst in connection with the search for the mathematical criteria of chemical equilibrium and chemical spontaneity. The solution he proposed for predicting the equilibrium had been formulated, and its practical significance recognised, by chemists, for over a century. It had taken the form of experimental investigations designed to provide an exhaustive catalog of chemical affinity relationships. An old and puzzling question was why certain chemical reactions proceed while others do not; or more precisely, how far a given reaction will go before it reaches equilibrium. The more general problem may be stated as follows: Given the magnitude of the energy changes that accompany the transition of a system from one equilibrium state to another, is it possible - using only the first and second laws of thermodynamics - to calculate theoretically, for that transition, the quantity of maximum useful work, i.e., the Helmholtz available work function?

By 1900 it was known that the thermodynamic calculation of chemical equilibria, using thermal data alone (that is, heats of reaction, specific heats, and the thermal coefficients for both heats of reaction and specific heats), could not be carried through because of what Haber in 1904 called the thermodynamically indeterminate integration constant J that appears in the integrated form of the Gibbs-Helmholtz equation, $\Delta F = \Delta H + T(\partial \Delta F/\partial T)_P$ which relates the free energy change ΔF to the heat content or enthalpy ΔH and to the entropy change ΔS , the latter being expressed here in the form of the thermal coefficient of free energy change, since $(\partial \Delta F/\partial T)_P = -\Delta S$. The Gibbs-Helmholtz equation, derived from considerations of the first and second laws, was expressed in slightly different form by Gibbs (1875-1879) and Helmholtz (1882-1883), but we will not take time to emphasise those distinctions here.

From the Gibbs-Helmholtz equation, $\Delta F = \Delta H + T(\partial \Delta F/\partial T)_P$, and the general expression of ΔH as a function of temperature, $\Delta H = \Delta H_o + \alpha T + \beta T^2 + \gamma T^3 + \dots$, and the Kirchhoff law of 1858 that relates to the specific heat changes at constant pressure ΔCp , viz., $\Delta Cp = (\partial \Delta H/\partial T)_P$, it can readily be shown that the free energy equation takes the integrated form: $\Delta F = \Delta H_o - \alpha T \ell n T - \beta T^2 - \gamma/2\ T^3 - \dots + JT$. In this equation, ΔH_o is the heat of reaction at absolute zero; that is, the integration constant in the integrated form of the Kirchhoff law. It can be evaluated empirically from the knowledge of ΔH at any temperature. Likewise, the heat-capacity coefficients, $\alpha, \beta, \gamma, \dots,$ can be calculated from calorimetric data. The only remaining difficulty encountered in putting this equation into practice is that the integration constant J cannot be evaluated calorimetrically and must be obtained from the knowledge of ΔF at some temperature.

From the mid-1880's until the publication of Nernst's theorem in 1906, this integration constant J became the focus of a genuine dilemma within chemical thermodynamics. It essentially meant that the task of predicting the equilibrium conditions for chemical reactions, was, as heretofore, an experimental problem and not one that could be approached theoretically. The challenge apparently was straightforward: namely, to invent more ingenious physical techniques to overcome the analytical difficulties associated with the determination of ΔF . This problem had plagued all the investigators: Arrhenius, van't Hoff, Ostwald, Le Châtelier, Haber, T.W. Richards, G.N. Lewis, and, of course, Nernst. The solution that Nernst proposed in 1905 relates to the way in which the J of the integrated Gibbs-Helmholtz equation is interpreted and also to the practice of thermochemistry prior to the 1880's.

According to the Thomsen-Berthelot principle, the driving force of a chemical reaction, the free energy ΔF , is equated simply with the heat of reaction ΔH from calorimetric measurements. That is, it was assumed that $\Delta F = \Delta H$. For any given process, say a reaction represented by the chemical equation $Y \rightleftharpoons Z$, this principle predicts that the magnitude of ΔH is a direct measure of the driving force of the reaction. For exothermic reactions, where $\Delta H > 0$, the formation of Z from Y is favored. For endothermic reactions, where $\Delta H < 0$, the formation of Y from Z is favored. When $\Delta H = 0$, no reaction takes place. The application of this principle gives almost the right experimental answer most of the time, but not invariably.

In retrospect it seems that this should have been self-evident. It could have been argued, logical considerations suggest, that the course of spontaneous processes necessarily must be governed by both the first and second laws of thermodynamics and not just the first as the Thomsen-Berthelot principle implies. Manifestly, that principle owed its long-standing practical successes to the fact that the entropy changes in most chemical reactions is quite small in comparison with ΔF and ΔH . On these premises it became evident that probing into the theoretical significance of the magnitude and sign of $\partial \Delta F / \partial T$ might furnish the clue to an explanation of the discrepancies that the thermochemists had tried so diligently, although rather arbitrarily, to bring into line with the Thomsen-Berthelot principle.

In his treatise on theoretical chemistry of 1893, Nernst had noted that this principle was surprisingly accurate for solids. He proceeded to show that for special cases, such as ideal gases and dilute solutions, for which $\Delta H = 0$, ΔF and therefore $\partial \Delta F / \partial T$ also should approach zero. In 1894 Nernst discussed the Gibbs-Helmholtz equation for the free energy of mixing of concentrated solutions and found a close fit between ΔF and ΔH , plotted as a function of temperature. In Göttingen, between 1894 and 1905, Nernst carried out experimental work with gaseous reactions in order to test the Gibbs-Helmholtz equation over a wide range of temperature. His analyses of the high-temperature equilibrium conditions for the formation of nitric oxide and his work on the synthesis of ammonia were motivated by the search for methods of nitrogen fixation from the air that would provide a potential source of nitrates for fertilizers and explosives. These were specific practical instances of the application of classical thermodynamic principles to the computation of chemical equilibria. Their study, and the inherent theoretical difficulties connected with computing the driving force of chemical reactions, mentioned above, led Nernst to prove the deeper thermodynamic significance of free energies, heats of reaction, and specific heats, especially in the low temperature range.

Other investigators were working along similar lines at about this time. In 1884 in Comptes rendus..., in his comprehensive 225-page paper on chemical equilibrium, under the heading "Constante d'integration," Le Châtelier recognised the problem in connection with his own equilibrium studies on cements and blast furnace reactions. In 1888 he explored these problems systematically in a memoir published in Annales des mines. Nernst later remarked that he had not seen Le Châtelier's work, and, in fact, it is unlikely that he would have examined this periodical in connection with his thermodynamic studies.

At Harvard, G.N. Lewis (1899) and T.W. Richards (1902) carried out calorimetric and galvanic cell measurements to determine what happens to ΔF and ΔH at low temperatures. Richards found that the thermal coefficient of electromotive force for most galvanic cells $\partial E/\partial T$ approaches zero at low temperatures. By graphing ΔE and ΔH as a function of temperature and extrapolating to absolute zero he concluded that $\partial \Delta F/\partial T < 0$, since $(\partial \Delta F/\partial T)_p = -n\mathcal{F}(\partial \Delta E/\partial T)_p$, that $\partial \Delta H/\partial T > 0$, and that both of these terms become equal to zero as $T \to 0$.

Richard's paper was a suggestive contribution to physical chemistry at the time and was the subject of van't Hoff's special memoir in the Boltzmann Festschrift of 1904. Haber was influenced by the discussions of both Richards and van't Hoff and was particularly impressed that for some reactions at low temperatures on Richards' graph the extrapolated values of ΔF and ΔH practically overlap. Nevertheless, in spite of the stimulus provided by Richards' work, it is doubtful that he saw the intrinsic implications for thermodynamics of vanishing ΔF and ΔH values at absolute zero. Nernst was later to reject Richards' priority claim to the third law and imply that Richards' work did not reveal even an intimate acquaintance with the second law.

In 1904, Haber attacked head-on a problem that Nernst solved - or thought he had solved - a year later. He concluded that if Kopp's law (the additivity of atomic heat capacities) holds for reactions between solids, the integration constant J and therefore the entropy change at absolute zero must have zero value. Lacking knowledge about specific heats at these low temperatures, Haber felt bound to leave open the possibility that the integration constant might have a small finite value owing to deviation from additivity for atomic heat capacities at low temperature. Thus Haber was led to discuss at some length what he called the "thermodynamically indeterminate constant." Nernst later simply postulated the validity of Kopp's law near absolute zero.

In his Gasreaktionen (1905) Haber took up the problem of the integration constant in relation to gas reactions. As Planck would later do, Haber explained that the nature of the integration constant should be expressible in terms of heat capacity and entropy constants characteristic of the components of the gaseous reaction. In particular he stressed the importance of knowing the variation of these properties with temperature in the vicinity of absolute zero. Because of insufficient heat capacity information, he simply left open the question of what happens with gases at absolute zero. For gaseous reactions involving no change in the number of molecules he concluded, however, that the integration constant, if not equal to zero, was probably quite small. His experimental data could support this conclusion, and so he adopted it as a guide in setting up his free-energy equations. By the end of the year Nernst had announced his Wärme-Theorem, as he then called it. Haber immediately recognised its immense importance.

Nernst's theory began with the Gibbs-Helmholtz equation. Like Le Châtelier, Lewis, Richards, and Haber, Nernst noted that all the quantities in this equation can be obtained by thermal measurements, with the exception of the integration constant J . Thus, although the form of Nernst's equation was somewhat different, there was nothing new in it. He knew, of course, that for chemical reactions involving gases the sum of the specific heats of the reactants in general is not equal to the sum of the specific heats of the products, since the degrees of freedom may differ. By considering only condensed phases and by assuming that Kopp's law would apply he equated the specific heat sums of products and reactants. Accordingly, he reasoned that ΔF and ΔH do not differ markedly at room temperature, nor even at somewhat higher temperatures; that the ΔF curve is not likely to start at a steep angle at absolute zero. He recognised that the simplest assumption would be for the ΔF and ΔH curves to run together and become tangent to one another as $T \rightarrow 0$. The important point about Nernst's hypothesis is that, since $\Delta F - \Delta H \rightarrow 0$, as $T \rightarrow 0$, the integration constant J is known and there is no longer a need to seek special techniques for determining ΔF ; this can now be accomplished calorimetrically.

Having postulated the heat theorem for condensed phases Nernst turned directly to the question of the determination of the integration constant for gaseous systems. Crediting Le Châtelier and Haber for having recognised the fundamental problem, he proceeded without fanfare or apologies to treat gaseous reactions at sufficiently low temperatures so that all of the constituents could be considered to be in the condensed state.

This was just the beginning of Nernst's investigations into chemical thermodynamics. From this time forward, and for the next eight years, he directed all of his efforts to furnishing empirical support for the theoretical soundness and practical utility of the heat theorem. By 1907 his theoretical views on the subject were essentially complete – as can be seen from his Silliman lectures at Yale University. His thesis was that the heat theorem manifestly revealed new truths about the relation between chemical energy and heat, but he suggested that it would prove useful beyond the solution of problems of chemical equilibrium and spontaneity. He was quick to emphasise that a great deal of experimentation would be necessary in order to decide whether the theorem represents an approximate principle or an exact law of nature similar to the first and second laws. Irrespective of the theoretical status of the theorem, Nernst maintained that approximations based upon the theorem would provide answers to the all-important question about the driving force for chemical reactivity. In this he was right. Still, he suggested that chemists work over the whole field of thermochemistry from a new point of view, namely to measure heats of reaction, specific heats, and their temperature coefficients over the entire experimentally feasible temperature range. Thus, Nernst was led to declare that all that thermodynamics can contribute to chemistry is already implied in the Gibbs-Helmholtz equation, provided that the heat theorem be used to furnish an interpretation of the mode of behavior of nature in the vicinity of the absolute zero of temperature.

The clue to the theoretical analysis of the problem for Nernst lay in the approximate success at ordinary temperatures of the Thomsen-Berthelot principle. The solution that he provided was to assume that this principle was not only approximate at ordinary temperatures but also in the neighborhood of absolute zero, so that ΔF and ΔH, plotted as a function of temperature, approach each other asymptotically with contrary slopes. Nernst concluded that this theorem would settle the question by making it possible to predict the thermodynamic stability of any stoichiometrically correct chemical equation, based on calculations derived from thermal data alone.

During the next decade, virtually the entire facilities and personnel of the physical chemistry institute of the University of Berlin were organised into a huge work program to experimentally test Nernst's Wärme-Theorem (1906). The immediate consequence of Nernst's new idea was that radically different thermochemical techniques were put into practice to elucidate chemical equilibrium. These involved the determination at very low temperatures (in fact as close to zero as possible) of the specific heats and thermal coefficients of specific heats for the constituents of the chemical reactions under investigation. In a series of seven papers published between 1910 and 1914 Nernst and his co-workers (F. Koref, F.A. Lindemann, and F. Schwers) presented impressive experimental evidence to support Nernst's theorem based on the electrical measurement of electrically induced temperature changes. The rigorous test of the validity of the theorem was soon seen to be an enormously challenging experimental undertaking. To approach this objective the Nernst group constructed ingenious electrical and thermal devices, developed a vacuum calorimeter, and built a small hydrogen liquefier to achieve temperatures low enough to be able to extrapolate safely to absolute zero. Nernst tackled all these problems imaginatively and successfully, and step by step came increasingly to believe that his hypotheses should be elevated to the rank of a bona fide law of thermodynamics.

Although unaware of it when he announced his heat theorem, Nernst had laid the foundation for the connection between chemical thermodynamics and the quantum theory set out five years earlier by Planck. Not long after the Nernst heat theorem had become established as a method to predict chemical equilibrium it was seen that quantum statistical calculations of entropy constants for gases became accessible. As a follow-up to his four classic papers of 1905, Einstein treated the specific heat of solids as a problem in quantum mechanics. In Nernst's 1906 paper there had been no mention of Planck's quantum ideas of 1900, nor had he in 1907 taken seriously Einstein's suggestion that quantum theory predicted vanishing heat capacities for solids at absolute zero. During Nernst's visit to Einstein in Zurich in March 1910, the two men discussed the extent of agreement between the Einstein theory of specific heats and the experiments being conducted at Nernst's institute in Berlin. In 1911 Einstein wrote Michele Besso that his theory of specific heats had celebrated true triumphs, since Nernst had experimentally confirmed virtually everything that his theory predicted.

Apparently both Nernst and Einstein gloried in the turn of events: Nernst, realising that his precious heat theorem was linked with and in agreement with predictions from quantum theory; and Einstein, that his revolutionary quantum conceptions were receiving experimental backing from Nernst's work. Working with Lindemann in 1911, Nernst showed that Einstein's specific-heat equation was in agreement with his data, except for certain systematic deviations. They also showed that a revised Einstein formula was in still closer agreement with the experimental information except at the very lowest temperatures.

The experimental evidence for the quantum theory was one of the central themes for discussion at the first Solvay Congress in Brussels in early November of that same year. Nernst had taken the initiative for setting up and organising the sessions and seeing to it that the leading physicists would be there. The twenty-two participants included Lorentz, Planck, Rubens, Sommerfeld, Wien, Jeans, Einstein and Lindemann. In a communication on the theory of specific heats Einstein discussed the empirical formula for the thermal energy of solids of Nernst and Lindemann. The following year the issue was clarified theoretically by Max Born and Theodore von Kármán. About the same time, in 1912, Peter Debye who had been Sommerfeld's assistant and was then lecturing on thermodynamics in Zurich, independently presented his theory of specific heat at the Physikalische Gesellschaft in Bern. Born and von Kármán had reached their results by a different route than Debye; both had built on the foundations laid by Einstein. For some time these formulations proved satisfactory. Of unique significance was Debye's derivation of the T^3 law that gives for the lowest temperatures the proportionality between atomic heats and the third power of the temperature - a relationship that was seen to fit the empirical data very well.

Planck markedly enhanced the Nernst theorem by putting it into the form in which it most generally has been given ever since; stressing that it was a major extension of the second law because it permits the calculation of absolute entropies. He also pointed out - but in rather cautiously worded statements - that the entropy form of the Nernst theorem necessarily means that the third law, like the second, is intrinsically connected with probability, atomistics, and statistical implications.

Planck's formulation of the heat theorem in terms of vanishing entropies at absolute zero did not appeal to Nernst, who considered it inappropriate and intuitively unclear. The paradox inherent in Nernst's position was that his own formulation, in which there was no reference to entropies, was unwieldy and theoretically less elegant. Unquestionably Nernst's peculiar reluctance to employ the entropy concept was frowned upon by most of his colleagues in the Berlin scientific community.

In 1912 Nernst formulated his heat theorem in terms of the theoretical unattainability of absolute zero. According to this principle, it is impossible to build a caloric machine that will allow a substance to be cooled to absolute zero. From this negative assertion Nernst concluded that the thermal coefficients of all physical properties for solid bodies should vanish in the approach to absolute zero. The

properties of bodies that subsequently were investigated as tests of
the third law included studies on thermal expansion, surface tension,
magnetic and dielectric polarisation, and the thermoelectric properties
of matter. More indirect, and therefore less convincing were the
studies in fluidity, solutions, mixed crystals, frozen-in states, and
crystallographic transformations - all designed to demonstrate the disappearance of physical properties as $T \rightarrow 0$; that is, to show, in the
Planck formulation, that $\lim_{T \rightarrow 0} \Delta S = 0$.

Nernst's peculiar way of enunciating the principle of the unattainability
of absolute zero was questioned by scientists, even while the principle
itself was recognised to be plausible and theoretically significant.
Rather than furnish a logically acceptable argument for the principle,
Nernst had thought to demonstrate the consistency of the principle with
the impossibility of a perpetuum mobile. In order to steer clear of
formulating his ideas in terms of entropy-temperature diagrams, Nernst
presented his "proof" in the form of a Carnot cycle. He showed that
a perpetuum mobile of the second type results from taking absolute zero
as the lowest temperature of the cycle, thus demonstrating that the
attainment of absolute zero is theoretically impossible. The way in
which the unattainability of absolute zero was deduced from arguments
based on the disappearance of specific heats as $T \rightarrow 0$ was rather
obscure. In 1913 his proof was challenged by Einstein, who reasoned
that Gedanken-experimente should be possible in principle even if not
in practice and that this was not the case for Nernst's formulation.
Nernst stuck to his guns, and so did Einstein.

One of the most serious difficulties confronting Nernst was that classical kinetic gas theory predicts that the heat capacity at constant
volume does not tend to zero as $T \rightarrow 0$, as the heat theorem demands,
but reaches limiting values of $3/2\ R$ for monatomic gases, and $5/2\ R$
for diatomic gases, and so forth. Thus, even if the Nernst theorem had
given a fairly acceptable interpretation of the heat capacities of
condensed phases at low temperatures, it seemed likely that gases might
have to be excluded from the theorem. Of course this would have been
very detrimental to Nernst's attempt to elevate the heat theorem to the
status of a general law of thermodynamics. Fortunately for Nernst, the
developments in quantum mechanics just prior to World War I gave Sackur
(1912), Tetrode (1912), and Stern (1913) the means to calculate directly
the entropy differences for monatomic gases, and later, the same for
more complex molecules. The calculations agreed tolerably well with
the experimental results. Theory predicted and experiment confirmed
the falling-off of heat capacities at low temperatures.

Nernst interpreted these developments as a step toward confirming his
theorem. In 1914, he proceeded heroically, using some quite primitive
arguments and despite the incredulity of many scientists, to postulate
a state of degeneracy (Entartung) for gases. Subsequent advances in
quantum mechanics showed Nernst to be on the right track. The Bose-
Einstein and Fermi-Dirac statistics both confirmed the gas degeneracy
idea. The progress of quantum theory notably justified Nernst's idea,
if not the reasons for enunciating it, when it was shown that electrons
in metals present an example of degeneracy at much higher temperatures

and that the same principle suffices to account for the interior physical conditions of stars at very high temperatures and pressures.

Let me conclude with a summary statement. For Nernst, the clue to the theoretical solution to the search for the criteria of chemical equilibrium and spontaneity lay in the approximate success, at ordinary temperatures, of the Thomsen-Berthelot principle. His original goal had been to find a practical means of computing chemical equilibria from thermal measurements. The search for the mathematical criteria for chemical equilibrium and chemical spontaneity led Nernst to suggest a new interpretation of the Gibbs-Helmholtz equation. The bold heat-theorem that he postulated generated new problems, both theoretical and experimental, that no one had anticipated. Free energies, heats of reactions, and specific heats, were shown to have deep theoretical significance for the structure of matter in the low temperature range. Quantum theoretical formulations for the heat capacities of solids in the vicinity of absolute zero were seen to be coupled with probabalistic, statistical and atomistic implications. Questions about the theoretical unattainability of absolute zero, degeneracy states for gases at low temperatures, and frozen-in equilibrium states became the topic of an extensive and polemical literature.

The period we have treated in this paper represents only the first chapters of a more complete history of third law considerations that extends well into the 1940's. After World War I the problems connected with the heat theorem were taken up anew by Nernst's students and colleagues. By that time, Nernst's scientific interests were so far removed from third law formulations that he was satisfied to pursue other problems in physical chemistry that were closer to what he considered to be the new frontiers of his discipline, e.g., chemical kinetics and astrophysics.

A physicist by training, Nernst early in life had been fired with enthusiasm for physical chemistry in his contacts with Ostwald, Aarhenius and van't Hoff. His complete theoretical command of the subject matter of physical chemistry was unparalleled. He was a superb craftsman with keenly developed technical skills and an imaginative grasp of what was experimentally feasible. We might characterise Nernst as a physicist, turned chemist, who was quick to seize on novel ideas no matter what their source. This certainly was the case in regard to his approach to chemical thermodynamics and in his formulation of the third law. Much more remains to be said about the fate of the third law after 1914; but that has another history within another historical context.

W. J. Hornix

THE THERMOSTATICS OF J. WILLARD GIBBS
AND 19TH CENTURY PHYSICAL CHEMISTRY

Gibbs's famous and notorious paper 'On the Equilibrium of Heterogeneous
Substances' (EHS) [1] is considered an extremely important contribution
to physical chemistry because it supplies the fundamental structure of
chemical thermodynamics and some of its important applications. The
first objective of EHS is to develop a theory of equilibrium and stability conditions which includes heterogeneous substances and systems
of variable composition. This aim is fulfilled in the first part of EHS
which was published in 1876. The second part which appeared in 1878
modifies the theory to cover complications which arise through aniso-
tropic strain in solids and surface effects, and ends with a short discussion of the thermodynamics of the galvanic cell.
I will restrict myself to the first part of Gibbs's paper and confront
his results with work of Kirchhoff, Massieu, Guldberg, Debray, Troost,
Hautefeuille, Moutier and Horstmann.

On open systems

The starting point of EHS is a formulation of definitions of the energy
and entropy of a system in terms of 'heat' and 'work': "the difference
of the values of the energy for any two states represents the combined
amount of work and heat received or yielded by the system when it is
brought from one state to the other, and the difference of entropy is
the limit of all possible values of the integral $\int \frac{dQ}{t}$,....." (Sci Pap.
p 55). That these definite descriptions are uniquely instantiated is guaranteed by the first and second laws. For homogeneous systems both laws are
unified in the concise statements $dU = TdS - PdV$. This equation holds only
for closed systems; the treatment of systems with variable composition
demands a more general expression. Gibbs's solution is straightforward
and simple: the "complete value" of the differential of the energy is
$dU = TdS - PdV + \sum_i \mu_i \, dM_i$ where the "potential" μ_i is defined by
$\mu_i \equiv \frac{\partial U}{\partial M_i}$ (Sci Pap. p 63) [2]. This extension of the energy function is
physically not self evident at all, because energy and entropy are only
defined for systems without material exchange with the environment. A

314

physical justification is given by Gibbs only in passing: "We may choose independently for each simple substance the state in which its energy and entropy are both zero.... The values of the energy and entropy of any compound body in any particular state will then be fixed. Its energy will be the sum of the work done and heat expended in bringing its components, from the states in which their energies and entropies are zero into combination and to the state in question, and its entropy is the value of the integral $\int \frac{dQ}{t}$ for any reversible process by which that change is effected" (Sci Pap. p 85). This is a statement of an experimental procedure, such as Kirchhoff already used in 1858 to determine, among other things, the energy function for the solution of a salt in water and which resulted in an expression $U(t,P,m)$ for 1 gram of salt in m grams of water at t° Celsius [3]. An energy function $U(t,P,m) = U(T-273, P, M_{water}/M_{salt})$ is however quite different form a function $U(S,V,M)$.

In his early papers of 1873 Gibbs was already aware of the usefulness of entropy and volume as independent variables and in particular of certain advantages of the relation between volume, entropy and energy for the characterization of a system [4]. Probably Gibbs found in Kirchhoff's papers the justification of the extension of the energy function to open systems, and recognized in $U(S,V)$ the most promising candidate for such an extension, which leads to the well known equations.

On fundamental equations

The importance of $U(S,V)$ and $U(S,V,M_1,...M_n)$ lies in their information content: they contain all the thermodynamical information about a system. If U is a known function of $S,V,M_1,...M_n$, then a very large class of properties of the compound considered can be determined," we may say in general, all its thermal, mechanical, and chemical properties, so far as active tendencies are concerned...." (Sci Pap. p 86). Gibbs calls an equation giving such a function: a fundamental equation and lists the following, which are related as given

$U = U(S,V,M_1,...M_n)$

$F = F(T,V,M_1...M_n)$ $\qquad$ $F = U - TS$

$H = H(S,P,M_1...M_n)$ $\qquad$ $H = U + PV$

$G = G(T,P,M_1...M_n)$ $\qquad$ $G = U - TS + PV$

He recognises F, H and G as Legendre transforms of U by demonstrating
the defining properties in geometrical language for the simplified case
of a system of invariable composition (Sci Pap. p 116).

These considerations are clearly the final clarification and generalization of the insights which led to Gibbs's preference for U(S,V) in the
papers of 1873 and to Massieu's preference for the equations
$\psi(V,T) = S - \frac{U}{T}$ and $\psi'(P,T) = S - \frac{U + PV}{T}$ which the latter in 1869 named
"fonctions caractéristiques" because "on peut en tirer tòutes les
propriétés du corps que l'on considère dans la thermodynamique" [5].

A special property of the fundamental equation $U = U(S,V,M_1,...M_n)$
is its homogeneity of the first order which we know, according to Gibbs,
a priori (Sci Pap. p 86). From this property follows the "Gibbs-Duhem
equation"

$$SdT - VdP + M_1 d\mu_1 + ... + M_n d\mu_n = o$$

which equation generates another fundamental equation: $P = P(T,\mu_1...\mu_n)$.
The latter contains all the thermodynamical information related to,
what Gibbs calls, "the phase" of the system.

These abstract considerations naturally invite the formulation of specific concrete fundamental equations. Gibbs himself derives the fundamental equations for ideal gases starting from the definition of an
ideal gas through PV = aT ; dU = cdT and U = cT + E where a,c and E are
constants. For mixtures of ideal gases he derives analogous equations
supposing that by approximation $P_{mixture}$ $(T,\mu_1...\mu_n) = \sum_i P_i (T,\mu_i)$ the
so called "Gibbs Dalton approximation". The Gibbsian fundamental equations
for pure ideal gases F = F(T,V,M) and G = G(P,T,M) differ from Massieu's
"fonctions caractéristiques" for ideal gases of 1869. The difference is
a consequence of the use of different methods of derivation and of the
choice of reference states for energy and entropy [6].

On equilibrium and stability

The nucleus of EHS is the theory of equilibrium and stability. The principle on which the theory is founded reads: "For the equilibrium of any
isolated system it is necessary and sufficient that in all possible variations of the state of the system which do not alter its energy, the
variation of its entropy shall either vanish or be negative" (Sci Pap. p 56)
Or concisely: for all possible variations $(\delta S)_{U,V,M} \leqslant o$, which is

316

equivalent to: for all possible variations $(\delta U)_{S,V,M} \geqslant 0$.

With methods derived from Lagrangian mechanics it follows from the energy-minimum-principle firstly, that for heterogeneous systems a necessary condition of equilibrium is, that for existing homogeneous parts, the temperature, the pressure and the potentials must be constant throughout the system; secondly, that for substances A_1, A_2, etc., related by a reaction equation $\Sigma n_i A_i = 0$ in equilibrium there holds $\Sigma n_i \mu_i = 0$.

A third consequence is about the stability of an equilibrium: supersaturated solutions, superheated water, etc. are familiar examples of "practically unstable equilibrium". The following stability criterium is proved: if for existing parts of a system the temperature, pressure and potentials are T,P, and $\mu_1,\mu_2...\mu_n$ respectively, then a sufficient condition for stable equilibrium is, that it is impossible to form new parts out of the same substances for which

$U' - TS' + PV' - \mu_1 M_1' - ... - \mu_n M_n' < 0$ (accented letters relate to the new parts).

The entropy maximum principle and the energy minimum principle are considered by Gibbs as theorems which have to be proved starting from the first and second law. The proofs given form a first stumbling-block for the student of EHS (Sci Pap. pp 58 - 62). In the abstract of the paper by the author the central position of the principles is justified by noting that the energy minimum principle (and the equivalent free-energy-minimum principle $(\delta F)_{T,V,M} \geqslant 0$) "may be regarded as extensions of the criterion employed in ordinary statics to the more general case of a thermodynamic system" (Sci Pap. p 355). The quantities $-U$ and $-F$ may be regarded as force functions for constant entropy and constant temperature respectively. This, and the methods Gibbs uses, show the relationships between Gibbsian thermodynamics and Lagrangian statics.

Another set of stability conditions follow from the consideration of neighborhoods of states on the equilibrium surfaces defined through the fundamental equations. These 'local' stability conditions are expressed in an analytical form e.g. $\Delta U - T\Delta S + P\Delta V - \mu_1 \Delta M_1 - ... - \mu_n \Delta M_n > 0$ where $\Delta U = dU + d^2U + ...$ etc. and from which follows:

$$\left(\frac{\partial^2 U}{\partial S^2}\right)_{V,M} = \left(\frac{\partial T}{\partial S}\right)_{V,M} > 0 \; ; \; \left(\frac{\partial^2 U}{\partial V^2}\right)_{S,M} = -\left(\frac{\partial P}{\partial V}\right)_{S,M} > 0 \text{ and}$$

$$\left(\frac{\partial^2 U}{\partial M i^2}\right)_{S,V,M_j} = \left(\frac{\partial \mu i}{\partial M i}\right)_{S,V,M_j} > 0.$$

These conditions can be concisely summarized in geometrical language when we consider the graphical representation of the fundamental equation $U = U(S,V,M_1 \ldots M_n)$: the representing surface appears to be convex downwards everywhere. Similar conditions of local stability are expressed in terms of the other fundamental equations. If some of the conditions are not satisfied phase transitions or phase separations will occur. Critical states are also defined within this framework. The above stability conditions are already formulated in the 1873 paper on geometrical representation of the thermodynamic properties for the simpler case of substances whose equilibrium states can be represented by a surface in V,S,U space (Sci Pap. p 39 - 43). They are expressed in geometrical language. In EHS the generalized problem is initially treated in analytical language. Geometrical considerations are relegated to an illustrative chapter. The stimulus for the exercise of the 1873 paper seems to be the consideration of the continuity of the liquid and the gaseous state [7]. The generalization of the problem in multidimensional chemical thermodynamics, and the example of generalized analytical mechanics, probably was the incentive for the development of the analytical thermodynamics of the 1876 paper [8].

The phase rule

Sandwiched between the statement of the general and the local stability conditions we find the short chapter in which the concept of 'phase' is defined and the famous phase rule is stated: "A system of r coexistent phases, each of which have the same n independently variable components is capable of n + 2 - r variations of phase" (Sci Pap. p 96). Gibbs gives only a few examples "An example of n = 1 and r = 3 is seen in the coexistent solid, liquid and gaseous forms of any substance of invariable composition. It seems not improbable, that in the case of sulphur, and some other simple substances, there is more than one triad of coexistent phases, but it is entirely improbable that there are four coexistent phases. An example of n = 2 and r = 4 is seen in a solution of a salt in water, in contact with vapor of water and two different kinds of crystals of the salt" (Sci Pap. p 97).

Other applications are connected with the generalized Clausius - Clapeyron

equation which Gibbs derives, and certain consequences, now known as
the Gibbs - Konovalow rules (Sci Pap. pp 99 - 100).

Gibbs and Guldberg

An investigator of the origin of Gibbs's phase rule has to question its
connections with the theory of indeterminate chemical compounds, which
Guldberg explains in the Acta of the Scientific Society of Christiania
of 1870 [9]. Chemical reactions of compounds of a determinate composition
and the transformations of solutions or liquid mixtures of an indeterminate composition can, according to Guldberg, be treated within the same
thermodynamical theory. The starting point of this theory is, that for
every state of aggregation of a chemical compound, there exists a pair
of "fundamental functions" which determine the thermodynamical properties. For compounds of determinate composition these functions are: the
specific volume $v(P,T)$ and the specific energy $u(P,T)$. For indeterminate
compounds, consisting of two components in variable proportion $\frac{y}{x}$, the
fundamental functions are $v(P,T,\frac{y}{x})$ and $u(P,T,\frac{y}{x})$.

All compounds have certain limits of existence. At these limits the
indeterminate compounds separate into their constituent parts, e.g. water
evaporates or freezes out from a salt solution. The limits of existence
of a solution of a salt in water can be represented in the following
diagram, with coordinates P,T and $\frac{y}{x}$ (figure 1): abfd is the freezing
surface, abec the vaporization surface and acgd the saturation surface.
Guldberg derives general and approximated equations for these surfaces
e.g. the approximated equation for the vaporization surface is identical
with the equation Kirchhoff derived in 1858 for the vapor pressures of
salt solutions [10].

The intersection ab of the freezing surface and the saturation surface
defines the conditions under which simultaneously ice and salt separate,
leaving the composition of the solution unchanged. At constant pressure
this happens when the temperature is a minimum. The intersection P =
constant (fig. 2) shows at this temperature the composition of the
"cryohydrate" (this name, and the interpretation as a chemical compound
of determinate composition is due to Guthrie [11].
Guldberg shows that the data about vapour pressures, freezing points
etc. of salt solutions, mainly supplied by Wüllner and Rüdorff, can be
explained within the framework of the theory.

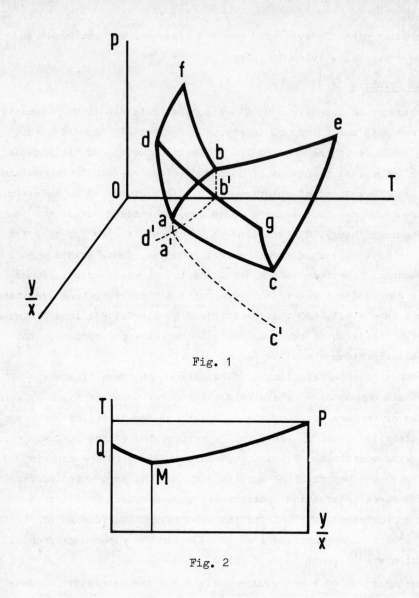

Fig. 1

Fig. 2

There are sufficient ressemblances to ask whether Gibbs read Guldberg's paper, without mentioning it. Gibbs is, certainly in EHS, not very communicative about his knowledge of the thermodynamical literature, and his theory is, compared with that of Guldberg, original enough to justify not mentioning the latter, supposing that a very restricted citation policy is used.

It is however also possible that Gibbs did not see Guldberg's paper, but considering the same data in the same spirit, gave a solution which necessarily shows similarities. The latter hypothesis is supported

by the circumstance that Guldberg's paper was published in a journal
of restricted circulation, and thus could easily be overlooked. As a
matter of fact it was overlooked by almost everyone, and consequently
did not have any effect on the development of thermodynamics.

Although the most important sources of Guldberg's paper: Clausius,
Kirchhoff, Wüllner and Rüdorff are also mentioned by Gibbs, the use of
these sources is so different that a direct influence of Guldberg is improbable. The main result of Guldberg's paper is a theory of salt solution. Gibbs mentions salt solutions only in passing. The differences
between the two papers are equally worth mentioning with the similarities.

Allotropy

The thermodynamic theory of equilibrium and stability is relevant for
the theory of allotropic modifications. Research on allotropy and on the
related subject of dissociation are at the centre of interest of the
physical chemists in the late sixties and the seventies. The main centre
of activity in both cases is certainly the École Normale Supérieure.

A first glance at dissociation theory: Debray, in his famous researches
on the dissociation of calcite and of sodium phosphate hydrates in the
late sixties, confirmed Deville's earlier view that there is a close
analogy between dissociation and vaporization. Debray showed that the
above dissociations are characterized through equilibrium pressures of
CO_2 and H_2O respectively, which increase with the temperature. Horstmann
in 1869 showed for the case of the dissociation of sal ammoniac and
Peslin in 1871 for the case of calcite that the pressure-temperature
curve follows the Clausius Clapeyron equation and that the heat of
dissociation can be calculated from the P,T curve. Clearly a chemical
process, dissociation, followed the same laws as a physical process:
vaporization [12].

There are similarities between the above dissociations and the transformation of white phosphorus into red phosphorus, of cyanic acid into
cyanuric acid and cyamelide, and of cyanogen into paracyanogen. The
research on these transformations by Troost and Hautefeuille and by
Lemoine between 1868 and 1874 revealed the existence of a vapour
pressure at equilibrium, which is a function of the temperature only [13].

321

White phosphorus is at higher temperatures quickly transformed into red phosphorus. The maximum vapour pressure of the system diminishes till a final value is reached. The vapour of phosphorus can be condensed and always gives the white modification. Troost and Hautefeuille explain the phenomena as follows: The molecular constitution of red phosphorus is different from that of white phosphorus; the transformation is consequently a chemical reaction. The transformation of liquid white phosphorus into vapour is a physical process: vaporization. The initial vapour pressure is the maximum vapour pressure of white phosphorus. The transformation of red phosphorus into vapour is a chemical process, because the vapour consists of molecules of the white modification. The transformation is an equilibrium reaction: the final vapour pressure ("tension de transformation") is the equilibrium pressure of the chemical reaction: white phosphorus (gas) $\rightleftharpoons$ red phosphorus (solid). This pressure is a function of the temperature only, according to contemporary dissociation theory.

Moutier, one of the leading thermodynamicists in Paris, considered in 1874 the transformation of allotropic modifications and the vaporization of a substance as identical processes. The equilibrium vapour pressure of the transformation of white phosphorus into red phosphorus is the vapour pressure of red phosphorus and consequently follows the Clausius Clapeyron equation. The characterization of allotropic modifications through the distinction of the vapour pressure and the transformation tension is, according to Moutier, not to the point: in general two modifications of a substance (e.g. water and ice) at the same temperature (e.g. 0° centigrade) will have different vapour pressures [14].

It will be clear that Gibbs could have clarified the discussion considerably: from the phase rule it follows that the system phosphorus (1), phosphorus (g), phosphorus (s) must be instable in general; the different fundamental equations for liquid (white) phosphorus and solid (red) phosphorus generate different vapour pressures at the same temperature and the unstable modification has the greatest vapour pressure; thermodynamics explains the phenomena considered quantitatively without deciding between explanations at the molecular level. However, Gibbs does not make any allusion to the problems of allotropy except for that halve sentence about sulphur (Sci Pap. p 97) and nothing is found which could be interpreted as an intentional clarification of the problems

the chemists struggled with. It is probable that Gibbs did not follow
the discussion. If he had, presumably he would have made some clarifying statements, at least in passing, which could have aroused the
interest of some of the discussants in his thermodynamics.

On homogeneous gas reactions

A most interesting application of the new thermodynamics to chemistry
is Gibbs's treatment of homogeneous gas reactions. He considers a gas
mixture which in equilibrium consists of M_1 gram A_1, M_2 gram A_2 and M_3
gram A_3. The equilibrium reaction is written volumetrically as
β_1 volume A_1 + β_2 volume A_2 = 1 volume A_3.
From the equilibrium condition $\sum_i n_i \mu_i = o$ and the expression derived
for the fundamental equation $P = P(T, \mu_1, \ldots \mu_n) = \sum P_i(T, \mu_i)$ for ideal
gas mixtures it follows that

$$\ln \frac{M_1^{\beta_1} M_2^{\beta_2}}{M_3 \cdot V^{\beta_1 + \beta_2 - 1}} = \frac{A}{a_3} + \frac{B}{a_3} \ln T - \frac{C}{a_3 T} \text{ and}$$

$$\frac{d}{dT} \ln \frac{M_1^{\beta_1} M_2^{\beta_2}}{M_3 \cdot V^{\beta_1 + \beta_2 - 1}} = \frac{BT + C}{a_3 T^2} = \frac{Q_v}{a_3 T^2}$$

$$\text{and } \frac{d}{dT} \ln \frac{M_1^{\beta_1} M_2^{\beta_2} P^{\beta_1 + \beta_2 - 1}}{M_3 (a_1 M_1 + a_2 M_2 + a_3 M_3)^{\beta_1 + \beta_2 - 1}} = \frac{Q_p}{RT^2}$$

where a_1, a_2 and a_3 are the constants in the equation PV = aT for 1 gram of
A_1, A_2 and A_3 respectively, and A, B and C are constants related to the
constants defining the ideal gases considered, such that BT + C can be
interpreted as the heat evolved at constant volume (Q_v) in the formation
of 1 gram A_3 out of the gases A_1 and A_2 and Q_p is the heat evolved at
constant pressure for this reaction (Sci Pap. pp 168 - 172). Gibbs takes
considerable care to verify that these equations are in agreement with
available experimental data. According to Ostwald, this is because Gibbs
here finds the only possibility to test his theory [15]. In EHS he discusses data on the vapour density of peroxide of nitrogen as a function
of temperature and pressure as measured by Deville and Troost (1867),
and Playfair and Wanklyn (1861). Supposing that peroxide of nitrogen
dissociates according $N_2O_4 \rightleftharpoons 2NO_2$ he derives a theoretical relation
between the vapour density of the gas mixture, the temperature and the

pressure which "exhibits a somewhat striking agreement with the results of experiment" (Sci Pap. p 364). In the 1879 paper "On the vapor densities of peroxide of nitrogen, formic acid, acetic acid and perchloride of phosphorus", the same exercise is done more thoroughly: vapor density data are collected from the literature over many years and compared with the theoretical equations. Discrepancies between the calculated and determined values are critically discussed by evaluating possible errors connected with the different experimental methods used [16] Thus Gibbs contributed to the solution of the problem of abnormal vapour densities which was a matter of intensive research and discussion in his days.

The chemists of the seventies hesitated between a physical explanation - non ideal behaviour of the vapours concerned - and a chemical explanation: the existence of a polymerization process. For the case of peroxide of nitrogen Hautefeuille and Troost decided in favour of the physical explanation, because at low temperatures and low pressures the densities are normal. If one considers dissociation as the counteraction of the chemical force of affinity by the physical force of heat, one will not expect a pronounced influence of the pressure on the relative densities in case of the chemical explanation. We have seen that Gibbs opted for the chemical explanation. But he did more: he explained the experimental data quantitatively on the basis of the dissociation hypothesis and the hypothesis of ideal behaviour, within the framework of generalized thermodynamics.

A reward followed soon: in the Annales de Chimie et de Physique of 1882 his contribution was discussed by the chemist Lemoine, who however was not convinced of the great merits of Gibbs's solution, in comparison with that of Horstmann. Lemoine concluded that the theories of Gibbs and Horstmann differ only in certain terms "d'une importance médiocre" which are either neglected or maintained [17].

Gibbs and Horstmann

Gibbs's involvement with the dissociation problem and the nature of his solution pose the question of his obligations to Horstmann, particularly with respect to the latter's 1873 paper in Liebig's Annalen on the theory of dissociation. Horstmann's theory is based on the entropy maximum principle of Clausius for equilibrium: $\delta S = o$ with respect to the possible

variations of the system, and on a second principle of Clausius: the
existence of the disgregation function. The latter principle transforms
the equation $TdS = dU + dW$ via $TdS = dH + dI + dW = dH + dL$ into
$TdS = dH + TdZ$ or $dS = \dfrac{dH}{T} + dZ$.

dH is the increase of the real heat content, H being a function of the
temperature only; dI is the heat used for internal work, also a state
function; $dL = dI + dW$ is the total work done, internal and external,
or that part of the heat TdS which is transformed into work. In his
sixth treatise on thermodynamics (1862) Clausius stated the new principle
$dL = TdZ$ where dZ is another state function called "Disgregation". The
equation $dS = \dfrac{dH}{T} + dZ$ explains, according to Clausius, the entropy in
molecular terms [18].

Horstmann, who explained in his paper of 1870 'for the chemists' entropy
in terms of heat content and disgregation, translated in 1873 the already
dubious disgregation equation for the case of chemical reactions in a
still more dubious way: $dS = \dfrac{Qx}{T} + dZ$ where Q is the heat of reaction,
x is the variable which measures the progress of the reaction: the
proportion of the molecules transformed, and Z the disgregation of the
system. The total disgregation is supposed to be the sum of the disgrega-
tions of the components of the system $Z = \sum_i Z_i$, For gaseous components
$Z_i(T,V) = Z_i^O(T) - R \ln \dfrac{V}{Vo}$, an equation again taken from Clausius. For
solid components $Z_j = Z_j^O(T)$. On this basis Horstmann derives for the
reaction $A_1(solid) \rightleftharpoons A_2(solid) + A_3(gas)$ e.g. $CaCO_3 \rightleftharpoons CaO + CO_2$:
$\ln P = \dfrac{Q_v}{RT} + constant$, and for the homogeneous gas reaction
$n_1A_1(gas) \rightleftharpoons n_2A_2(gas) + n_3A_3(gas)$:
$$\ln \frac{P_2^{n_2} P_3^{n_3}}{P_1^{n_1}} = \frac{Q_v}{RT} + constant.$$

The theory was confirmed by the experimental data of many reactions.
Horstmann's dubious theoretical exercise thus produced remarkable results which established his reputation as the first who applied thermodynamics succesfully to chemistry [19].

Why did Gibbs not mention Horstmann's work? It is difficult to believe
that he did not see it or that he was not interested. It is even probable that Gibbs got acquainted with Horstmann personally during his stay
in Heidelberg in 1868/9. And certainly he would have been highly interested in Horstmann's 1873 paper: in fact, Gibbs derivation of the equili-

brium condition of homogeneous gas reactions does the same job, starting
from the same general condition $\delta S = o$, resulting in analogous equations.
But the job is done in a much better way. Probably Gibbs read Horstmann's
papers and decided not to mention them because, being of retiring disposition, he did not want to be provocative. The historian however might
well consider, that Horstmann's 1873 paper was a stimulus to Gibbs to
do a better job.

NOTES AND REFERENCES

1. J. Willard Gibbs, "On the Equilibrium of Heterogeneous Substances",
   Trans. Conn. Acad. (Oct., 1875 - May, 1876) 3: 108 - 248 and (May,
   1877 - July, 1878) 3: 343 - 524. Reprinted in The Scientific Papers
   of J. Willard Gibbs. H.A. Bumpstead and R.G. Van Name, eds.
   (New York 1906; reprinted 1961) Vol I, pp 55 - 353. This reprint
   will be noted as Sci. Pap., and page references will be to this
   edition.

2. I will use a modern notation for thermodynamic quantities, except in
   literal quotations.
   Concerning the equation $dU = TdS - PdV$ see the article of Martin J. Klein
   in these Proceedings.

3. G. Kirchhoff, "Ueber einen Satz der mechanischen Wärmetheorie, und
   einige Anwendungen desselber". Ann. Phys. u. Chem. (1858) 103: 178 -
   206. The "Wirkungsgrösse W" (= -U) of 1 gram of salt in m grams of
   water at t^o Celsius is (p. 201):

$$W = -k\left(m\gamma t + \int_o^t c'dt\right) + R(a + t)^2 \frac{\partial}{\partial t} \int_o^m dx \ln \frac{\mu}{a + t} , \begin{cases} o < x \leqslant \alpha: & \mu = \mu_1(t) \\ \alpha \leqslant x & \mu = \mu(t,m) \end{cases}$$

where k = mechanical equivalent of heat

> y = specific heat at constant volume of water vapour
>
> c' = specific heat of solid salt
>
> -a = temperature in degrees Celsius of the absolute zero
>
> $\mu_1 = \mu_1(t)$ = maximum vapour pressure of a satured solution of
>   the salt
>
> $\mu = \mu(t,m)$ = maximum vapour pressure of a solution of 1 gram
>   salt in m grams of water
>
> α = quantity of water needed to dissolve 1 gram of salt

4. J. Willard Gibbs, "Graphical methods in the Thermodynamics of Fluids;, Trans. Conn. Acad. (April - May, 1873) $\underline{2}$: 309 - 342, Sci. Pap. 1- 32 "A Method of Geometrical Representation of the Thermodynamic Properties of Substances by Means of Surfaces". Trans. Conn. Acad. (Dec. 1873) 2: 382 - 404. Sci. Pap. 33 - 54.
   See in particular Sci. Pap. pp 20 - 28, pp 33 - 35.

5. F. Massieu, "Sur les functions caractéristiquès des divers fluides". Comptes Rendus. Acad. Sci. Paris. (1869) $\underline{69}$: 858 - 862, 1057 - 1061. In Gibbs's early papers there is no indication of acquaintance with Massieu's results. Gibbs compares Thomson's surface in V,T,P-space with an equilibrium surface in V,S,U-phase (Sci. Pap. p 34). Massieu considers "fonctions caractéristiques" ψ and ψ' in the variables V,T and P,T respectively.

6. F. Massieu, op. cit. neglects in the derivation of $\psi(V,T)$ and $\psi'(P,T)$ constants which arise when the reference state is taken into consideration. This is corrected in his: "Mémoire sur les fonctions caractéristiques des divers fluides et sur la théorie des vapeurs". Mém. Sav. Étrang. (1879) $\underline{22}$: (no 2) 1 - 92, and leads to results which are comparable with Gibbs's equations.

7. See Martin J. Klein's contribution in these Proceedings.

8. EHS might properly be characterized as the development of thermodynamics as a branch of analysis, analogous with Lagrange's characterization of the "Mécanique Analitique" in the "Avertissement" of the first edition e.g. "On ne trouvera point de Figures dans cet Ouvrage . Les méthods que j'y expose ne demandent ni constructions, ni raisonnemens geómétriques ou méchaniques, mais seulemens des opérations algébriques, assujetties à une marche régulière et uniforme. Ceux qui aiment l"Analyse, verront avec plaisir la Méchanique en devenir une nouvelle branche..." J.L. Lagrange Mécanique Analitique Avertissement 1788, Paris 1965.

9. C.M. Guldberg, Forhandlinger: Videnskabs - Selskabet. Christiana (1870, 11 Febr., 25 Mar.): p 1 - 43 with a summary in French.

Translated as "Beitrag zur Theorie der unbestimmten chemischen Verbindingen: in Thermodynamische Abhandlungen über \Molekulartheorie und chemische Gleichgewichte von C.M. Guldberg R. Abegg ed. Ostwald's Klassiker No 139 Leipzig 1903, pp 27 - 70.

10. C.M. Guldberg (transl. R. Abegg) op. cit. p 51; G. Kirchhoff op.cit. p 199.

11. F. Guthrie, On salt solutions and attached water, Phil. Mag. 4e ser. (1875) 49: 1-20, 206 - 218, 266 - 276, 5e ser. (1876) 1: 49 - 60, 354 - 369, 446 - 455. (1876) 2: 211 - 225.

12. H. Debray, "Recherches sur la dissociation". Compt. rend. (1867) 64: 603 - 606; (1868) 66: 194 - 197.
A. Horstmann, "Dampfspannung und Verdampfungswärme des Salmiaks" Berichte: (1869) 2: 137 - 140; reprinted in Abhandlungen zur Thermodynamik chemischer Vorgänge von August Horstmann. Ostwald's Klassiker no 137, Leipzig 1903, pp 3-7.
M. Peslin, "Sur les lois des tensions de dissociation des composés chimiques". Ann. chim. phys. (1871) 24, 208 - 214.

13. L. Troost, P. Hautefeuille, "Sur la production du paracyanogène et sa transformation en cyanogène", Compt. rend. (1868), 66: 735 - 738,
-, "Lois de la transformation du paracyanogène en cyanogène et la transformation inverse", Compt. rend. (1868) 66: 795 - 799.
-, "Lois de la transformation de l'acide cyanique en ses isomères, et de la transformation inverse, Compt. rend. (1868) 67: 1345 - 1348.
G. Lemoine, "Etudes expérimentales sur la transformation réciproque des deux états allotropiques du phosphore", Compt. rend. (1871) 73: 797 - 801, 837 - 841.
L. Troost, P. Hautefeuille "Recherches sur les transformations allotropiques du phosphore", Compt. rend. (1873) 76: 76 - 80.
-, "Etude sur le phosphore", Compt. rend. (1874) 78: A useful review can be found in the Encyclopédie chimique M. Frémy ed. Paris 1882 , in the article:G. Lemoine, "Etudes sur les equilibres chimiques" Fasc. 2 p 69.

14. J. Moutier, "Recherches sur les vapeurs émises à la même température par un même corps sous deux états différents". Ann. Chim. Phys. (1874) 5e ser. 1: 343 - 373. The strength of the theory is shown by a calculation of the heat of transformation of white phosphorus into red phosphorus from the vapour pressures of white and red phosphorus.

15. W. Ostwald, Lehrbuch der Allgemeinen Chemie, Verwandtschaftslehre Erster Teil. Leipzig 1896 - 1902, p 139 - 140.

16. J. Willard Gibbs, "On the Vapor-Densities of Peroxide of Nitrogen, Formic Acid, acetic acid, and Perchloride of Phosphorus.", American Journal of Science, (1879) ser. 3, 18: 277 - 293, 371 - 387. Reprinted in Sci. Pap. pp 372 - 403.

17. G. Lemoine, "Théories des equilibres chimiques". Ann. Chim. Phys. (1882) 26: 289 - 405.
A. Horstmann, "Theorie der Dissociation". Ann. Chem. Pharm. (1873) 170: 192 - 210.

18. A. Horstmann, op. cit.
See also: A. Horstmann, "Ueber den zweiten Haupsatz der mechanischen Wärmetheorie und dessen Anwendung auf einige Zersetzungserscheinungen". Ann. Chem. Pharm. (1870) Suppl. 8: 112 - 132.
R. Clausius, "Ueber die Anwendung des Satzes von der Aequivalenz der Verwandlungen auf die innere Arbeit", Pogg. Ann. (1862) 116: 73
Reprinted in: R. Clausius Abhandlungen über die Mechanische Wärmetheorie I Braunschweig 1864, pp 242 - 296 (with addition).
See also on disgregation:
Edward E. Daub, Atomism and Thermodynamics, Isis (1967) 58: 293 - 303.
Martin J. Klein, Gibbs on Clausius, Historical Studies in the Physical Sciences (1969) 1: 127 - 149.

19. W. Ostwald, op. cit. p 111 - 114.

THE EARLY PAPERS OF J. WILLARD GIBBS:
A Transformation of Thermodynamics

Josiah Willard Gibbs introduced himself to the scientific public in
1873 with two papers in the Transactions of the Connecticut Academy
of Arts and Sciences. (1,2) These two works on geometrical methods
in thermodynamics, written by an unknown author and published in an
obscure provincial journal, had but few readers at the time of their
appearance. Later on, when their author's genius had been recognised
by his scientific peers on the strength of his great monograph on
heterogeneous equilibrium, when he was being compared with Lavoisier (3)
and Lagrange (4) in France, with Clausius (5) in Germany, and with
Kelvin (6) in England, and was known at home (at least to some) as
America's leading scientist, (7) these two early papers came to be
looked upon as preparatory essays for the masterpiece that followed
them. Gibbs himself wrote that their "previous perusal (was) very
well adapted to diminish the difficulties which students may feel in
reading the latter," (8) that is, the work on heterogeneous equilibrium.
Many years later one of Gibbs's former students, Edwin B. Wilson, agreed
with another of that small, select company, Henry A. Bumstead, that the
early papers were "of importance not so much for any place they made for
themselves in the literature as for the preparation and viewpoint they
afforded the author as groundwork for his great memoir." (9) It had
become virtually impossible for scientists to see Gibbs's early papers
as anything other than preparatory - for their author or for their
readers.

Yet the Gibbs who wrote those papers on geometrical methods in thermodynamics in 1873 certainly did not look at his work that way. Neither
did the one contemporary reader of whose existence we are certain,
James Clerk Maxwell. In February 1875, when Maxwell called this work
a "most important American contribution" by means of which "problems
which had long resisted the efforts of myself and others may be solved
at once," (10) and spoke of it to Charles Sanders Peirce "in terms of
warm laudation," (11) he was evidently not referring to its preparatory role. The interest and significance of Gibbs's early papers
do not depend on what their author did next. Their interest and
significance are to be found by studying these papers and the context
in which they were written, by trying to understand the problems that
prompted Gibbs to write them and the methods that he used in solving
those problems. When we do so we find that Gibbs was concerned with
problems his contemporaries were thinking about, although his laconic,
mathematician's style makes that far from obvious. We find, too, that
his approach and methods were very much his own, from his first papers
on, and that they constituted nothing less than a transformation of
thermodynamics.

Gibbs began his first paper, "Graphical Methods in the Thermodynamics
of Fluids," (1) by referring to the "general use" and "good service" of

geometrical methods in representing thermodynamic propositions. Such
representations had, however, been restricted to diagrams whose recti-
linear coordinates represent pressure and volume, and he proposed to
discuss a variety of alternatives, "preferable ... in many cases in
respect of distinctness or of convenience." This beginning suggested
a work of largely didactic or expository value, rather removed from
whatever might have been the current concerns of scientists already
actively involved in thermodynamic research. Gibbs went on to list
the quantities relevant to his discussion: volume v , pressure p ,
absolute temperature t , energy ε , and entropy η, all of them
functions of the body's state, as well as the work done W and heat
received H , by the body in passing from one state to another. These
quantities were related by the equations,

$$d\varepsilon = đH - đW ,\tag{1}$$

$$đW = pdv ,\tag{2}$$

$$đH = td\eta ,\tag{3}$$

The first and third of these equations express the definitions of the
state functions energy and entropy whose existence is required by the
first and second laws of thermodynamics, respectively, while equation
(2)· is just the mechanical expression for the work done by an expanding
fluid. (12) Eliminating the work and heat leads directly to the
equation

$$d\varepsilon = td\eta - pdv ,\tag{4}$$

which Gibbs referred to as the differential form of "the fundamental
thermodynamic equation of the fluid," the equation expressing the energy
as a function of entropy and volume.

Now Gibbs used hardly any more words than I have just used in stating
these matters, as though he, too, were simply reminding his readers
of familiar, generally known truths, and writing them down only to
establish the notation of his paper. That, after all, is what one
might expect as the modest starting point of a first scientific paper,
especially from one "of a retiring disposition." (13) Could there have
been any doubt or disagreement that these were the essentials of thermodynamics?

To answer this question one must look beyond Gibbs's paper to the
general state of thermodynamics in 1873. Almost a quarter of a
century had gone by since Rudolf Clausius set the subject on its
proper, dual foundation and William Thomson endorsed and developed
the idea that there were two basic laws of thermodynamics. Clausius
explored the second law of the theorem." (14) He was convinced that
that "real nature" had been found with the help of his analysis of
transformations, first for cyclic processes in 1854 and then for the
general case in 1862, when he introduced the concept of disgregation.
(15) It was not until 1865 that Clausius invented the word entropy
as a name for "the transformational content of the body." (16) The
physical meaning of entropy had already been determined, as far as

Clausius was concerned; it was simply the sum of the body's disgregation and the transformational content of the heat it contained. Entropy had the great advantage of making possible a terse, if portentous, form of the second law: "The entropy of the universe tends towards a maximum." I would emphasise that Clausius saw entropy as a convenient summarising concept, rather than as the basic concept for understanding the second law. It is no accident that he introduced the term in a paper entitled "On Several Forms of the Fundamental Equations of Thermodynamics Convenient for Application," and that he began that paper by contrasting its purpose with his previous explorations of the meaning of the second law. (Exactly the same limited summarising role is assigned to the entropy in Clausius's lecture to the Association of German Scientists in 1867, a lecture in which he tried to make the meaning of the second law clear to a broad scientific audience.) (17) The principal goal of the 1865 memoir - deriving a number of experimentally useful consequences of the thermodynamic laws - is reached without any use of the entropy function, or the internal energy function, for that matter.

Clausius had evidently not changed his mind about the status of both energy and entropy in 1875, when he reworked his thermodynamic articles into a treatise that could be used as a textbook. (18) Although he showed how the two laws of thermodynamics guarantee the existence of the two state functions, energy and entropy, Clausius proceeded to eliminate these quantities from his working equations as soon as possible (quite in contrast to Gibbs's procedure described above). Work and heat, the two original thermodynamic concepts, remained at the center of his thinking, although he did devote a chapter to the experimental determination of energy and entropy.

Clausius's word, entropy, did enter the thermodynamic literature in English, but only by misappropriation. Peter Guthrie Tait, who needs no introduction to an Edinburgh audience, liked Clausius's "excellent word," but proceeded to use it in his Sketch of Thermodynamics (1868) as the name of quite another concept: "It would only confuse the student," he wrote, "if we were to endeavor to invent another term for our purpose." (19) Tait, acting rather like Humpty Dumpty, chose to make entropy mean available energy, and when Maxwell wrote his Theory of Heat a few years later, he followed his friend Tait in this new usage. (20)

When Gibbs wrote that "geometrical representations of propositions in the thermodynamics of fluids are in general use and have done good service," he was referring to something more than mere illustrations. In the years since Clapeyron had introduced the indicator diagram (or pressure-volume diagram) into his exposition of Carnot's ideas, (21) that diagram had been used in a variety of ways. Rankine's long paper to the Royal Society of London in 1854 showed the power of this method as he used it for "the solution of new questions, especially those relating to the action of heat in all classes of engines," and for presenting "in a systematic form, those theoretical principles which are applicable to all methods of transforming heat to motive power by means of the changes of volume of an elastic substance." (22) As an indication of how far Rankine could go in this way, I refer to his geometric derivation of the general equation for the difference between

the specific heats at constant pressure and constant volume. (23)
Rankine's methods were used extensively in his Manual of the Steam
Engine, and were also taken up in other textbooks, such as the widely
used one by Gustav Zeuner. (24) Another famous user of geometrical
proofs in this field, as well as in others, was Maxwell, whose "four
thermodynamic relations" were derived that way in his Theory of Heat.
(25)

Gibbs had already demonstrated his abilities along this line in his
Yale doctoral dissertation, "On the Form of the Teeth of Wheels in
Spur Gearing," a work recommended by its modern editor only to those
who have "a natural friendliness for the niceties of geometrical
reasoning." (26) Thermodynamics provided a new domain for his geometrical explorations. He was concerned with finding "a general
graphical method which can exhibit at once all the thermodynamic
properties of a fluid concerned in reversible processes, and serve
alike for the demonstration of general theorems and the numerical
solution of particular problems." Gibbs considered the general
properties of any diagram in which the states of the fluid were
continuously mapped on the points of a plane. The thermodynamic
properties of the fluid could then be determined from the geometrical
properties of the several families of curves connecting states of equal
volume, of equal temperature, of equal entropy, and so forth. Since
the equations relating work to pressure and volume, and heat to
temperature and entropy (equations (2) and (3) above) are of exactly
the same form, the temperature-entropy diagram must share many of the
useful features of the familiar pressure-volume diagram. Gibbs
pointed this out, as well as the additional advantage that Carnot's
ideal cycle is always a rectangle in the temperature-entropy diagram,
regardless of the nature of the working substance.

The real advantage of the temperature-entropy diagram, however, is that
it "makes the second law of thermodynamics so prominent, and gives it
so clear and elementary an expression." Gibbs meant that although
there is no formal difference between representing the work done in
a process as area under its curve in the pressure-volume diagram, and
representing the heat exchanged in the process as the area under its
curve in the temperature-entropy diagram, the former depended only
on the mechanical definition of work. The latter, however, was "nothing
more nor less than a geometrical expression of the second law of thermodynamics in its application to fluids, in a form exceedingly convenient
for use, and from which the analytical expression of the same law can,
if desired, be at once obtained." (27)

Gibbs did not consider the temperature-entropy diagram worth an
extended discussion, and devoted only three of his thirty-two pages to
it. The diagram that interested him most, the one whose "substantial
advantages over any other method" called for more detailed consideration,
was that in which the coordinates of the state point were the entropy
and volume of the body. (28) Its importance was already suggested by
the form of the fundamental thermodynamic equation in which entropy and
volume determine the energy. Such a diagram has a variable scale
factor, that is, a variable ratio of the work done in a small cyclic
process to the area enclosed by that cycle in the diagram. In this

it contrasts with constant scale diagrams like both pressure-volume and temperature-entropy. But what might be a difficulty, or at least an awkwardness, for engineering purposes was a definite advantage if one's goal was to grasp the properties of matter at equilibrium.

Some features of the diagrams, that is to say some aspects of the families of curves representing thermodynamic properties, are independent of the choice of coordinates. Gibbs also examined these invariant features, and especially the order and character of the crossings of such curves at any point of the diagram. He indicated that the crossings might be of higher order, involving a tangency as well, and commented in a footnote: "An example of this is doubtless to be found at the critical point of a fluid. See Dr Andrews, On the continuity of the gaseous and liquid states of matter." (29)

This footnote by Gibbs calls for further comment. It even suggests the answer to a historical question I have not raised yet. In 1873, when Gibbs wrote his first paper, the paper under discussion, he was thirty-four years old and had already been for two years Professor of Mathematical Physics at Yale. (30) He had evidently been appointed to this position on the strength of his brilliance as a student, his demonstrated abilities as an engineer, his three years of postdoctoral study at Paris, Berlin, and Heidelberg, and the faith of the Yale intellectual community in which he had grown up. What drew his attention to thermodynamics as the subject of his first scientific inquiries? Why should this untried, but mature and independent thinker, as he at once proceeded to show himself, have selected this particular set of problems to begin with? It is true that in 1872 the pages of the Philosophical Magazine carried a good many lively, pointed, and even angry words on the subject of thermodynamics, as Clausius and Tait debated the history, which meant to them the priorities, of the discovery of the second law. (31) Gibbs could hardly have avoided noticing this dispute, in which the names of Maxwell and Thomson were often mentioned, and it may well have prompted some reading on his part. (There is no evidence that he had studied thermodynamics systematically before then, neither in his American or European lectures nor in his reading.) But controversy repelled rather than attracted Gibbs, and his papers do not follow up any of the themes of the Clausius-Tait debate. We must look elsewhere for the origins of his interest in thermodynamics, and the footnote referring to Andrews points the way.

In 1869 Thomas Andrews presented to the Royal Society the results of almost a decade of experimental work establishing the continuity of the two fluid states of matter - liquid and gas. (32) This work was the high point of his scientific career, and Andrews was well aware of it, writing to his wife: "I really begin to think that Dame Nature has at last been kind to me, and rewarded me with a discovery of a higher order than I ever expected to make." (33) His careful measurements of the isotherms of carbon dioxide established the existence of a critical temperature: if the gas were compressed isothermally below this temperature it would eventually begin to liquefy and become a two-phase system with a visible surface of separation between gas and liquid. Further compression would lead to complete transformation of gas into liquid, the liquid then being almost incompressible. In sharp contrast to this behavior an isothermal compression of the gas at a temperature above critical never led to two phases, although the density would eventually take on values appropriate to the liquid state. Above the critical temperature there was no distinction between liquid and gas. It was always possible to pass from a state clearly liquid to one that was equally clearly gas without ever going through the discontinuous two-phase region. These remarkable properties were by no means peculiar to carbon dioxide. They "are generally true of all bodies which can be obtained as gases and liquids," as Andrews confirmed by studies on some half-dozen substances. (34) He carefully refrained from any theorising about the implications of his discovery, attempting neither a kinetic-molecular explanation nor a thermodynamic analysis.

Even if Gibbs missed Andrews's Bakerian Lecture when it appeared in the Philosophical Transactions, he would have read a discussion of its contents in Maxwell's Theory of Heat, (20) published in London in 1871 and in New York the following year, and referred to soon afterward by J. D. van der Waals as "the little book which is surely to be found in the hands of every physicist." (35) Andrews's discovery - a new, surprising, and general feature of the properties of matter, as yet quite unexplained - would be just the sort of thing to attract Gibbs's attention, as that promising new professor of mathematical physics sought for a suitable subject on which to reflect. Despite his mathematician's instincts he preferred "the study of the problems set us by nature." (36) If I am right in thinking that Gibbs turned to thermodynamics with the questions raised by Andrews's results in his mind, then one can see why he might well have chosen to emphasise just those aspects of the subject that he did.

Gibb's second paper, (2) finished a few months after the first, provides evidence in support of this suggestion. Although this second paper might seem to be only a continuation of the first, merely extending the discussion of geometrical methods from two dimensions to three, one does not have to read very far to see that Gibbs is doing something different. He is no longer exploring a variety of alternative representations; his concern now is not with methods as such but rather with the phenomena to be explained. His problem was to characterise the behavior of matter at equilibrium, to determine the nature of the equilibrium states of a body that can be solid, liquid, or gas, or some combination of these, depending on the circumstances. In this paper Gibbs goes directly to the single three-dimensional representation indicated by the fundamental equation, where the three rectangular coordinates are the energy, entropy, and volume of the body, and the equilibrium states constitute a surface in this space. While there had been extensive use of two-dimensional diagrams, if only the pressure-volume diagram, before Gibbs, there was only one precedent for using a three-dimensional representation of equilibrium states, and Gibbs showed his awareness of it in his first footnote reference.

James Thomson, William's older brother and former collaborator in thermodynamic matters, had introduced the thermodynamic surface in pressure-volume-temperature space to assist him in thinking about

Andrews's experimental results. (37) Thomson, who was Andrews's colleague at Queen's College, Belfast, was aware of his findings and trying to interpret them as early as 1862, though he did not publish his ideas until 1871. Thomson's "chief object" in his paper was to argue that Andrews had not gone far enough in his claim that the liquid and gas phases are continuous with one another. Below the critical temperature Andrews's isotherms included a straight line segment parallel to the axis of volume. This segment of the isotherm, in which the pressure did not change, represented the states where gas and liquid could co-exist in proportions varying from all liquid to all gas.

A crucial question about these isothermal curves that neither Thomson nor van der Waals could answer was: where must the straight line segment be drawn? Or in other words, what is the condition that relates the pressure and temperature at which gas and liquid can co-exist? What is the condition for the equilibrium of the two phases? Van der Waals discussed the long history of this problem and the inadequacy of the attempted solutions in his thesis. (38) Not even Maxwell, who included a discussion of James Thomson's ideas along with Andrews's results in his "elementary" textbook on heat in 1871, had been able to resolve this difficulty.

What was missing from all these attempts was nothing less than the second law of thermodynamics. Gibbs, who started from the thermodynamic laws, settled the question in his usual brief and elegant manner. (39)

Referring to his treatment almost twenty years later in a letter to Wilhelm Ostwald, who had proposed to do a German translation of Gibbs's papers, he states:

> "It contains, I believe, the first solution of a
> problem of considerable importance, viz: the
> additional condition (besides equality of tem-
> perature and pressure) which is necessary in order
> that two states of a substance shall be in
> equilibrium in contact with each other. The
> matter seems simple enough now, yet it appears
> to have given considerable difficulty to
> physicists. ... I suppose that Maxwell
> referred especially to this question when he
> said ... that by means of this model, problems
> which had long resisted the efforts of himself
> and others could be solved at once." (8)

Maxwell expressed his appreciation of Gibbs's thermodynamic surface by constructing a model of it for water and sending a cast of it to Gibbs. (40) He also included a fourteen-page discussion of the Gibbs surface in the 1875 edition of his textbook, giving more details of its proper-ties than Gibbs had. (41) In that same year Maxwell developed an alternative form of the equilibrium condition - the Maxwell construction - which states that the horizontal, two-phase, portion of the isotherm must cut off equal areas above and below in the van der Waals-Thomson loop. (42) The proof involved a direct application of the second law carried out with what may be called genuinely Maxwellian ingenuity.

Clausius independently arrived at Maxwell's result five years later by a slightly different argument. (43) When van der Waals told him of Maxwell's prior publication, Clausius seemed a bit annoyed, (44) chiefly because Maxwell had not included his work in the latest version of his Theory of Heat, whose successive editions had acquired something of the status of a review journal. Clausius had, however, apparently missed Maxwell's discussion of the Gibbs surface, and there is no sign that he ever read any of the reprints Gibbs regularly sent to him.

Gibb's analysis of the critical point provides another striking demonstration of the power of his methods. (45) His analysis of the geometry of the thermodynamic surface in the neighborhood of the critical point led him to several sets of conditions satisfied at this point. Among these were the following two pairs,

$$\left(\frac{\partial p}{\partial v}\right)_t = 0 \ , \qquad \left(\frac{\partial^2 p}{\partial v^2}\right)_t = 0 \ ; \qquad (10)$$

$$\left(\frac{\partial t}{\partial \eta}\right)_p = 0 \ , \qquad \left(\frac{\partial^2 t}{\partial \eta^2}\right)_p = 0 \ . \qquad (11)$$

Although the first pair of conditions was implicitly present in the discussions by Thomson and van der Waals, as already mentioned, their methods did not even suggest looking for conditions of the second type. (46) Only Gibbs's new approach to thermodynamics made it possible to see the real significance of the critical point, as the limit of the boundary between the domains of stable and unstable behavior. (47)

Gibbs wrote relatively little science during the thirty years of his professorship, and he was even more reticent when it came to expressing his views about science. Such remarks as he did make are scattered through his occasional writings - prefaces, letters, even obituary notices. Among them there is a statement that recurs several times and expresses something Gibbs saw as essential about his way of doing science.

There is a letter to Lord Rayleigh, for example, written on 27 June 1892, in which Gibbs tells of his plans to write on statistical mechanics. In the preface to the book he wrote on statistical mechanics some years later he restated the same point made in that letter; namely, that the novelty of his ideas lay in their manner of presentation, not in their substance. (48) Indeed, he had made a very similar remark in 1881 when he wrote to accept the Rumford Medal awarded him by the American Academy of Arts and Sciences at Boston for his monograph on heterogeneous equilibrium, (49) indicating that even in his first works Gibbs had already found a new standpoint from which he saw thermodynamics more clearly and to greater effect than it had ever been seen before. (50)

1. J.Willard Gibbs, "Graphical Methods in the Thermodynamics of Fluids," Trans. Conn. Acad. 2 (1873), 309-342. Reprinted in The Scientific Papers of J.Willard Gibbs, H.A. Bumstead and R.G. Van Name, eds. (New York, 1906; reprinted 1961), Vol. I, pp. 1-32. This reprint will be noted as Sci. Pap. and page references will be to this edition.

2. J.Willard Gibbs, "A Method of Geometrical Representation of the Thermodynamic Properties of Substances by means of Surfaces," Trans. Conn. Acad. 2 (1873), 382-404. Sci. Pap. I, pp.33-54.

3. Henri Le Châtelier, Préface du traducteur in J.Willard Gibbs, Equilibre des systèmes chimiques (Paris, 1899) p.vi.

4. Pierre Duhem to J.W. Gibbs, 29 May 1900. (Yale University Library)

5. Wilhelm Ostwald to J.W. Gibbs, 9 August 1891. Reprinted in Aus dem wissenschaftlichen Briefwechsel Wilhelm Ostwalds, H.-G. Korber, ed. (Berlin, 1961) Vol. I, p. 99.

6. Joseph Larmor, "Josiah Willard Gibbs, 1839-1903," Proc. Roy. Soc. 75 (1905), 292.

7. Wolcott Gibbs to J.W. Gibbs, 19 April 1902. (Yale University Library). Wolcott Gibbs was then President of the National Academy of Sciences to which Willard Gibbs had been elected in 1879.

8. J.W. Gibbs to W. Ostwald, 27 March 1891. Reprinted in op. cit., Note 5, pp. 97-98.

9. Edwin B. Wilson, "Papers I and II as Illustrated by Gibbs's Lectures on Thermodynamics" in A Commentary on the Scientific Writings of J. Willard Gibbs, F.G. Donnan and Arthur Haas, eds. (New Haven, 1936), Vol. I, p. 19. Bumstead's remarks are in his "Biographical Sketch" of J.W. Gibbs, Sci. Pap. I, pp. xii-xiv.

10. James Clerk Maxwell, "On the Dynamical Evidence of the Molecular Constitution of Bodies," Nature 11 (1875), 357-359, 374-377. Reprinted in The Scientific Papers of James Clerk Maxwell, W.D. Niven, ed. (Cambridge, 1890; reprinted New York, 1965), Vol. II, p. 426.

11. Charles Sanders Pierce, Review of The Scientific Papers of J. Willard Gibbs, The Nation, 84 (1907), 92.

12. Gibbs did not use the notation đ to denote inexact differentials. This notation was introduced by Carl Neumann, Vorlesungen über die mechanische Theorie der Wärme (Leipzig, 1875), p. ix.

13. See H.A. Bumstead, in Sci. Pap. I, p. xxiii. Bumstead's phrase was used to characterise Gibbs's scientific personality by Pierre Duhem

in his tribute to Gibbs, Josiah Willard Gibbs à propos de la publication de ses mémoires scientifiques (Paris, 1908).

14. Rudolf Clausius, "On a Modified Form of the Second Fundamental Theorem in the Mechanical Theory of Heat," Phil. Mag. 12 (1856), 81-98.

15. R.Clausius, "On the Application of the Theorem of the Equivalence of Transformations to the Internal Work of a Mass of Matter," Phil. Mag. 24 (1862), 81-97, 201-213. For an analysis of disgregation see M.J. Klein, "Gibbs on Clausius," Historical Studies in the Physical Sciences 1 (1969), 127-149 and also Keith Hutchison, "Der Ursprung der Entropiefunktion bei Rankine und Clausius," Annals of Science 30 (1973), 341-364. Both papers contain additional references.

16. R.Clausius, "Über verschiedene für die Anwendung bequeme Formen der Hauptgleichungen der mechanischen Wärmetheorie," Pogg. Ann. 125 (1865) 353. English translation in R.Clausius, The Mechanical Theory of Heat, T.A. Hirst, transl. (London, 1867), pp. 327-365.

17. R.Clausius, "On the Second Fundamental Theorem of the Mechanical Theory of Heat," Phil. Mag. 35 (1868), 405-419.

18. R.Clausius, Die mechanische Wärmetheorie Vol. I (Braunschweig, 1876).

19. P.G. Tait, Sketch of Thermodynamics (Edinburgh, 1868), p. 100.

20. J.C. Maxwell, Theory of Heat (London, 1871), p. 186. Reading Gibbs's first paper persuaded Maxwell to correct the error he had "imbibed" from Tait. See my paper cited in Note 15. Tait's misusage of "entropy" can also be found in G.Krebs, Einleitung in die mechanische Wärmetheorie (Leipzig, 1874), pp. 216-218.

21. Emile Clapeyron, "Mémoire sur la puissance motrice de la chaleur," Journal de l'École Polytechnique 14 (1834), 153-190. The earlier history of the indicator diagram is discussed in D.S.L. Cardwell, From Watt to Clausius (Ithaca, N.Y. 1971), pp. 80-83, 220-221. Cardwell remarks that the diagram had been a closely guarded trade secret of the firm of Boulton and Watt, and that John Farrey, the English engineer, only learned about it in 1826 in Russia, presumably from one of Boulton and Watt's engineers working there. Since Clapeyron was also in Russia during the late 1820's he may have acquired his knowledge of the indicator diagram in the same way.

22. W.J.M. Rankine, "On the Geometrical Representation of the Expansive Action of Heat, and the Theory of Thermodynamic Engines," Phil. Trans. Roy. Soc. 144 (1854), 115-176. Reprinted in W.J. Macquorn Rankine, Miscellaneous Scientific Papers, W.J. Millar, ed. (London, 1881), pp. 339-409. See also K. Hutchison, op.cit., Note 15.

23. W.J.M. Rankine, op. cit. Note 22. Papers, pp. 359-360.

24. Gustav Zeuner, Grundzüge der mechanischen Wärmetheorie (Leipzig, 1966 2nd ed.)

25. J.C. Maxwell, Theory of Heat (London, 1877, 5th ed.), pp. 165-169.

26. Gibbs's thesis was first printed in The Early Work of Willard Gibbs in Applied Mechanics, L.P. Wheeler, E.O. Waters and S.W. Dudley, eds. (New York, 1947). The thesis is on pages 9-39. The quotation from Everett O. Waters's commentary is on p.43.

27. J.W. Gibbs, op.cit. Note 1, p. 11.

28. J.W. Gibbs, op.cit., Note 1, pp. 20-28.

29. Ibid., p. 30 fn.

30. For details of Gibbs's life and career see the official biography: Lynde Phelps Wheeler, Josiah Willard Gibbs, (New Haven, 1952, 2nd ed.).

31. See my paper, op.cit. Note 15, for a discussion of this controversy and further references.

32. Thomas Andrews, "On the Continuity of the Gaseous and Liquid States of Matter," Phil. Trans. Roy. Soc. 159 (1869), 575-590. Reprinted in The Scientific Papers of Thomas Andrews (London, 1889), pp. 296-317. Further references are to this reprinting.

33. Ibid., p. xxxi.

34. Ibid., p. 315.

35. J.D. van der Waals, Over de continuiteit van den gas-en vloeistof-toestand (Leiden, 1873), p. 81.

36. In June 1902 Gibbs advised his former student, Edwin B. Wilson, who was leaving for a year's study in Paris, to take some work in applied mathematics. "He ventured the opinion that one good use to which anybody might put a superior training in pure mathematics was to the study of the problems set us by nature." E.B. Wilson, "Reminiscences of Gibbs by a Student and Colleague," Scientific Monthly 32 (1931), 210-227. Quotation from p. 221.

37. James Thomson, "Considerations on the Abrupt Change at Boiling or Condensing in Reference to the Continuity of the Fluid State of Matter," Proc. Roy. Soc. 20 (1871), 1-8. Reprinted in James Thomson, Collected Papers in Physics and Engineering (Cambridge, 1912), pp. 278-286. References will be to this reprinting. Related papers and unpublished notes by Thomson are to be found at pp. 276-277 and pp. 286-333. Although Gibbs does not refer to them, he may well have read Thomson's papers to the British Association for the Advancement of Science in 1871 and 1872 (pp. 286-291, 297-307) in which the triple point is first named and discussed.

38. J.D. van der Waals, op.cit., Note 35, pp. 120-121.

39. J.W. Gibbs, op.cit., Note 2, pp. 35-38.

40. Photographs of this surface are included in the works cited in Notes 9 and 30 among others. See also A.L. Clark and L. Katz, "Thermodynamic Surfaces of H_2O", Transactions of the Royal Society of Canada 32 (1939), section III, 59-71.

41. J.C. Maxwell, op.cit., Note 25, pp. 195-208. Tait's comments on Maxwell's Theory of Heat are very much to the point. "In intrinsic importance it is of the same high order as his Electricity, but as a whole it is not an elementary book ... In the little book we are discussing there is matter enough to fill two or three large volumes without undue dilution (perhaps we should rather say, with the necessary dilution) of its varied contents." P.G. Tait, "Clerk-Maxwell's Scientific Work," Nature 21 (1880), 317-321.

42. J.C. Maxwell, op.cit., Note 10, p. 425.

43. R. Clausius, "On the Behavior of Carbonic Acid in Relation to Pressure, Volume, and Temperature." Phil. Mag. 9 (1880), 393-408. See particularly pp. 405-407.

44. R. Clausius, "On the Theoretic Determination of Vapour-Pressure and the Volumes of Vapour and Liquid," Phil. Mag. 12 (1881), 381-390. See particularly the long footnote on p. 382.

45. J.W. Gibbs, op.cit., Note 2, pp. 39-49.

46. When Kamerlingh Onnes and Keesom wrote their review article on equations of state in 1912 they described equation (10) as having "probably been first applied for determining the critical point" by J. Douglas Hamilton Dickson, in Phil.Mag. 10 (1880) 40, and then found independently by Onnes himself in 1881. See H. Kamerlingh Onnes and W.H. Keeson, "Die Zustandsgleichung," Comm. Phys. Lab. Leiden 11 (1912), supp. 23.

47. A relatively recent discussion of phase equilibrium and critical points which axiomatises Gibbs's theory and extends it can be found in Laszlo Tisza, "The Thermodynamics of Phase Equilibrium," Annals of Physics 13 (1961), 1-92. Reprinted along with other relevant papers in L. Tisza, Generalised Thermodynamics (Cambridge, Mass., 1966). See also Robert B. Griffiths and John C. Wheeler, "Critical Points in Multicomponent Systems," Phys. Rev. A2 (1970), 1047-1064.

48. J.W. Gibbs, Elementary Principles in Statistical Mechanics (New Haven, 1902), p.x.

49. J.W. Gibbs to the American Academy of Arts and Sciences, 10 January 1881. Quoted in full in L.P. Wheeler, op.cit., Note 30, pp. 88-89.

50. I thank the National Science Foundation for supporting this work through Grant SOC 73-05419.

Dr. Edward E. Daub

SOURCES FOR CLAUSIUS' ENTROPY CONCEPT: REECH AND RANKINE

INTRODUCTION

Gibbs claimed, in his notable review of Clausius' scientific work in 1889, that if we define thermodynamics in the words of Maxwell as "a science with secure foundations, clear definitions, and distinct boundaries," then that science of thermodynamics dates from the publication of Clausius' first memoir in 1850.[1] And yet, at that date, only one of the two fundamental concepts of classical thermodynamics had taken definitive form, the internal energy U.

The entropy concept remained undefined until 1865 when in his ninth memoir Clausius phrased his memorable maxims: "Die Energie der Welt ist constant; Die Entropie der Welt strebt einen Maximum zu." Gibbs made these two principles the cornerstones for his thermodynamics of heterogeneous equilibria[2] and, without Clausius' calling explicit attention to the irreversible increase of entropy, it is difficult to imagine thermodynamics ever escaping the narrow confines of steam engine cylinders for the wider reaches of phase equilibria as it did with Gibbs. It is not Clausius' first memoir, therefore, that marks the laying of the foundations. It is the ninth.

In this paper I will examine the initial steps which Clausius took in developing his understanding of the integral $\int dq/T$: in 1854 when he first discussed the integral and in 1856 when he fully developed his analysis of irreversibility. I will suggest the significant influence of two major figures in the history of thermodynamics upon these developments: the influence of Reech on Clausius' proof that $\int dq/T = 0$ for a reversible cycle and the influence of Rankine for the irreversible case where $\int dq/T > 0$.

THE REVERSIBLE CASE: $\int dq/T = 0$

Clausius' Curious Cycle and His Errors in Logic.

In developing his proof that $\int dq/T = 0$ for a reversible cycle, Clausius analyzed a cycle that is distinctively different from the traditional Carnot cycle. Instead of the simple combination where a pair of isothermal curves cut a pair of adiabatic curves to carve out the pressure-volume diagram, Clausius combined three isothermal curves with three adiabatic ones. This curious cycle shown in Figure 1 cost Clausius two errors in logic.

The cycle traced the following sequence of processes.[3] The gas began in contact with the heat reservoir at the highest temperature t and absorbed a quantity of heat q in expanding isothermally at that temperature (fa). After expanding adiabatically to a lower temperature t_1 (ab), the gas expanded isothermally and absorbed a quantity of heat q_1 from a second heat reservoir at that temperature (bc). After falling to the lowest temperature t_2 in a second adiabatic expansion (cd), the gas rejected heat q_1 to a third reservoir during an isothermal compression at that temperature (de). The gas finally completed its cycle via an adiabatic compression from t_2 to t (ef). By the first law for a cycle the net work w equaled the net heat q.

342

Clausius demonstrated that the relation between q and q_1 must be universal for all substances given the temperatures t, t_1, and t_2. If that were not the case, then the cycle could be made to violate the natural behavior of heat.[4] He then set out to determine the universal functional relationship between these quantities.

To do so Clausius interpreted the cycle in terms of two apparent transformations of heat: the transformation of heat q at temperature t into work w, and the transformation of heat q_1 at temperature t_1 into heat q_1 at temperature t_2. Each supposed transformation cost Clausius an error in logic.

Consider first the imagined transformation of heat q into work w at the temperature t. The net work w in the cycle obviously equals the net transformation of heat q $(q + q_1 - q_1)$ by the first law of thermodynamics. Clausius, however, treats w not just as the net work performed in the cycle but as the amount of work performed in the transformation of q into work during the isothermal expansion at the temperature t. This could only be true if the substance in the cycle were a perfect gas. Thus, his claim to a universal transformation of heat q into work w at the temperature t is a figment of the imagination.

In no way can Clausius be saved from this logical error. He claimed an equivalence between the two supposed transformations which he had singled out from the cycle. By equivalence he meant that each single transformation could replace the other with no other change in the universe.[5] Thus, he explicitly assumed that $q = w$ for the transformation at the temperature t. Since that is true only for perfect gases, the validity of his theorem of the equivalence of transformations is likewise so restricted. It is not the universal theorem that he claimed.[6]

This logical error does not invalidate Clausius' proof that $\int dq/T = 0$, however, because the relationship which is derived for perfect gases must be true for all substances. But the proof is far more complicated than it need to have been due to the second logical error that Clausius committed. This error derived from his other imagined transformation of heat, from q_1 at t_1 to q_1 at t_2.

The proof begins as follows. Because the two supposed transformations are equivalent, they should each represent an equal equivalence value. Their signs, however, will be opposite because it is actually the reverse of the one transformation that replaced the other. The transformation of heat q into work w at temperature t will be $-q\,f(t)$ where $f(t)$ is some function of the temperature; the transformation of heat q_1 at temperature t_1 to heat q_1 at temperature t_2 will be $q_1\,F(t_1,t_2)$ where $F(t_1,t_2)$ is some other unknown function of the two temperatures. For the cycle the sum of these equivalence values will be zero:
$$-q\,f(t) + q_1\,F(t_1,t_2) = 0.[7]$$
But Clausius committed a logical error in assigning different temperature functions to the three isothermal processes. What logical reason could he give for attributing a function $f(t)$ to the first isothermal process and then linking the other two with a different function $F(t_1,t_2)$? Since the other two processes were equally isothermal, he should logically have been able to write immediately that
$$-q\,f(t) - q_1\,f(t_1) + q_1\,f(t_2) = 0;$$
and that generally for any reversible cycle, if we take $f(t) = 1/T$,
$$\int dq/T = 0.$$

His proof is overly complicated because he had first to establish that

$$F(t_1,t_2) = f(t_2) - f(t_1)$$

which he could have stated at the very beginning. The source for these two logical errors by Clausius laid not in the cycle itself. The cycle merely expressed the concepts that Clausius brought to the analysis. He stated them clearly at the outset. Immediately after titling the section, "Theorem of the equivalence of transformations," he wrote:

> Carnot's theorem, when brought into agreement with the first fundamental theorem, expresses a relation between two kinds of transformation, the transformation of heat into work, and the passage of heat from a warmer to a colder body, which may be regarded as the transformation of heat at a higher, into heat at a lower temperature.[8]

So these slips in reasoning, rare for the usually rigorous Clausius, arose because he conceived a cycle to represent these two ideas.

I want to suggest that the source of these two ideas was a paper published by the French engineer Ferdinand Reech in 1853. But before doing so, I will complete Clausius' demonstration that $\int dq/T = 0$ for any reversible cycle. As indicated earlier, Clausius spent the major share of the argument on the imaginary function $F(t_1,t_2)$.

In order to establish the nature of this function, Clausius imagined a second cycle which was identical with the first cycle except that the transformation of heat $\underline{q}'$ into work $\underline{w}'$ occurred at a different high temperature $\underline{t}'$. Thus, for this cycle the equivalence of transformations would require that

$$-q' f(t') + q_1 F(t_1,t_2) = 0.$$

Subtraction from the analogous equation for the first cycle yielded the relation

$$-q f(t) + q' f(t') = 0.$$

Clausius then argued that this relation would represent a combination of the two cycles in which the second cycle operated in reverse, thus causing the two exchanges of heat q_1 between temperatures $\underline{t}_1$ and $\underline{t}_2$ to cancel each other. For this combined cycle, Clausius said, the principle of equivalent transformation should again apply and could be expressed as

$$-(q-q') f(t) + q'F(t,t') = 0$$

In combination with the previous equation, this established the form of the unknown function

$$F(t,t') = f(t') - f(t).$$

Upon introducing the symbol $1/T$ for $f(t)$ and generalizing the principle to any reversible cycle, Clausius established that

$$\int dq/T = 0.[9]$$

The derivation obviously was made cumbersome by the unnecessary function $F(t_1,t_2)$. Although that function did not explicitly appear in Reech's analysis of the theory of heat, the idea which led Clausius to formulate the function did.

Reech's Influence on Clausius' Thermodynamic Thought

Reech published a long and difficult paper in 1853.[10] There he sought to derive equations of such generality that both the original theory of Carnot and the new theory established by Clausius and Thomson could be satisfied as special cases.

Reech began by deriving the traditional equation for a Carnot cycle.

It related the work $\underline{w}$, the heat $\underline{q}$, and the two temperatures for the cycle $\underline{t}'$ and $\underline{t}$ under the caloric theory assumption that heat was conserved.

$$w = q \int_{t}^{t'} dt/C.^{11}$$

C was the universal Carnot temperature function introduced by Clapeyron.

He then abandoned the caloric theory assumption that heat must be conserved and, by imagining two Carnot engines operating between the two temperatures, one of them as a refrigerator, he proposed two conceivable situations. In the first case, he said, the combined cycles would result in no net change of heat at the temperature $\underline{t}$, and in the other case in no net change at $\underline{t}'$. (Thomson's statement of the second law, of course, would not let him get away with either case.) Reech now argued that the heat $\underline{q}$ and the net work done in the first case could only depend on some function of the temperature $\underline{t}$ because no change in heat occurred at the temperature $\underline{t}'$. Similarly in the second case, the heat $\underline{q}'$ and the net work done could only depend on some function of the temperature $\underline{t}'$. By this reasoning Reech established his fully general equation for work in a cycle,

$$w = q'T(t') - qT(t) = \int_{t}^{t'} \frac{d}{dt} [qT(t)]dt.$$

Reech now demonstrated that his generalized equation could satisfy either the original theory of Carnot or the new mechanical theory of heat. If the Carnot theory and the conservation of heat were true, the equation would reduce to

$$w = q \int_{t}^{t'} dt/C.$$

If the mechanical theory of heat and the equivalence of heat and work were true, the equation would reduce to

$$w = G(q'-q)^{12}$$

where G is a constant. Since the equation could thus satisfy both competing theories, Reech believed that he had found a valid generalized equation for the behavior of heat.

Reech now speculated about various possible kinds of behavior which heat might manifest in satisfying his generalized equation. He made several such speculations. One for example was that heat $\underline{q}'$ is destroyed at temperature $\underline{t}'$ and that heat $\underline{q}$ is created at temperature $\underline{t}$. It was another speculation, however, that brought forth ideas which closely parallel the two transformations imagined by Clausius. Reech rewrote his equation in the following form:

$$w = (q'-q)T(t') + q[T(t') - T(t)],^{13}$$

and he interpreted the equation to mean that a quantity of heat $(q'-q)$ was destroyed at $\underline{t}'$ and that another quantity $\underline{q}$ was transferred from $\underline{t}'$ to $\underline{t}$. The parallel with Clausius' two transformations of heat seems too obvious for comment. I would contend, therefore, that Clausius was influenced by reading Reech's paper and that this influence accounts for the curious cycle and the mode of argument that Clausius adopted to prove that $\int dq/T = 0$ for reversible cycles.

I do not mean to imply that Reech's speculation was equivalent to Clausius' theorem of the equivalence of transformations. Reech did not associate his supposed destruction of heat $(q'-q)$ at $\underline{t}'$ with the production of an equivalent amount of work because in 1853 he was still unable to accept the mechanical theory of heat. It was only in 1856, after he had read a French translation of Clausius' 1854 paper, that Reech saw clearly how to reconcile the theory of Carnot and the new theory.[14]

Reech's paper, however, and the way he concocted functions that border closely on the entropy function, make it quite clear that the entropy function for reversible cycles would not long have escaped good thermodynamic reasoning when governed by both the first and second laws. Rankine did so before Clausius in 1854.[15] But Reech's paper also makes it quite clear that it was unique for Clausius to extend his theorem to irreversible cycles.

Reech, for example, in speculating about his generalized function $q\mathcal{T}(t)$, conjectured that it might be thought to measure the motive force of heat at the temperature t. He found this line of thought blocked, however, when he considered the irreversible case. Suppose, said Reech, that an amount of heat passes between two temperatures t' and t, not via a reversible engine cycle, but by direct radiation or conduction. This would mean, if we were to assume that $q\mathcal{T}(t)$ measures the motive force of heat, that one source of heat would have lost more motive force than the other had gained. A conclusion, Reech said, which should singularly shock our good sense.[16] Clausius did not find himself so intimidated by that prospect, and he alone of the founders of thermodynamics fully explored and developed the irreversible case.

THE IRREVERSIBLE CASE: $\int dq/T \geq 0$

Keith Hutchison, in reviewing and comparing the development of the entropy function in the thermodynamics of Clausius and Rankine[17], considered three of Clausius' papers to be the most significant: the 1854 paper on equivalence values, the 1862 paper on the disgregation function, and the 1865 paper which introduced the concept of entropy.[18] He disregarded Clausius' 1856 paper on the steam engine,[19] perhaps because it does not at first appear to have any enduring significance for thermodynamics.

That paper, however, may turn out to be more significant than the others in understanding the genesis of the entropy concept and for two reasons. In that paper Clausius first fully developed his thoughts on the $\int dq/T \geq 0$ in irreversible cycles, a topic only touched upon in 1854. Moreover, in that paper Clausius admitted having read Rankine's 1854 paper, and in Rankine's 1854 paper can be found the one case where Rankine applied his analysis to an irreversible process and indicated that his thermodynamic function (equivalent to entropy) increased.

Clausius on Irreversibility in 1854 and 1856.

In 1854 Clausius spent but a few paragraphs on the irreversible case. Since he had adopted a negative value for the transformation of heat at a lower temperature to the same quantity of heat at a higher temperature, the sum of all transformation values in an irreversible cycle could never be negative for that would violate his statement of the natural behavior of heat. Thus, the sum for any cyclical process could only be positive, and he called this an uncompensated transformation. After mentioning several kinds of uncompensated transformations, Clausius concluded with the suggestion that his previous discussion had already made clear how to calculate uncompensated equivalence values.[20] He did not elaborate this point until he discussed irreversibility in 1856.

In his 1856 paper on the steam engine Clausius indicated that in order to measure the equivalence value for any particular uncompensated transformation, it was necessary to conceive a reversible process for returning the substance to its original condition. This would constitute a small cycle to which his equation $\int dq/T > 0$ would apply.[21] He then argued that because the reversible return of the substance to its original condition could not contribute to the uncompensated transformation, the integral for the cycle must be equal to the value for the initial uncompensated transformation.[22] Thus, only in 1856 did Clausius clearly state how to calculate the uncompensated transformation.

In addition to this explanation of the equivalence value for an uncompensated transformation, Clausius developed one other new dimension in his thinking about irreversibility, namely, that irreversibility leads to a decrease in net work. To do so he considered an engine cycle in which the heat q_1 was added to the substance at various temperatures while all of the heat q_0 rejected by the substance occurred at the constant absolute temperature T_0. The amount of heat q_0 rejected would vary between the reversible and irreversible cycles; the amount of heat q_1 added would not. In this cycle the value for the uncompensated transformation N would be

$$\int_0^{q_1} dq/T + q_0/T_0 = -N$$

where the negative sign takes account of a change in Clausius' convention for assigning positive and negative values to the quantities of heat. (In 1854 he referred the values to the additions and subtractions of heat from the reservoirs; now he referred them to the substance in the cycle.) By combining this equation with his statement of the first law, $q_1 + q_0 = w$,[23] Clausius obtained the expression for the net work in the cycle,

$$w = [q_1 - T_0 \int_0^{q_1} dq/T - NT_0]$$

If N is zero, the reversible case, then the net work is a maximum. If N is positive, the irreversible case, then an amount of work NT_0 is lost due to irreversibility.[24]

Thus, in his 1856 paper on the steam engine, Clausius developed two significant new dimensions in his understanding of irreversibility. He conceived the thought experiment of returning a substance by a reversible process to its original condition in order to calculate the uncompensated transformation (entropy increase) for any particular irreversible change occurring in the substance. He also developed an equation to show how irreversible processes cause a loss in net work. Both dimensions have become integral parts of the thermodynamic analysis of irreversibility and of the entropy concept. We need therefore to account for their origins.

We might simply assume that the first development, the method for calculating the value of an uncompensated transformation, represented a mere generalization of an idea apparently obvious to Clausius in 1854, when he commented on the free expansion of a perfect gas as follows:

> ... when a vessel filled with air is suddenly connected with an empty one -- a portion of air is then propelled with great velocity into the empty vessel and again comes to rest there. It is well known that in this case just as much heat is present in the whole mass of air after expansion as before, even if

differences have arisen in the several parts, and therefore
there is no heat permanently converted into work. On the other
hand, however, the air cannot again be compressed into its
former volume without a simultaneous conversion of work into
heat.[25]

Thus, Clausius might have been implying that the equivalence value for
the recompression of the air would measure the uncompensated transformation in its free expansion. We could therefore conclude that no
development occurred between 1854 and 1856, that Clausius merely
explicitly generalized in 1856 an idea already eminently clear to him
in his discussion of free expansion in 1854.

But were his ideas eminently clear? When Clausius discussed the
free expansion of air, he claimed to be illustrating a class of uncompensated transformations in which "a force, in doing mechanical work,
has not to overcome an equal resistance, and therefore produces a perceptible motion, ... the vis viva of which afterwards passes into heat."[26]
And such a conversion of work into heat would be a positive transformation.
But when he discussed free expansion as an example of this class, Clausius
explicitly stated that the gas after its expansion contained just as
much heat as it did before. Thus, no net heat was generated.

In a free expansion, heat first generates vis viva which then in turn
regenerates heat. Would not these two opposite transformations compensate for each other? Could Clausius clearly identify an uncompensated
positive transformation in the free expansion?

Phrasing the issue somewhat more philosophically we could ask: "Where
is the value of the uncompensated transformation before it is made
evident by returning the gas to its original condition?". After 1865
Clausius could reply: "It is the entropy change". But before 1865
Clausius did not have the concept of entropy as a function of the state
of the substance. True, he had proved that $\int dq/T$ was a perfect differential for reversible cycles, but this did not hold for a cycle which
included an irreversible transformation. Thus, he could not yet claim
a change in some function of the state of the substance.

It is highly significant that Clausius in 1854 did not assign a
specific symbol to the quantity $\int dq/T$ for reversible cycles even though
he identified dq/T as a perfect differential. This contrasts sharply
with his identifying the function U in 1850. Immediately after demonstrating that a certain cluster of terms involving partial differentials
of heat was a perfect differential, he replaced that cluster with a
differential dU. He further stated that U "comprises the sensible heat
and the heat necessary for interior work ... [and] is a function of v
and t, and is therefore determined by the state of the gas."[27]

There were probably two reasons why Clausius did not make such an
immediate identification of dq/T with a function of the state of the
substance. His proof that dq/T was a perfect differential held only for
a reversible cycle. Moreover, he had no immediate conceptual understanding
of $\int dq/T$ in 1854 as he did of the function U in 1850. Nonetheless in
1856, when he made the method for calculating values for uncompensated
transformations explicit, Clausius implicitly assumed such a function.
For in justifying his method for making that calculation he said, with
regard to the reversible return of the substance to its original condition,
"the uncompensated transformation involved in the given change of
condition could not have been increased by the above reduction [back to
the original condition], [because it] was executed in a reversible

manner."[28] In making that statement for a single process, and not for
a cycle, Clausius required the implicit assumption that something in the
reversible return matched the transformation value occurring in the heat
reservoir, namely, the entropy change in the substance.

I believe that we can account both for this development in Clausius'
thinking and for the development in terms of lost work, by introducing
the hypothesis that Clausius gained insights from reading Rankine's
treatment of the free expansion of a gas. In that treatment Rankine
quantitatively identified the work that was lost and clearly indicated
that his thermodynamic function $\underline{F}$, which is equivalent to entropy,
increased.

Rankine's Influence on Clausius' Thermodynamic Thought

In his paper on the steam engine in 1856, Clausius noted that he had
read Rankine's paper, "On the Geometrical Representation of the Expansive
Action of Heat, and the Theory of Thermo-dynamic Engines." He said that
the Philosophical Transactions for 1854, which contained the paper,
appeared in 1855 and that he was surprised to see that Rankine had derived
certain equations identical to his own. For this aspect of his work,
therefore, he claimed no priority; and he concluded by saying that reading
Rankine had assured him of the validity of his own approach.[29] The
context in which Clausius made this admission dealt solely with his
analysis of a reversible steam engine cycle. Did reading Rankine influence his analysis of the irreversible cycle?

Unfortunately, no definitive answer can be given to that question
because no existing manuscripts show to what extent Clausius had developed
his thoughts about the steam engine prior to his reading of Rankine. We
can only make conjectures from the comments which Clausius occasionally
made in the detailed summary he made of Rankine's paper, a manuscript
that is fortunately preserved at the Deutsches Museum.[30]

Clausius gave Rankine's paper no mere cursory reading.[31] He sketched
every diagram and occasionally added one of his own. He followed every
derivation closely, even supplying the missing steps. He generally
summarized the text, but frequently translated entire sections, especially
those in the first part of the paper which presented the fundamental
principles of Rankine's thermodynamics. Finally, and most importantly,
he sometimes made personal comments on the text and made comparisons
with his own ideas.

Thus, at the point where Rankine defined his thermodynamic function $\underline{F}$
in terms of the heat $\underline{H}$ and the total actual (molecular kinetic) energy
$\underline{Q}$ (proportional to the absolute temperature) as $F = \int dH/Q$, Clausius wrote
in red ink:

dH/Q ist nach meiner Bezeichnung dq/T von welchem Ausdruck ist
schon in dem erwähnten Heft, „Theoretische Bemerkungen über
die Wärme" zu Anfang bewiesen habe, dass er ein vollständige
Differential ist, und ganz ähnlich Entwicklung gemacht habe
wie die vorstehenden.[32]

Significantly, he cited no references to his own ideas when he came to
Rankine's treatment of irreversible free expansion, and his extensive
comments there suggest that he was thinking afresh about the problem.
He also followed Rankine's argument in close detail.

Rankine began his discussion of free expansion by saying that, in the
previous analysis he had assumed that the full mechanical force of an
expanding substance was transmitted to objects outside of the substance,

to a piston, for example, which then provided power to run a machine.
(In our terminology, he had been discussing reversible thermodynamics.)
Now, however, Rankine continued:

> ... suppose that as much as possible of the motive power
> developed by the expansion is expended in agitating the
> particles of the expanding substance itself, by whose mutual
> friction it is finally reconverted into heat (as when compressed
> air escapes freely from a small orifice); and let us examine
> the properties of the curves which, on a diagram of energy,
> represent the law of expansion of the substance ... which may
> be called Curves of Free Expansion.[33]

Rankine, thus, considered a different free expansion than that which
Clausius briefly discussed in 1854. Rankine chose as his model the
Joule-Thomson experiments where a gas continuously flows through a porous
plug; Clausius chose the traditional free expansion of a compressed gas
from one container to an evacuated one. The Joule-Thomson free expansion
is the more complicated model.

In order to interpret Rankine's curves for free expansion, it is first
necessary to understand the method by which Rankine represented quantities
of heat geometrically. His key analytic tool was the curve of no transmission, identical with what we now call an isentropic curve ($q=0$ and
$dS=0$). He assumed that these curves, if indefinitely extended, will
finally merge with what he called the Curve of Absolute Cold, where the
substance is wholly devoid of heat.[34] Thus, if on a pressure-volume
diagram two curves of no transmission are drawn from the end points of
any given curve, which represents some real process involving heat and
mechanical work, and the two curves are extended indefinitely, then the
two curves of no transmission together with the original curve will
enclose an area which constitutes a cycle.

This enclosed area, by analogy to a Watt indicator diagram, would
represent the net work performed in the imagined cycle. Moreover,
according to the mechanical theory of heat, the net work in that cycle
msut equal the net heat exchanged during the cycle because heat and work
are equivalent. But the only heat exchanged in Rankine's imagined cycle
is the heat involved in the original process. Therefore, the area that
is enclosed between any given curve and two curves of no transmission
which are drawn and indefinitely extended from its end points, geometrically
represents the heat exchanged in the process designated by the given
curve.[35]

So whenever Rankine analyzed a thermodynamic process, he always
represented the quantities of heat by drawing indefinitely extended
curves of no transmission. Thus, it was inevitable that he would draw
such curves for his analysis of free expansion. Furthermore, by the time
he had reached this discussion in his paper, he had defined his thermodynamic function F and had indicated that each curve of no transmission
has its own constant value of F, i.e., $dF = 0$, the function that Clausius
realized was equivalent to his perfect differential dq/T.

Rankine depicted the process of the Joule-Thomson experiment as shown
in Figure 2, where FF is the curve of the free expansion from pressure
P_G to pressure P_H, and the curves GM and HN are the curves of no transmission indefinitely extended.[36] The area enclosed by their indefinite
extension MGHN, Rankine stated, "represents the heat which would have to
be communicated to the substance, if the motive power developed were
entirely transferred to external bodies." The area V_GGHV$_H$, he continued,

"represents that motive power." The analysis became complicated by the fact that the gas performs work $P_H H V_H O$ when it leaves the vessel and has work $P_G G V_G O$ done upon it as it enters. By taking account of these amounts, Rankine established that the area $P_G G H P_H$ ($\int V dP$) was the amount of the expansive power which,

> ... by the mutual friction of the particles of the expanding
> substance, is entirely reconverted into heat, and is exactly
> sufficient (by the definition of the curve of free expansion)
> to render the communication of heat to the substance unneces-
> sary; from which it follows that this area is equal to the
> area MGHN.[37]

Rankine only stated the proof verbally in a very long paragraph as was his practice with all the geometrical demonstrations in this paper. Clausius outlined each step of the argument with an appropriate equation. When he reached the final conclusion, however, that area $P_G G H P_H$ equals area MGHN, he did not exactly paraphrase Rankine's interpretation quoted above. Clausius simply stated the meaning of this equation in terms of a reversible process, namely, that the area MGHN represented the heat which must be added to the substance if the substance performed its maximum work and followed the same curve GH.

> $P_G G H P_H$
> Dieses ist also die Wärme welche der Substanz mitgetheilt
> werden musste wenn sie bei der Ausdehnung ihre volle Arbeit
> thäte und doch derselben Kurve GH folgen sollte; und diese
> Wärmemenge wird anderweitig dargestellt durch
> MGHN.[38]

The two statements together suggest the key for understanding the equivalence values for uncompensated transformations, and Clausius joined the two when he appended to his notes an additional analysis of the traditional free expansion into a vacuum. He sketched a diagram for such a process as shown in Figure 3. He called it the curve of totally free expansion because the substance performed absolutely no external work.

> Sei F'F' eine Kurve ganz freier Ausdehnung, welche eine etwas
> andere Lage hat, wie FF, so ist die äussere Arbeit 0 und könnte
> sein $V_G G H V_H$. Diese letztere ist also die Wärmemenge die
> mitgetheilt werden musste wenn bei voller Arbeit dieselben
> Kurve entstehen sollte also
> MGHN = $V_G G H V_H$ ...[39]

In sketching that diagram and in making that comment, Clausius showed his new understanding of the uncompensated transformation that he had only briefly discussed in 1854. For by sketching that diagram he showed that he understood that Rankine's thermodynamic function F increased in the expansion, and by making that comment he showed that this increase would have been the same amount if the expansion had occurred reversibly along the same curve. Therefore, he now understood adiabatic uncompensated transformations more fully. He knew that they were accompanied by an increase in the thermodynamic function F. He also knew that by reversibly compressing the substance to its original condition, this change in the thermodynamic function would be nullified and replaced by an equivalent value in terms of the transformation of work into heat via the reversible compression.

So we can account for the first new development in Clausius' under-standing of irreversibility in 1856 through the influence of Rankine's ideas. But we can also further find in Rankine's diagram for the Joule-

Thomson free expansion a perfect model for the second development, the
equation for lost work in a cycle,

$$w = q_1 - T_0 \int_0^{q_1} dq/T - NT_0.$$

It is only necessary to look at the diagram while keeping the behavior
of steam in an engine in mind rather than the behavior of compressed air
passing through a tiny orifice. We know that Clausius thought about the
steam engine while reading Rankine's paper because the second half of
the paper analyzed steam engine performance, and Clausius occasionally
drew comparisons with equations to be found in his own notebook "Zur
Theorie der Dampfmaschinen."[40]

If we look now at Figure 2 keeping the steam engine in mind, then $P_G G$
corresponds to the vaporization of water into saturated steam at the
constant pressure P_G and its entrance into the cylinder. The curve GM
represents the indefinite ideal reversible adiabatic expansion of the
steam. In an engine the expansion only continues to the point where GM
intersects $P_H H$, after which the steam passes out of the cylinder and is
condensed at the constant pressure P_H and its saturation temperature.
The curve GH corresponds to the most extreme case of irreversible steam
expansion in which all the work of expansion is dissipated (for example,
when the steam initially enters the dead space in a cylinder), after
which the steam leaves to be condensed again at constant pressure.
Rankine's diagram clearly shows that the amount of heat that would be
rejected to the condenser in the irreversible case is greater than that
rejected in the reversible one because the volume of the wet saturated
steam remaining after irreversible expansion is greater. This volume is
directly proportional to the amount of saturated vapor that remains.[41]
Moreover, since the condensation of steam occurs at constant temperature,
the amount of the additional heat rejected to the condenser will be
directly proportional to the increase in the function F during the
irreversible expansion. This additional heat, therefore, will represent
the work lost due to irreversibility because the net work must be reduced
by that additional amount.

This interpretation of Rankine's diagram perfectly matches Clausius'
model for the work lost due to irreversibility. In Clausius' model, the
amount of heat q_1 added to the substance is always the same, and it is
added at various temperatures. This matches the process of vaporizing
water from the condenser temperature to the temperature of saturated
steam. Furthermore, Clausius' model stipulated that the amount of heat
q_0 rejected by the substance only occurred at the constant temperature
T_0 and that this amount would vary between the reversible and irreversible
cases. This stipulation again is fully matched by the above interpretation
of the diagram. And finally, in Clausius' equation for lost work, the
amount lost due to irreversibility is NT_0, and N is obviously the change
in Rankine's function F in the diagram.

Thus, we could account for both of the new dimensions that Clausius
introduced into his understanding of irreversibility in 1856 by viewing
them as extensions of ideas which Clausius culled from reading Rankine's
1854 paper. This hypothesis in no way detracts from Clausius' achievement
because it remains true that Clausius alone followed out the full implications of irreversibility, in sharp contrast to Rankine who avoided the
irreversible case except for his isolated analysis of the Joule-Thomson
experiment.

CONCLUSION

But one final puzzle remains. Why did Clausius take 10 years before he made explicit what he had learned from Rankine? By 1855 Clausius knew that Rankine had defined a thermodynamic function which was equivalent to his theorem of equivalent transformations in reversible cycles. He also knew that this thermodynamic function, defined for reversible processes, increased in irreversible ones, which confirmed his theorem of uncompensated transformations. And yet Clausius did not define entropy as a property of a substance nor state the unique characteristic of entropy to increase in irreversible processes until his ninth memoir in 1865. Clausius remained bound to his ideas of equivalence of reversible transformations and the non-compensation of irreversible ones.

Here I believe that Rankine must again be given credit for assisting in the development of the full entropy concept, although paradoxically, it contradicted his own perspective. In 1862 he speculated that the apparent dissipation of heat that we observe in the world might probably be reversed in the far reaches of space. At the bounds of the universe, Rankine imagined, the radiant energy that continually disperses itself might be brought back to its original power by reflection into foci of heat so intense that it could even sever the chemical bonds of matter.[42]

This supposition directly contradicted the fundamental axiom upon which Clausius had built his whole theory of heat, namely, that heat cannot spontaneously, (von selbst) move from a lower to a higher temperature. Clausius, therefore, felt compelled to refute Rankine's conjecture.[43] But without this challenge from Rankine, Clausius might never have severed his thinking from the limited domain of equivalent and uncompensated transformations in particular cycles and processes. He extended his ideas to the fate of the universe, "Die Entropie der Welt strebt einen Maximum zu," to meet the challenge of Rankine, who conceived the existence of entropy as a property of a substance but never fully explored its irreversible increase.[44]

ACKNOWLEDGMENTS

I wish to express my thanks to the National Science Foundation for providing the funds which made my trip to the Deutsches Museum possible, to Dr. Jurgen Teichman and Archivist R. Heinrich at the Deutsches Museum for their cordial welcome and kind help, and to Ms. Elsie Benson for helping me to decipher Clausius' script.

NOTES

1. J. Willard Gibbs, The Scientific Papers, Volume 2, Dover edition, "Rudolf Julius Emanuel Clausius," 261–267, p. 262.
2. J. Willard Gibbs, The Scientific Papers, Volume 1, Dover edition, "On the Equilibrium of Heterogeneous Substances," 55–353, p. 55.
3. Donald Cardwell has outlined the main points of the proof without any critical analysis in his book, From Watt to Clausius, London, 1971, pp. 264–266.
4. In his 1850 paper, "On the Moving Force of Heat and the Laws of Heat which May be Deduced Therefrom," Clausius did not single out beforehand the premise on which he was to base his proof of Carnot's theorem (the identical behavior of all substances in the reversible Carnot cycle). He first argued that two engines with different efficiencies could be so arranged that the only outcome of their operation would be the transfer of heat from the reservoir at the lower temperature to that at the upper temperature. He then stated that this was impossible because it contradicted the natural behavior of heat.

   Hence by repeating both these alternating processes, without expenditure of force or other alteration whatever, any quantity of heat might be transmitted from a cold body to a warm one; and this contradicts the general deportment of heat, which everywhere exhibits the tendency to annul differences of temperature, and therefore to pass from a warmer body to a colder one.

   Rudolf Clausius, The Mechanical Theory of Heat, first edition, edited by T. Archer Hirst, London, 1867, p. 45. Hereafter designated by MTH.

   In his 1854 paper, "On a Modified Form of the Second Fundamental Theorem in the Mechanical Theory of Heat," he did state the principle beforehand and spelled it out in considerably more detail.

   This principle, upon which the whole of the following development rests, is as follows: — Heat can never pass from a colder to a warmer body without some other change, connected therewith, occurring at the same time. Everything we know concerning the interchange of heat between two bodies of different temperatures confirms this; for heat everywhere manifests a tendency to equalize existing differences of temperature, and therefore to pass in a contrary direction, i.e. from warmer to colder bodies. Without further explanation, therefore, the truth of the principle will be granted.

   MTH (1854), pp. 117–118.

   Clausius' most succinct statement of the principle, "Heat cannot of itself pass from a colder to a warmer body," first appeared in his 1862 paper, "On the Application of the Theorem of the Equivalence of Transformations to Interior Work," MTH p. 215. In one of his many controversial exchanges with Peter Guthrie Tait, Clausius made the mistake of claiming this statement of the principle for his 1850 memoir. Tait thus managed to salvage a debating point while losing the main contest. See my "Entropy and Dissipation," Historical Studies in the Physical Sciences 2(1970):321–354, pp. 333–336.

5. Clausius' choice of the word equivalent was somewhat unfortunate
    since the two transformations did not actually replace each other.
    The reverse of the one could replace the other.

    > Of the two transformations in such a reversible process
    > either can replace the other, if the latter is taken in
    > the opposite direction; so that if a transformation of
    > the one kind has occurred, this can again be reversed, and
    > a transformation of the other kind may be substituted with-
    > out any other permanent change being requisite thereto.

    MTH (1854), p. 122.
    This reverse equivalence becomes clear when Clausius assigns
    positive and negative values to the two "equivalent" transformations.
6. I refer here to Clausius' second statement of the theorem and not
    to his first. His first statement is always correct.

    > In all cases where a quantity of heat is converted into work,
    > and where the body effecting this transformation ultimately
    > returns to its original condition, another quantity of heat
    > must necessarily be transferred from a warmer to a colder
    > body; and the magnitude of the last quantity of heat, in
    > relation to the first, depends only upon the temperatures
    > of the bodies between which heat passes, and not upon the
    > nature of the body effecting the transformation.

    Ibid. p. 116.
    His second statement is correct only for perfect gases.

    > If two transformations which, without necessitating any
    > other permanent change, can mutually replace one another,
    > be called equivalent, then the generation of the quantity
    > of heat Q of the temperature t from work, has the
    > equivalence-value

    $$\frac{Q}{T}$$

    > and the passage of the quantity of heat Q from the temperature
    > t_1 to the temperature t_2, has the equivalence-value

    $$Q \left[\frac{1}{T_2} - \frac{1}{T_1} \right],$$

    > wherein T is a function of the temperature, independent of the
    > nature of the process by which the transformation is affected.

    Ibid. pp. 125-126.
7. Ibid. pp. 123-124.
8. Ibid. p. 116.
9. Ibid. pp. 124-129.
10. Ferdinand Reech, "Theorie Generale des Effets Dynamiques de la
    Chaleur," Journal de Mathematiques, 18 (1853) 357-568.
11. Reech used S for work. I am using a consistent set of symbols.
12. Ibid. pp. 364-371.
13. Ibid. p. 371.
14. Ferdinand Reech, "Récapitulation Trés-Succincte des Recherches
    Algébriques faites sur la Théorie des Effets Méchaniques de la
    Chaleur par Differents Auteurs," Journal de Mathématiques (1856) 21:
    58-75, p. 62.
15. W.J.M. Rankine, "On the Geometrical Representation of the Expansive
    Action of Heat, and the Theory of Thermo-dynamic Engines,"

Philosophical Transactions 144 (1854), 115–175.

16. Reech (Note 11) pp. 376–377.

17. See Keith Hutchison, "Der Ursprung der Entropiefunktion bei Rankine und Clausius," Annals of Science 30 (1973) 341–364.

18. Ibid. p. 352.

19. R. Clausius, "On the Application of the Mechanical Theory of Heat to the Steam-Engine," 1856, MTH, pp. 136–207.

20. MTH (1854) pp. 133–134.

21. In the paper on the steam engine Clausius changed his convention for assigning positive and negative values to quantities of heat. In 1854, the heat reservoirs served as his reference system. Positive quantities represented additions of heat to the reservoirs, negative quantities represented removals. The equation for the equivalence of transformations expressed this underlying convention clearly, $-q\ f(t) - q_1\ f(t_1) + q_1\ f(t_2) = 0$. In 1856, Clausius made the substance in the engine cycle his reference system, so that his conventions for quantities of heat were directly reversed. Therefore, his statement of the theory for irreversible cycles now became $\int dq/T < 0$. I prefer to keep consistently with his first convention since it has become common practice in thermodynamics.

In order to prevent confusion, Clausius continued to employ the symbol N for the value of an uncompensated transformation. N must therefore always be positive. In 1854 he wrote $\int dq/T = N$; in 1856 he wrote $\int dq/T = -N$. Thus, his convention that uncompensated transformations are positive is not changed by his change in convention for heat quantities.

22. MTH (1856), pp. 142–144. In a note added in 1864, Clausius illustrated his discussion with the calculation of the uncompensated transformation in the free expansion of a perfect gas. Ibid. p. 144.

23. Clausius' statement of the first law for a cycle was $q = Aw$, where A is a proportionality constant, the heat equivalent of a unit of work. I have simplified the equations by assuming that all quantities are consistently in mechanical units.

24. Ibid. pp. 145–146. Tait fumbled this derivation miserably in his Thermodynamics and thus gave Clausius an opportunity to give him (and Thomson) a lesson in thermodynamics. See my "Entropy and Dissipation" (Note 4), pp. 338–342.

25. MTH (1854), p. 134.

26. Ibid. pp. 133–134.

27. Ibid. p. 29. By using the disgregation function which he introduced in 1862, Clausius was able to make an analogous interpretation for entropy in 1865. See Martin J. Klein, "Gibbs on Clausius," Historical Studies in the Physical Sciences 1 (1969) 127–150, especially pp. 135–141.

28. MTH (1856), p. 143.

29. Ibid. Footnote, p. 161.

30. Deutsches Museum manuscript, Hs. 6459

31. And Clausius' manuscript permits no cursory reading. He writes with the old slashing German penmanship and generally omits all vowels from his words. See the comments made by Ivo Schneider, "Clausius' erste Anwendung der Wahrscheinlichkeitsrechnung im Rahmen der atmosphärischen Lichtstreuung," Archive for History

of Exact Sciences <u>14</u> (1974) 143–158, in Notes 23 and 24, p. 150.

32. Clausius' manuscript (Note 30), p. 16.

33. Rankine (Note 15), p. 130.

34. <u>Ibid</u>. p. 117.

35. <u>Ibid</u>. pp. 117, 119.

36. <u>Ibid</u>. p. 130. Hutchison (Note 17) mentioned briefly Rankine's discussion in footnote 35. "Die Unterscheidung zwischen umkehrbaren und nichtumkehrbaren Vorgängen wird ... kurz erwähnt." p. 349.

37. <u>Ibid</u>.

38. Clausius manuscript (Note 30), p. 23.

39. <u>Ibid</u>.

40. For example, with reference to an equation for heating water, he noted: "Entspricht der Gleichung (3) auf Seite 6 meines Heftes, Zur Theorie der Dampfmaschinen'." Clausius manuscript (Note 30), p. 52.

41. Both Rankine and Clausius independently demonstrated in 1850 that saturated steam would condense during an adiabatic expansion.

42. W.J.M. Rankine, "On the Reconcentration of the Mechanical Energy of the Universe," <u>Philosophical Magazine</u> (1862), reprinted in <u>Miscellaneous Scientific Papers</u>, London, 1881, pp. 200–202.

43. R. Clausius, "On the Concentration of Rays of Heat and Light, and on the Limits of its Action," 1864, MTH, pp. 290–326. See especially p. 292.

44. Keith Hutchison credits Rankine with the discovery of the existence of entropy, Clausius with the discovery of the law of its increase. "Seine [Rankine's] Entdeckung bestand in der Erkenntnis des Daseins der Entropie, Clausius' Entdeckung gagegen in der Aufstellung des Gestezes ihrer Zunahme." Hutchison (Note 17) p. 363.

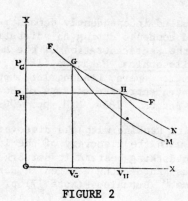

FIGURE 1

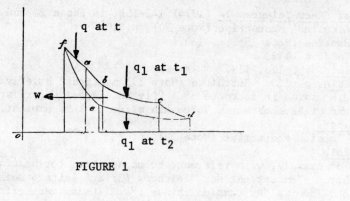

FIGURE 2

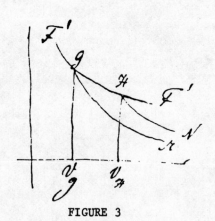

FIGURE 3

U.Hoyer

CONSIDERATIONS ON CARNOT'S MECHANICAL EQUIVALENT OF HEAT

In 1878, half a year after Robert Mayer's death, the younger brother of Sadi Carnot presented to the French Academy of Sciences a number of posthumous notes, which obviously had escaped the burning of Sadi's infected manuscripts in 1832. Perhaps the most interesting of these notes reads as follows [1] :

"According to some ideas which I have formed on the theory of heat, the production of a unit of motive power necessitates the destruction of 2.70 units of heat."

The units of heat and motive power which Sadi Carnot employed in his "Reflections on the motive power of fire" from 1824 were 1000 calories and 1000 mkg respectively, so that the foregoing relation can be read as

(1) $$1000 \text{ mkg} = 2.70 \text{ kcal}$$

or

(2) $$1 \text{ kcal} = 370 \text{ mkg} .$$

This is Carnot's mechanical equivalent of heat. It nearly coincides with the equivalent published by Robert Mayer in 1842, namely [2]

(3) $$1 \text{ kcal} = 365 \text{ mkg} .$$

Carnot did not tell us how he obtained the cryptic

result (1) which, of course, repeatedly attracted interest during the past century. The simplest conjecture is that his procedure was identi= cal with Robert Mayer's method. Indeed, Henri Poincaré[3] and Louis Décombe[4] suggested that Carnot might have calculated the value (1) from the difference of the heat capacities under constant pressure and constant volume c_p and c_V of air. However, Décombe's suggestion was challenged immediately afterwards by Camille Raveau[5] who advocated a different method, similar to the one I shall discuss subsequently. Although there are some logical difficulties in Decombe's deduction (as well as in Raveau's), the assumption that Carnot proceeded like Mayer cannot be rejected as improbable. Nevertheless, I think it is worthwhile here to discuss a method different from Mayer's reasoning,[6] since it seems to be more in line with Carnot's original calculations given in the "Reflections". Afterwards I shall inquire into its relation to Mayer's calculation.

I suppose that Carnot's method simply consisted in equa= ting the work done under the isotherm of air to the heat absorbed thereupon. This suggestion, which was first put forward by Ernst Mach in 1896[7], immediately encounters a serious difficulty, namely the question: Whence could Carnot have obtained the quantity of heat absorbed on the isotherm? For in Carnot's time this quantity had not yet been measured. When Carnot, on writing the "Reflections", first met with this problem, he still held the conviction that heat is con= served in any closed cycle. He therefore calculated the amount of heat in question from the cycle given in figure 1. This cycle con=

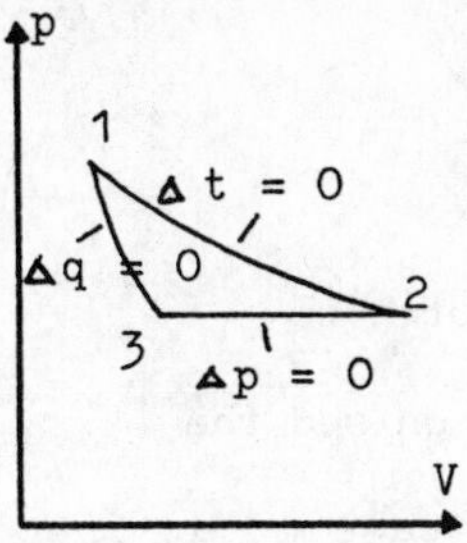

sists of three curves, one isothermic, one adia=
batic and one isobaric. Then the law of conser=
vation of heat yields

figure 1

$$(4) \qquad \int_1^2 dq = \int_1^3 dq + \int_3^2 dq \quad .$$

Since $\int_1^3 dq = 0$, Carnot obtained

$$(5) \qquad \int_1^2 dq = \int_3^2 dq = m \int_3^2 c_p \, dt \quad .$$

By this ingenious consideration Carnot was able to reduce the un=
known quantity $\int_1^2 dq$ to the specific heat at constant pressure
which had already been measured at his time.

However, even when working on the "Reflections",
Carnot had difficulties in reconciling the theory of heat conser=
vation with certain experimental results. From his posthumous notes
we know that Rumford's experiments were the major stumbling-stone.
Presumably it was this problem that made him give up the idea of
conservation of heat in favour of the assumption that heat and mechanical work are quantities of the same dimension. Therefore, Carnot
must have realized one day that equation (5) can no longer be applied
for the determination of the heat absorbed on the isotherm. This is
the main objection to be raised against all derivations of Carnot's
equivalent that have hitherto been published. However, Carnot
certainly realized that the calculation of the efficiency of a Carnot
machine working with air between infinitesimal temperature differences
based on the conservation of heat yielded surprisingly satisfac=
tory results. The values of air and water at $0\,^\circ$ C were

$$\eta \text{ (air)} = 1.395 \text{ mkg/kcal} \quad \text{and}$$

(6)

$$\eta \text{ (water)} = 1.290 \text{ mkg/kcal}$$

The differences which were less than 10% could be attributed to experimental errors. Furthermore, Carnot presumably recognised the reason for this remarkable agreement, namely that in case of an in= finitesimal process the work gained disappears at a higher rate than the heat absorbed. If air is the working substance, then heat is

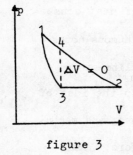

represented by the column under the isotherm (fig.2). In other words: Carnot may well have been aware of the fact that conservation of heat, although untenable in the general case of finite processes, is still applicable, if due restriction is made to infinitesimal processes.

figure 2

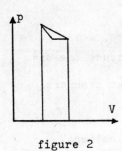

figure 3

He may then have proceeded as follows: Consider the cycle shown in figure 1 and completed by the line 3-4 drawn under constant volume (see figure 3). Then

$$(7) \quad W = \int_1^2 p \, dV = - \int_4^1 p \, dV + \int_4^2 p \, dV \,,$$

whence as a first approximation it can be deduced that

$$(8) \quad \alpha = \frac{1}{V} \frac{\partial V}{\partial t} \bigg|_{\Delta p = 0} \quad \text{and}$$

$$(9) \quad = - \frac{1}{V} \frac{\partial V}{\partial t} \bigg|_{\Delta q = 0}$$

On the other hand, from equation (5) we find in first approximation

362

$$(10) \qquad \int_{1}^{2} dq = m \, c_p \Delta t + \ldots$$

On the assumption that the work gained is equivalent to the heat consumed, it finally follows that

$$(11) \qquad\qquad m \, c_p = p \, V \, (\alpha + \beta) \qquad .$$

This equation is exact. All infinitesimal or small quantities have disappeared from it.

The experimental values which Carnot employed in 1824 for air at 0° C were

$$(12) \qquad m = 1 \text{ kg} \;, \; p = 10.40 \; 10^3 \text{ kg/m}^2 \;, \; V = 0.77 \text{ m}^3 \;,$$
$$c_p = 0.267 \text{ kcal}, \; \alpha = \frac{1}{267} \text{ centigrade}^{-1}, \; \beta = \frac{1}{116}$$

centigrade^{-1}.

Insertion of these values in equation (11) yields Carnot's mechanical equivalent of heat (1).

The numerical value of the coefficient α Carnot adopted from Guy-Lussac's experiments on the expansion of air, and c_p he owed to measurements of Delaroche and Bérard. β is perhaps the most interesting number here. In a footnote to his calculations in the "Reflections" Carnot added that the value of β appearing in (12) was due to Poisson and agreed satisfactorily with experiments made by Clément and Desormes on the expansion of air into rarified air and also with Gay-Lussac's experiments on the quantity $k = c_p/c_v$ of air. Since Carnot did not give an exact reference to his source for Poisson's number, I originally supposed that there was a connection to Poisson's famous paper on the adiabatic theory from 1823. Since then I have learned from Dr. Fox "Caloric Theory of Gases"[8] that Poisson had in 1808 already deduced this number from his theory of the propagation of sound and that Clément and Desormes in 1819 had been able to verify his result by measuring $\beta = 1/114$. The relation of β to Gay-Lussac's and Welter's experiments will be discussed presently.

If we insert the following modern values in equation (11), namely

(13)
$$m = 1 \text{ kg} \, , \, p = 10.34 \ 10^3 \text{ kg/m}^2 \, , \quad V = 0.773 \text{ m}^3$$

$$c_p = 0.240 \text{ kcal}, \quad \alpha = \frac{1}{273.2} \text{ centigrade}^{-1} \, , \quad \beta = \frac{1}{109.3}$$

centigrade^{-1}, we obtain the modern equivalent

(14)
$$1 \quad \text{kcal} \quad = \quad 427 \quad \text{mkg} \quad .$$

Therefore I contend that Carnot might have obtained the modern value of the mechanical equivalent of heat, if he had had better measurements at his disposal. Moreover I think that this result is a convincing illustration of the fact that the theory of heat conservation is a first approximation to thermodynamics of reversible processes.

I should like now to discuss the relation of the method presented here to Robert Mayer's procedure.

In the "Reflections" Carnot calculated the proportion $k = c_p/c_V$ for air from the coefficients α and β, and found

(15)
$$k = 1.44 \, ,$$

whereas Gay-Lussac's and Welter's experimental result was

(16) $k = 1.3748$.

Again the agreement between theory and experiment was very convincing. The reason is, of course, that the theory of equivalence of heat and work yields the same result $k = \alpha + \beta/\beta$ as does the theory of conservation of heat. This can easily be seen from the fact that the mechanical work, represented by the triangles 1-4-3 and 4-2-3 in figure 3 again disappears at a higher rate than the heat absorbed on the curves 1-2 and 3-2, provided the process is an infinitesimal one.

Insertion of $k = \alpha + \beta/\beta$ in (11) yields

(17) $m (c_p - c_V) = \alpha \, p \, V$.

This is Robert Mayer's relation. Therefore we see that the two methods discussed here are intimately connected with each other. Although distinctly different in their points of departure, they are mathe= matically equivalent and lead to the same result. Thus, as long as Carnot's original calculation does not reappear, there is no reason for excluding Mayer's method, as was suggested by Raveau.

It should be mentioned here that the foregoing calculations presuppose that the total amount of heat absorbed on the isotherm is consumed solely for the production of mechanical work, i.e. that there is no loss of heat due to inner work done on expan= ding the gas. From Carnot's posthumous notes it is evident that he knew Gay-Lussac's famous experiment on the adiabatic expansion of air into a vacuum. This experiment shows, as is well known, that no change of temperature occurs. If inner work had to be done on expanding the gas, then the air would at the end of the process be

cooler than at the beginning. From the absence of such an effect
Carnot could draw the conclusion that his calculations rested on a
sound basis.

The discovery of the equivalence of heat and work, of course,
deeply affected the socalled second law of thermodynamics which Carnot
had discovered in 1824. For his reasoning at that time had been based on
the assumption that heat is conserved in any closed cycle. This assumption was no longer valid now. However, Carnot's posthumous manuscripts
also testify that he tackled the problem at the right point. He wrote:[9]

"If one produces motive power by transfer of heat from a body
A to a body B, is then the quantity of heat which arrives at B (if it
is not the same which was taken from A, if a part of it really is
consumed to produce motive power), is this quantity the same, whatever
body we apply to create motive power?"

This is the decisive question which later led Clausius and
Thomson to a revision of the second law of thermodynamics. So we may
conclude in saying that Carnot is not only the founder of thermodynamics,
but that he almost finished the classical theory of heat.

References:

1) Reflections on the Motive Power of Fire by Sadi Carnot and other
   Papers on the Second Law of Thermodynamics by É. Clapeyron and
   R. Clausius. Ed. E. Mendoza, New York 1968, p. 68.

2) Robert Mayer, Bemerkungen über die Kräfte der unbelebten Natur.
   Liebigs Annalen der Chemie und Pharmacie, Bd XLII, 1842. Cp.
   Die Mechanik der Wärme in gesammelten Schriften von Robert Mayer.
   Ed. Jacob J. Weyrauch, Stuttgart 1893, 3rd ed. pp. 23-30.

3) Henri Poincaré, Thermodynamik, Berlin 1893, p. 37.

4) Louis Décombe, Sadi Carnot et le principe de l'Equivalence. Son calcul de l'équivalent mécanique de la chaleur reconstitué à l'aide de données exclusivement empruntées aux "Réflexions sur la puissance motrice du feu". Comptes rendus, t. 168, 1919, pp. 268-271.

5) Camille Raveau, Comment Carnot a calculé l'équivalent mécanique de la chaleur. Un document inédit. Comptes rendus, t. 168, 1919, pp. 549-552.

6) Ulrich Hoyer, How Did Carnot Calculate the Mechanival Equivalent of Heat? Centaurus, vol. 19, 1975, pp. 207-219.

-, Das Verhältnis der Carnotschen Theorie zur klassischen Thermo= dynamik. Archive for History of Exact Sciences, vol. 15, 1976, pp. 149-197.

-, La théorie de Carnot. Première et seconde approximations de la thermodynamique.   Sadi Carnot et l'essor de la thermodyna= mique, Paris 1976, pp. 221-228.

-, Didaktische Aspekte der Carnotschen Theorie. Der mathematische und naturwissenschaftliche Unterricht (MNU), 30.Jg, 1977, pp. 84-90.

7) Ernst Mach, Die Principien der Wärmelehre -historisch-kritisch entwickelt, Leipzig 1896, p. 245.

8) Robert Fox, The Caloric Theory of Gases from Lavoisier to Regnault. Oxford 1971, p. 86 and 144.

9) Sadi Carnot, Réflexions sur la puissance motrice du feu et sur les machines propres à développer cette puissance. Paris 1878, p. 93.

Symposium 7.   PROBLEMS OF SOURCE MATERIALS IN THE HISTORY OF SCIENCE

M.T.d'Alverny

ARCHIVES AND HISTORY OF SCIENCE

Archives,in a general sense means a collection of documents,of
public or private origin,preserved as records.In many countries
the term is used too for a depository of documents. These
documents are normally handwritten (or, in our century, typed).
A distinction must be precised to begin with.The archival material
is different from the book,which is a text intended to be read and
used for educational(or religious,literary,scientific,etc)purposes.
The fact that in Antiquity and in the Middle Ages the book is handwritten is the source of an equivocal meaning.In the United States
for instance,the Division of the Library of Congress called "Manuscripts Division" is a depository of documents:papers,diaries,corres-
pondences.What is usually called "manuscripts" in Europe,viz.hand-
written books is preserved in the rare books Department. We must
remark,however,that in many European libraries,the Departments of
Manuscripts,primarily the sanctuary of handwritten books are frequently repositories of private documents,especially correspondence.
The great depositories of Archives are mainly established to preserve
public documents,but it so happens that they keep,and even acquire,
many private documents,mostly from institutions,but sometimes from
families and individuals.The task of the historian of science who
intends to carry on a research with any kind of source material
requires some explorative imagination and patience,the more so
because catalogues,inventories,indexes,are often still in the making.
This research is essential;this is why our Commission,when we were
asked to precise our aims decided to give the following description:
Sources for history of science;manuscripts and documents.Among our
members we had and still have historians especially devoted to the
study of archival documents.We remember with regret our late colleague
L.Rosenfeld,who gave us precious information on the present condition
of contemporary Archives of scientific papers.He was the director
of the Niels Bohr Archives in Copenhagen and had begun to edit the
works of Bohr,but did not live long enough to fulfill his plans.
The study and discussion of contemporary Archives is the topic of
an other part of our Symposium;we must precise,anyhow,that this topic
is not foreign for our members.
We shall first indicate the kind of information that the historian
of science can find in archival material,be it preserved in public,

institutional or private depositories,or in libraries,from the Middle
Ages onwards. Medieval archives,secular and ecclesiastical,have
not been much used by historians of science,who are mostly concerned
with the research for manuscript texts and eventually instruments.
The biographical data that can be found in documents is not devoid
of interest however;it is difficult to understand the development
and transmission of scientific knowledge if the men involved are
unknown.In recent years,the publication of documents from Spanish
ecclesiastical archives has allowed to follow the career of several
translators of important scientific texts working in Spain in the
XIIth and early XIIIth cent. From the XIIIth cent.onwards the University archives are an important source for the biography of
scientific scholars;they have been used as well as ecclesiastical
archives and Public Records by A.B.Emden to produce his excellent
biographical registers for Oxford and Cambridge. Medieval documents
are a precious source of information to investigate the development
of medicine,surgery and technology.Archival material has been used
by our late member,E.Wickersheimer for his:Dictionnaire biographique
des medecins en France au Moyen Age; the author of the Supplement,
D.Jacquart,could find much additional material in inventories of
Archives and in recent publications of documents,like Les Chartes du
Forez(edd.G.Guichard;G. de Neufbourg;E.Perroy).The same kind of
investigation was carried on by Ch.H.Talbot and E.A.Hammond when
they wrote The Medical Practitioners in Medieval England.Archival documents contain also references to what is termed "illegal practice
of medicine",sometimes with picturesque details.It would be rewarding
to explore the legal suits,mostly ecclesiastical,in which physicians,
surgeons and midwives are involved-even eventually the trials of
sorcererd and witches.A register preserved in the Archives nationales
provides us with a good instance of this statement.(Registre MM.266;
XVIIIth cent.copy of earlier deeds).It contains ecclesiastical sentences dating XIVth century against several men and women who practice medicine and chirurgy without licence of the Faculty.The comp
plaints are issued from the part of the procurator of the Dean of
the Faculty of medicine of Paris(between circa 1312-1331).

 A number of practicians are named,some are qualified: "empeiricus"
and "empeirica"(quack).The most remarkable piece is the trial of
a woman,Iacoba Felicie,in 1322(Chart.Univ.Par.II,nº811,p.255-267).
From her sayings;Jacquette or Jacqueline was about thirty;not at
all the doting old woman warrying around mysterious recipees.The reports of the witnesses called in court is very favourable;men and

women relate how they were cured by her;several had tried first
official approved physicians,who did not help them.These reports
contain some details on Jacqueline's way of examining sick persons}
(rather in agreement with the traditional principles),and on the
treatment,which consisted mostly of potions with herbs.The defence
put forward by Jacqueline with courage and boldness includes interesting
remarks,particularly the statement that it is more appropriate for
a sick woman to be examined and treated by a woman.It seems,however,
that Jacqueline had many male patients too.
In the same XIVth century the Archives of the "Corona de Aragon"
preserved in Barcelona present some deeds of the learned kings of
Catalonia and Aragon,showing their interest in astronomy and astrology.
We feel that our still unadequate appreciation on the use and impact
of astrology might be improved by the exploration of Medieval documents.

   In the Renaissance period,then in the XVIIth and XVIIIth centuries,
Public Record offices,ecclesiastical Archives,archives of Universities,
scientific institutions and Academies furnish material for the biography of scientists and for the history of the development of the
sciences.We should rather say:could furnish material,for these depositories are not much explored for our special purpose.The present condition of the catalogues accounts in part for this lack of interest:
a number of collections are imperfectly catalogued,or even uncatalogued.
 A fundamental kind of documents,especially in the XVIIth and XVIIIth
centuries,but still important for the history of science up to our days
are the correspondences of scientists.One of the most typical cases
is the correspondence of Marin Mersenne,exchanging letters with the
learned men in Europe. A Colloquium was held two years ago (May 1975)
at Chantilly on this topic:Les correspondances,leur importance pour
l'historien des sciences et de la philosophie.Problèmes de leur édition.
(published:Revue de Synthèse,3e ser.n°81-82,1976).These private documents are sometimes deposited in Public Archives or Libraries,but
they may be yet in private hands.Hence the risk of destruction or
damage,plus the difficulty of obtaining permission to consult them.
This remark applies to the papers, notebooks, diaries of scientists; if
they were attached to a University or Academy, or scientific Institution,
there is some chance for their preservation and eventual publication,providing that they bequeathed their papers,or that their
heirs were generous.The oldest and most respectable institutions,
notably the Academies,dispose rarely of a numerous staff for cataloguing the scientific papers or correspondence,still less of funds
for editing their holdings.The preliminary work and the subsequent

publication require frequently outside help and international collaboration. Some years ago our Commission, at the request of our colleague L.Belloni, sponsored the publication of a catalogue of the manuscripts documents of the "Fondo Halleriano"(Albrecht von Haller) preserved in Milan, Biblioteca Braidense; this catalogue was prepared by G.Vergnano   Pecorella and published with a grant of the IUHPS. Our colleague M.D.Grmek devoted several years to the investigation of the important collection of documents concerning Claude Bernard preserved at the Collège de France(including drafts of his works, notes, projects, lecture notes, records of experiences, letters); he edited the "Catalogue des manuscrits de Claude Bernard"in 1970. Claude Bernard is a good case, for his papers were carefully preserved in the institution where he taught and worked. The fate of the papers of scientists left in private hands is unpredictable; the research is adventurous and generally difficult. Professor Henry Guerlac, who presided in 1960 a conference held in Washington on "Science manuscripts"(the records were edited in Isis 53, 1962), and has a wide experience, delivered during the present Symposium a lively report on the fate of the Lavoisier papers, in which he was involved.

The second part of this Symposium is dedicated to XXth century Archives, with an important American contribution , for contemporary scientific papers are collected and preserved with great care and efficiency in American Universities and Institutions. I should like to present a short note concerning XVIIIth and XIXth century scientific documents preserved in some Institutions of the United States and in the Library of Congress .A generous collaboration allowed us to collect information, especially from the American philosophical Society, of which the records are kept since 1769, and where collections of papers, reports, diaries, notebooks, correspondences of scientific interest have accumulated. A "Guide to the Archives and manuscript collections of the American philosophical Society" compiled by the present Librarian, Dr. W.J.Bell, and his assistant, M.D.Smith, was published in 1966. The accessions of the last years are described in the annual Reports of the Library. The conditions of preservation and accessibility of the documents are excellent. The tradition of scientific interests initiated with Benjamin Franklin is one of the distinctive features of the American philosophical Society; a large part of the

holdings come from members of the Society or correspondents of
Benjamin Franklin:Lucien Bonaparte,Darwin,Humphrey Davy,
Duhamel du Monceau,Gianbattista Beccaria,Giovanni Fabroni,
Alexander von Humboldt,John Le Conte, the geologist Charles
Lyell,Heinrich Muhlenberg,etc. It can be considered that this
repository is one of the most important for the history of
XVIIIth and XIXth century science. It is also rich in new
accessions for XXth century history of science,but this will be
said in the second part of the Symposium. A "Guide to science
and history of science resources in the Philadelphia area",
compiled by L.D. Gundry is a publication of the American philosophical Society produced in mimeogram,1970. It contains short
but precise notices on the Institutions of Philadelphia and
vicinity; those in possession of scientific documents are the
Academy of natural sciences founded in 1812; these papers
concern primarily the scientists connected with the Academy(cf.
A Guide to the mss. collections in the Academy of natural sciences
of Philadelphia by Maurice and Venia Phillips);the College of
Physicians of Philadelphia,founded in 1787,with a collection
of documents dating mostly XVIIIth and XIXth cent.;the Hugh Mor-
ris Library(Univ.of Delaware),founded 1834),particularly important
for the history of chemistry;the Samuel Paley Library,Temple Univ.
preserving papers and correspondence of Richard Owen and Andrew
Pritchard,among others.
  The Library of Congress preserves in the Manuscripts Division
many collections of documents;a number concern scientists and
scientific institutions.The larger collections have finding aids.
A list of these holdings was communicated to us by one of the
librarians,R.S.Wilkinson,in charge of the scientific documents.
Most of these collections are XXth century,for instance the papers
of Robert Oppenheimer;there are however notes and letters dating
XVIIIth and XIXth cent.,including a collection of Benjamin Franklin's papers(the most important part being in the Library of
the American philosophical Society).New accessions are described
in the reports of the Manuscripts Division published in the
Quarterly Journal of the Library of Congress.
  The Smithsonian Institution(Washington D.C) has Central Archives

with considerable collections of scientific papers.A "Preliminary
Guide to the Smithsonian Archives"was published in 1971.Part of the
holdings include papers of scientists and scientific institutions
of the XIXth cent.and some dating XVIIIth and even XVIIth cent.
Among the earlier collections we note:Meteorological records
1848-1879(information going back to 1820);Topographical Data
project(W.L.Nicholson papers) circa 1873-1882;the papers of Joseph
Henry(1797-1878);Luis Berlandier(1826-1851);Joseph Sexton(1799-1873)
etc.The Smithsonian Institution recently acquired the Dibner collection including books and papers relating to the early history
of electricity and magnetism;moreover some manuscripts concerning
Galileo,Harvey,Newton.(cf. a note by B.J.T.Dohls on the Newton
mss. at the Smithsonian Institution,Isis 68,1977,p.105 sqq.).
The Dibner Library of the history of science in the Smithsonian
Institution.History and Technology-was opened in October 1976.

The National Library of medicine at Bethesda preserves mostly
papers of modern physicians in its Archives;among the holdings are
a small number of earlier documents,for instance XVIIIth cent.medi-
cal diaries.The papers are classified,indexed and cross-indexed.
Many Universities possess in their Archives holdings of scientific
interest.At the request of our colleague H.Guerlac,Miss Ellen Wells
sent us a considerable documentation on the manuscript and archival
resources in the History of science collections preserved at Cornell University.The papers of American scientists date XIXth and
XXth century;a very interesting collection of manuscripts and documents in French date XVIIIth and early XIXth cent.This includes
mainly the "Lavoisier papers",acquired some fifteen years ago.
More recently,the History of science collections could buy from
a bookdealer the scientific Archives of prince Baldassare Boncom-
pagni,including the correspondence of this distinguished historian
of mathematics with other European scholars during the XIXth cent.
and the information that he collected on scientific manuscripts.

Other important holdings of the Cornell Library are the Hollister
collection on the history of engineering,and the papers of Ezra
Cornell(1807-1874). The activity of the departments of Cornell
University are well represented:papers of specialists of agriculture
forestry,botanics,horticulture,biochemistry,bacteriology,apiculture
poultry,husbandry,zoology,entomology;reports of the Departments of
Physics ,Chemistry,Entomology,Micology,Veterinary medicine,Plant-
breeding,Plant pathology,Zoology .There are papers and records of

special institutions:New York State College of agriculture,of Veterinary medicine;Laboratory of ornithology.The Cornell medical Archi
ves collect the records of the New York hospital founded in 1771
and of the institutions affiliated with this Center;the documents
include students notebooks and medical case books.
(A preliminary Guide to the medical Archives has been prepared and
is available upon request:Adele Lerner,Medical Archivist, N.Y.Hospi
tal-Cornell medical Center,1300 York avenue,New York 10021)

THE LAVOISIER PAPERS

The story of the Lavoisier papers deserves a brief account. His
published writings - his books and articles, together with a number of
hitherto unpublished papers - were printed in the Oeuvres de Lavoisier,
brought out in six volumes, under the auspices of the Ministry of Public
Instruction, from 1864-1893. The first four volumes were edited by the
instigator of the project, the distinguished chemist, J-B. Dumas,[1] and
the last two by Edouard Grimaux who brought out in 1888 the first edition
of his well-known biography of Lavoisier.

When the project for the Oeuvres was first discussed, the bulk of
the Lavoisier papers were in the possession of M. Léon de Chazelles,
a collateral heir of Mme. Lavoisier (the Lavoisiers, be it remembered,
had no children). Generously, M. de Chazelles presented the Academy of
Sciences in 1846 with twenty boxes (cartons) of Lavoisier's scientific
papers, and entrusted to François Arago, for reasons that are not clear,
the volumes of Lavoisier's laboratory notebooks began in 1773. All but
one of these registres eventually found their way to the Archives of
the Academy of Sciences. The missing registre, for a time thought to
be lost, had in fact been given to the city library of Perpignan by Arago,
an instance of misguided local patriotism (Arago was born in Roussillon).

The publication of the Oeuvres left unaccounted for a mass of correspondence, several notebooks - including the registres - and a mass of
drafts, academic reports, records of Lavoisier's early geological excur-
sions, and snippets of various kinds. A considerable part of this material
was retained by M. de Chazelles, although he gave some to the city library
of Clermont-Ferrand in his native Auvergne. Much of this material was
borrowed in the 1880's by Edouard Grimaux when he was preparing his biography of Lavoisier. Some came from the Academy of Sciences and some
from M. de Chazelles.

The Dreyfus affair interrupted the relations between Grimaux - an
ardent Dreyfusard - and the Chazelles family, firmly on the side of the
Church and the Army. At the death of Léon de Chazelles, the Lavoisier
material in his possession was divided between an unmarried daughter and

a Mme. de Chazelles residing in the Sarthe. The latter obtained Lavoisier's scientific instruments, some papers, and a substantial portion of Lavoisier's personal library.[2] Only in the 1930's was a considerable mass of papers returned by Grimaux's heirs to the sister who lived in the château of Aigueperse, near Riom in Auvergne, where - as a young graduate student in 1939 - I was privileged to examine them.

During the German occupation of France in World War II, Mlle de Chazelles became active in the resistance movement; her activities at last discovered, she was shipped to a detention camp in Germany where she died. For a time there was a fear, on the part of Lavoisier scholars, that her Lavoisier papers might have fallen into enemy hands or been destroyed. It turned out, however, that the papers joined the family archives of the Chabrol family in the muniment room of a chateau in Auvergne, a third sister having married the then Comte de Chabrol.

In 1956 the material in the possession of the surviving Mme de Chazelles was put up for sale.[3] The scientific instruments were acquired by the French Government for the Conservatoire des Arts et Métiers, and many of the books and some papers and memorabilia were purchased by a well-known Paris bookseller, M. Lucien Scheler. Largely (but not exclusively) through M. Scheler, a collection of Lavoisier books and manuscripts had gradually been assembled by Denis I. Duveen, a chemical manufacturer of New York City. These books, manuscripts, and bibelots, acquired from a New York bookseller through the generosity of the Honorable Arthur Dean, now form the substantial core of the Lavoisier Collection at Cornell University. This collection, enlarged by later acquisitions and the the addition of works of Lavoisier's scientific contemporaries, also includes a punch-card index to the papers in the Archives of the Academy of Sciences. It should be continued to encompass such other sources as the few documents in the Arsenal Library in Paris, the notarial papers in the Minutier Central, the papers in the Archives Nationales, dealing with Lavoisier's trial and execution during the French Revolution, and a lone document in the manuscript room of the Bibliothèque Nationale.

M. René Fric, a chemist employed by the Michelin Tire Company, became interested in the Lavoisier papers as early as 1925. A committee of the

Academy of Sciences to supervise the publication of an edition of the Lavoisier Correspondence (as Volume VII of the Oeuvres de Lavoisier) was established in 1947 and M. Fric was put in charge of the publication. In 1949 he published, under the sponsorship of UNESCO, a short-title, preliminary list of Lavoisier letters. The first fascicle of the Correspondance appeared in 1955, the second in 1957, and the third in 1964. It has to be confessed that M. Fric's editing was, to put it mildly, idiosyncratic. Many well-known personalities are annotated at excessive length; little-known items are passed by with no comment; and many important letters (readily accessible to M. Fric) are omitted altogether. Before his death, M. Fric completed another fascicle of the Lavoisier Correspondance, and this (I understand) is at the moment being proof-read at the Academy of Sciences. M. Fric had also, to my personal knowledge, completed, or nearly completed, a typed transcript of Lavoisier's Registres de laboratoire. This should by all means be published, and should (if it does not already) include the volume given by Arago to the City Library of Perpignan.

Notes

1. The need for an edition of Lavoisier's works was first strongly urged by J-B. Dumas in lectures delivered at the College de France in 1836, the year of the death of Mme Lavoisier. These lectures were published under the title Lecons sur la philosophie chimique.

2. Mme Lavoisier had cleared her husband's name in 1798; all his books and papers were returned to her in 1808. A good part of Lavoisier's personal library was put up for sale at the death of Mme Lavoisier in 1836. A catalogue of this sale has survived.

3. Souvenirs de Lavoisier, Paris, Hotel Drouot, Salle No. 6, 7 Mars, 1956.

Pr Guy BEAUJOUAN

LA PREPARATION D'UN "GUIDE DE L'HISTOIRE DES SCIENCES DANS LES ARCHIVES DE LA REGION PARISIENNE".

Dès 1950, j'ai fait, au Congrès international d'Histoire des Sciences d'Amsterdam, une communication sur L'histoire des sciences aux Archives Nationales de Paris. Ayant, depuis, abandonné la conservation des archives pour assumer une direction d'études d'histoire des sciences, je me suis trouvé, ces dernières années, chargé d'un Guide de l'histoire des sciences et des techniques dans les archives de la région parisienne. Le présent rapport rendra compte de cette expérience; il tentera aussi d'en dégager quelques réflexions visant à la sauvegarde et à la meilleure accessibilité des archives scientifiques.

Entendons-nous bien sur le mot "archives". En principe, les archives sont constituées par l'accumulation de documents résultant normalement du fonctionnement d'une quelconque institution. Cette définition exclut les manuscrits rassemblés par des collectionneurs ou des bibliothécaires (cf. la monumentale série du Catalogue général des manuscrits des bibliothèques publiques de France).

En France, depuis le fameux décret du 21 juillet 1936, les papiers des services publics civils doivent, en principe, être versés aux Archives Nationales ou Départementales. Les Archives Nationales ont, à cet effet, une Section contemporaine; elles comportent, en outre, un service susceptible de recueillir des archives économiques et privées. En 1969-70 fut créée, sous la présidence de Pierre Renouvin, une Commission des archives de l'enseignement.

Mais c'est surtout par la mise en valeur des documents scientifiques du passé qu'il est possible de faire comprendre aux savants actuels la nécessité de rendre un jour consultables leurs propres archives. Lorsque, en février 1965, la Commission d'équipement culturel du Ve Plan s'est préoccupée, hélas sans lendemain, de la sauvegarde des archives techniques françaises, il est clairement apparu que l'établissement d'un guide des archives techniques consultables était bien le premier pas vers une meilleure conservation des archives techniques présentes et à venir: cf. A. BIREMBAUT, Quelques réflexions sur les problèmes posés par la conservation et la consultation des archives techniques françaises, dans Archives internationales d'histoire des sciences XIX, 74-75 (janvier-juin 1966), p. 21 - 102.

LA PREPARATION DU "GUIDE".

Sous la présidence de M. Robert Courrier, secrétaire perpétuel de l'Académie des Sciences, la Commission d'histoire des sciences du Comité des travaux historiques et scientifiques tenta, en 1971, la mise en chantier d'un Guide de l'histoire des sciences et des techniques dans les archives de la région parisienne. Sous l'égide du Pr Huard, cette entreprise a obtenu pendant trois ans (1973-75) le label du Centre National de la Recherche Scientifique, sous forme de "recherche coopérative sur programme", mais avec des crédits véritablement dérisoires (10.000 FF par an) et, par surcroît, chargés de contraintes assez lourdes.

Il ne s'agit évidemment pas d'un inventaire systématique, mais plus simplement d'un guide relativement sommaire conçu à la même échelle que, par exemple, celui de Mlle Mireille RAMBAUD, Les sources de l'histoire de l'art aux Archives Nationales (Paris, 1955): revue donc des principales cotes intéressant l'histoire des sciences et des techniques, avec indication systématique des inventaires et, si nécessaire, de la bibliographie facilitant les recherches. En rappelant brièvement la formation et la structure du fonds considéré, on s'efforce d'y expliquer la présence ou l'absence de tel ou tel type de documentation.

L'avancement du travail s'est traduit par plusieurs colloques au cours desquels furent présentées les ressources qu'offrent à l'histoire des sciences un certain nombre de dépôts. Ont ainsi été obtenus et multicopiés les avant-projets de plus d'une vingtaine de notices. L'expérience a montré qu'il était vain d'imposer aux coauteurs des directives trop contraignantes sauf, lors de la publication, pour la "carte d'identité" de chaque dépôt.

LES ARCHIVES NATIONALES DE PARIS.

Les Archives Nationales de Paris font l'objet d'un travail plus poussé à la faveur notamment des importants dépouillements que j'ai pu personnellement effectuer à l'époque où j'y étais conservateur. Pour compléter notre collection d'inventaires imprimés, j'ai fait xérocopier, dans de nombreux inventaires épuisés ou inédits, les pages concernant l'histoire des sciences: ceci constitue une sérieuse source de documentation dont, même après la parution du Guide, l'utilité restera grande.

Il n'est évidemment pas possible de détailler ici l'extrême variété des renseignements fournis à l'histoire des sciences par cette énorme masse de fonds généralement non scientifiques. Cf. G. BEAUJOUAN, L'Histoire des sciences aux Archives Nationales de Paris, dans Archives internationales d'histoire des sciences III, 13 (1950), p. 874-881.— M.-J. IMBAULT-HUART, Sources de l'histoire de la médecine aux Archives Nationales de 1750 à 1822, dans Revue d'histoire des sciences XXV (janvier-mars 1972), p. 45-53. Les recherches seront, du reste, prochainement facilitées grâce à une publication claire, systématique et à jour: Archives Nationales. Etat général des fonds (Paris, à partir de 1978, 4 vol.)

Pour la Section Ancienne (avant la Révolution), l'historien des sciences sera utilement guidé vers des séries qu'il n'aurait peut-être pas eu l'idée de consulter. Ne pouvant ici tout énumérer, citons à titre d'exemples:
K. "Monuments" historiques (comptes, hôpitaux, métiers, sorcellerie).
M. Mélanges (ordres hospitaliers, universités, collèges, affaires militaires et maritimes, hôpitaux, papiers scientifiques notamment ceux du P. Truchet).
O 1. Maison du Roi (services médicaux de la Cour, inventions et encouragements aux savants, institutions scientifiques comme l'Observatoire et le Jardin du Roi, machine de Marly, pompe de Chaillot, etc.).
T. Séquestres, surtout T 160, papiers de Louis Henri Duchesne (plusieurs cartons concernent la Société libre d'émulation de France pour l'encouragement des arts, métiers et inventions).— T 1116. Cassini, etc.
X 1. Parlement civil, enregistrement des privilèges ancêtres de nos brevets d'invention (voir aussi la série E. Conseil du Roi).
Dans les séries judiciaires (X, Y, Z): informations sur les médecins attachés auprès des juridictions et sur l'état sanitaire des prisonniers.

A la Section Moderne, une énorme masse de documents provient des ministères et des administrations en dépendant (cf l'Etat sommaire des versements

dont la table alphabétique a paru en 1972). Voir notamment les sous-séries F 4.
Comptabilité (de divers établissements scientifiques, pensions).— F 8. Police
sanitaire.— F 10. Agriculture.— F 12. Commerce et industrie (très importante
pour les techniques au XVIIIe siècle).— F 13 et F 21 pour les bâtiments
civils abritant des institutions scientifiques.— F 14. Travaux publics.—
F 15. Hospices.— F 17. Instruction publique.— F 20. Statistiques (notamment
médicales et hospitalières, épizooties, observations météorologiques).—
F. 70. Ministère d'Etat du Second Empire (Académies, encouragements, etc.).
— F 80. Algérie.

En ce qui concerne la sous-série F 17, la plus importante pour lui,
l'historien des sciences a beaucoup à puiser dans le tout récent volume de
M.-E. ANTOINE et S. OLIVIER, Inventaire des papiers de la division des sciences et lettres du Ministère de l'Instruction publique, t. I (Paris, 1975).
Le tome 2, promis pour 1978, va inventorier les archives des Caisses de Recherches Scientifiques qui ont précédé l'actuel C.N.R.S.

Il y a maints autres documents concernant l'histoire des sciences et des
techniques sous la Révolution (C, D, A F II, F 7, W), sous l'Empire
(A F IV, O 2) et même sous la Restauration (O 3 et B B pour les troubles
provoqués par l'emploi des machines). Des autographes de savants et de médecins se trouvent dans la série A A (surtout A A 40 à 46 et 63 à 65). Mais il
faut, plus encore, attirer l'attention sur divers fonds remis aux Archives
Nationales: A J 2. Maison Nationale de Charenton.— A J 15. Muséum d'histoire
naturelle.— A J 16. Académie de Paris.— A J 18. Académie de Strasbourg
(documents récemment renvoyés à Strasbourg).— A J 52. Ecole des Beaux-Arts.—
A J 61. Ecole Normale Supérieure.

En quelques lignes, nous ne pouvons rendre compte ici des dépouillements
effectués pour les factums et prospectus de la série A D, pour les archives
privées (A P: cf. Etat des fonds par Ch. de TOURTIER-BONAZZI et S. d'HUART.
Paris, 1973), pour les dossiers techniques des archives d'entreprises (A Q),
pour les acquisitions diverses de A B XIX, pour enfin les "cartes et plans"
qui comportent aussi des dessins techniques.

Pour ce qui est du Minutier central, sera établie, grâce à la collaboration
de Mme Jurgens, une typologie des documents notariaux concernant les sciences,
les techniques et la médecine, notamment aux XVIIe et XVIIIe siècles (contrats
d'apprentissage et d'enseignement, marchés de soins pour des malades, conventions pour l'exploitation de secrets techniques ou pour la mise en valeur
d'eaux minérales, informations sur les manufactures, inventaires de bibliothèques, de pharmacies, de collections scientifiques, etc.). Les recherches
du professeur américain Roger Hahn illustrent bien ce que le Minutier peut
apporter à l'histoire sociale du milieu scientifique français du XVIIIe siècle.

Dans les archives de la Marine déposées aux Archives Nationales, plusieurs
séries touchent directement l'histoire des sciences et des techniques:
— avant 1790 (cf. E. TAILLEMITE, Les arch. anc. de la Marine, Paris, 1961):
C 2, ingénieurs officiers de santé.— D 1, constructions navales.— D 2, travaux
hydrauliques.— G, mémoires scientifiques (cf. catal. des mss. de la Marine).
— après 1790 (cf. E. TAILLEMITE, dans Mémoires de l'Académie de Marine, année
1964, p. 189-230): C C 2, personnels civils et corps associés.— D D 1, constructions navales.— G G 1, mémoires et projets.— A quoi s'ajoute le très important dépôt du Service central Hydrographique dont certains documents remontent
au XVIIe siècle, par exemple des journaux de bord.

Etroitement lié à la Marine se trouve enfin le fonds des colonies jusqu'en
1815 (cf. E.TAILLEMITE, dans Gazette des archives, 3e trim. 1964, p. 93-116):

documents concernant la géographie, les sciences naturelles et la médecine
tropicale (surtout dans la collection Moreau de Saint-Méry, pour les XVIIe
et XVIIIe siècles).

NOTICES SUR DIVERS AUTRES FONDS D'ARCHIVES INTERESSANT L'HISTOIRE DES SCIENCES
(en dehors des Archives Nationales).

Lorsqu'une institution verse normalement ses papiers aux Archives Nationales,
elle apparaît, à ce titre, dans notre guide des Archives Nationales, sans
qu'il y ait théoriquement lieu de lui consacrer ensuite une notice particulière. Si en effet la préparation du Guide doit provoquer une salutaire sensi-
bilisation à l'intérêt des archives scientifiques, il serait catastrophique
que cette prise de conscience conduise des services publics à retenir leurs
papiers au lieu de les confier aux Archives Nationales, comme c'est leur devoir.

Faute de place, nous ne pouvons guère fournir ici qu'une assez sèche énumération des notices provisoires déjà rédigées: pour chacune, le nom du rap-
porteur apparaîtra ci-dessous, entre parenthèses, après la mention de
l'institution considérée.

SECTION OUTRE-MER administrativement rattachée aux Archives Nationales
(Mme Monique POULIQUEN). Documentation postérieure à 1815, continuant donc
celle déposée aux Archives Nationales (cf. ci-dessus). Les indications concernant l'histoire des sciences sont assez dispersées: exemple typique de
l'utilité du Guide.

ARCHIVES DE PARIS ET DE L'ANCIEN DEPARTEMENT DE LA SEINE (Mme Françoise
JENN): surtout intéressantes pour les techniques plus ou moins liées à
l'urbanisme.

UNIVERSITE DE PARIS (Mme Paule RENE-BAZIN). Pour avant 1940, les papiers
de l'Université de Paris vont, en principe, à la série A J 16 des Archives
Nationales. Pour la période postérieure une mission des Archives Nationales
tente d'organiser un préarchivage dans les nouvelles universités. Sauf exceptions (Institut de Physique du Globe, par exemple) les papiers de laboratoires échappent malheureusement encore à ces mesures de sauvagarde. M. Jean
THEODORIDES a attiré notre attention, toujours dans la mouvance de l'Université de Paris, sur l'intérêt que présentent, pour l'histoire de la biologie
et de la zoologie à la fin du XIXe siècle, les papiers de Henri LACAZE-DUTHIERS
(laboratoires de biologie marine de Banyuls-sur-mer et de Roscoff). Quant à
la FACULTE DE MEDECINE (Mme M.-J. IMBAULT-HUART), certains de ses documents
les plus anciens sont à la bibliothèque de la Faculté. Le reste de ses archives (assez malmenées) doit rejoindre la série A J 16 des Archives Nationales.

ACADEMIE DES SCIENCES (M. P. BERTHON): procès-verbaux depuis 1666, dos-
siers de seances, dossiers personnels, prix et fondations, plis cachetés.
Des documents détournés se trouvent dans le fonds Libri de la Bibliothèque
Nationale (nouv. acq. fr. 3133 à 3153).

COLLEGE DE FRANCE (Marguerite OSWALD): archives de l'administration centrale du Collège. Les laboratoires y apparaissent, mais ils gardent leurs
propres papiers. Le fonds Claude Bernard fait l'objet d'un inventaire imprimé dû au Dr M. D. GRMER (Paris, 1967).

MUSEUM D'HISTOIRE NATURELLE (M. Yves LAISSUS): archives écartelées entre
la série A J 15 des Archives Nationales, les services de la Direction du
Muséum et la bibliothèque. Distinction floue entre papiers des laboratoires
et papiers personnels des chercheurs.

OBSERVATOIRE DE PARIS (Mme Geneviève FEUILLEBOIS). Voir surtout M. G.
BIGOURDAN, Inventaire général et sommaire des manuscrits de l'Observatoire

de Paris, dans Journal for the History of Astronomy VI (1975), p. 72-74 et aussi le catalogue de l'exposition Observatoire de Paris: trois siècles d'astronomie (Paris, 1967).

CONSERVATOIRE DES ARTS ET METIERS (M. Jacques PAYEN). Papiers de Vandermonde et de Cl. P. Molard, récompenses pour la promotion des arts et de l'industrie (1791-1796), vie du Conservatoire pendant la période 1794-1840: classement méthodique en 40 séries.

SERVICE DES INSTRUMENTS DE MESURE (M. l'ingénieur divisionnaire L. MARQUET).

ECOLE NORMALE SUPERIEURE. Ses papiers ont été versés aux Aechives Nationales où ils constituent la série A J 61. Mais, depuis, M. Petitmengin a retrouvé les archives du laboratoire de chimie, en fait celles de Henri Sainte-Claire-Deville: manuscrits, cahiers de laboratoire (1843-1880), correspondance reçue d'illustres savants entre 1855 et 1881 (inventaire par M. CAZACU). Un travail est en cours sur les archives de la "Bibliothèque de mathématiques".

ECOLE DES MINES (Paulette ENJOLRAS): Archives administratives de l'Ecole, registres d'analyses du Bureau d'essais (1826, 1830), comptes-rendus de voyages des élèves (1826-1887). Ajouter au catalogue imprimé en 1914, nombre de manuscrits datant du XVIIIe S. et du début du XIXe.

ECOLE POLYTECHNIQUE (M. A. MOREAU): registres des divers conseils, dossiers par promotions. Pour les origines, cf. M.-E. ANTOINE, Les archives de l'Ecole Polytechnique, dans Bulletin d'histoire économique et sociale (année 1967), p. 51-62.

SERVICE HISTORIQUE DE L'ARMEE DE TERRE (M. Claude DEVOS).

SERVICE HISTORIQUE DE L'ARMEE DE L'AIR (Mlle Marie-Hélène DEGROISE). Remarquer la série A concernant l'aviation pendant la guerre 1914-18 et, dans la série B, la correspondance technique avec la maison Dewoitine (1923-26).

ARCHIVES DE L'ARTILLERIE (Mlle Chr. LACOMBE): d'un très grand intérêt scientifique et technique pour la période 1795-1910 (Ecoles, études et expériences, inventions et projets).

DEPOT DES FORTIFICATIONS aux Archives et bibliothèque du Génie (Mlle LACROCQ): depuis l'époque de Vauban, plus de 1100 volumes manuscrits (atlas, cours des écoles du Génie, etc.).

CEDOCAR = Centre de documentation de l'Armement (Ingénieur général RENAUD) et CENTRE D'ARCHIVES DE L'ARMEMENT à Chatellerault (C F CHOPARD). Dans ce dernier dépôt, nombreux fonds d'archives de services, établissements publics ou sociétés privées ayant travaillé ou travaillant pour l'armement (chaque année, état sommaire multicopié).

VAL DE GRACE (médecin général A. FABRE): archives concernant le Service de Santé militaire de 1700 à 1920; fonds spécial pour la guerre 1914-18.

ASSISTANCE PUBLIQUE (M. Marcel CANDILLE). Archives remontant au XIIe S.; informations assez dispersées concernant notamment la démographie sanitaire. Ce rapport a été publié: M. CANDILLE, Contribution éventuelle du fonds de l'Assistance publique à Paris à l'histoire des sciences et des techniques, dans Société française d'histoire des hôpitaux, Bulletin 31 (1975), p. 40-48.

INSTITUT PASTEUR (M. Henri PERRIER): archives administratives de l'Institut, papiers scientifiques des sociétés dont l'Institut Pasteur est le siège, archives historiques du Musée Pasteur. Mais divers services et laboratoires restent maîtres de leurs papiers.

MANUFACTURE NATIONALE DE SEVRES (Mme Tamara PREAUD): techniques et secrets de la porcelaine depuis le XVIIIe siècle.

Parmi les autres établissements conservant des archives susceptibles

d'intéresser l'histoire des sciences, il en est un certain nombre pour lesquels
des notices sont en préparation ou, au moins, des contacts ont été pris.
— Archives du Ministère des Affaires Etrangères (Mlle Paulette Enjalran).
— Archives de la Marine (MM. Joël Audouy et Erik Le Maresquier).
— Institut National de la Propriété Industrielle: sur l'importance de ses
collections du XIXe siècle, voir Ministère de l'Industrie et du Commerce.
Brevets d'invention français 1791-1902: un siècle de progrès technique
(Paris, 1958). Depuis la loi de 1902 prescrivant la publication intégrale
des brevets, les originaux sont habituellement détruits au bout de trente ans.
Cependant, ceux des années 1944-47 sont actuellement déposés dans l'annexe
des Archives Nationales sise à Fontainebleau.
— Institut géographique national.
— Institut de France et Académie Française.
— Académie de Médecine. Une partie de ses archives se trouve actuellement,
pour y être classée, dans l'annexe des Archives Nationales à Fontainebleau.
— Ecole des Ponts et Chaussées (Mme Anne Samba).
— Ecole Vétérinaire d'Alfort (Mlle Grison).
— Musée de l'Air (M. Y. Kayser).
— Bureau international des Poids et mesures.
    D'autres démarches et prospections restent à entreprendre (Bureau des
Longitudes, Institut d'Optique, sociétés savantes, petits musées parisiens
consacrés à des savants, Chambre de Commerce, dossiers techniques d'entre-
prises n'ayant pas déposé leurs papiers aux Archives Nationales, etc... etc...

REFLEXIONS INSPIREES PAR CETTE EXPERIENCE.
    Ces réflexions pourraient s'inscrire sous trois rubriques: efforts de sen-
sibilisation, obtention de moyens, rêve d'une meilleure coordination des
bonnes volontés.
    Il faut d'abord mieux mettre en valeur ce que les archives apportent à
l'histoire des sciences: ainsi pourra-t-on espérer convaincre les actuels
chercheurs scientifiques qu'ils s'insèrent dans une histoire et que cette
histoire aura besoin de leurs archives pour être écrite. Réapprenons tous
que nous nous situons dans une continuité!
    Si une telle prise de conscience évite d'irréparables destructions, elle
ne doit pas, non plus, se traduire par un trop possessif amour des documents:
à long terme, la meilleure sauvegarde reste la conservation dans un grand
dépôt public comme les Archives Nationales.
    Les avantages d'une certaine concentration sont manifestes, mais ils ont
des limites. S'il est, en effet, relativement aisé de récupérer les archives
de la gestion bureaucratique des centres scientifiques, il s'avère en revanche
très difficile de sauver les papiers des laboratoires: les chercheurs consi-
dèrent trop souvent de tels documents comme leur bien personnel. Or, lorsque
le souci de leur gloire l'emporte sur la négligence ou sur l'appréhension
d'apparaître à nu devant la postérité, les savants préfèrent naturellement
laisser leurs papiers à une institution scientifique prestigieuse, plutôt
qu'à un dépôt fourre-tout. Longue vie donc aux petits dépôts d'archives des
établissements scientifiques anciens et glorieux comme le Collège de France
ou l'Académie des Sciences! Je ne crois pas, pour la France, à la nécessité
de créer un centre spécial d'archives scientifiques et techniques.
    Les tâches archivistiques requièrent de l'abnégation et des moyens: elles
exigent plus que ne peut habituellement donner un universitaire même enthou-
siaste. Tout effort sérieux trouvera donc avantage à passer par l'organisa-

tion et le personnel des archives et bibliothèques, même si leurs moyens apparaissent misérables par rapport aux énormes tâches qui leur incombent. Insistons, au passage, sur le profit qu'il y aurait à faire entrer un plus grand nombre de jeunes historiens des sciences dans le personnel des Archives et bibliothèques.

La préparation de notre "guide" n'a reçu du C.N.R.S. qu'une aide chiche et mesquine; l'achèvement se fait difficilement et non sans amertume. Ceci n'encourage évidemment guère à étendre l'entreprise. Le Comité des Travaux Historiques et Scientifiques a cependant décidé de mettre au programme des prochains "Congrès des Sociétés Savantes" la question des sources de l'histoire des sciences dans les archives de province (celles non couvertes par notre Guide).

D'autre part, il faudra, de plus en plus, prendre en considération l'immense documentation scientifique multicopiée qui se situe à mi-chemin entre archives et bibliothèques. Tout au plus veux-je rappeler que ces problèmes de documentation scientifique font, à l'Unesco, l'objet du projet UNISIST: j'ai suggéré d'y intégrer l'histoire des sciences, au moins dans la mesure où ses sources fournissent des données utilisables par les savants et les techniciens actuels (astronomie, climatologie, prospection minière, épidémiologie et "pathocénose", écologie, etc.).

Au sein même de l'Union Internationale d'Histoire des Sciences, une meilleure coordination des commissions ne serait pas inutile, si l'on veut au moins — peut-on faire plus ? — reconnaître les bonnes orientations et les recommander avec une certaine autorité morale.

V.A.Volkov, Yu.Kh.Kopelevich, B.V.Levshin

ARCHIVES SERVING THE HISTORY OF SCIENCE IN THE USSR

Sometimes the work of the historian of science is compared to that of paleontologist. Both have to make use of scattered, incomplete materials to reconstruct the entire picture of a more or less remote epoch in science. The very nature of the historic research requires continuous scanning of new documents which make it possible to fill in the essential gaps in our knowledge or to change fundamentally the existing concepts concerning the historic past or, finally, to create new basis for broad historic and scientific generalisations (1).

The most ancient depositories of the written relics on the USSR territory date back to the middle of the first millennium B.C. There is some information on the existence of archive depositories of Kiev Russia Princes, in the states of Middle Asia. There appeared in the XVIth century in Moscow the Tsar Archive. But it was only at the time of·Peter the First that in the new state institutions created by him the historic documents were separated from the current paperwork and concentrated in the archives (at the same time the words 'archive' and 'archivist' came into use). In 1724 of the documents of the abolished Posol'skii Prikaz they made "The General Archive of Old State Affairs" - the first historic Russian archive (it was called later "The Moscow Archive of the Ministry of Foreign Affairs". The most important act for the history of science in the USSR was the creation in 1728 of the Archive in the St Petersburg Academy of Science. But besides this archive, historians of science make use of the materials of other archives that appeared during XVIIIth and XIXth centuries, namely those at the Depot of Maps, Berg-Board, Military, Land-Survey Administration, the Geographical Society, the architectural and engineering institutions, the individual ministries and committees.

The October Socialist Revolution introduced radical changes in the organisation of the archives in our country. The decree of June 1, 1918 signed by V.I.Lenin and named "O reorganizatsii i tsentralizatsii arkhivnogo dela v RSFSR (2)" subordinated all the archives of the former state offices to a single supervisory organ - The Main Administration of Archives. All of them together constituted the Unified State Archive Fund. Its individual parts were later used to form 11 central state archives of the USSR subordinated to the Main Archival Administration of the Council of Ministers of the USSR (GAU). Those include the most intensively used by the historians of science Central State Archive of Ancient Acts (TsGADA), Central State Historical Archive of the USSR (TsGIA SSSR), Central State Archive of the Navy of the USSR (TsGAVMF SSSR), Central State Archive of the October Revolution, High Organs of State Government and Organs of State Administration of the USSR (TsGAOR SSSR), Central State Archive of the National Economy of the USSR (TsGANKh SSSR), Central State Military History Archive of the USSR (TsGVIA SSSR), Central State Archive of Film, Photo and Phonographic Documents of the USSR (TsGAKFD SSSR). In some cases, especially when one studies the biographical materials concerning scientists, the researchers use the central archives of the Union republics, state and county archives which are governed by the Archival administrations of the Councils of Ministers of the republics and the Archival Departments of the Executive Councils Committees of the Workers' Deputies. The documents of the republican and local archives are of special significance in the formulation of the history of development of science in individual regions of the Soviet Union. In addition to the state archives belonging to the system of GAU SSSR certain department archives are also preserved; those include the archives of the academies of sciences of the Union republics, archives of the Geographical Society of the USSR, of the All-Union Geological Fund, the Central Cartographical Fund, the depository of the Hydro-Meteo-Service Administration and some others.

Of considerable significance for the acquisition of Soviet archives was Lenin's decree of July 29, 1919 concerning the abolishment of the right of private property over the archives belonging to deceased Russian writers, composers and scientists and deposited at that time in the libraries and museums (3). Certainly, this decree did not abolish the right of property of the scientist and his (her) heirs over the personal documents associated, in particular, with the scientific creativity.

The edict of the State Archival Fund of the USSR including all the documents that are on state depository in our country, which has been asserted by the decree of the Council of Ministers of the USSR of August 13, 1958 (4), foresees stocking of this fund by the personal archives that arrive at state depositories as a gift or on some other lawful grounds. The owners of personal archives give them to state archival institutions, libraries, museums which guarantee for them the unlimited right to use the transferred material. They also enjoy the right to allow or not to allow other persons to work with the documents of the fund.

The Soviet archives conduct enormous work on the use of documents deposited in them; one can judge the scale of this work from the following figures; during a five-year period (1971-1975) only the archival institutions of RSFSR enabled 81,000 researchers to become acquainted with 2.3 millions of files and 455,000 microcopies of documents (5).

The largest depository of documents concerning the history of science and technology in the USSR is the Archive of the USSR Academy of Sciences. The documents deposited in it reflect the history of Russian science, culture and education over the period of XVIII - XX centuries, accumulated as a result of the scientific and organisational activity of the Academy of Sciences, its institutions and expeditions, as well as the personal funds of academicians and other outstanding Russian scientists.

At present the Archive of the Academy of Sciences of the USSR contains above 1,000 funds with the total amount of 725,000 depository units. (6) The scientific correspondence is represented in the Archive on an especially broad scale. The academicians' correspondence on scientific topics was regarded in the XVIIIth - the beginning of the XIXth centuries as the property of the Academy and was collected in the Archive in a special fund which is a unique collection of documents about the scientific life of its time. Later the correspondence of scientists arrived in the Archive after their demise as a part of their personal funds. Thus, for instance, the fund of the academician-mineralogist V.I. Vernadsky deceased in 1945 contains about 16.5 thousand of letters from 1887 correspondents (7).

The use of the riches of the academic Archive for the historic and scientific research started as early as in XVIII century. But the number of the scientists working in the Archive before the October Revolution was small - not more than 30 per year, and nearly all of them were academicians. After 1917 the circle of the Archive readers gradually expanded. The scientific and publication activity of the Archive itself also intensified. From 1917 up to the beginning of the Great Patriotic War (World War II), in other words, over 23.5 years the reading rooms of the Archive saw about 1.5 thousand of readers, on the average about 80 a year; in 1957-1967 in the same reading rooms worked from 300 to 400 readers a year (8). In 1973 the number of readers reached 420 (among them 30 readers from abroad), in 1976 - 462.

This rapid expansion of the use of the Archive documents demonstrated not only the growing interest to the history of science in our country but the intensive information activity of the Archive which opens its riches before the masses of scientific workers and provides the key to their use.

From 1933 the Archive of the Academy of sciences of the USSR began publishing its works. The first publication was the description of materials on depository in the archive, which could be considered the first archival guide. By 1941 the Archive has already published 4 volumes of works. They contained a description of the manuscripts of the great Russian scientist M.V. Lomonosov (Proceedings, vyp. 3, 1937), a description of the materials of expeditions conducted by the Academy in XVIII - XIXth centuries (Proceedings, vyp. 4, 1940), a description of the scientific correspondence of the Academy of Sciences over 1766-1782 (Proceedings, vyp. 2, 1937). Every new issue attracted the attention of scores of researchers to the development of a new complex of materials. Thus, the issue "The Manuscript Heritage of M.V. Lomonosov" helped many authors in writing research papers concerning the life and work of M.V. Lomonosov. This description served as the basis for the preparation of 10-volume complete works of M.V. Lomonosov (1940-1965), publication of six volumes of collected papers and materials "Lomonosov" and for publications of documents devoted to this scientist. The description of M.V. Lomonosov's manuscripts was also used for creating the expositions of M.V. Lomonosov's memorial museum, for organisation of radio and TV broadcasts. The same Proceedings series publishes the volumes of "Reviews" that unify the descriptions of many funds. The six volumes of "Reviews" that have been published up to the present time provide a description of about 600 funds. The Proceedings of the Archive devoted to individual scientists who left large manuscript heritage open the richest material to the scholars. Let us give a few examples. The scientific description of I.I. Mechnikov's fund (1960) includes his manuscripts on zoology, embryology, phagocytosis, the theory of immunity, the medical microbiology, etc., the records of experiments, drawings, journals and note books, his letters to Russian and foreign scientists, scientific societies and organisations. The Appendix contains I.I. Machnikov's letters to the well-known Russian zoologist and anthropologist A.P. Bogdanov.

A similar description has been compiled about K.E. Tsiolkovsky's fund (1966). The publication gives the names of his manuscripts on aeronautics, aviation and rocket technology, astronomy and natural science and also on the issues concerning the problem of mastering the near-Sun space; it contains the descriptions of drawings and computations. The biographical materials, Tsiolkovsky's notebooks are present, as well as his vast correspondence; there is a reproduction of his "Album of Space Travels" which was never published before and which is devoted to the questions of behaviour and state of human in interplanetary space. This album is of special interest because many assumptions of the scientist were confirmed during the space flights. The above description of K.E. Tsiolkovsky's fund is widely used not only by those who investigate the documents of the pioneer of space research but it served as the documental foundation during organisation of a number of departments in the Space Navigation Museum of K.E. Tsiolkovsky in Kaluga and for creation of films and TV shows about K.E. Tsiolkovsky.

The description of Leonard Euler's manuscripts (1962) published in the Archive Proceedings formed the basis of a large publication work about the fund of this outstanding academician of the XVIIIth century and of numerous research papers that appeared in Soviet and Western journals and volumes of collected papers. As a result of further study of Euler's epistolary heritage, in 1967 an annotated index of his correspondence preserved all over the world was published - a description of more than 3 thousand letters (2273 of them being from the Archive of the Academy of Sciences of the USSR) with a brief abstract of the contents of each letter. This book served as a foundation for a new, completed description published together by the Soviet and Swiss Eulerologists in Basel in 1975, which in its turn will form the basis for the multivolume edition in Euler's correspondence in the last series of his "Opera omnia". A part of this correspondence which reflects the activity of and connections between them St Petersburg and Berlin Academies - 872 letters in all - was published in Berlin by the Soviet and DDR scientists in three volumes of the series "Die Berliner und die Petersburger Akademie der Wissenschaften im Briefwechsel Leonhard Eulers" (1959, 1962, 1976). The same group published again the incomplete XIXth century edition that lacked commentaries of Euler's correspondence with Christian Goldbach (Berlin, 1965).

One of the peculiar features of the Archive of the Academy of Sciences of the USSR operation during the last decades inasmuch as the publication of its funds' descriptions is concerned is the fact that in addition to the Archive's personnel the distinguished specialists in corresponding sciences are invited to do this work, and this allows to penetrate deeper into the contents of the manuscripts and estimate their significance.

Giving on a large scale its materials to researchers, the Archive of the Academy of Sciences of the USSR conducts a consistent publication work itself. Thus, for instance, Archive published the "Letters of A.O. Kovalevsky to I.I. Mechnikov" (1955). The correspondence covering the period of 1866-1900 and containing 185 letters throws some light for the first time on certain important aspects of the scientific activity of these scholars. The second publication - "Letters of V.G. Khlopin to V.I. Vernadsky" (1961) reveals the creative plans of these outstanding scientists; it contains the little known information about their life and scientific activity, on the development of radiochemistry and radium industry in the USSR.

In connection with the preparation for the 100-year anniversary from V.I. Lenin's birthday Archive published a collection of documents "Lenin and the Academy of Sciences" (1969) edited by the academician P.N. Pospelov. Those who compiled this collection managed to unify both the complete Lenin's heritage that was published before and the unpublished documents, letters and memoirs of scientists and statesmen who reveal Lenin's role in the attraction of the Academy of Sciences to cooperation with the Soviet rule and in the organisation of scientific research.

Archive has published by the 250th anniversary of the USSR Academy of Sciences a two-volume collection "The Academy of Sciences of the USSR. Personnel" (1974) that contains a list of 4096 members of the Academy of Sciences, with pictures of scientists, and also a collection of documents "The Statutes of the Academy of Sciences of the USSR" (1974) including all regulations that led the Academy of Sciences over the period of its activity.

In 1975 the Archive of the Academy of Sciences of the USSR and the Archive of the Academy of Sciences of DDR published a joint collection of documents "Russian-German Scientific Connections between the Academy of Sciences of the USSR and the Academy of Sciences of DDR. 1700-1974".

Beginning with 1967 the Archive of the Academy of Sciences of the USSR publishes a handbook "The Topics of Research on the Documents of the archives of the USSR Academy of Sciences". The purpose of this handbook was to inform the interested institutions and persons about the work that started on the basis of the archival documentation of the Academy of Sciences of the USSR. It is especially valuable because it complements essentially the data of the organs of scientific information giving also the information on the research work that is not included in the State Plan. The publication of such handbooks in the Soviet Union guarantees the legal aspects of the studies on the history of science as well since it protects the priority of scientists in the formulation and development of a given topic. This edition acquaints the reader with monographs, theses, scientific reports, textbooks, films, dramas, expositions, diploma works, fiction books, the authors of which made use of the materials of the archives of the USSR Academy of Sciences, its Siberian department and its branches. The above handbooks make it possible to form a judgement concerning the ever growing use of the archival documents with the scientific, cultural, educational and other objectives in mind. Thus, while the first issue contained a systematic list of 1752 topics on various scientific and cultural disciplines and included information on 1602 researchers studying the documents in 31 archives of the Academy of Sciences of the USSR in 1967, in 1976 the amount of research works increased up to 2104, and that of the scientist - to 2039 persons. About a third of this number constitutes the studies on the history of science.

From 1971 the Archive of the Academy of Sciences of the USSR publishes a bibliography handbook "The Archives of Academies of Sciences of Socialist Countries". The first issue of this handbook (1971) contains a description of more than 1,000 publications of the archives of Academies of Sciences of the USSR, PRB, PRH, DDR, PPP and ChSSR. This is a retrospective annotated bibliography (the first issue contains the bibliography over the period 1917-1968, with the Soviet part constituting 552 titles). The second issue (1975) gives the bibliography over the period 1969-1972.

The Archive of the Geographical Society of the USSR is the oldest and the only one in the country special geographical archive; at present it has on depository more than 60 thousand of manuscripts and other documents. Here one can find manuscripts on geography, history and the associated sciences, documents of numerous expeditions organised by the society, and also manuscripts on folklore, ethnography and linguistics of the people of the USSR and foreign countries. Among the materials of the Archive one can single out the vast fund of the Society's office the acquaintance with which recreates the history of the Geographical Society over the 130 years of its existence.

The personal funds of the famous Russian travelers and scientists - P.P. Semenov-Tyan-Shansky, N.M. Przheval'sky, N.N. Miklukho-Maklai, P.K. Kozlov, G.Ya. Sedov, V.A. Obruchev, L.S. Berg, Yu.M. Shokal'sly etc. - represent an outstanding scientific value.

Of extreme interest is the collection of the rarest photographs, negatives, slides systematised according to parts of the world and countries, the collection of the pictures of the Geographical Society members and other scientists, collections of drawings, etchings, lithographs and post cards.

The most ancient materials of the Archive (acts, manifestos, petitions, edicts) date back to XVI-XVIIIth centuries, but the major part of them pertains to the middle of the XIXth century.

The Archive has published a description of the personal funds of Russian geographers and travelers (1971), as well as a description of collections of manuscripts (1973).

The Central State Historical Archive of the USSR (TsGIA SSSR) in Leningrad represents the largest in the USSR depository of documents of the high organs of the state power and administration of Russia covering the period from XVIIIth century up to 1917; it also keeps the numerous documents on the history of Russian science and technology. Several reference editions have been published that help to find the proper direction in the documental materials on the history of science and technology. Among these editions there is the archival guidebook (1956) annotating the materials of the scientific societies and, in particular, of the Russian technological Society, and also of the personal funds. Among the topic reviews published by the Archive of the processing industry in Russia the first issue of which (1957) covers the first half of the XIXth century, and the second (1962) - the period from 1864 till 1914. The review describes the files and documents concerning the technological innovations, the equipment of the industrial plants and the improvement of the industrial technology, the production and introduction into industry of new devices. There is information on the organisation of industrial exhibitions in Russia, as well as on participation of scientific institutions and scientists in international exhibitions.

The review on the history of medicine and health care in Russia (1958) contains data on the work of clinics, medico-biological scientific research institutes and societies and the outstanding scientists N.V. Sklifosovsky, V.M. Bekhterev and so on belonging to them.

The review of the material on the history of communications (1966) discusses the development and improvement of various types of communications (postal communications, telegraph, telephone, etc.).

Leningrad State University together with TsGIA SSSR and the State Historical Archive of the Leningrad oblast' published in 1961 a review of archival documents "Materials on the History of Leningrad University, 1819-1917", and in 1965 - a collection of papers devoted to the 250th anniversary of one of the oldest enterprises of our country Okhta Chemical Group (9).

Of considerable interest from the point of view of studying the history of Russian science and technology is the personal catalogue of the college teachers compiled by the workers of TsGIA SSSR (according to the funds of the Ministry of People's Education). The catalogue takes into account the reports on the scientific travels abroad, autobiographies, lists of publications, the documents on professorship appointments, comments on scientific research papers, etc. They have also completed the preparation of the catalogue "Science and Scientific Organisations", reflecting the documents about the activity of scientific societies, committees and scientific institutions, about participation of Russian scientists in international scientific congresses, about the publication and censorship of scientific literature, and so on. The archival catalogue reflects in detail the history of giving privileges for innovations. When one works with the archival documents, it is important to know whether one or another document is introduced in science for the first time or has been used already by the previous investigators. To clarify this question, there is a very useful bibliographical index published by TsGIA SSSR: "The Documents of TsGIA SSSR in the Papers of Researchers". Its first issue (1960) covers the papers and publications in Russian in periodical and continuous editions over 1917-1957; the second (1966) also takes books into account, chronologically it follows the first one (1958-1962).

The funds of the Central State Archive of Ancient Acts (TsGADA) has on depository interesting documents concerning the organisation of scientific expeditions in XVII-XVIIIth centuries, including the Second Kamchatka expedition of V. Bering (1732), the work of M.V. Lomonosov, L. Euler, Zh.N. Delil', A.K. Nartov, I.P. Kulibin and other distin-

guished scientists. The Archive published a collection of documents
"Discoveries of Russian Travelers and North Sailors of the XVIIth century
at the North-East of Asia" (1951).

Funds of the Central State Military History Archive of the USSR (TsGVIA
SSSR) contain documents on the history of aeronautics, aviation, engineering, construction of fortresses and engineering objects, production of
Russian side and fire-arms. In 1956 Glavarkhiv SSSR and TsGVIA SSSR
together with the Soviet National Unification of the historians of
science and technology published a collection of documents on the history
of aeronautics and aviation in Russia up to 1907, and in 1966-1972 - six
issues of collections of documents and materials "Aviation and Aeronautics in Russia in 1907-1914".

The Central State Archive of the Navy of the USSR (TsGAVMF SSSR) keeps
the funds of the central apparatus of pre-revolutionary Navy, Naval
colleges, research institutions, shipyards, admiralties and scientific
expeditions. The archive also has on depository the documents concerning
the activity of the outstanding Russian admirals, researchers, sailors
and scientists.

In 1965 TsGAVMF SSSR published a collection of documents "The Submarine
Construction in Russia. 1900-1917" which covered 104 documents reflecting
the major stages of development of projects and construction of submarines
in the pre-revolutionary Russia. The book gives for the first time the
original schemes, drawings and photographs of submarines.

The documents of certain pre-revolutionary institutions, collections of
manuscripts, autographs of scientists, etc. are concentrated in the
Manuscript Departments of V.I. Lenin State Library of the USSR, the State
Public Library of M.E. Saltykov-Shchedrin, the State Historical Museum,
the Republican and Oblast' museums and libraries, the memorial museums
(the Scientific Archive-Museum of D.I. Mendeleev at Leningrad State
University, K.E. Tsiolkovsky State Museum of Space Travels in Kaluga,
and so on). There are guidebooks for the researchers working in these
institutions. Besides, V.I. Lenin State Library of the USSR, for instance,
publishes from 1938 "Zapiski Otdela Rukopisei Gosudarstvennoi Biblioteki
im. V.I. Lenina" (the current 37th issue was published in 1976) the
purpose of which is to familiarise the scholars with the primary sources
on depository, as well as with the new material or the documents discovered during the fund processing and representing the scientific
interest.

The information on the personal funds of scientists kept in more than
300 state archives, libraries and museums is placed in the summary
index "The Personal Archival funds in the State Depositories of the
USSR"(Vol. I-II, Moscow, 1962, 1963) prepared by Glavarkhiv SSSR, V.I.
Lenin's State Library of the USSR and the Archive of the Academy of
Sciences of the USSR.

The unique sequences of films and photographs of the polar explorer
Sedov's expedition to the North Pole (1912), ice travels of icebreakers
"Sibiryakov", "Malygin", "Yermak", "Sadko", "Krasin", "Chelyuskin", the
construction of first Soviet electric power stations and plants, the

film documents about the celebration of the 200th anniversary of the Academy of Sciences of the USSR (1925), films of I.V. Michurin (1929), K.E. Tsiolkovsky (1931), meetings of the academician I.P. Pavlov with the British writer H. Wells (1935), photographs and documentary film sequences about the work of the Academy of Sciences of the USSR sessions, of its institutes and scientists and many other film and photographic documents are on depository in the Central State Archive of Film and Photo Documents of the USSR. There is an essay-guidebook over the Archive (1961).

Numerous documents about the history of Soviet science and industry are kept in the Central State Archive of the October Revolution, High Organs of State Government and Organs of State Administration of the USSR (TsGAOR SSSR), the Central State Archive of the National Economy of the USSR (TsGANKh SSSR) and the Central State Archive of RSFSR (TsGA RSFSR).

TsGAOR SSSR keeps documents that throw light on the activity of the high organs of the state government and the state administration of the USSR concerning the supervision of high and intermediate special education and the training of scientific personnel.

The funds of TsGANKh SSR contain materials reflecting the history of creation and development of the socialist industry and giving an idea of the scientific and technological progress in the USSR. Many documents tell about the important scientific and technological under-takings of the first years of the Soviet government, including the development of the plan of electrification of Russia, chemisation of the national economy, elucidate the history of creation of the leading branches of the USSR industry.

In the fund of the Main Administration of the Scientific, Museum and Arts Institutions (Glavnauka) which is kept in TsGA RSFSR there are documents about the establishment of the Soviet science: the reports of Glavnauka and scientific research institutes, journals and records of proceedings of the sessions of the institutes' Scientific Councils, documents concerning the preparation of scientific and technological personnel, etc.

An efficient way of informing the scientific community and the researchers about the documents on the history of science and technology kept in the State archives and the simultaneous bringing of these documents into the scientific turnover was found in the form of publishing the collections of documents which are prepared by archives together with the interested scientific organisations. Let us list a few of them.

In 1956 the Institute of Economics of AN SSSR and TsGAOR SSSR published a collection of documents "The Development of Electrification of the Soviet Country. 1921-1925".

A.F. Joffe Physical and Technical Institute and the Archive of the Academy of Sciences of the USSR published the correspondence between the academician A.F. Joffe and a Dutch theoretical physicist F.

Erenfest (10) that contained interesting information about the life and work of two outstanding physicists, about the development of physics research in the USSR.

The Institute for the History of Science and Technology of the Academy of Sciences of the USSR together with the Archive of the Academy of Sciences of the USSR prepared collections of documents "The Organisation of Science during the first Years of the Soviet Government (1917-1925)" (1968) and "The Organisation of the Soviet Science in 1926-1932" (1974). The collection makes use of the materials of many Soviet archives, including TsGAOR SSSR, TsGANKh SSSR, TsGA SSSR. The publication of the above collections initiated the creation of the documentary basis for studying the diversified problems of the State supervision of science.

Glavarkhiv SSSR, the Institute of Marxism-Leninism at the Central Committee of the Communist Party of the USSR, the Institute of History of the USSR and the Institute of Economics of the AN SSSR together with TsGANKh SSSR published a series of documentary collections in four volumes "The History of Industrialisation of the USSR, 1926-1941" (1969, 1970, 1971, 1973). There are also a number of collections of documents concerning the history of industrialisation of the major economic regions of the USSR.

On May 21, 1964 the Council of Ministers of the USSR adopted a resolution "On Centralisation of Keeping the Scientific and Technological Documentation and on organisation of its Wide Use" (11). This resolution the aim of which was to guarantee the safety and the organisation of the use of documents that have been formed as a result of the activity of the R&D and other institutions, organisations and enterprises created the Central State Archive of the Scientific and Technological Documentation of the USSR concentrating the technical documentation of the all-Union significance; it also created the network of central state archives of scientific and technological documentation of the Union republics and the departments of scientific and technological documentation of the State archives of the autonomous republics, regions and counties.

A number of research institutions, in particular, the All-Union Scientific Research Institute for Documentation and Archival Affairs (VNIIDAD), organised in Moscow in 1966, and also the chairs of the Moscow State Historical and Archival Institute involved in the study of archives are engaged in the development of the problems of keeping and using the documentary materials of the archives. The Chair of the Scientific and Technological Archives of this Institute published a textbook which presented the principles of organisation of keeping the scientific, technical, film and photographic documents in the archives (acquisition of archives, the systematisation, description and taking into account documents, the scientific reference apparatus of the archives, etc.). Various questions of the use of the documentary materials of the archives with scientific purposes in general and in the interests of the history of science and technology in particular are elucidated in the journal "Sovetskiye Arkhivy" ("Soviet Archives") which is being published by Glavarkhiv SSSR from 1966. Earlier the journals "Krasnyi Arkhiv" (1922-1941), "Arkhivnoye Delo" (1923-1941), "Istoricheskii Arkhiv" (1955-1962), "Voprosy Arkhivovedeniya" (1959-1965) were published.

In November 1976 in Moscow the All-Union Scientific Conference was held that was concerned with the questions of scientific and information activity of the State archives of the USSR, with summarising the work for the last five-year period and outlining the paths of further improvement of the organisation of the overall use of documents (12).

To make a more efficient practical and scientific use of documents, VNIIDAD works on the development of an automated system of scientific and technical information on the documents of the State Archival Fund of the USSR. The system will allow to quickly obtain the retrospective information concerning science and technology in Russia and in the USSR which is contained in the documents of all State archives in the USSR.

In order to help the researchers working in the Archives, the USSR conducts the development of the theory and methods of archival research, publishes textbooks on archival heuristics, for example, "The Work of a Researcher with the Archival Documents" by L.E. Shepelev (1966) and his "Archival Search and Investigation" (1971).

* * *

The beginning of systematic research on the history of science in the USSR is associated with the names of academicians V.I. Vernadsky, V.L. Komarov, P.P. Lazarev, V.A. Steklov and others who participated in the work of the Committee on the Study of the History of Science, Philosophy and Technology (from 1922 - the Committee on the History of Knowledge) that was formed in 1921 with V.I. Vernadsky's initiative, at the Academy of Sciences. Since 1927 the Committee has published 10 issues of "Proceedings" and several monographs. In "Proceedings" they presented on a broad scale the publications of documents concerning the history of science and technology. In 1932 the Committee was used as the base for forming the Institute of History of Science and Technology. The creation of the Committee on the History of Knowledge, and then of the special Institute took the history of science out of the sphere of the amateur interests of individual scientists, put it on the grounds of collective work, gave it State organisation. The study of documents on the history of physics, mathematics, mechanics, biology and other sciences spread widely. The development of the methodological problems on the history of science and technology started, the Institute organised the publication of a periodical "Arkhiv Istorii Nauki i Tekhniki" (in 1933-1936 six issues were published).

In 1938, instead of a single institute, the Committees on the history of individual branches of science and technology were organised at the Departments of AN SSSR, and at the Presidium of AN SSSR Committees were formed on the development of the history of the Academy of Sciences of the USSR, on the development of the scientific heritage of M.V. Lomonosov, D.I. Mendeleev, A.M. Butlerov and others.

In 1945 inside the system of AN SSSR a new Institute of the history of science was organised (from 1953- the Institute for the History of

Science and Technology) whose first director was the President of AN SSSR the academician V.L. Komarov, and after his demise in 1946 - the correspondent member of AN SSSR Kh.S. Koshtoyants.

The first All-Union Conference on the history of science (December 1946) and the session of the General Meeting of AN SSSR, devoted to the problems of history of Russian and Soviet science (January 1949), as well as the first All-Union Conference on the History of Technology (April, 1952) were important events for the development of the historical and scientific research in the USSR.

As was stressed in the resolutions of the above conferences, the study of the primary sources, the archival materials not published before will exercise a favorable influence on the development of Soviet science and technology.

With the initiative and the editorship of S.I. Vavilov in 1948 they started publication of collections of documents "The Scientific Heritage". The task of these collections was to publish the unknown scientific works, journals, memoirs and correspondence of the outstanding Russian and foreign scholars. In 1948-1961 four volumes of "The Scientific Heritage" were published.

The first two volumes contained individual papers and letters of M.V. Lomonosov, I.P. Kulibin, K.M. Ber, D.I. Mendeleev, I.I. Mechnikov, V.O. Kovalevsky and A.O. Kovalevsky, P.N. Lebedev, R. Guk, A.M. Butlerov, V.I. Vernadsky, K.F. Rul'e. The third volume (1956) was devoted to the scientific heritage of I.M. Sechenov; it included his unpublished articles, correspondence and documents, the fourth (1961) - the letters of Russian chemists to A.M. Butlerov. Then the publication of "The Scientific Heritage" stopped. In 1976 Presidium of AN SSSR made a decision of publishing "The Scientific Heritage" again.

From 1954 to 1962 45 volumes of "Proceedings" of the Institute for the History of Science and Technology were published, and from 1956 (four times a year) they published a periodical "Voprosy istoril yestestvoznaniya i tekhniki" (By 1977 55 issues were available) which regularly presents new archival documents and materials on the history of science and technology.

The use of new manuscript materials, including the unpublished ones, made it possible for the Soviet historians of science to open up the new aspects in the work of M.V. Lomonosov, L. Euler, M.V. Ostrogradsky, D.I. Mendeleev, A.M. Butlerov, K.F. Rule'e, I.M. Sechenov, I.P. Pavlov, K.N. Ber and other scientists, to identify more completely the significance of discoveries of Russian scientists for the development of the world science.

Thus, the many-year study of the scientific heritage of D.I. Mendeleev allowed the academician B.M. Kedrov to decipher and attach dates to a number of manuscripts of the great Russian chemist that are kept in the scientific archive-museum of D.I. Mendeleev at the Leningrad University. The results of archival search were published in the documentary collections "D.I. Mendeleev. The New Materials on the History of

Discovery of the Periodic Law" (1950) and "D.I. Mendeleev. The Scientific Archive, Vol 1. The Periodic Law" (1953). In 1955 B.M. Kedrov (together with T.N. Chentsova published a book "Browner - the Associate of Kendeleev. To the 100th Anniversary of B. Browner" that contained new documents on the scientific contacts between the Russian and the Czech chemists. During the next years B.M. Kedrov prepared and published the monographs: "The Day of a Great Discovery" (1958), "A Philosophical Analysis of the First Works of D.I. Mendeleev on the Periodic Law of 1869-1871" (1959), "The Microscopic Anatomy of the Great Discovery" (1970) and so on, in which he made much more precise the path of Mendeleev to the discovery of the periodic law, arrived at historical, philosophical and psychological conclusions that are important for considering the problem of the scientific discovery in general.

The study of a number of documentary sources that were not published before allowed the Soviet scientists also to revise the previous erroneous and much too low estimates of the level of development of science in Russia in XIth-XVIIth centuries. The effort in this direction was summarised in the collective works: "The History of Science in Russia (from the most ancient times to 1917)" (in four volumes, 1957, 1960, 1962); "The History of Mathematics in Russia" (in four volumes, 1966-1970); "Essays on the History of Technology in Russia" (in four volumes; two volumes are published: 1973, 1975), in the monographs: A.P. Yushkevich, "The History of Mathematics in Russia before 1917" (1968), P.M. Luk'yanov, "The History of Chemical Crafts and Chemical Industry in Russia before the End of the XIXth Century" (in six volumes, 1948-1965) and other books of Soviet historians of science and technology.

The use of manuscripts on the history of natural sciences, in particular, of mathematics in the countries of the East made it possible to develop a new, more objective conception of the development of mathematics in this region which opposed the Europe-centrist point of view that was spread previously (13).

In 1964-1976 a series consisting of six volumes "Russian Geographic Studies in the XIXth- the beginning of the XXth centuries" was published, in which the expedition activity of Russian scientists and travelers concerned with the study of land and waters of the World Ocean, especially of the Russian territory with its adjacent seas, was revealed with the help of documents; this activity was insufficiently studied earlier.

There is a deep and comprehensive description of the activity of the Academy of Sciences, based on the archival documents and covering the whole history of its existence in the three-volume "The History of the Academy of Sciences of the USSR" (1958, 1964); the third volume corresponding to the period from 1917 to 1967 is at present in print. For the 50-year anniversary of the Soviet government the Institute for the History of Science and Technology of AN SSSR prepared and published in 1967 the eleven-volume series "The Development of Science and Technology in the USSR over 50 years" and later "The Essays of Development of Technology in the USSR" in five volumes (1968, 1970, 1971, 1974, 1976).

These fundamental editions make use of the documentary sources to tell about the role of social transformations in the development of science and culture, about the achievements of the Soviet science and technology.

In 1970-1975 the generalising collective works on the world history of mathematics (in three volumes), mechanics (in two volumes), physics (in two volumes) based on the deep study of the primary sources were prepared and published. At present the work is nearing its completion on the last volumes of the General History of Chemistry (in 5 volumes) and the General History of Technology (in 4 volumes).

The traditional directions in the work of the historians of science include publishing of scientific-biographical essays the authors of which try to identify the diversified connections between the biography of a scientist and his work. Studies of this type allow to go deep into the problems of scientific creativity, to recreate to some extent the image of the scientist's personality, to become acquainted with the historical and social conditions under which the scholar lived and worked. From 1961 the Academy of Sciences of the USSR publishes a special series "The Scientific and Biographical Literature" (14). The authors of this series are distinguished scientists, historians of science and researchers whose sphere of work is close to that in which the main characters of their books laboured. At present this series published more than 200 biographies of scientists. The authors of many biographies used widely the archival documents, for instance, in the descriptions of life of A.M. Butlerov, I.P. Kulibin, M.V. Lomonosov, M.V. Ostrogradsky, A.G. Stoletov, V.Ya. Struve, V.I. Vernadsky, K.E. Tsiolkovsky, M.S. Tesvet and other scientists. The materials of Soviet archives are employed also in certain biographies of foreign scientists, for example, J. Berzelius, J. Kepler, K. Linnaeus, A. De Candolle, W. Ostwald.

In our review we tried to present in a more or less complete manner the main depositories of documents on the history of science and technology as well as their publication activity, we made an attempt to show how the Soviet historians of science and technology make use of the documentary riches of the archives in their research. There is still a large piece of work facing us on the identification of the documentary materials on the history of science and technology, especially in the republican and county archives, their description and catalogization. This will be promoted by further work of Soviet archives on the improvement of the scientific-reference apparatus, intensification of their information activity, the measures directed at a fast introduction into the scientific turn-over of the newly arriving materials. It appears to us that the possibility of creating the "living archive" is extremely important; this means interviewing the living outstanding scientists on their path into science and the development of the corresponding methods of interview; it also means the collection of letters, the laboratory journals and other materials having historical significance. The rapid development of the new scientific trends makes it necessary for the archives to expand the types of documentation deposited in them, to watch the appearance of its new types and to plan their storage.

Notes

(1)   See: S.R. Mikulinsky. The Contemporary State and the Theoretical
      Problems of the History of Science as a Science - "Voprosy
      filosofii", 1976, No. 6.

(2)   "Dekrety Sovetskoi vlasti", Vol. II, Moscow, 1959, pp. 383-385.

(3)   "Dekrety Sovetskoi vlasti", Vol. V. Moscow, 1971, p. 412.

(4)   "Sobranie Postanovlenii Pravitelstva Soyuza Sovetskikh Socialist-
      icheskikh Respublik", 1958, No. 14, p. 112.

(5)   See: "The Materials to the All-Union Scientific Conference on
      Scientific and Information Activity of State Archives of the USSR",
      Moscow, 1976, p. 7.

(6)   B.V. Levshin. The Archive of the Academy of Sciences of the USSR
      (for the 250-year jubilee of the AN SSSR), - "Sovetskie arkhivy".
      1974, No. 2, p. 23.

(7)   Ibid. p. 23.

(8)   T.I. Lysenko, B.A. Malkevich. The Use of Documentary Materials of
      the Archive of the Academy of Sciences of the USSR over the Years
      of Soviet Rule, - Academic Archives of the USSR over the 50 years
      of Soviet Rule, Moscow, 1968, p. 91.

(9)   "Okhta Chemical Group. 1715-1965. Essays, Documents, Memoirs".
      Leningrad, 1965.

(10) Erenfost-Joffe. The Scientific Correspondence (1907-1933).
      Leningrad, 1973.

(11) "Sobraniye Postanovlenii Pravitelstva Soyuza Sovetskikh
      Sotsialisticheskikh Respublik", 1964, No. 9, p. 65.

(12) See: "The Materials to the All-Union Scientific Conference on
      the Problems of Scientific and Information Activity of the State
      Archives of the USSR", Moscow, 1976.

(13) See: A.P. Yushkevich. The History of Mathematics in the Middle
      Ages (1961), and also papers in collections "From the History
      of Science and Technology in the Countries of the East" (in three
      issues, 1960-1963); Physical and Mathematical Sciences in the
      East (in two issues, 1966, 1969), etc.

(14) See: Z.K. Sokolovskaya. 200 Scientific Biographies, - "A
      Bibliographical Reference Book", Moscow, 1975. Besides, the
      biographies of scientists are also published in "Popular Science
      Series" of AN SSSR and by the Publishing House.

      "Molodaya Gvardiya" in its series "The Life of Remarkable People"
      (from 1933). Beginning from 1933 and up to the present moment

"LRP" has published about 500 books, a considerable part of which is devoted to the description of life and work of scientists (see "40 years of ZhZL. The Catalog. 1933-1973" Moscow, 1974).

M.L.Righini Bonelli and T.Settle

ITALIAN ARCHIVES AND THE HISTORY OF SCIENCE AND TECHNOLOGY

In Italy there are few archives that specialise in the history of
science or technology; the great majority of them, here as elsewhere,
do not normally distinguish among the several historical disciplines.
So the possible fruits as well as the difficulties of archival research
are about equal for all. Italian collections are immense and not at all
completely explored or catalogued. People working in relatively standard subjects have little difficulty finding the sources they need,
simply by consulting the scholarly apparatus of the most recent monographs. But someone trying to work up a new subject or trying to exploit
a new point of view may well spend months in fruitless searching. Since
there is no way at present for that fruitless work to be registered,
others coming later to the same subject may substantially repeat the
experience. Given the sheer magnitude of materials on one shelf or
another and the usual difficulties of finding the funds and personnel
with which to do what would be required, it has been (and perhaps will
continue to be) impossible to put the house completely in order.

Italy does have a Code pertaining to archives, both public and private:
La Legge sugli Archivi (Ministero dell' Interno, Direzione Generale
degli Archivi di Stato, Roma, 1963). This Code sets standards for the
organisation of public archives, norms for the collecting and preserving
of materials, requirements regarding the making and updating of inventories, and so on. Needless to say, the work implicitly mandated by
the provisions is going on; essentially it is an endless task. The Code
also has sections dealing with archives still in the hands of private
families. It requires the preservation of documents of "notable
historical interest" and provides mechanisms for ascertaining this
quality; and it specifically forbids the alienation of these resources
(See: La Legge, Title II, Chapter II). Unfortunately it provides no
sanctions for the illegal sale of manuscripts and books, and one knows
that many important collections have been dispersed to the world and.
for a while, lost to scholarship. Nor, of course, does the Code
protect against the vicissitudes of fortune: war, earthquake (Messina,
1908), flood (Florence, 1966), and fire (Archivio di Stato, Naples,
making research into Francesco Fontana and Neapolitan astronomy very
difficult), all of which can destroy whole collections or put them out
of use for many years.

For the archives that are available, however, there are the usual card
indexes of published materials and catalogues of manuscripts. Many of
these, unfortunately, lack depth of subject indexing and cross-referen-
cing, and many of the latter have a tendency to be less thorough with
regard to scientific and technical materials than with literary and
"humanistic" ones. In addition, many of the more useful manuscript
catalogues exist only in manuscript themselves and at the archive in
which the collection is housed. One important case in point is the
Galileiana Collection at the Biblioteca Nazionale Centrale of Florence.
Antonio Favaro's handwriten inventory of it is available for consultation

there, but of course this does not help the individual who may have need
of it at some distant location. It is true that part of it has been
edited and published (Angiolo Procissi, "La Collezione Galileiana della
Biblioteca Nazionale di Firenze, Vol. I 'Anteriori' - 'Galileo',"
Indici e Cataloghi, n.s. V, Instituto Poligrafico dello Stato, Roma,
1959); unfortunately, the rest and by far the most of it has not yet
been put in print. One can order a microfilm of this remainder, but
the research libraries of the world have not generally done so.

State archives have some characteristic problems of their own. The
Archivio di Stato, for instance, has shelves of inventories and inven-
tories-of-inventories (in manuscript, typescript, and print) but no
comprehensive name and subject index for the whole collection.
Exploring the Archivio is tantamount to educating oneself in the
administrative and political history of Florence and Tuscany. While
for many purposes this is a very worthwhile thing to be achieved, it is
an impediment to those who might have very specific items to find and
could well delay acquiring the more comprehensive view. To date the
best introduction and guide to the Archivio is in Giulio Prunai's
"Firenze (Secolo XII - 1800)" (in: Acta Italica: Piani particolari di
pubblicatione, 6, Editore Antonio Giuffre, Milano, 1967). In the
course of presenting detailed plans for the publishing of certain of
the currently valid inventories, Prunai does summarise the administrative history of Florence and briefly describe the contents of those
particular inventories. (Similar Piani have been issued or are in
preparation for the State Archives of other Italian Cities; consult
the publisher.) But the volume itself lacks an analytical index; and
not to be published, apparently, are such inventories as those pertaining to the Academia del Disegno or the Conventi Soppressi, to cite
two very important fonti. In sum, even when this whole (and very
worthy) project is completed, there will still be great gaps in our
detailed knowledge of the Archivi.

Consulting private archives in Italy is as chancy an affair as it is
anywhere. Some families are cordial and hospitable, others not; some
have well done inventories and some not. For some collections there
are printed catalogues such as those for the Gondi, Guicciardini and
Da Verrazzano archives published by Roberto Ridolfi in the late 1920's
and early 1930's. Perhaps the greatest difficulty in using the private
sources, however, is the lack of any national or union listing of the
collections.

It is clear, then, that in spite of the existence of real riches there
are many impediments to doing archival research in Italy, especially
for the foreigner who has not grown up in the system and initially
has no "sixth sense" of where to go to look for things. With respect
to the history of science and technology in particular, it would be
extremely useful for everyone if there were in existence a "centrale"
which kept and published records of where individual items,
collections and other types of sources could be found. The institutional bases already exist (for instance, the Domus Galileiana in Pisa
or the Instituto e Museo di Storia della Scienza in Florence, to
name two possibilities), and the idea is certainly not new; what is
needed is both the staff and the money.

There are two publications that are important, each in its own way, which could seed on-going projects. One is the Annuario delle Biblioteche Italiane (3 Volumes, Roma, 1956 + 1959) published by the Ministero della Pubblica Instruzione. This current edition lists most, but not all, of the non-private libraries in Italy and Vatican City and for each provides basic information about the size of the collections, numbers of incunabula, numbers of codexes, etc. It gives a brief summary of the history of each library, and the sources of its major fonti, and it lists the published and unpublished catalogues. Each entry also includes the bibliography citing works about the library and its collections. While in its present state it is far from an exhaustive or even satisfactory guide to Italian archives, it is useful for scholars in the exploratory stages of an investigation. Moreover, a new edition of the Annuario is being prepared. It will list some of the libraries missed in the current edition. Hopefully its historical and bibliographical sections will be systematically expanded, and it will include the analytical name and subject matter index not present in the current version. If such improvements were made, the work would be a necessary tool for the researcher and could provide one point of departure for building a national catalogue of fonti and their contents.

The second item is the Atti del Primo Congresso Internazionale di Ricognizione delle Fonti per la Storia della Scienza Italiana:I Secoli XIV - XVI, (Firenze, Barbera, 1967). The Congress took place in September of 1966 and ellicited many impressive papers and much important discussion about the state of scholarship in the history of science for the period indicated, all united by the themes of our knowledge (or lack of it) of the sources and what might be done to improve that knowledge. In the final session Eugenio Garin very ably expressed the "sense of the meeting" and, among other things, listed a number of suggestions about what be done immediately and easily. That the suggestions were obvious to anyone acquainted with the problems does not make them any less appropriate. One set of suggestions amounts to that of making a detailed inventory according to name, subject, language, source, location, period, fortuna, etc. of manuscripts, codexes, books and collections pertaining to the history of science. This would not be something that could be done in one day, or even ever finished completely. But with modern, computer assisted indexing and information retrieval techniques, something quite substantial could be achieved in a relatively short time. Interestingly, the editor of the Atti, Carlo Maccagni, has already seen to it that it has a complete index of the names mentioned in the papers and discussions and a location list for all the manuscripts cited. These Atti and similar Atti of other such congresses could be another point of departure for an inventory of fonti.

Of course, one thing that would be most helpful, and not just for archival research, would be an annual bibliography devoted to the Italian literature in the history of science and technology, the sort of bibliography published by Isis, for instance, or Technology and Culture or the Bulletin Signaletique. As one knows, the annual Italian production is huge; much of it is in places difficult of access; and it is not at all reliably reflected in the non-Italian bibliographies.

406

Zofia Wardeska

ARCHIVES POLONAISES, SOURCES DE L'HISTOIRE DES SCIENCES.

Dans divers types d'archives se trouvent maintes sources importantes pour l'histoire des sciences: Archives Nationales divisées en Archives Centrales et Archives de Voîvodies; Archives des universités et des écoles supérieures, celles des Sociétés Savantes et de différents organismes et institutions politiques. Les archives d'Eglise constituent un groupe à part.

Nous voudrions présenter, d'une façon plus détaillée, des Archives relativement jeunes, créées à Varsovie en 1953, et disposant aujourd'hui de filiales à Cracovie, à Posnań et à Zabrze: les Archives de l'Académie Polonaise des Sciences. Ce sont précisément ces archives qui ont pour but de rassembler les fonds concernant l'histoire des sciences et des techniques en Pologne, notamment pour les XIXe et XXe siècles. Les archives de l'Académie Polonaise des Sciences possèdent 2400 mètres de rayonnages; elles conservent trois catégories de fonds:

1º Actes des sociétés savantes et des instituts qui fonctionnaient avant la création de l'Académie Polonaise des Sciences, par exemple, les actes de la Société Savante Cracovienne (1815-1872), de la Société Savante Varsovienne (1907-1951) et de l'Académie Polonaise des Sciences et des Lettres, P.A.U. (1873-1952).

2º Actes de l'Académie Polonaise des Sciences proprement dite: documents scientifiques et administratifs de l'Académie depuis 1953; documentation concernant l'organisation des recherches scientifiques.

3º Manuscrits constituant l'héritage scientifique des savants polonais décédés: matériaux très précieux pour les historiens des sciences, car on y trouve souvent des textes d'ouvrages scientifiques non publiés, des documents représentant l'activité scientifique, didactique ou sociale de leurs auteurs, la correspondance des savants ou des données biographiques. Cette collection est continuellement complétée par des dons ou des achats. Citons quelques exemples: l'héritage scientifique du mathématicien W. Sierpiński, du physicien L. Wertenstein renfermant des lettres de Marie Curie-Skłodowska, d'Irène et de Frédéric Joliot-Curie, d'E. Rutherford et de P. Langevin; les papiers ayant appartenu au physicien M. Wolfke et comportant, entre autres, des lettres de L. de Broglie et d'Einstein. Tous ces documents sont conservés par une équipe spécialisée, classés et rendus accessibles aux chercheurs.

Les Archives de l'Académie Polonaise des Sciences disposent également d'un Service de microfilm et de photocopie. Le Service possède aussi une assez riche photothèque des savants polonais.

La Bibliothèque des Archives compte actuellement environ 15600 volumes; elle réunit surtout des imprimés concernant l'histoire des sciences et des techniques polonaises, ainsi que des éditions de documents complétant parfois les collections des Archives. La collection de médailles et de plaques commémoratives se révèle souvent utile pour les recherches d'histoire des sciences.

Les Archives possèdent un système d'information qui permet de trouver assez rapidement les matériaux conservés dans d'autres archives et collections manuscrites des différentes bibliothèques: c'est l'Inventaire Central des Sources de l'Histoire des Sciences et des Techniques Polonaises qui

compte ca 32000 fiches, plus l'index des noms (environ 40000 fiches). Autre
aide supplémentaire pour les chercheurs: le Répertoire des savants polonais,
fournissant les données biographiques des savants (du XVIe jusqu'au milieu
du XXe siècle), avec double classement alphabétique et selon les disciplines
scientifiques. Un guide des collections des Archives de l'Académie Polonaise
des Sciences (Przewodnik po zespołach i zbiorach Archiwum PAN) a été publié
à Wrocław en 1965; une nouvelle édition va paraître en 1978. Il renseigne
sur l'ensemble des fonds des Archives. Un Bulletin annuel des Archives a
apporté, jusqu'à présent, 57 inventaires de fonds des Archives.

SOME GENERAL PROBLEMS AND BRITISH EXPERIENCE

Charles Weiner's paper did not arrive before I wrote mine so that I could not take account of his points, and I apologise for this. I should say however that I am speaking about sources for the history not only of 20th century science, but of technology and medicine. I agree with Charles Weiner that sources for the history of science form part of much wider questions of science but I propose now to take more limited questions. Firstly I shall note the categories of sources which may interest historians of science, technology and medicine; secondly I shall point to some general problems in the relationship of historians to their sources; and thirdly I shall say something about 20th century sources for the history of science in Britain.

First then the categories of sources. These seem to me to consist of

1) written records of all kinds which survive from the past: a heading under which I include such things as engineering drawings, technical and non-technical reports, and all kinds of statistical returns. This category can be subdivided into a) primary records: the private records of persons or organisations, along with printed contemporary documents such as the printed proceedings of bodies like Parliament, together with newspapers, journals, etc. Under b) comes the secondary material - which is self-evident in the form of printed books.

2) consists of artefacts: scientific or medical instruments, machines and plant, and buildings - what is called scientific and industrial archaeology. This includes the territory of the museums but goes further because it also includes objects, equipment and buildings which are still, or were recently, in use - some of them of enormous size.

3) consists of graphic material which again I take to be largely but by no means exclusively the province of museums and libraries.

4) consists of films or sound recordings made by individuals, companies or broadcasting organisations.

All these four categories are concerned mainly with material created not for posterity's historical interests but for the contemporary purposes of a person, or an organisation, or a broadcasting company.

5) however, is different. It consists of records created currently for historical purposes. These may be written records, such as the autobiographical material collected by the chairman of this symposium or taped interviews of all kinds (including of course the typed scripts) which come under the heading of oral history.

Before I deal with British sources for the history of 20th century science, technology and medicine within these five categories, I should like to make three general points about the relationship between history and its sources. Two concern historiography and one is practical. The first problem is that the methodology of history, like that of science, consists of the interplay of hypotheses with evidence. But the historian's evidence is very different from the scientist's: it consists of what happens to have survived from the past. This is a very random business and often we cannot even assess the completeness of evidence because we do not usually know what once existed. Moreover we cannot even be sure that we have seen all that has survived. So the scope for falsifiability is a matter of chance. As for 20th century material, we are all aware of the warnings that most important transactions are now carried on by telephone and that documents alone give a very misleading picture. If all this suggests that history must be so misleading that we would be better without it - the Henry Ford view that history is bunk - I suppose that we can only say that a society without history is like someone who has lost his memory and a bad memory is better than nothing.

This point leads on to my second historiographical point. Everyone yearns to theorise and generalise about categories of people, whether they be scientists, engineers, novelists or painters: that way lies scientific or anyway sociological respectability. I view with misgivings attempts to generalise about the scientists, doctors, or engineers of the past simply on the basis of the sources we are so busy collecting. There is a great danger that the evidence that does survive - the records for example of the people who did not throw away their correspondence or school reports will be given too much weight and the evidence that went into the bonfire too little. My own experience in checking oral evidence of surviving scientists and engineers with surviving documents of their past activities also underlines the fallibility of some Category 5) material.

The practical problem in all this for the archivist is: what do you keep and what, if anything, do you throw away? Twenty-five years ago archivists felt rather confident. For example as far as government paper was concerned, you kept the papers showing the evolution of policy at the higher levels. As for scientists' own papers, you kept only the papers of the greatest scientists showing the evolution of their big and pure ideas. But historians' interests have changed and will go on changing. And along with historians you have to reckon with sociologists, political scientists and economists. Nothing is too small - say even the exchange of Christmas cards - to escape these academic nets. The correspondence of lesser scientists may be more interesting than that of great scientists. Moreover the ubiquity of the computer means that all kinds of material previously thrown away or sampled because it could not be used in its entirety can now be processed in its entire, and often vast, bulk. I sometimes think that it is now impossible to look at any piece of paper - except multiple copies - and say, "no historian will ever want to look at that". The leaflet of the American Institute of Physics gives a very good guide to the papers historians might want to look at. The only snag is that it doesn't have a section on what you can throw away. Is the answer nothing? Yet if this is true the problems of cataloguing, storage and cost become truly enormous until the time when the spoken word takes over so completely that the historian of later years goes out of business. Similarly there is no limit to the possibilities of oral history and the collection of autobiographical material. What is the right expenditure of effort here?

I will now go back to British sources in the categories of records I gave at the beginning. There is not time to give a careful description of the records available in this country under these headings but I have encouraged people with specialist knowledge either to come to this symposium and contribute to formal or informal discussion or - if they could not come - to write notes about their work. First primary documents. The main subdivisions here are, I think, papers of public organisations, businesses, and private individuals. Public organisations include a large variety of government or quasi-government bodies such as the nationalised industries and the National Health Service. The Public Record Office was unable to send someone to talk about the records available there about science, engineering, or medicine but they have produced a note about records on science. We also have with us Lorna Arnold who until recently combined her historical work on atomic energy with the post of records officer for the Atomic Energy Authority, the largest of the public science-based organisations. She has brought with her a copy of the excellent record guides used in the Authority while she is also able to talk knowledgeably and generally about the public records. Medical records form another large sector of the public records and we have, here, a collection of notes about them compiled after a recent one-day conference in London on the subject.

There are, I shoud add, many public bodies which are financed from public funds but are not subject to the Public Records Act. These include bodies of interest to historians of science such as the Science Research Council or the universities or the British Broadcasting Corporation. All these make their own archive arrangements.

Business archives are very diverse in the care given to them, and a Business Archives Council exists; a note about it is available. Mrs Symons, the archivist of the Institution of Electrical Engineers, who is here today, has been dealing with records of electrical firms as well as public utilities and can talk about the problems. The Scottish Record Office is very active and knowledgeable about e.g. shipbuilding firms' records.

The records of individual scientists, engineers and medical men are spread, of course, among many different repositories: university libraries, the Royal Society, the Royal Institution and other learned societies. Churchill College at Cambridge should be mentioned as an especially important collection of the papers of 20th century scientists. The Historical Manuscripts Commission will be publishing next year a Guide to the Papers of British Scientists 1600-1940, which can be consulted now in the Commission's office. This takes further the guidance produced on microfiche some years ago by MacLeod and Friday.

I will say rather more about the Contemporary Scientific Archives Centre at Oxford, not because I am the honorary director but because it embodies some new principles which make it especially cost-effective and which may therefore be interesting to others. A Joint Committee of the Royal

Society and the Historical Manuscripts Commission had been worried in the 1960s that so little was being done to preserve the papers of eminent contemporary scientists, engineers and medical men. With a few notable exceptions, university libraries had few archives staff and could not easily deal with technical papers. At first the idea was to create a science archive but at a general discussion meeting at the Royal Society in 1969 Michael Hoskin of Cambridge had a bright idea: we needed not a specialist archive but a specialist service which would identify papers to be preserved, collect them and bring them to a centre where they could be sorted, listed and boxed. They would then be handed over to existing libraries for permanent preservation. Catalogues would be kept at central points but dispersal of the collections meant that no expensive buildings would be needed.

This formula was aacepted, fund-raising began, and in 1973 it was possible to open shop, albeit on a shoe string. Neither the Royal Society nor the Historical Manuscripts Commission could run the Centre so I was asked to take it under my wing at Oxford. Two members of staff were recruited - Mrs Alton and Mrs Weiskittel, and they still comprise the total staff for all purposes, including typing and driving to collect and deposit collections. They had no previous experience in archives work or science but they have done a splendid job. In less than 4½ years they have catalogued and deposited 54 collections. A further 10 collections are in process and they have numerous inquiries in hand. 28 different libraries have accepted collections. The quality of their work is as impressive as the quantity as the specimen catalogues here will show. The Centre's reports are also available. Their work shows how much can be done in how short a time with how few resources if the right people are found. I should add that at present the Centre is financed by grants from the Royal Society, the British Library and the Council of Engineering Institutions. Mrs Alton and Mrs Weiskittel are available today to talk about the Centre.

So much then for written records. I have asked Dr Lindsay Sharp of the Science Museum in London to come prepared to talk about categories 2) and 3) - artefacts and graphic material. On category 4), films and sound recordings, British sources are probably exceptionally rich because of the existence of the British Broadcasting Corporation and to a lesser extent, the National Film Archive. Under the Corporation's charter, its terms of reference are restricted to producing programmes and do not include running archives to serve historians whether of science, war, sport, fashion, linguistics or anything else. But it has built up immensely valuable archives to serve its own programming needs and, conscious of their importance to the outside world, it has recently set up an Archives Advisory Council to go into the whole question. No report has yet been issued but the B.B.C. has kindly produced a note about its archives for this symposium.

In category 5) - archives material created currently for historical purposes - Britain to my knowledge has very little to report compared with the very important work conducted in the United States by our chairman, by the American Institute of Physics, by Professor Weiner, and other historians of science, technology and medicine. Individual British historians have undoubtedly taped interviews for the purposes of their own books or other purposes; thus there is a good deal of taped material on the history of molecular biology. But I know of very few systematic attempts here to commission autobiographical material from scientists, engineers and medical men themselves nor to tape their recollections. We clearly lag a long way behind the United States in this field and should no doubt try to catch up. But the usual questions arise: who should do this and who will finance it? Any ideas will be gratefully received. It will be equally interesting to know of work going on in other countries besides the United States.

A.Hermann

A REVIEW OF GERMAN SOURCE MATERIAL

I have, in my own studies, extensively used the material collected by
American scholars, especially the material of the Sources for History
of Quantum Physics. History of Physics in the 20th century can not any
longer be written without this material. With my survey I now will also
help scholars from other countries to use our sources.

Our professor at college used to say: "Differentiation is a labour,
integration an art." I adopt this and say: To find published sources
is a labour, to find unpublished sources is an art. Thus I shall merely
mention the printed sources. By far not the largest, but the most
important German library for history of science is the "Bibliothek des
Deutschen Museums" in Munich. Here you find about 600,000 books and
volumes of journals. You find there a card catalogue for secondary
literature in our field. So for example you look up Bohr or Printing
machines or relativity theory and find a lot of articles in journals
written about this topic. This index is restricted to the journals in
possession of the Deutsches Museum. It includes about 40,000 cards.

Let me now indicate the nature and location of the unpublished source
material. There are two different keys to the unpublished source
material, the first of which is the bibliographies. There are six
bibliographies of Nachlässe (literary estates):

A)   Federal Republic of Germany

1)   Die Nachlässe in den deutschen Archiven (= Verzeichnis der schriftl.
     Nachlässe in dt. Archiven und Bibliotheken. Bd. 1). Bearbeitet von
     Wolfgang A. Mommsen. Boppard 1971.

2)   Die Nachlässe in den Bibliotheken der Bundesrepublik Deutschland
     (= Verzeichnis d. schriftl. Nachlässe in dt. Archiven u. Bibl. Bd.
     2). Bearbeitet von Ludwig Denecke. Boppard 1971.

B)   Switzerland

3)   Repertorium der handschriftlichen Nachlässe in den Bibliotheken und
     Archiven der Schweiz (= Quellen zur Schweizer Geschichte. Neue
     Folge.   IV Abtl.: Handbücher. Bd. VIII). Bearbeitet von Anne-
     Marie Schmutz-Pfister. Bern 1967.

C)   German Democratic Republic

4)   Die Nachlässe in den Wissenschaftl. Allgemeinbibliotheken
     (= Gelehrten- u. Schriftstellernachlässe in den Bibliotheken
     der Deutschen Demokrat. Republik. Tl. 1). Berlin 1959.

5)   Die Nachlässe in wiss. Instituten u. Museen und in den allgemeinen
     Bibliotheken (= Gelehrten- u. Schriftstellernachlässe in den
     Bibliotheken der Deutschen Demokrat. Republik. Tl. 2). Herausgegeben

von Hans Lülfing und Ruth Unger. Berlin 1968.

6) Nachträge, Ergänzungen, Register (= Gelehrten- u. Schriftsteller-
   nachlässe in den Bibliotheken der Deutschen Demokrat. Republik.
   Tl. 3). Herausgegeben von Hans Lülfing und Horst Wolf. Berlin
   1971.

D) Austria

   A bibliography for unpublished Nachlässe doesn't exist.

The second key to unpublished letters in German archives and libraries is
a special documentation service, which the Preußische Staatsbibliothek in
Berlin, in Western Berlin, had established the last years. So you may
write to the Staatsbibliothek Berlin, c/o Tilo Brandis, 1000 Berlin-
Dahlem, Archivstraße 14, enquiring about letters, let's say from Planck
and you will get a list of all Planck letters in libraries and archives
in the Federal Republic and in the German Democratic Republic.

Now I should characterise the most important archives, of which there are
five in all:

1) The Deutsches Museum, where you will find artefacts (physical sources)
   such as Philipp Lenard's cathode ray tubes or the equipment, with
   which Paul Knipping and Walther Friedrich discovered the diffraction
   of x-rays in 1912, pursuing an idea of Max Laue.

   The Address: Deutsches Museum, Postfach, 8000 München 26

2) The Preußische Staatsbibliothek which holds a lot of letters
   especially from scientists who worked in Berlin for some time. They
   have parts of the Nachlass of Carl Runge, of Friedrich Schmidt-Ott,
   of Johannes Stark, a lot of Planck letters and are just taking over
   (I hope) the Nachlass of Max Born.

   The Address: Staatsbibliothek Preußischer Kulturbesitz, Archivstraße 12-14, 1000 Berlin 33

3) The Staats- and Universitätsbibliothek Göttingen, where the Nachlässe
   of most of the famous Göttingen Professors, i.e. David Hilbert and
   Felix Klein are preserved - a programme has just been initiated to
   edit the letters of Georg Christoph Lichtenberg.

   The Address: Archiv der Georg-August-Universität, Wilhelmsplatz 2,
   3400 Göttingen

4) The recently-founded "Bibliothek und Archiv der Max Planck-
   Gesellschaft" in Berlin. As is well known the Max Planck Society
   was founded in 1911 as the Emperor William Society, that it has now
   about 50 research institutes and that an important part of German
   science was done in the institutes of the society. So a central
   archive is of course extremely valuable. It will get, and has
   already received, a great many administrative documents from the
   institutes, but also scientific correspondence including notes and

letters from Walther Bothe, Otto Hahn, Richard Kuhn and Peter
Debye. A project of collecting material - restricted to the Max
Planck institutes - is currently in progress.

The new institute has been housed in a charming location, the
former Otto Warburg institute, which provides wonderful research
facilities. So I advise you all to come and do research there.
We have a keen hope: that this archive for the history of the Max
Planck society will develop one day to a real Max Planck Institute
for History of Science.

C. Weiner

SOURCES FOR HISTORY OF 20th CENTURY SCIENCE: PROGRESS AND PROBLEMS

Historians and archivists have been creating time capsules that will reveal what we think is important and symbolic in our culture, institutions and lives. These messages for the future focus on many aspects of mid-twentieth century life. The role of science in this world will be prominently portrayed as a result of the proliferation of efforts in the past two decades to preserve the historical record of the work of scientists and of the institutions which nurture them. How accurate, balanced and complete will this portrayal be? What has influenced the perceptions that historians of science have about their responsibilities in the present and for the future? I will attempt to explore these questions in this paper in the hope of stimulating critical discussion and action. I will focus on some highlights of the work that has been done in the United States, since other participants in this symposium will talk about programs in other countries.

I. Origins and Motivations

In 1957, the year of Sputnik, a note appeared in the history of science journal, ISIS, calling for an organized and systematic effort to collect and preserve the personal papers of American scientists. Gerald Gruman, its author, declared that ". . . we historians of science have the serious responsibility of preserving the materials necessary for future scholarly research. Our duty in that regard probably outweighs anything else we might do. . . . We ought not to let our interest in the distant past absorb us so completely

that we miss the beautiful opportunities before our eyes."[1] The focus of
historians of science on "the distant past" was clearly evident in the
issue of ISIS in which Gruman's plea appeared: of the scholarly articles
and notes, none made it into the 20th century, and only one note went beyond
1800.

A year later, in June 1958, Gruman amplified his call beyond the small
group of historians of science to the much larger audience of scientists, in
an invited editorial in Science magazine, entitled "Preserving the Stuff of
History."[2] The basic message was the same: valuable source materials
essential for understanding of the history of science were being destroyed,
and a systematic effort to preserve them was urgently needed. This effort
should involve scientific institutions, scholarly societies, and national and
local archives.

The appeal made public what many individuals had been saying privately.
It had special resonance in the post-Sputnik period of international scientific competition when American political and scientific leaders felt it
necessary to point to their country's past contributions to science and technology. Gruman emphasized that during the period since the late 19th century
to the post-World War II years, American science had "won a position of
world leadership,"[3] and that the historical record of this "heroic age of
American science"[4] would be of special interest to future historians. Whatever the motivations, the climate was receptive. However, it takes more
than editorials to mobilize support and to launch projects.

A key role was played at this point by Nathan Reingold, who joined
the staff of the Library of Congress in 1959. As a scholar in the history
of American science, he was able to turn the attention of the Manuscripts
Division of this great national library to collecting the personal papers of

leading American scientists. Just before joining the Library of Congress,
Reingold decided to use the occasion of Gruman's editorial to publicize the
need for a policy and program that would take into account the rapid growth
of the scientific enterprise and the changes in the nature of its documentation.
He was also concerned with problems relating to the organization and accessibility of 20th-century science collections. Reingold convinced the History
of Science Society to back him for a conference, and a small grant to support
it was made by the National Science Foundation.[5]

The Conference on Science Manuscripts was held in May 1960 at the Cosmos
Club in Washington, D.C., attended by forty-eight historians, archivists and
scientists from universities, government agencies, scientific societies,
scholarly associations, libraries and archives. During the two days of
discussions, there was clear agreement that something had to be done and that
the effort would require the cooperation of the scientific community, historians
and archivists. What should be saved? Where should collections be deposited?
How should they be organized? Where do the new techniques of oral history
fit in? What priority should be given to publication of archival documents?
These questions were explored in the light of the special problems presented
by 20th-century science and its historical sources. All the discussions pointed
to the need to provide continuity beyond the conference, perhaps in the form
of a joint national committee or commission, as originally suggested by Gruman.
But when the conference was over, it was not clear which of the participating
groups was willing to take the initiative and assume the basic responsibility.

In the immediate aftermath of the conference, Reingold kept the issue
alive. He amplified its message by organizing sessions on science manuscripts
at meetings of the American Association for the Advancement of Science, and
by attempting to get other organizations and institutions involved. The

419

conference proceedings were published as a special issue of ISIS in March 1962, with an appendix describing budding British efforts along the same line.[6] Although no national commission or committee was established, the conference helped to heighten consciousness of the special needs and opportunities in the field of 20th century science manuscripts, and it provided legitimation and respectability to individuals and collecting institutions who were interested in exploring this area.

During the same period, the late 1950s to the early 1960s, history-minded physicists began to take steps in the same direction. A major impetus to this effort was the passing of many of the heroes of 20th century theoretical physics. The feeling is exemplified in this statement by John A. Wheeler, himself an important contributor to mid-20th century theoretical physics: "The bell has been tolling; time is short. Einstein died in 1955, von Neumann in 1957, Pauli in 1959, and Schroedinger in 1961."[7] What could be done to preserve the personal records and recollections of the physicists who had participated in the intellectual adventures of relativity and quantum mechanics in the first third of the 20th century? This question was discussed in a series of informal conversations beginning in 1958. Among those involved were the physicists Edward Purcell, Gerald Holton, Charles Kittel and Wheeler. Historians of science Harry Woolf, Hunter Dupree and Thomas Kuhn were brought in to the discussions in 1960. By July 1961, these efforts culminated in a grant from the National Science Foundation for a project on the history of quantum physics directed by Kuhn, and jointly sponsored by the American Physical Society and the American Philosophical Society.[8] At about the same time, the American Institute of Physics sponsored a project on the History of Recent Physics in the United States, also supported by the National Science Foundation. The aim of these projects was to involve physicists, historians and archivists in large-scale systematic efforts to locate, preserve and catalogue correspondence and other archival material that would illuminate the development of 20th century physics and its institutions.

Why were physicists so supportive of these efforts? In addition to a general interest in the past of their discipline, some prominent physicists shared these motivations voiced by John Wheeler:

> Complementarity and indeterminism, the quantum principle and relativity, together constitute the major scientific innovations of our time. They are key concepts in a revolution in scientific theory without parallel in the last three hundred years. Of all the advances that the mind of man has ever made towards a complete harmonious account of human experience, none is more revolutionary. None has penetrated more deeply into the inner working of the machinery of the universe.[9]

Given this belief, the development of such concepts certainly had to be documented, studied and understood. And for further justification, Wheeler argued that no physicist "can serve society with full effectiveness until the past has sprung into intense and unfolding drama before his eyes: the great men, the great struggles, the great ideas. These historical insights are not for scientists alone, but also for analysts of the creative process, and makers of government and university policy towards science."[10] The message was clear: physics had passed through a heroic, golden age, and since World War II it had demonstrated its powerful practical applications in warfare and industry. A major historical task perceived by many scientists was the need to explain the peculiar success of this endeavor, and the preservation of historical source materials would make it possible.

Enthusiasm of physicists for historical activities in the early 1960s coincided with increased government financial support for science. The growth of history of science as a well-defined academic discipline was also stimulated by the new largess for science, which provided a small overflow for historical studies. At the same time, some of the individuals coming in to the history of science as a professional discipline were attracted to issues in more recent science. Their motivations for studying the history of science may have differed from those of the scientists, but many of the questions they wanted to ask coincided with the interests of scientists. Together they set out to find and preserve the source materials essential for such studies.

II. Results of the Past Two Decades

What has happened since the early 1960s? The record is full and rich, and there is no need to recite it here. I will mention only a few highlights of the activities in the United States.

The project on "Sources for History of Quantum Physics," mentioned above, accomplished its major goals during the period 1961-1964. The project staff located, cataloged, and microfilmed letters and manuscripts relating to the development of quantum physics. Many of the materials were in the possession of European physicists and their families. Now these documents are preserved on more than 100 reels of microfilm and are available for research use at the American Philosophical Library in Philadelphia, the Bancroft Library of the University of California at Berkeley, the Center for History of Physics at the American Institute of Physics in New York, the University of Minnesota in Minneapolis, and the Niels Bohr Institute in Copenhagen. Also available as an important supplement to these written sources, are the transcripts of oral history interviews with about 100 individuals who participated in the development of quantum physics. Some additional materials were collected in a final stage of the project, which terminated its activities in 1966.[11] The collected source materials have helped to open up a new field of scholarship in the history of science.

The American Institute of Physics' Project on History of Recent Physics in

the United States was launched in 1961 with support from the National Science
Foundation. By 1965 it had become a permanent division of the American
Institute of Physics, and since then, as the Center for History of Physics,
the AIP program has continued its documentation efforts and has served a large
number of users as a research resources center and an information clearing
house. Through its newsletter and its projects over the last decade the
Center has stimulated and assisted archival activities in scores of institutions
and has enlisted the continuing support of physicists, historians and archivists.
Past projects of the Center included efforts to document the history of nuclear
physics, and current projects are concerned with the development of astrophysics
and the special archival problems of the large postwar national laboratories.
The Center's oral history program, and its catalog of sources for history of
physics and astronomy, are continuing activities. The basic financial support
of the Center by the American Institute of Physics is supplemented by contributions from individual physicists and by grants for specific projects. Heading
the Center is Spencer Weart, director, and Joan N. Warnow, associate director.
For many years there has been wide agreement that the AIP Center has had an
enormous positive impact on the documentation of twentieth century physics.[12]

Other major U.S. projects of more recent origin include the Bancroft
Library's project to document the development of the electronics industry and
the history of nuclear physics and chemistry at Berkeley, described by
Arthur Norberg in this symposium, and the Berkeley survey of physics sources
reported on by John Heilbron.

In the last few years there has been life in the biological sciences as
well. A major national project to document the history of biochemistry and
related fields is in full swing. Sponsored by a joint committee of the American
Academy of Arts and Sciences and the American Philosophical Society headed by

the distinguished biochemist John Edsall, the project has been very successful in locating and cataloging biographical information, personal papers and institutional records of a large group of significant biochemists and laboratories. David Bearman, who chairs this symposium, is responsible for the day-to-day work of the project at the American Philosophical Society Library in Philadelphia, and he works with Robert Kohler to accomplish its aims. The major support for the project comes from the National Science Foundation and the National Endowment for the Humanities.

In addition to these projects, active programs to collect papers of scientists have been underway at the Library of Congress and the American Philosophical Society Library. A number of university archives have made a serious commitment to preserve the papers of their own science faculty, including those at the University of Illinois, California Institute of Technology (reported on by Judith Goodstein at this symposium), and most recently, Massachusetts Institute of Technology where the new archival program is linked to an oral history program.

This partial list of activities is impressive. They have been made possible by financial support from scientists (notably the physicists), university budgets, the National Science Foundation, the National Endowment for the Humanities, and to a lesser extent, private foundations. The yield on these investments has been substantial when measured in terms of the rich resources that are now available for research and that otherwise might have been lost or destroyed. Included are collections of personal papers, oral history interviews, manuscript autobiographies, department and laboratory histories, and photographs. Catalogs, newsletters, sophisticated retrieval techniques, and an active information network among cooperating individuals and institutions--they are all part of a system that increases awareness of the need to preserve source materials and the opportunities to use them.

424

Uses of the collected resources have ranged from doctoral dissertations and monographs to popular articles and educational curriculum materials.

The science archives activities since the early 1960's have had other far ranging effects, including the raising of consciousness within the scientific community about the need to preserve source materials, the ways historians of science use documents, and the standards of historical scholarship. Historians and scientists have been brought into closer contact, and more historians are deliberately broadening the audience for their work to include scientists. Many, but not enough, collecting institutions have recognized their responsibilities in this area and have overcome their reluctance to tackle the problems involved in preserving the archival record of twentieth century science. Finally, a number of promising young historians have received unique on-the-job training by participating in these projects.

III. Needs for the Future

It will be clear to anyone looking back on the twentieth century that a tremendous spurt of archival activity took place in the 1960's and 1970's. Much has been done in the last two decades; it is a major accomplishment, and more of the same is needed. But before we congratulate ourselves, we ought to think about our responsibilities to the future. We have entered the last quarter of the twentieth century, and we should think seriously about what has to be done in the next two decades. What will the historical record be like when we are at the turn of the century? We should also think about the historical questions that will be asked about twentieth century science, and about who will be asking them. Certainly, all of our efforts should not be focused exclusively on providing source materials to help answer the questions of a relatively small group of historians of science.

Even from our own contemporary perspective, the gaps are clear. We need to make special efforts to preserve and make available the records of the government agencies and foundations that support science and the international scientific organizations and institutions where much of contemporary scientific activity takes place. In many cases, such files are destroyed everyday or are neglected and scattered.

We have made considerable progress in documenting the history of physics, and are beginning to move into certain areas of biology. But there is more to science than these fields. What sources will be available for study of the development of chemistry and its applications in the last half of this century? How adequate will the source materials be for documenting communications technology, nuclear engineering, metallurgy and materials science, and computer science? Are we to exclude the record of the applications and social impact of these fields, or of the responses of non-scientists and non-technologists to the changes in their lives brought about by these developments? We should bring our experience to bear on aspects of the history of science that may not engage many historians now, but that are likely to be of interest in the future as central issues in twentieth century science. I cannot imagine an adequate account of our times that would not pay serious attention to such developments as the consumer and environmental movements, the public response to the development of nuclear energy, and the increasing concern with government regulation of science and public participation in scientific-political issues. Certainly, we have more than enough to do in the special area of preserving the papers of scientists. These existing efforts need all the help they can get. But we should also be aware of neglected fields and issues and should devote at least some of our efforts to encourage and assist in the documentation of these important aspects of science.

My point is that we should begin planning for the future. I hope that we can begin such a discussion in this symposium. My own agenda would include efforts for closer coordination of projects' in various institutions, fields and countries to help identify gaps. We should also explore how to broaden the base of advisors and form alliances with other groups beyond historians of science and scientists. We need advice and cooperation from engineers, economists, sociologists, political scientists, public interest science advocates, humanists in general, science writers, and critics of science and technology. They can help inform us of their view of what is important about twentieth century science and technology, and where the source materials are. We ought to consider ways to use archival materials for educational purposes, not only for public understanding but to help meet the increasing need for informed public participation in decisions regarding science. This would also help to demonstrate novel uses of source materials and would increase appreciation of the need to preserve documents, tapes, and pictorial sources. Efforts are also needed to investigate appropriate techniques for retrieval of information from large collections for a variety of studies from different disciplinary perspectives.

Planning for the future should also include consideration of how to obtain more accounts of the day-to-day experience of individual scientists. Qualitative information of this kind, whether obtained through well-researched oral history interviews or from specially written autobiographies, can provide unique personal perspectives and much needed information to supplement the written record. They can also help identify and locate written materials that should be preserved. The experience obtained using this approach during the past fifteen years should be of great help in planning for the future.

Finally, I want to emphasize the need to get involved in documenting contemporary history. The efforts of the MIT Oral History Program to document the current recombinant DNA controversy have already produced a significant collection of correspondence and related documents, on-the-scene audio and video tapes of major events, and oral history interviews with about 80 individuals who have played significant roles in the United States and Europe in this major case of science and society interaction. The project was initiated in the spring of 1975, and is scheduled to complete its work in 1978. It is supported by the MIT Oral History Program and by a joint grant from the National Science Foundation and National Endowment for the Humanities. A significant portion of the collection is already available in the MIT Archives and is being used by scholars from a variety of disciplines. Following a contemporary scientific issue while it is still developing leads to relevant source materials that might otherwise escape notice. In addition, the perspectives gained can be useful in approaching the documentation of less recent developments. True, there are risks as well as benefits in doing contemporary history, and they must be understood and minimized. However, to repeat the words of the ISIS editorial 20 years ago, "We ought not to let our interest in the distant past absorb us so completely that we miss the beautiful opportunities before our eyes."[13]

NOTES

[1] Gerald Gruman, "Saving Our Primary Source Materials," ISIS, Vol. 48, No. 152 (June 1957), p. 184.

[2] Gerald Gruman, "Preserving the Stuff of History," Science, Vol. 127, No. 3313 (June 27, 1958).

[3] Ibid.

[4] Ibid.

[5] Nathan Reingold, personal communication.

[6] The Conference on Science Manuscripts, ISIS, Vol. 53, No. 171 (March 1962).

[7] John Archibald Wheeler, "Preface," in T. S. Kuhn, J. L. Heilbron, P. Forman, and L. Allen, eds., Sources for History of Quantum Physics (American Philosophical Society, Philadelphia, 1967), p. vi.

[8] Ibid., pp. vi-viii.

[9] Ibid., p. v.

[10] Ibid., p. vi.

[11] For a report on the project and an inventory of materials collected in its initial three years, see Kuhn, Heilbron, Forman and Allen, eds., op. cit.

[12] For a brief account of the approach of the AIP project in 1970, see Charles Weiner, "Resource Centers and Programs for the History of Physics," in Stephen G. Brush and Allen L. King, eds., History in the Teaching of Physics, Univ. Press of New England (Hanover, New Hampshire, 1972), pp. 47-62.

[13] Gruman, op. cit.

SCIENCE ARCHIVES DEVELOPMENTS IN THE WESTERN UNITED STATES

The Bancroft Library, the special collections and research library on
the University of California's Berkeley campus, has a regional focus,
and contains material - books, manuscripts, photographs, and paintings -
largely related to western North and Central America, from Alaska to
Panama, from the Rocky Mountains to the Pacific Ocean. The Library's
program for a history of science and technology archive at Berkeley is
organised also along regional lines and has both academic and industrial
foci.

The organisation of the College of Chemistry in 1912 was the first in a
series of important events in the history of twentieth-century science
in the western United States. Over the next two decades, this department contributed greatly to physical chemistry, and some of its staff,
notably Gilbert Newton Lewis, helped lay the foundation for a scientific
research tradition at Berkeley. The invention and development of the
cyclotron, and later advances in linear accelerators, occurred in and
around Berkeley and Stanford University. As historians have noted so
often, World War II accelerated the pace of scientific and technological
developments everywhere. At Berkeley, separate laboratories were established for biochemistry, radiochemistry, molecular biology, medical
physics, and a number of others. From 1950 an increasing number of
Nobel prize winners have resided on the campus. They and others have
not only been involved in research, but have played increasingly important
roles in national and international science, and participated in
formation of government science policy. The acquisition of the personal
papers of these scientists, which we are vigorously pursuing, is consistent with the program outlined in the early 1960s discussed earlier
by Professor Weiner.

The Library's program for the acquisition of these papers can be applied
to any academic discipline or economic development. It includes the
analysis of the significant historical events in the field of interest
and identification of the important persons connected with these events.
We began with two fields: first nuclear physics, which focussed on
the work of the University of California Radiation Laboratory
established by Ernest O. Lawrence in 1931. We chose an arbitrary
cutoff date of 1960, which is two years after Lawrence's death. (Our
hope is that the growing importance of the archive and the momentum
established in these early years will encourage the later deposit of
papers from physicists active in the post-1960 years, leaving us freer
to pursue detailed studies of other fields.) Analysis of events in the
field yielded a list of names - both persons on and off campus. With
those on the campus, or in institutions with no archival program, I
request a meeting at which I outline the goals of our program and why
I believe their papers are important. Usually, if papers exist, the
client and I examine them and I note which papers have historical
relevance. Of course, there is a need for clairvoyance here. It is
necessary to anticipate the needs of not only historians of science,

but of historians-at-large, sociologists, psychologists, political
scientists, and others.

Collections acquired in the first two years of the program, 1973
through 1975, were mostly personal papers of scientists and engineers.
Information obtained from these collections, as well as that obtained
from related and previously acquired collections in our Manuscripts
Division and the University Archives, helped us to understand the
strengths and weaknesses of the Library's holdings in the history of
science and technology. For example, the University Archives contains
a fine collection of documents concerning scientific activities on the
campus from the perspective of the President's and Dean's offices.
One weakness found was in administrative records of departments and
laboratories, especially for the period since 1945. Casual visits to
other campus buildings, plus discussions with campus figures, led to
the belief that this weakness could be rectified in some areas. Thus,
in the last two years, 1975 through the present, we emphasised organisational papers, both academic and industrial organisations.

The acquisition of such collections - personal papers of scientists
and academic department records - has proceeded apace, and libraries
have a respectable selection of correspondence with all sorts of
persons, research notebooks, manuscript drafts, and photographs. But,
the solicitation of personal accounts, either written autobiographies
or tape-recorded interviews, has not been pursued vigorously. Our
program has not neglected this aspect. Over the past four years we have
recorded interviews with 45 scientists and engineers, varying from 1 to
12 hours of tape. The interviews contain information on technical
developments, details surrounding discovery or innovation, educational
background, religious and political outlook, social structure of science,
and any involvement in government science. Emphasis on these latter
topics varies from person to person.

The early emphasis in interviewees was on individuals, but as time
passed emphasis shifted from individuals who pioneered to groups
involved in important subject development. The emphasis on subjects
is necessitated by the complexity of scientific and engineering growth
following World War II. Interviews obtained contain discussions of
nuclear physics, especially accelerator development, new elements,
radio isotope research, atomic weapons, photosynthesis, biochemistry,
microwave equipment, electrical engineering related business, persons
involved and funding for all of these. Usually, a person interviewed
fits into more than one of these categories.

Preparation for these interviews is extensive, and therein lies one of
the two serious problems connected with interviewing, not to mention
further the quality of conversations. It is superfluous to repeat in
an interview what can be found in the printed literature or available
correspondence. Hence, we examine both of these and develop our questions
from them, taking care to fill out details that are not present in
other sources. As you can imagine, this requires significant research
time on the part of the interviewer.

The second problem involves the staff time transcribing these tapes, putting the text in a form to allow the researcher to maximise his or her efficiency when reviewing them. This raises the cost. Various mechanisms can be used to keep such costs reasonable, but it is diffi- cult. Nevertheless, we should promote further such programs.

This interview program serves as a useful device for approaching engineers, those associated with the second field of foremost interest to us. Concurrent with the search for physicists' papers, we have analysed the growth of the electronics industry in northern California. This important center of radio communications had its origin in hydro- electric power development and ship-to-shore communication required by an expanding shipping industry. Besides the inherent interest in the growth of this center and its technology, there are interesting conn- ections between this and local nuclear physics equipment developments. For example, several electronic engineers provided advice and made devices for use in powering the cyclotrons. A number of our collections provide insight into this crossing of the interface between science and technology.

We proceeded with this field in the same manner as outlined above for the acquisition of paper and interviews of scientists, and we encountered two major problems. First, engineers outside academia seem not to be as ready paper savers as those inside. What they keep has intrinsic importance, but it tends largely to be information on processing or maintenance of equipment. As members of an organisation that has proprietary interests their correspondence and research notebooks remain in the firm's control for eventual shredding when the patents run out. (Though I should add in fairness that not all companies act this way. Some have developed important archival programs.) Usually, the engineers are quite willing to offer names of persons who might know where the material can be found. Here we confront our second problem. Directed to the storage warehouse for records one is suddenly faced with an over- whelming amount of material to search through for the desired documents - say 11,000 boxes of paper. This is an unparalled opportunity for preserving documents useful to many interdisciplinary studies; but it is also an enormous task. With the electronics companies, we have faced this problem often. Proper analysis of the material requires certain knowledge of that company's growth and product structure. Two criteria we apply to such companies is: Did their contributions advance tech- nology or cross the interface of science and technology? If they simply derive their products from the industry we will need a closer look at their effects on the region before deciding to accept or reject their papers.

While in the pre World War II period electronics related materials were produced in small quantities, the lack of preservation by the engineers limits the amount of material that can be retrieved. Sometimes this has meant accumulating historical documents piece by piece, and from the strangest sources. In this way we have gathered an archive for the history of radio communications on the Pacific Coast, related mostly to technical and economic developments. Concerning our larger task, we have transferred records of a few companies and continue to argue the criteria of choice for the larger collections. Space is always at a premium, and to take large collections of marginal material is not a good policy. Hopefully my next report will include a model for solving this problem of large quantities of organisational material in order to make it usefully available to scholars.

In closing I would like to add that while we have emphasised the physical sciences and electronic engineering, we have not neglected other sciences. Several important collections in the life sciences were sought and obtained over the past few years, adding strength to our holdings in these areas. In June and July we surveyed the geology/ paleontology activity of several persons now at Berkeley, and in the Fall we intend to do the same for medical physics. As the number of collections increases the time delay between acquisition and efficient access by scholars increases, but we are trying valiantly to cope with this problem. Because there is so much to be done, we constantly cooperate with other archives, always keeping in mind that it is the scholarly community we must all serve.

J.R.Goodstein

SCIENCE ARCHIVES DEVELOPMENTS IN THE WESTERN UNITED STATES

Caltech is a young school. In 1966 it celebrated its 75th anniversary. Two years later, the Institute appointed an archivist and formally opened its archives to scholars. The Caltech archives serve a dual purpose: they are the permanent records of the institution and the activities of those associated with it. The records of the past, in Caltech's case, included school publications; books published by the faculty; medals, trophies, and honorary degrees awarded to its professors; and finally, the papers of its past presidents. These materials formed the backbone of the Caltech archives.

In 1968, we also made the decision to make the history of science since 1900 the central focus of the Caltech archives. Our program to create a university science archives which would document the growth of science and technology in our own century is now in its 10th year. We are grateful to the organizers of this symposium for the opportunity to give a brief report on the first milestone of this enterprise.

The question we are frequently asked is: who provides the archives with its source material? The answer, in general, is Caltech's faculty and administrative officers. Ten years ago, the archives had two major research collections: the scientific correspondence and personal papers of physicist Robert A. Millikan and astronomer George Ellery Hale.   In the spring of 1977, the number of

research collections passed the 50 mark.   Today the archives house

approximately half-a-million documents.   Documents is a vague word.

By documents I mean correspondence, laboratory notebooks, reference

reports, unpublished manuscripts, autobiographical memoirs, lecture

notes, drafts of talks, even income tax returns.

       Let's take just a minute or two and put some people into

this archival laundry list.   The Millikan papers, for example,

include two extremely interesting oil-drop experiment notebooks;

there are also several drafts, in Millikan's hand, of his auto-

biography--parts of which, I might add, differ from the final published

version.   In the Howard Percy Robertson papers, there is a long and

rather personal exchange of letters between the American cosmologist

and the Polish physicist, Leopold Infeld.   To read Infeld's letters

is to gain a keener appreciation of events and personalities later

described in his public autobiography, Quest.   Infeld's colleague,

H. P. Robertson, served as a referee for Reviews of Modern Physics

in the 1930s; Robertson's short, direct, and sometimes acid comments

have been preserved along with student notebooks from his Göttingen

days, as well as his extensive and colorful correspondence with other

relativists and mathematical-physicists from around the world.   Eric

Temple Bell, the historian and mathematician, destroyed his scientific

papers, but saved his correspondence with science-fiction publishers

and fans: to them, he was John Taine, a raconteur of science fantasy.

The archives contain many of the stories Bell, alias Taine, published

in the pulp magazines from 1918 on, plus the small batch of correspond-

ence he wished historians to consult in the future.   Bell, who died in 1960, remains to this day, one of the few National Academy of Sciences members without an official Academy biographical memoir. There are in theoretical physicist Paul Epstein's papers letters from Arnold Sommerfeld, Albert Einstein, Wolfgang Pauli and Sigmund Freud. Epstein served as physics colloquium chairman at Caltech for many years; he made a point to take very careful notes on what his speakers, including Paul Ehrenfest and Erwin Schrödinger, said in Room 201 in the Norman Bridge Laboratory of Physics (East Bridge). These items, too, are preserved in the Epstein papers. Russian-born and Leiden-trained, Epstein brought with him to Pasadena in 1921 family letters written in the 1880s, and the habit of exchanging scientific information with colleagues who remained in Europe. His papers are a fruitful starting-point for anyone interested in the transmission of quantum physics from the old world to the new one. I chose my people examples from the physical sciences; my apologies to the biologists in the audience.

Before I mention some projects in progress now, let me just sketch in, very quickly, some other components of the Caltech archives. In addition to written records, we also house

-- Millikan's scientific library, also Epstein's (as well as his reprint collection)

-- about 5,000 photographs, both candid and formal, of scientists. The photographs date from the latter part of the 19th century and include scenes in the laboratory, at international meetings, and during off-duty hours; the photograph of Einstein riding a bicycle   is, of course, well known.

-- a tape-cassette library which contains a number of unusual
recordings: Linus Pauling's 75th birthday party, Max
Delbrück and Murray Gell-Mann's Nobel Prize dinner and
Richard Feynman's classroom lectures on freshman and
sophomore physics.

In several instances, the Caltech archives have been offered
important related collections from off-campus sources. We presently
hold the papers of a pioneering American physicist Frederick Bedell;
the nineteenth-century family correspondence of American chemist
Edward Morley; and in xerox form, the extensive correspondence of
the Italian mathematician and relativist, Tullio Levi-Civita. (The
originals are deposited in the National Academy of the Lincei in
Rome).

Less than 10 percent of our research collections are un-
processed; the rest are organized and available for use by scholars.
The Hale papers have long been available to scholars on microfilm;
the microfilm edition of the Robert A. Millikan collection will be
available shortly. This project, now nearing completion, has been
financed by a grant from the National Historical Publications and
Records Commission.

Caltech's research collections are supported, in part, by
a grant from the National Endowment for the Humanities. The Endowment
is currently providing support for the processing of the presidential
papers of Lee A. DuBridge (who succeeded Millikan as Institute head
in 1946) and the scientific correspondence and personal papers of
aerodynamicist Theodore von Kármán. The Kármán collection will be

finished and available for use in the late fall; the DuBridge papers
we expect to complete a year from now.

Kármán's papers are a natural starting-point for investigating a wide variety of topics including 1) the racial persecution
of scientists in Germany and Italy, 2) the rise of the aircraft
industry in the United States, 3) the role of the Guggenheim Foundation
in the development of the aeronautical sciences, 4) the collapse of
the Tacoma Narrows bridge, 5) early helicopter designs, 6) the growth
of aeronautics in China and Japan before World War II, 7) the development of the Aachen Aerodynamic Institute between 1912 (when Kármán
became Director) and 1930 (when he accepted a similar position at the
newly established Guggenheim Aeronautical Laboratory at Caltech, 8)
international cooperation in science--because Kármán practiced what
most people only preach--, 9) the problem of jet propellants in 1939,
and above all, 10) the role of scientists in government, in Kármán's
case, his association with the military, particularly the Air Force.

The involvement of scientists in matters of state presents
a problem of some importance for the community of scholars interested
in the recent history of science.   After 1945, for example, it
becomes increasingly difficult to document with precision just what
scientists were up to in Washington.   These governmental files, in
general, are classified; such files have a tendency to disappear from
the historian's (and ultimately the public's) view.   Such files, if
they arrive in the archives at all, contain only confirmation of
travel plans to Washington, and related per diem expense vouchers.

Let me close this review of the Caltech archives by relating
an incident to you involving the governmental files of Richard

438

Chace Tolman, professor of both physics and chemistry at Caltech for many, many years, and who, during World War II, served among other things, as scientific advisor to General Groves, the Army head of the Manhattan District Project.   After the war, he became scientific advisor to Bernard Baruch at the United Nations.   At the time of his death, in 1948, Tolman was also chairman of the Declassification Board of the Atomic Energy Commission and a consultant, besides, to the Commission.   Shortly afterward, a local AEC agent went all through Tolman's files at Caltech and removed everything relating to the atomic energy program.   The material taken included unclassified as well as highly classified files, documents and correspondence.   The material was catalogued and delivered to the Office of Santa Fe Directed Operations at Los Alamos, and that was the end of the story until last spring.   In April, 1977, we wrote to the AEC and requested the return of the Tolman AEC material in order that we might incorporate it with the rest of his papers in the archives.   This story does not have a happy ending.   Only last month I learned that the records of Tolman's involvement with the AEC were largely destroyed in 1961, apparently because the government ran out of storage room at Los Alamos just then.   No list exists of what was destroyed.   On the other hand, some of his Declassification Board files--about 12 folders of reports--did turn up, some still classified.   We will now go ahead and order copies of the unclassified material.

This is a story without heroes or villains.   No one quarrels with the government's need to restrict access to certain records, or to safeguard those records, as it did, just after Tolman's death.

The quarrel is over what was subsequently done to those records--in
1961.   It is safe to say, I think, that the yet-to-be written history
of the atom in our century will be poorer because of a senseless
bureaucratic decision taken only 15 years ago.

Symposium 8.   RELATIONS BETWEEN THEORIES OF HEREDITY AND EVOLUTION, 1880-1920

INTRODUCTION TO SYMPOSIUM 8

The choice of topic represents an explicit attempt to escape from the domination of international meetings by the Darwin and Mendel research industries and to distance ourselves from conventional presentations of the evolution of genetics as an independent science which only became integrated with the theory of evolution in the 1930s. This neo-darwinian synthesis is generally pictured as a result of endeavours that can be traced back to the seminal paper of R. A. Fisher in 1918. This orthodox view will be examined critically by a study of R. A. Fisher's motivation and achievement in 1918.

In the nineteenth century theories of evolution and of heredity were linked through the question of the inheritance of variations. Varying degrees of emphasis were given to variations acquired during the life of the individual and allegedly inherited. Extreme positions were taken by Weismann against such inheritance and by Spencer in support of it. Frederick Churchill expresses the view that facts alone could not decide the issue, but that behind their opposing stances lay Weismann's sympathy for the naturalist tradition and Spencer's for the physiological and reductionist tradition. This, according to Churchill, may well account for the enthusiastic welcome given to Weismann's ideas in Britain where the naturalist tradition flourished, whereas the German community, with its great strength in cytology, was strongly critical.

In his commentary on Churchill's paper, Ernst Mayr pointed to the rapidity with which de Vries and others rebutted Weismann's claim of the isolation of the germ plasm on cytological grounds. Also Weismann's view of cellular differentiation as a distribution of qualitatively different factors to different cells was seen already in the 1880s as a fundamental error. Continuing acceptance of the inheritance of acquired characters by botanists and zoologists until the 1930s was partly due to the satis-factory explanation it offered of continuous variation over climatic gradients.

Michael Ruse, in his commentary, asked how important a figure Herbert Spencer was by the time of his controversy with Weismann, and as a corollary, whether the controversy chosen was an important one. Ruse felt that medical opinion of the 1880s and 1890s might well have supported the continued acceptance of the inheritance of acquired characters. A pangenetic conception of heredity seems to have been behind the assumption then current in medical circles that frequent masturbation impoverished the physique and intellectual powers. Wallace's support of Weismann, on the other hand, might have arisen from Wallace's studies of the poly-morphic butterflies whose striking variety of forms seemed not explicable in terms of the inheritance of acquired characters.

Brigitte Hoppe added to the historical context of the discussion by drawing some parallels and contrasts between Weismann's Das Keimplasma, eine Theorie der Vererbung (1892) and Nägeli's Mechanisch-physiologische Theorie der Abstammungslehre (1884). Although a prominent exponent of the cytology of the 1840s Nägeli did not acquire all the new facts of the changed cytology of the 1880s, and cytology did not play an important role in his Abstammungslehre.

Hugo de Vries' contributions to theories of heredity and evolution have found a place in the history of science because of his rediscovery of the Mendelian theory and laws of inheritance and his introduction of the mutation theory of evolution. Heimans' paper aims to retrieve de Vries from this "shadow of Mendel" image, and to exhibit the remarkable prescience of his hypothesis of intracellular pangenesis (1889) and its role as a conceptual scheme for his ensuing experimental programme. Lindley Darden sounded a note of caution on de Vries' anticipations of our modern understanding of the relations between chromosomes and genes. She felt that de Vries' mutation theory has important implications for the question of the relations between theories of heredity and evolution.

One thing the history of the period between 1880 and 1920 can teach us is that one and the same theory might have been both a theory of heredity and a theory of evolution, as indeed de Vries' mutation theory was. It just turned out that the world is such that two different theories were required and then their relations had to be sought. As we look back at Darwin's theory of natural selection, we see the lack of a theory of heredity as an incompleteness in his theory. Darwin's extensive catalogues of variations provided evidence for its existence but a reasonable question arises: what is the cause of variation and how is variation inherited? Darwin tried unsuccessfully to supplement the theory of natural selection with his theory of pangenesis in order to resolve that incompleteness. But others, including the neo-Lamarckians and de Vries, saw a theory of the causes of variation, not as filling an incompleteness in Darwin's theory of natural selection, but as an alternative theory. Theirs were theories of heredity that were at the same time theories of the transformation of species. If, indeed, acquired characters were inherited and a suitable hereditary mechanism could have been given to explain that inheritance, then that mechanism could also have served to give rise to well-adapted species without the mechanism of selection. Lamarck himself had used such inheritance to explain the diversity and adaptedness, though not the extent of complexity, of all life in his own theory. Curiously, both Lamarck and the neo-Lamarckians were more concerned to give a physiological mechanism for explaining how the characters were acquired than to give a hereditary mechanism for explaining how they are transmitted to offspring. Nonetheless, had such a theory of the inheritance of acquired characters been provided it would have rivalled both Darwin's theory of natural selection and his theory of pangenesis as a unified account of heredity and the origin of species. To be sure, Darwin's pangenesis did provide a mechanism for the inheritance of acquired characters, but Darwin saw it as merely another source of variation and not as a substitute for selection.

Hugo de Vries, on the other hand, provided, in the mutation theory, a theory that was at one and the same time a theory of heredity and a theory of species transformation. He did not, however, provide an explanation for the inheritance of acquired characters which he believed did not occur. Professor Heimans has correctly pointed out that the mutation theory is merely an extension of intracellular pangenesis; this is an important point missed by historians who treat de Vries' pangenesis as his theory of heredity and the mutation theory as a separate theory of evolution.

The unification accomplished by the mutation theory is an elegant one. Varieties were distinguished from species on a hereditary basis, namely different types of changes in pangenes occurred in each case. And such changes could even be detected experimentally: varieties showed segregation; species did not. Wouldn't it have been convenient if a simple genetic test could distinguish species and varieties? The problem of an adequate species definition would have been solved so simply. Molecular geneticists would, most likely, have been able to find the molecular changes that resulted in a progressive mutation and thus the molecular differences between species. Our whole study of species change would have been on a much more exact molecular basis than is possible to-day. Unfortunately, the world is not so simple. De Vries' elegant and practical mutation theory did not withstand the test of further experiments. The mutations that Thomas Hunt Morgan and other geneticists investigated were smaller scale changes that did segregate and did not, with one jump, give rise to new species.

Darden noted de Vries' de-emphasis of the need to explain adaptation: De Vries' mutations were internally caused with no necessary connection to the environment in which the species was found. De Vries advances this feature of his theory as an answer to the problem plaguing selection theories. In The Mutation Theory he notes that many characters which separate species are not of "particular advantage in the struggle for existence, but are as a rule useless and inconsiderable." (Op. cit. vol. i, p. 65). His own theory accounts for this fact: "useless, but not dangerous mutations must appear as often as useful ones, and have almost as much likelihood as these of persisting." (Ibid.) He does concede a small role to selection: species, once they have arisen by mutation, may be eliminated by selection if they are unfit for their environment. Thus, species, not individuals, are the units of selection; furthermore, he stresses, selection does not create, it only eliminates. (Op. cit. vol. i, p. 199).
The failure of de Vries' mutation theory has much to teach us about the growth of scientific knowledge. The emergence of separate fields of science such as evolutionary studies of selection in populations and genetic studies of heredity in individuals are a result of the way the world happens to be. It might have been the case that the study of heredity and the origin of species were united and one mechanism within one theory might have explained both. But empirical evidence did not support such unified theories as neo-Lamarckism or the mutation theory. Mutation and selection are two separate processes leading to the origin of species. Genetics emerged as a separate field of study and later had to be synthesized with evolutionary studies. Hence, we are to-day discussing their relations.
Secondly, de Vries' theory provides an example of a disagreement between scientists attempting to explain the same phenomena as to important aspects of that phenomena. In this case adaptation was an important item that Darwin included in his theory's domain; no theory of the origin of species would have been adequate, in Darwin's eyes, if it had not explained adaptation. In sharp contrast, de Vries emphasized that his theory could explain characters that lacked adaptive value. This dispute is still alive to-day with the debate about the existence and

importance of neutral mutations. We learn from this dispute that
phenomena are not simply out there, ready to be observed and seen in
need of explanation. Their importance and very existence can be a
matter of legitimate scientific debate. Adaptation is perhaps more
problematic than other empirical items since it requires a judgment
about the relation between an organism and its environment; whether
a character is adaptative or not may be difficult for the observer to
determine. The fact remains that this shifting of importance of items
to be explained is a prevalent aspect of the evaluation of competing
theories and needs the attention of historians and philosophers of
science.
A third lesson can be learned from the fate of de Vries' theory: the
simplest, most unified, most practical theory may not turn out to be
correct. De Vries' mutation theory was an elegant account which
proposed a single mechanism for hereditary change and species change.
Furthermore, it provided a practical, experimental test to distinguish varieties and species. But these simple and practical aspects
of the theory turned out to be incorrect. Consequently, simplicity,
unity or practicality are not necessarily reliable criteria for
empirical adequacy, contrary to claims often advanced by scientists
and philosophers.

Darden concluded that the legacy left by de Vries' mutation theory "was
not the concrete proposals about the origin of species but the search
for the genetic basis of variation and the use of experiments to tackle
the problems."

In his paper on the neo-darwinian synthesis, Bernard Norton examines
R. A. Fisher's putative role as biometric-Mendelian synthesiser. He
sees Fisher's 1918 paper "not simply as an attempt to resolve certain
'internal' problems within genetics, but also, and predominantly, a contribution to the hereditarian social ideology of eugenics." He goes on
to show that the sense in which Fisher did synthesise Darwinism, biometry and Mendelism does not relate simply and directly to the questions
of the early pre-synthetic British Mendelians. In his commentary, Jonathan Hodge agreed with Norton that

we are bound to be misled if we read Fisher's 1918 paper as part, ultimately, of a larger movement eventuating in this new synthesis of
the 1930s. Norton suggests, to be sure, that there is 'a strong
sense' in which Fisher did synthesize Darwinism [i.e. roughly, grad-
ualistic selectionism] biometry and Mendelism'. Moreover, he also
implies that such a synthesis was eventually to constitute the core
of the new synthesis in the sense of the books by Dobzshansky, Huxley,
Mayr and others. But he has sensed that to join Provine and others
in (a) seeing Fisher as primarily resolving the biometrician-Mendelian
debate and in (b) seeing that resolution as central to the synthetic
theory is inevitably to risk misreading Fisher in 1918 as a knowing
contributor to the problems later addressed by the synthetic theorists.

Hodge set out three contrasts introduced by Norton:

(a) between _internal_ and _external_ determinants of scientific change
(b) between _reasons_ and _motivation_ and (c) between the _science of_

genetics and the ideology of eugenics. For he is suggesting that
Fisher's thinking in the 1918 paper was guided not only by scientific
reasonings internal to genetics but also by ideological motivation
external to science. My own view is that in this particular case -
the early Fisher and his 1918 paper - the distinction between reasons
and motivation is useful, but that the other two contrasts, between
the internal and scientific and the external and ideological determinants of Fisher's problem-situation, are not applied quite so appropriately here.

In making a new start, let us characterise a branch of applied science
that we might call population-genetical engineering. This applied
science is essentially the study of good breeding. It is thus normative, but not restricted to the human species (The word "eugenics"
could be used to name this general science of good breeding, but it
seems that it has usually been used only in connection, specifically,
with human breeding). Someone working out the best breeding strategy
for enhancing egg production in chickens, milk yield in cows or rust
resistance in wheat is combining normative judgments as to what are
desired ends with technical judgments as to what are effective means
to those ends. So the problems here are to decide on the ends and to
discover how to maximise the achieving of those ends within various
biological and economic constraints.

Now, obviously, unless human heredity is altogether unlike animal and
plant heredity in almost every way, and it plainly is not, then the
findings as to means of the general populational-genetic engineer may
well be applicable, in principle though quite possibly indeed probably not in practice, to our own species. Moreover, the applicability of these findings to the human species will be independent of
any particular choice made of the specific eugenetical goals for mankind. To accept that the breeding strategies used to raise milk yield
in cows can be applied to humans is to leave open the question of
whether they should be used in attempts to raise average height or to
lower it, to raise mean I.Q. test scores or to lower them.

We can see now how Fisher's paper could at one and the same time have
(a) been motivated by the specific ideological eugenist ends to which
he explicitly subscribed throughout his career and which are most
familiar from the later chapters of his 1930 book and (b) consisted,
almost entirely, of reasonings which were directly applicable to population genetic engineering problems as to the best means for breeding
strategies in animals and plants no less than in humans. Moreover, we
can see that for Fisher himself the paper must have been construed
in this twofold manner. For, of course, given that human and animal
and plant heredity were taken to often work in the same way - most
especially to be Mendelian - then any treatment of the theory of population genetics which could contribute to the motivating goals would
have to consist of reasonings applicable to animals and plant breeding
as well. I would, then, want to modify Norton's description of
Fisher's paper as "predominantly... a contribution to the hereditarian
social ideology of eugenics;" I would suggest, instead, "predominantly
a contribution to hereditarian principles of general populational genetic engineering."

Hodge turned to Fisher's introduction of his 1918 paper to expose its
connection with earlier work.

His opening sentences (a) acknowledge that several attempts have already been made to interpret the empirical findings of biometry according to Mendelian inheritance (b) explain that his own attempt will
trace these findings to a wider range of Mendelian postulates than before, he will try, that is, "to ascertain the biometrical properties
of a population of a more general type than has hitherto been examined,
inheritance in which follows this [the Mendelian] scheme" and (c) expresses the hope that this will make possible "a more exact analysis
of the causes of human variability." In the two paragraphs closing
his introduction, he explains how he is going beyond existing attempts
along these lines both in generality of analysis and in success in
ascribing variance to heritable factors.

He thereby shows us, in effect, how his specific hereditarian eugenic
goals and his general aims as a Mendelian interpreter of biometry are
mutually reinforcing because crucially consilient. They are crucially
consilient, particularly, on the decisive matter of dominance. And
why should dominance be so decisive? As Fisher recounts, Yule had
shown that the assumption of incomplete dominance would allow a Mendelian to explain why Pearson found the parental and fraternal correlations to be too low on the standard Mendelian assumptions. Then
he stresses that Yule has shown the similarity in the effects of dominance, in general, and of the environment in reducing the correlations between relatives. But, says Fisher, Yule has stated that the
two reudctions, the one caused by dominance, the other by the environment are identical. But luckily, Fisher says, there is a remarkable
exception to this statement not discovered by Yule. Luckily, because
it is an exception "which enables us, as far as existing statistics
allow to separate them and to estimate how much of the total variance
is due to dominance and how much to arbitrary outside causes". I.e.
how much is due to dominance and how much to non-additive genetic and
environmental factors.

Here we have, then, a major clue to Fisher's whole paper and indeed to
several subsequent ones too. His great achievement, as he sees it, is
to separate variance due to dominance from largely-but not entirely-
environmentally caused variance, and thereby to claim the former for
heredity, for nature as opposed to nurture. We have here one source
at least for his lifelong preoccupation with dominance - in the 1922
paper on the dominance ratio for example.

These general points suggested by the centrality of dominance variance
for Fisher's early work, allow us to appreciate and reinforce without
reservation Norton's perceptive remarks about Fisher after 1918, and
more especially his valuable thesis which I take to assert the following:

1. Contrary to legend the British Mendelians before Fisher were explaining some continuous phenotypic variation as due to discontinuous
Mendelian genotypic differences.
2. So Fisher and Bateson did not disagree on this question of a syn-
chronic variational continuity and discontinuity (Norton's discontinuity 1.).
3. Where Fisher and Bateson did disagree was over diachronic phylogenetic and speciational discontinuity.
4. This issue concerning phylogenetic and speciational discontinuity
was not one that Fisher's 1918 paper or even his 1930 book (which is

explicitly on natural selection and not evolution by means of natural selection) attempted to resolve. For, of course, theories of the evolution of dominance, of multifactorial inheritance and balanced polymorphism leave such issues in evolutionary biology untouched. This thesis (which I welcome) suggests once again that Fisher was both a eugenist and a Mendelian biometrician but not an evolutionary biologist.

The first two decades of this century also saw Johannsen's formulation and defence of his genotype theory. Roll-Hansen commented on Johannsen's guarded recognition in 1923 of the important contributions made by the chromosome theory of heredity. Hereditary factors, he considered to be "local conditions", and those experimentally demonstrated were
"expressions for local deviations from the original ('normal') constitutional state in the chromosome". If this is the case, argued Johannsen, then all the impressive achievements of experimental Mendelian genetics concern only "chromosomal irregularities, disturbances and diseases" and has contributed nothing to our "understanding of the 'normal' constitution of natural biotypes", he concludes. For Johannsen this means that the origin of species is as mysterious as before: "The Problem of Species, Evolution, does not seem to be approached seriously through Mendelism nor through the related modern experiences in mutation". This defence of the genotype theory is continued in the third and last edition of Johannsen's great textbook Elemente der exakten Erblichkeitslehre (1926).
Though Johannsen was persistently critical of "Darwinism", his genotype theory implied reform of current "Darwinist" theories of evolution rather than wholesale rejection. The transformation of species was an unquestionable presupposition for Johannsen. It was the problem of the hereditary mechanism, in particular of hereditary change, that led to disagreement. Johannsen's view of "scientific Darwinism" was that by accepting "the fact of mutation Darwinism can save large parts of the theory of selection". He also pointed out that mutations "can be quite small, and in so far approximate continuity". In general Johannsen's genotype theory represented a large step toward the neo-Darwinian theory of evolution that emerged in the nineteen-twenties. Johannsen's genotype theory contributed considerably to a clearer distinction of genetics from evolutionary and embryological studies. Inspired by Claude Bernard's methodology of biology Johannsen strove to establish the study of heredity as an experimental discipline independent of evolutionary studies. Already in 1896 he underlined that while every theory of evolution presupposes a theory of inheritance, the study of heredity does not depend on theories of evolution.
The main weakness of Johannsen's theory with respect to evolution was the neglect of genetic recombination. He saw its importance, for instance in the hybridization experiments of Nilsson-Ehle, but apparently the theory of genetic types did not give an adequate grasp of the phenomenon. Here the chromosome theory of the Morgan Group proved superior.

In concluding the symposium Professor Ghilarov referred to his teacher Academician Ivan Schmalhausen, the first "who tried successfully to link the controversy of the phenotype and genotype in micro-evolution" when he

449

introduced the concept of phenocopies. Ghilarov commented on new possibilities of hereditary change which modern science has revealed and he looked forward to a new and fruitful synthesis between genetics, evolutionary studies, ecology and the biology of development.

Dr. Frederick B. Churchill

THE WEISMANN-SPENCER CONTROVERSY OVER THE INHERITANCE OF ACQUIRED CHARACTERS

In the published synopses of this symposium Dr. Norton remarks with approval that historians of biology are moving beyond the detailed examination of Darwin to the fate of Darwin's theory, "once it passed from his hands." I agree completely with the fitness of this trend, for I believe we will not greatly improve our understanding of the Darwinian debates of the 1860s and 1870s until we gain a broader perspective. I feel it appropriate to congratulate the program committee for resisting the temptation of focussing once again on Darwin's generation and for encouraging us to forge into the twentieth century. We need not forget our love affair with Darwin; who knows, we might even enhance it thereby.

My assignment is to examine the controversies over the inheritance of acquired characters which took place during the last two decades of the nineteenth century. The period and the task form a natural stepping stone into our century with its rediscovery of Mendel, its debates between the Mendelians and biometricians and the rise of classical chromosomal genetics. The decades between 1880 and 1900 present a period of an extraordinary profusion of explanations of evolution, which in the final analysis distill down to a multitude of theories about variation and heredity. For this reason it is perhaps appropriate to describe the period as the age of "over-rationalism." It is a period, too, when experimental embryology and cytology became the vanguard disciplines of biology; perhaps for this reason we should call it the age of "classical microscopy." I shall argue here, however, that there is more to this period than either of these exaggerated appellations encompass.

As for my specific task I quickly discovered how overwhelming the data was which proponents exploited in an effort to prove that acquired characters were inherited.[1] Medicine, pathology, experimental physiology, folklore, paleontology, natural history and human history all offered up their tribute to the Lamarckian alter. The advocates of transmission were clearly on the offensive in producing empirical examples to document their cause. Although much of this was of hearsay quality and must be understood as such today, some of it was not. Those who argued against such inheritance found themselves in the awkward position of proving a negative. This meant refuting the empirical proofs of their opponents or resorting to new principles which were not universally accepted or even understood.

The dilemma is well illustrated by the most famous experiments of the period, those of Brown-Séquard.[2] Over a period of three decades this Franco-American physiologist had

induced epileptic behavior and certain lesions by an incision
in the neck region of guinea pigs. Some offspring of the
animals displayed the same behavior or lesions without under-
going the initial incision. The experiments were difficult
to perform, yet Westphal and Obersteiner both claimed to
have reproduced them in the 1870s.[3] Romanes, after many
failures, reported success in the early nineties.[4] The
opponents of the transmission of such traits were forced
either to explain away the experiments on a priori grounds
or to devote considerable effort reproducing them and establishing experimentally a different causal sequence. For
Weismann and Ziegler, both of whom were committed neo-
Darwinians, the former strategy was sufficient.[5] For Romanes,
who believed in some inheritance of acquired traits, the
latter strategy alone sufficed.

How is the historian to understand such an affair? Short
of trying to reproduce the difficult experiments himself,
he must fall back on modern biology, which appropriately
proceeds unconcerned about such anomalies.[6] Modern genetics,
developed out of and amply established upon independent
grounds, has simply undercut the value of this kind of
evidence. Obviously what is of greater interest for the
historian is the context of this and similar disputes. We
need to come to grips with the prevailing change which
relegated Brown-Séquard's guinea pigs to the graveyard of
discarded data. I propose to do this by an analysis and
comparison of the ideas of August Weismann and Herbert
Spencer. Such a comparison is particularly appropriate for
this conference in Edinburgh, for it sheds some light on the
nature of biology in the British Isles.

AUGUST WEISMANN

In a longer version of this paper, I examine in detail
the development of the evolutionary ideas of both Weismann
and Spencer. To get on with the comparison I have assigned
myself, I must, however, limit the exegesis of their respective ideas to a hurried and somewhat harried summary. I turn
first to Weismann.

In 1864, nearly at the outset of his career as a zoologist,
Weismann suffered hyperemia of the retinas, an illness which
forced him to abandon for ten years his microscopical research.[7] As a young Dozent in Freiburg, Weismann turned to
the theoretical aspects of evolution, but unlike his exact
contemporary, Ernst Haeckel, he never presented biological
evolution as one aspect of a general cosmic process. The
research he carried out during the years of eye trouble may
be best described as natural history. It is characteristic
of this work, which included an extensive examination of the
protective and warning markings of sphinx catepillars, a deep
interest in seasonal polymorphism in butterflies, and a
brief examination of the curious metamorphosis of the Mexican
Axoloth, that Weismann constantly sought an evolutionary
explanation in terms of natural selection.[8] It is also
characteristic of this period in his life that Weismann

presented evolution in terms of changing and developing
natural populations. For example, in reference to the well-
known Steinheimer fossil snails, which presented a series of
interlocking paleontological types, Weismann declared that
"the constancy in a species doesn't begin suddenly but rather
gradually and it consists of the cross breeding of all
individuals."[9]

It is not easy to determine when Weismann abandoned all
recourse to the inheritance of acquired characters. Because
he never promoted the doctrine in the first place, there
was no spellbinding moment of grace and dramatic conversion.
In retrospect, Weismann claimed to believe in Lamarckian
inheritance in his earlier days and cited his 1883 pro-Rector's
address on "Heredity" as his first public challenge of such
a transmission.[10]

What is easier to assertain is the basis of Weismann's
public assault on the Lamarckian doctrine. His distinction
between the germ-plasm, on the one hand, and the somatoplasm,
on the other, increasingly articulated by Weismann during the
1880s, threw up a logical barrier between the hereditary
material and the environmentally malleable soma. A respite
between 1875 and 1886 from the hyperemia permitted Weismann
to follow and contribute to the exciting research on cell
division. With the assistance of his students, most notably
Ischikawa and Häcker, Weismann also proposed that a special
reducing division occurred during the process of gametogenesis.
This significant discovery led Weismann into further speculations about the fine structure of the germ-plasm. It
encouraged him to define life in terms of assimilating and
self-replicating units and to heap further doubt upon the
uncritical acceptance of the transmission of acquired traits.
Finally, his renewed interest in cytology allowed Weismann
to develop his grand synthesis, which appeared in the last
days of 1892 as the six hundred page volume, Das Keimplasma.
Havelock Ellis, with a keen eye for the controversial,
arranged for its nearly simultaneous publication in English.[11]

Others, as well as I, have considered Das Keimplasma a
comprehensive synthesis of hereditary, developmental and
evolutionary theory. Its nineteenth century readers, however,
did not find chapters dealing with phylogenies or the paleontological record; nor did Weismann make an effort to establish evolution on morphological, embryological, or taxonomic
grounds although these were subjects traditionally found in
volumes by Darwin, Wallace, Spencer, Haeckel, Eimer, and
Cope. Das Keimplasma presented a different kind of synthesis.
It was the evolution of germ-plasm which concerned Weismann,
and to some extent he reflected Carl Nägeli's mechanico-
physiological theory[12] of eight years earlier. But there is
an important difference here, too. Unlike Nägeli, Weismann
was actively promoting the neo-Darwinian cause, and when
considered on the germinal level, this meant the evolution
of the collective germ-plasms -- a population of genomes, to
use an anachronistic term. I feel that here is how Weismann's
deep interest in natural history exerted itself. In the 1870s

he had become absorbed in how different populations, be they
amphibians, snails, or polymorphic butterflies, evolved
through natural selection. After 1881 with his renewed microscopical research well underway, Weismann applied the same
population mentality to the germinal level. Natural selection
worked, in his mind, not on expressed traits distributed over
a varying landscape, but on combinations and recombinations
of germ-plasms -- hence Weismann's sustained and penetrating
analysis of reduction division and Amphimixis, as he called
the sequential loss and mixture of ancestral germ-plasms.
In contrast to earlier evolutionary syntheses, which were
concerned about generation in general, the Keimplasma of
1892 presented a theory of evolution through sexual reproduction.

Now, I don't want to be misunderstood. Weismann does not
deal with germ-plasm populations in a quantitative way be-
fitting 20th century genetics. This is perhaps his greatest
shortcoming and helped bring down the castle he constructed.
But all the germinal structures for which Weismann became
famous, the idants, ids, determinants, biophors, accessory
and reserve germ-plasms, and all the processes he attempted
to assign to them, qualitative as these were, made sense in
neo-Darwinian terms only when visualized as evolving populations of germinal constituents.

HERBERT SPENCER

It may seem inappropriate to contrast this seasoned biologist with the philosopher, Herbert Spencer. Spencer, however, was informed; he continually consulted with his fellow
members of the X-club, among whom numbered Huxley, Hooker,
and Lubbock. Furthermore, Spencer possessed an obsession
to articulate every alternative to and assumption in his
philosophy of nature. This trait renders his Synthetic
Philosophy an invaluable mirror of early Victorian science.
We do the history of biology a disservice when we fail to
exploit the conventional wisdom which can be found in
Spencer's writings.

With reluctance I not only skip over Spencer's earliest
writings on evolution,[13] but I abbreviate to a few aphorisms
the prolegomenon of Spencer's Synthetic Philosophy, the
First Principles of 1862.[14] First, as Medawar and Bowler
have pointed out anew,[15] Spencer used the term "Evolution"
in an embryological sense and applied it to all developmental
processes from the rise of the solar system to the ontogeny
of the chick. Second, Spencer maintained as a descriptive
generalization of evolution that homogeneity was an unstable
state and tended to evolve into heterogeneity. Third, he
was certain that the causal explanation for this evolutionary pattern must lie in the supposed circumstance that single
events seemed to produce multiple causes. Although Spencer
drew upon the first and second laws of thermodynamics, it is
clear that he was not conversant with the implications of a
general dissipation of energy. The world, as Spencer knew it,
existed in a perpetual and non-random flux on the molecular
level; this activity in the aggregate entailed evolution.

Within the covers of his First Principles Spencer remained
content to describe organic development, adaptation, and
heredity as the moving equilibrium of a single life cycle.
Yet the paradox even here was explicit: How was it that a
homogeneous germ consistently repeated the evolution of the
parental type? In common prose, how was it that like begat
like?

We must turn to Spencer's Principles of Biology, which
appeared in serial form between 1863 and 1867, to find the
particulars concerning life and heredity.[16] It is fortunate
that philosophers are at heart exhibitionists and don't blush
at exposing the undergarments of their science. Appropriately
we find Spencer being very direct in defining life as "The
continuous adjustment of internal relations to external
relations."[17] It was a dynamic or physiological definition
which contrasted with Weismann's. We find him also placing
a premium on the process of growth, which bespoke merely a
surplus of matter over expenditure, and on development,
which entailed simply an increase in structure. As fore-
shadowed in the First Principles, growth and development
together encompassed the essence of biological evolution.[18]
Furthermore, Spencer reinforced his reductionist presentation
when he drew the commonplace analogy of, on the one hand,
the growth of crystals through the selective assimilation of
chemicals from a medium with, on the other hand, the evolution
of the organism through the selective assimilation of special
"physiological" units from the body fluids.[19] Today we might
consider Spencer's perspective as a forerunner of bio-
energetics. In mid-nineteenth century it comprised a very
simple biological budgeting which consisted merely in the
reckoning of the intake and use of matter and "force".

Spencer's views of heredity flowed directly from his
simplified physiology. He claimed there existed an inverse
connection between offspring production and parental growth
and development. This inverse relationship, he felt, was
maintained throughout the plant and animal kingdoms; it was
maintained whether the reproduction occurred through budding,
fission, gemmation, or sexual pairing. As for sexual
reproduction or "gamogenesis," as Spencer called it, there
took place nothing unique. A mere "... coalescence of a
detached portion of one organism, with a more or less
detached portion of another," he remarked in one passage.[20]
The gametes, "have not been made by some elaboration, fundamentally different from other cells," he maintained in
another.[21] As cells, moreover, the gametes were "unspecial-
ized" and "... have departed but little from the original
and most general type."[22] Similar passages can be found in
Haeckel's writings.[23] Both downplayed the uniqueness of the
gametes even into the eighties and nineties when cytologists
were focussing on the peculiarities of gametogenesis; asexual
reproduction always remained their model for genesis.

As with Haeckel, this constellation of physiological and
reproductive concepts encouraged a cavalier acceptance of the
inheritance of acquired characteristics. Spencer was well

versed in the standard examples of purported use-transmission, including the rudimentary eyes of cave fish, the tendency for musical talent to run in families and, of course, Brown-Séquard's epileptic guinea pigs. This evidence would not pass muster today, but to call it undocumented folklore would be to miss its thrust. Blind cave fish and the Bach family there certainly were, and Spencer could not conceive of natural selection producing either example. His argument entailed excluding the alternative rather than demonstrating his own contention. The examples offered Spencer a proof of use-transmission through elimination.

Spencer added his so-called "physiological units" to make the process, he already assumed, intelligible, and he turned to a priori principles for his most conclusive demonstrations. A closing quotation from Spencer will give the flavor as well as the contents of his claims. "An organism," he reminded his readers,

being a combination of rhythmically-acting parts in moving equilibrium, it is impossible to alter the action and structure of any one part, without causing alterations of action and structure in all the rest; just as no member of the Solar System could be modified in motion or mass, without producing re-arrangements throughout the whole Solar System."[24]

What a magnificent statement, demonstrating both Spencer's holistic biology and his cosmic concerns! But we must now turn to the famous Weismann-Spencer controversy over the inheritance of acquired characteristics.

WEISMANN-SPENCER CONTROVERSY

A synopsis of the sequence of events will give you a feel for the intensity of the debate as it unfolded in England. Abstracts of Weismann's early essays on heredity began appearing in Nature from 1885 on. Although involved in other projects by this time, Spencer reviewed his ideas on organic evolution for a general audience in 1886. It was a becalmed weathervane of a paper which pointed backward to old ideas and away from the brewing storm. Spencer did not mention Weismann nor did he appear to be aware of the controversy then going on in Germany over the germ-plasm theory. The following year Weismann attended the Manchester meeting of the British Association, at which he delivered a formal paper on his work on polar body formation and introduced the audience to his newest ideas on reduction division. As significant as this research was in the long run for the germ-plasm theory, Weismann's participation on a panel the preceding day was unquestionably of more immediate interest. The topic for debate was "Are Acquired Characters Hereditary?" Weismann's co-panelists included Lankester and Poulton, who became neo-Darwinians, and Geddes and Hartog, who decidedly did not. For the next six years the columns of Nature were filled with claims and counterclaims about the adequacy of the "Weismann" position. Romanes, Vines, Hartog, Poulton, Wallace, Cunningham, and Mitchell were among the most frequent

contributors. The more popular and religiously oriented
Contemporary Review and The American Naturalist across the
Atlantic quickly picked up the controversy. Important
books by Wallace, Ball, Romanes, Poulton and a translation
of Eimer's Entstehung der Arten established the issue in the
English speaking world in a more substantial form. A collected edition of authorized translations of Weismann's early
essays on the germ-plasm, guided through the press by Poulton,
came out by the end of the decade; a second volume, including
the important essay on Amphimixis, appeared in 1892 -- the
same year, I'll repeat, as Das Keimplasma.
    This surge of literature reflected more than simply a
break-through in a recalcitrant scientific problem. The
confusion and profusion of reactions could only indicate
that deep assumptions had been shook; it was also the case
that the issue of heredity, then as now, had implications
which stretched beyond the scientific claims. But let us
turn directly to the Weismann-Spencer controversy which
followed, chiefly in the pages of the Contemporary Review.
Spencer's opening attack came in the Spring of 1893 in a two
part salvo entitled, "The Inadequacy of Natural Selection."[26]
Weismann responded that autumn with a two part volley of his
own, "The All-sufficiency of Natural Selection. A Response
to Herbert Spencer."[27] Spencer returned in December with
"A Rejoinder to August Weismann."[28] The following spring,
1894, Weismann travelled to Oxford to give the annual Romanes
Lecture; it consisted of a broadside entitled "The Effect
of External Influences upon Development."[29] The series'
benefactor, G.J. Romanes had contributed immensely to the
broader controversy; it was unfortunately one of the last
public affairs this disciple of Darwin attended before his
untimely death.[30] Spencer's reply, "Weismannism Once More"[31]
appeared in October of the same year, and Weismann's "Heredity
Once More"[32] followed twelve months later. A final one page
discharge to the editor by Spencer, also entitled "Heredity
Once More,"[33] ended the engagement. No one, I suspect, ever
got the parting shot on Herbert Spencer.
    It would be unproductive to follow the debate in detail,
but since the arguments for the most part focussed on particular biological phenomena which begged an unambiguous
explanation, some particulars are worth sampling.
    One of the central points of disagreement concerned the
issue of degeneration of structures during evolution; the
eyes and pigmentation of cave dwelling animals were favorite
examples. Weismann had sought an explanation in terms of a
relaxation of selective pressure, Panmixia as he called it.
It was an explanation which demanded a feel for a population
concept of species. At first, Spencer completely misunderstood the theory and, in discrediting it, claimed that the
nutrition saved in the process of an organ's degeneration
was insufficient to account for the evolution of the degenerate type. It was a telling misunderstanding, for it
revealed Spencer's propensity to think in terms of consumption and expenditure on the level of the individual -- that

budget mentality so often revealed in the First Principles. After his mistake was pointed out to him by Romanes, as well as Weismann,[34] Spencer persisted in the same mind-set but altered his example to an extreme case of degeneration, the rudimentary femur of a Razor-back whale. This vestigial bone weighed 1/900,000 of the total animals weight, he pointed out, how possibly could Weismann maintain that a relaxation of selection had permitted the femur to degenerate to such an insignificant remanent. Only a process of disuse and the repeated transmission of the consequent change could account for such a complete degeneration. It was an awkward example for Weismann to handle and forced him into a curious retreat.

The improbable co-adaptation of multiple parts was another repeated theme in the controversy. It was an argument commonly used by the special creationists in Darwin's day. For Spencer, of course, correlation, a feature common to most structures in all organisms, was an indicator of evolution his style. The adjustments and increase in the bones and neck muscles of the Irish elk which compensated for the increased size in antlers (Spencer called them horns!) was a favorite example. Paring the problem to size with his specially honed philosopher's knife, Spencer pointed out that: a) if organic parts adjust to one another in unison, the world would not have the multitude of variations so obvious to all; b) if the parts altered with complete independence of one another, the organism could not be considered the functioning whole it clearly is; c) if the parts varied independently but kept attuned to each other by the sole operation of natural selection, one would have, "to suppose that the prescribed end is somewhere recognized."[35] This Spencer refused to admit. The complex syllogism led Spencer to one conclusion only: "... either there has been inheritance of acquired characters, or there has been no evolution."[36]

Weismann's neo-Darwinian arguments for the correlation of parts were equally difficult to sustain. The selective value of a particular variation was not easy to demonstrate. One experiment cited by Weismann, had been performed by the Swiss naturalist August Forel. The hind tarsus of some insects has a finely structured notch, known as the strigil, which is used by the insect to cleanse the antennae. Forel had removed the tarsus and the insect's antennae had become dirty. Yet did this prove that such a beautifully correlated structure as the strigil, including its peculiar shape and comb, evolved through natural selection? Weismann recognized it didn't. At most he could only claim that this structure in the non-malleable exoskeleton of the insect ruled out the inheritance of functionally acquired characters. Weismann, no less than Spencer, had to rely on an argument by elimination to dispose of his opponent's theory. He had to depend upon a priori principles to establish his own as the only alternative. Thus we find Weismann admitting, "we know indeed, nothing at all but the chief foundation of the process; and therefore anyone who does not comprehend the logical necessity of the theory, or will not recognize it, can easily set aside the individual instances as untenable."[37]

Perhaps the most interesting set of arguments in the Weismann-Spencer exchange concerned neuter insects. Weismann brought the subject up first with respect to correlation of parts. With the colonial termites, ants, and hymenoptera there are worker castes which to a varying extent and often never participate in the production of or fertilization of eggs. The individuals of these castes possess highly modified structures designed for specialized tasks, e.g., the over-sized mandibles with soldier ants or the pollen collecting baskets with honey bees. These could not possibly have been developed through generations of functional adaptation and hereditary transmission. One could only argue that the modified castes had given the nest or hive an advantage and that natural selection operated on the colony rather than on the individual level. It was a good argument; Darwin had recognized its merits in the Origin of Species. Weismann's generation could cite more bizarre structures and habits to support it.

But it was not an invincible argument, in Spencer's mind; he found two outs. For the sake of brevity I'll turn only to the second of Spencer's ripostes. It had long been known that royal jelly, if given sufficiently early, could force worker bee larvae to turn into Queens. This worker caste, Spencer insisted, was clearly not determined by an inherited constitution but by nutrition. The social wasp colonies, furthermore, did not consist of well defined castes but of a gradation from large, fully fertile females to small neuters. This too, Spencer countered, could only be the result of a nutritional gradient.[39]

The lines were clearly drawn. Weismann could interpret the existence of social insect castes within the scope of his germ-plasm theory. There had to be determinants and ids for peculiar structures and sexes. These had been selected for in the course of evolution. To counter Spencer's emphasis on nutrition, Weismann also reported upon experiments in which he had partially starved developing blow-flies and found that the lack of nutrition affected the size of the adult but not the reproductive organs. "Poor feeding," he could analogize, "is not the causa efficiens of sterility among bees, but is merely the stimulus which not only results in the formation of rudimentary ovaries, but at the same time calls forth all the other distinctive characters of the workers."[40] Spencer, on the other hand, found Weismann's multiple ids and determinants too complex and too implausible to explain, let us say, the three worker castes and two sexes of a termite colony. It seemed that Weismann's argument was nothing more than an invocation of mystical potentialities in a hypothetical germinal line. Differences in nutrition by contrast were tangible and in some cases demonstrated efficient causes. Take the example of the honey ants, where the abdomen of certain individuals becomes a bloated bladder or food reservoir for the entire colony. For Spencer there was no need to interpret this phenomenon in terms of a hereditary constitution. "The life of the worker," he reminded us, "is mainly passed in taking in food and putting it out again. If the putting out is stopped its life will be mainly passed in taking in food. The receptacle will go on enlarging and it will eventually assume the monstrous form that we see."[41] No nicer example of the mid-century, matter and force accounting mentality could be found!

AFTER THOUGHTS ON THE CONTROVERSY

We have by no means exhausted the particulars of the Weismann-Spencer exchange, but we don't need to in order to agree with both men that further discussion was futile.

Let us see what of more general value we can extract from the controversy between Weismann and Spencer. In turning to Spencer first, there can be no doubt in our minds that his arguments were structured by the philosophical house in which he dwelled. Dr. Peel is probably right to insist that the edifice would have fallen had Spencer capitulated to Weismann.[42] There is no question that Spencer's conviction in use-transmission flowed from the Synthetic Philosophy and in the following ways: With the danger of stating the obvious, there was first a strong reductionist side to Spencer in the sense that he considered biological activities to be directly understood in terms of the raw data of physics and chemistry. His stance bore particularly on the heredity problem as we have seen, for it took merely a nutrition-and-force-accounting technology to explain on alternate pages of the same ledger: growth, development, reproduction, and evolution. Second, for Spencer the organism -- in fact, the entire cosmos -- was an integrated system. It was in keeping with his cosmology that he chose a physiological definition of life, viz., a continuously self-adjusting entity. For the same reason he disregarded the unique properties of gametogenesis. After all, if each part reflected the changes in the whole, it was illogical to claim a special dispensation for the germ-plasm. As for Spencer's "physiological units," these seem to be no more than molecules endowed with the capacity to reflect the whole. Third, Spencer constantly thought in terms of a contrast between heredity, on the one hand, and variation, on the other. This was an old antinomy, which can be found in Darwin and Haeckel. Heredity was considered a conservative process, a transmission of tried solutions to the eternal quest for improved adaptation. Variations by contrast represented new opportunities; they were the organism's way of responding to new challenges. Not all variations would be transmitted to offspring, but it went without saying that all constitutional elements started out as adaptations to external changes. Hence to deny the inheritance of acquired characters was indeed to repudiate evolution.

With Weismann we also find an elaborate system built on a number of premises. He, too, was a mechanist but not with the same reductionist zeal. The nutritional cost-accounting so dear to Spencer was suppressed for the most part and replaced by a qualitative reckoning of units in the germ-

plasm. As I have already suggested, Weismann thought of
evolution in terms of evolving populations of ids rather
than confounding the developmental and evolutionary processes.
Sexual reproduction, including the preparatory reduction
division, gained a favored position in the Weismann theory
that was maintained later in classical genetics. Second,
Weismann defined life very differently from Spencer. Still
within the mechanist spirit, he nevertheless perceived life
in terms of self-replicating rather than self-adjusting
units. To his contemporaries Weismann's was viewed as a
morphological approach and warranted the accusations of pre-
formationism. The ids, determinants and biophors were biological molecules lying intermediate between large chemical
molecules below and the cells above on the reductionist's
scale. Third, Weismann no longer clung to the assumption
that heredity and variations were conflicting processes.
Within the germ-plasm theory heredity fashioned rather than
opposed variations, and it followed perforce that natural
selection of certain variations brought about increasing
adaptation. In contrast to Spencer, then, adaptation was
the result of rather than cause of evolution.

THE NATURALIST'S IMPERATIVE

Given these three contrasts between Spencer and Weismann,
i.e., an extreme reductionism vs. a moderate mechanism; a
physiological vs. a morphological definition of life, and
the different understanding of the relation between heredity
and variation, it would seem natural to attribute these
debates and the general concern over the inheritance of
axquired characters to the emerging discipline of cytology.
According to this historical account, Weismann arrived at
his conclusions through his deep acquaintance with cytology.
The recently revealed mitotic process had encouraged a
morphological view of life, and the discovery of meiosis
promoted the uniqueness of sexual reproduction. Moreover,
Weismann's theory of the continutiy of the germ-plasm did
coincide with his attack on the transmission of acquired
traits. His opponents, not only Spencer, who was ignorant
of cytology, but botanists, who claimed Weismann was ignorant
of plant cytology, believed that the continuity of the germ-
plasm was the keystone to be dislodged in the Weismannian
fortress. Yet as historians we must tread carefully. In the
first place, Spencer adopted all the trimmings of the cell
theory from Schwann through Virchow to Haeckel; these
included the crystal analogies, the definition of life, and
the emphasis on growth and even overgrowth. It seems to me
clear that our historical account must make a strong distinction between the cell theories of the early and mid-century
and cytology of the last quarter of the century. Perhaps
no better indicator of such a difference can be found than
in Virchow's contemptuous dismissal of Weismann's theory as
mere "Cytomorphismus."[43] In the second place, if chromosomal
cytology was important to Weismann, it was not a sufficient
pre-condition, as evidenced by the cases of Oscar Hertwig,

Theodor Boveri, Wilhelm Roux, and T.H. Montgomery, all of
whom continued to believe in the inheritance of acquired
characters, and by E.B. Wilson, who as late as 1896, kept a
neutral stance.[44]

Nor can the rise of cytology be considered a necessary
pre-condition for discarding use-transmission! This brings
me to the most interesting of my conclusions. For the past
twenty years Ernst Mayr has insisted in print and in lecture
hall that the experienced naturalist comprehended the full
power of natural selection long before other biologists.[45]
His message is born out by the Weismann-Spencer debate. Recall that Weismann, originally trained as a descriptive
histologist and embryologist, was forced to spend ten years
away from the microscope. At that time he earned his creden-
tials as an excellent naturalist. We need not construe this to
mean that he turned explorer, as Darwin and Wallace had.
Museum studies, familiar also to Darwin and Wallace, were
sufficient to develop in Weismann a sensitivity to two
important considerations. First, he acquired an enormous
appreciation for the geographical distribution of varieties,
as his continued studies on the distribution of polymorphic
butterflies testify. Secondly, he developed the habit, as
is often said in modern ecology, of thinking like the animal
of his studies. His examination of the Axolotl shows this
in a magnificent way. What were the exact conditions of
the lakes on the Mexican plateau? he asked. How often did
the lakes dry up? What dangers were in store for the
Axolotl if it ventured on land? When the seasonal winds
picked up, how did this influence the muddy shores of the
lake? Weismann's studies, of the protective markings of
caterpillars show the same concerns. Along with all naturalists, Weismann became sensitive to the infinite diversity
of conditions, each of which presented slightly different
challenges to the organism. Such a sensitivity encouraged
Weismann to seek consistently the Darwinian explanation of
organic form long before he returned to the microscope and
worked out the continuity of the germ-plasm on Hydromedusae.

This brings us back to England and the curious circumstance that Weismann was more appreciated here than on the
continent. His influence is even more curious when one considers that it had been the German not English microscopists
who had largely developed the field of chromosomal cytology.
Let us look briefly at the most prominent English neo-
Darwinians who participated in the debates over the continuity
of the germ-plasm and the related theory of Panmixia, in
particular, and "use-inheritance", in general. It is worth
pointing out the old war horses of the Darwinian revolution,
and what is more the close friends and contemporaries of
Spencer, Huxley and Hooker had virtually retired from the
field. Huxley died in 1895, the year the Weismann-Spencer
debate ended; Hooker continued to work at Kew gardens till
the end of the century. He died in 1911 at the age of ninety-
four. I have yet to find any published recognition on their
part that they maintained an interest in heredity theory
after 1880.

By contrast the indefatigable Wallace, at the age of sixty-six, came out strongly for Weismann in his Darwinism of 1889. During the next decade he contributed many neo-Darwinian articles to Nature and other journals.[46] As for some of the most prominent younger neo-Darwinians: Thiselton-Dyer (1843-1928), son-in-law of Hooker and Director of Kew Gardens after 1885, had a deep interest in the geographical distribution of plants; E.R. Lankester (1847-1929), Linacre Professor of Zoology at Oxford and later Director of Natural History at the British Museum, possessed wide ranging interests in almost every sphere of zoology;[47] H.N. Moseley (1844-1891), Linacre Professor of Human and Comparative Anatomy, had spent four years as a naturalist on the Challenger expedition;[48] Raphael Meldola (1848/9-1915) was one of the few prominent coal-tar chemists in England,[49] and E.B. Poulton (1858-1943), Hope Professor of Zoology at Oxford, was author of the important neo-Darwinian book Colours of Animals (1890).[50] I'll stop here; the list is not intended to be comprehensive. It is nevertheless suggestive. Only Lankester out of this group could be considered a cytologist; he later became editor of the prestigious Quarterly Journal of Microscopical Sciences. All of them, however, were seasoned naturalists, and this included Meldola, who studied mimicry in butterflies as an avocation and became president of the Entomological Society. All of them, I would like to argue, would have possessed the same naturalist's sensitivity to the geographical distribution of varieties and the diverse challenges of life, found with Weismann and certainly shared by Wallace. They would have welcomed the all-sufficiency of natural selection once "use-transmission" had been challenged and once an explicit "ultra-Darwinian" explanation of heredity could be scientifically justified.

To what extent Galton's critical evaluation of the inheritance of acquired characters helped prepare the way for Weismann in England, I cannot judge. Despite its overconfident neo-Darwinian conclusion we get some clues as to the sequence of events from Wallace's autobiography:

Galton's experiments on blood transfusion with rabbits first staggered me, as it seemed to me to be the very disproof I had thought impossible. And later on, when Weismann adduced his views on the continuity of the germ-plasm, and the consequent non-heredity of acquired characters; and further, when he showed that the supposed transmission of such characters, which Darwin had accepted and which the hypothesis of pangenesis was constructed to account for, was not really proved by any evidence whatever; -- I was compelled to discard Darwin's view in favour of that of Weismann, which is now almost everywhere accepted as being the most probable, as well as being the most in accordance with all the facts and phenomena of heredity.[51]

In giving you this concentrated analysis of the Weismann-Spencer controversy, I run the risk of distorting the twenty years at the end of the century. The validity of "use-transmission" was on the minds of a great variety of biologists, and the historian must recognize that unique experiences shaped each differently. Yet there were overriding trends. Considering the cytological nature of Weismann's germ-plasm theory, it is natural to speculate on the role cytology played in promoting a neo-Darwinian position. As it turns out, cytology was neither a necessary nor a sufficient condition, and only when the assumptions of the older cell theory had been pared away, could it become even a contributing factor.

Moreover, as I survey the mounds of contentious articles and published letters produced during the period and find the American neo-Lamarckians, immersed -- as they were -- in paleontology, and German Lamarckians, familiar -- as they all surely were -- with the traditions of the cell theory, I cannot help but wonder whether both parties had not approached heredity with first principles not unlike Spencer's. The writings of Cope and Eimer suggest as much.[52] When I see how welcomed the germ-plasm theory was in England, I wonder, too, whether the naturalist tradition found in abundance here doesn't provide an essential experience for the recognition of the all-sufficiency of natural selection.

FOOTNOTES

1. The only attempt at a complete history of the concept of the inheritance of acquired characters is L.A. Blacher, The Problem of the Inheritance of Acquired Characters: A History of a 'a priori' and Empirical Attempts at its Solution. (Moscow: 1971, in Russian) I was fortunate to have access to a rough English translation; chapters 4-7 are pertinent to the subject of this paper. Other useful sources include: Vernon L. Kellogg, Darwinism To-day, (New York: 1908), Philip G. Fothergill, Historical Aspects of Organic Evolution (London: 1952), Yves Delage and Marie Goldsmith, The Theories of Evolution, trans. Andre Tridon (New York: 1913); William Platt Ball, Are the Effects of Use and Disuse Inherited? An Examination of the view held by Spencer and Darwin (Nature Series) London: 1890. Thomas Hunt Morgan, Evolution and Adaptation (New York: 1903). Richard W. Burckhardt, Jr., "Lamarckism in Britain and the United States," presented May 1974 at the Conference on the History of the Evolutionary Synthesis. American Academy of Arts and Sciences, Boston, Mass. To appear in the published proceedings of the conference.

2. For an English summary of his work see: Charles-Edouard Brown-Séquard, 'On the hereditary transmission of effects of certain injuries to the nervous system,' Lancet, 1875. 1:7-8. See also: J.M.D. Olmsted, Charles-Edouard Brown-Séquard, a nineteenth century neurologist and endocrinologist (Baltimore: 1946), pp. 168-195.

3. C. Westphal, "Ueber künstliche Erzeugung von Epilepsie bei Meerschweinchen," Berl. klin. Wschr., 1871, 18 Jg., 449-451, 461-463. Heinrich Obersteiner, "Zur Kenntniss einiger Hereditätsgesetze," Medizinische Jahrbücher, Jg. 1875, pp. 179-188.

4. George J. Romanes, Darwin, and After Darwin.II Post-Darwinian Questions, Heredity and Utility (Chicago: 1895) see particularly pp. 103-122.

5. August Weismann, "The significance of sexual reproduction in the theory of natural selection," in Weismann, Essays Upon Heredity and Kindred Biological Problems, 2 vols., ed. Edward B. Poulton, et al. (Oxford: 1891-92) vol. 1, pp. 319-332. Ernst Ziegler, "Können erworbene pathologische Eigenschaften vererbt werden und wie entstehen erbliche Krankheiten und Missbildungen?" Beitr. path. Anat., 1886, 1: 390-395.

6. In 1890 Ball expressed this view of the epileptic guinea-pigs. "Exceptional cases require exceptional explanations, and are scarcely good examples of the effect of a general tendency which in almost all other cases is so inconspicuous in its immediate effects." (n.1) p.35.

7. Ernst Gaupp, August Weismann sein Leben und sein Werk (Jena: 1917) pp. 4-29, Gloria Robinson, "Weismann, August Friedrich Leopold," DSB, 14:232-239. The author has also had access to a MS Autobiography by Weismann which he and Prof. Helmut Risler are preparing for publication.

8. August Weismann, "Ueber die Entstehung des rollendeten Insekts in der Larve and Puppe. Ein Beitrag zur Metamorphose der Insekten," Abh. Senckenbergischen Naturforschenden Gesellschaft, 1862-63, 4: 227-260. August Weismann, Studies in the Theory of Descent, trans., Raphael Meldola (London: 1882) This English edition contains a two page prefatory note by Darwin.

9. August Weismann, Ueber den Einfluss der Isolirung auf die Artbildung (Leipzig: 1872) p. 43. Ernst Mayr has very rightly pointed out that both Darwin and Weismann focussed on "intrapopulation variants" in an attempt to justify sympatric speciation by natural selection. In contrast, Moritz Wagner, who is the butt of Weismann's 1872 criticisms, referred to geographically distinct varieties while he insisted on the necessity and sufficiency of isolation in the evolutionary process. Both parties talked by one another and failed to see the multiplicity of factors involved. "Isolation as an Evolutionary Factor," Proc. Am. Phil. Soc. 1959, 103: 221-230.

10. August Weismann, The Germ-Plasm, A Theory of Heredity, trans. W. Newton Parker and Harriet Ronnfeldt (London: 1893) pp. 8-9.

11. August Weismann, Das Keimplasma. Eine Theorie der Vererbung (Jena: 1892) For the English translation see (n.10).

12. Carl Wilhem von Nägeli, Mechanisch-physiologische Theorie der Abstammungslehre (Munich and Leipzig: 1884).

13. Herbert Spencer, "The Development hypothesis," and "Progress: its law and cause" in Essays, Scientific, Political, and Speculative, 3 vols. (New York and London: 1910) Vol. 1. See also: J.D.Y. Peel, "Spencer, Herbert," D.S.B., 12: 569-572, _____, "Spencer and the neo-evolutionists," Sociology, 1969, 3: 173-191; Gilbert Charles Bourne, Herbert Spencer and Animal Evolution. The Herbert Spencer Lecture 1909 (Oxford: 1910); Raphael Meldola, Evolution, Darwinian and Spencerian. The Herbert Spencer Lecture 1910 (Oxford: 1910).

14. Herbert Spencer, First Principles of a New System of Philosophy (1st ed., 1862; New York: 1865).

15. Peter B. Medawar, The Art of the Soluble (London: 1967); Peter J. Bowler, "The Changing meaning of evolution," J. Hist. Ideas, 1975, 36: 95-114.

16. Herbert Spencer, The Principles of Biology, 2 vols. (London: 1884). All references will be to this later edition.

17. Ibid., vol. 1, p. 81.

18. Ibid., p. 133, ftn.

19. Ibid., pp. 151-152.

20. Ibid., p. 219.

21. Ibid., p. 220.

22. Ibid., pp. 219-222.

23. Frederick B. Churchill, "August Weismann and a break from tradition," J. Hist. Biol., 1968, 1: 91-112. See pp. 93-99.

24. Spencer, Principles of Biology, (n.16), Vol. 1, p. 255.

25. Herbert Spencer, "Factors of Organic Evolution," Nineteenth Century, 1886, 19: 570-589, 749-770.

26. Herbert Spencer, "The Inadequacy of 'natural selection,'"
    The Cont. Rev., 1893, 63: 152-166, 439-456.

27. August Weismann, "The All-sufficiency of natural selec-
    tion. A reply to Herbert Spencer." Cont. Rev., 1893,
    64: 309-338, 596-610.

28. Herbert Spencer, Cont. Rev., "A Rejoinder to Professor
    Weismann," 1893, 64: 893-912.

29. August Weismann, The Effect of External Influences upon
    Development (The Romanes Lecture 1894) London: 1894.

30. Ethel Romanes, Life and Letters of George John Romanes,
    (London: 1896) pp. 347-348.

31. Herbert Spencer, "Weismannism Once More," Cont. Rev.,
    1894, 66: 592-608.

32. August Weismann, "Heredity Once More," Cont. Rev., 1895,
    68: 420-456.

33. Herbert Spencer, "Heredity Once More", Cont. Rev., 1895,
    68: 608.

34. There took place an exchange between Spencer, Romanes,
    and Lankester on the mechanism of Panmixia in Nature,
    1890, 41 and 42. See also George J. Romanes, "A Note
    on Panmixia," Cont. Rev. 1893, 64: 611-12. George J.
    Romanes, Marcus Hartog, "The Spencer-Weismann Controversy,"
    Ibid., pp. 50-59.

35. Spencer (n.26) p. 29. This third alternative moreover
    required that one see the positive side of natural
    selection, an insight which called for the naturalist's
    feel for a large number of variations in space and time.

36. Ibid., p. 30.

37. Weismann (n.27) p. 321. Emphasis added.

38. Charles Darwin, On the Origin of Species (Facsimile of
    the First Edition; Cambridge, Mass.: 1964) pp. 235-242.

39. Spencer (n.28) pp. 5-20.

40. Weismann (n.29) pp. 29-39. Emphasis is Weismann's.

41. Spencer (n.31) pp. 603-604.

42. Peel, DSB (n.13).

43. Rudolf Virchow, "Anlage und Variation," Sber. preuss.
    Akad. Wiss., 1896, p. 527. For an interpretation of

Virchow's remarks see: Frederick B. Churchill, "Rudolf Virchow and the Pathologist's criteria for the inheritance of acquired characteristics," J. Hist. Med., 1976, 31: 117-148.

44. Theodor Boveri, Die Organismen als historische wesen (Wurzburg: 1906) p. 28, see also Fritz Baltzer, Theodor Boveri, Life and Work of a Great Biologist, 1862-1915, trans. Dorothea Rudnick (Berkeley and Los Angeles: 1967) pp. 130-142. Oscar Hertwig, Allgemeine Biologie (Jena: 1909) pp. 652-653. Hertwig documents his version of inheritance with a page-long quotation from Spencer's Principles of Biology! Thomas Harrison Montgomery, The Analysis of Racial Descent in Animals (New York: 1906) pp. 149-151. Edmund B. Wilson, The Cell in Development and Inheritance, (New York: 1896) pp. 329-330.

45. See particularly Ernst Mayr, (n.9) p. 226 and "Karl Jordan's contribution to current concepts in systematics and evolution," Trans. Roy. Ent. Soc. London 1955, 107: 45-66. Garland Allen has contrasted a naturalist tradition to the Drosophila school in America, "T.H. Morgan and the emergence of a new American biology," Quart. Rev. Biol., 1969, 44: 168-188, and Mark Adams has indicated the importance of the naturalist's perspective in Russia, "The Foundations of Population genetics: contributions of the Chetverikov school, 1924-1934," J. H. Biol., 1968, 1: 23-39, see especially p. 37. The concept of a "naturalist's tradition" needs further articulation and delineation.

46. Alfred Russel Wallace, Darwinism an Exposition of the Theory of Natural Selection with Some Applications (London: 1889), "Are Individually Acquired Characters Inherited?" Fortnightly Rev., 1893, 59 (ns 53): 490-498, 655-668.

47. Gavin de Beer, "Lankester, Edwin Ray," DSB, 8: 26-27.

48. The Times (London), 13 Nov. 1891, Fe.

49. The Times (London), 17 Nov. 1915, 5b.

50. Poulton, Edward Bagnall DNB, (1941-1950) pp. 687-689.

51. Alfred Russel Wallace, My Life, a Record of Events and Opinions, 2 vols. (New York: 1905), vol. 1, pp. 422-423.

52. E.D. Cope, The Origin of the Fittest, Essays on Evolution (New York: 1887); G.H. Theodor Eimer, Organic Evolution as the Result of the Inheritance of Acquired Characters According to the Laws of Organic Growth [!], trans., J.T. Cunningham (London: 1890).

HUGO DE VRIES AND THE GENE THEORY

As a student at the University of Leiden, Hugo de Vries read Darwin's "Origin of Species" of 1859, and became an enthusiastic adherent of the Theory of Descent. This brought him in conflict with the majority of leading scientists in the Netherlands, including his teachers of botany at Leiden, where de Vries' interest was concentrated on the infant science of plant-physiology.

Hugo de Vries' personal contact with Ch. Darwin began with a visit to Darwin's home in Down, where Darwin showed eager interest in de Vries' experimental phytophysiological studies on phytodynamics and growth processes, i.e. those pursued in Julius Sachs' laboratory at Würzburg.

The years 1876-1878 de Vries devoted, in charge of the Berlin ministry of agriculture, to an extremely intensive anatomical and chemical-physiological study of the development, from seedling to harvest-crop of some agricultural plants. The 150-page section of his report on Sugar-beets, furnished valuable indications for experimental pedigree-cultures for the continuous selection of strains with a higher percentage yield of sugar. (1)

In this way, after a decade of intensive and successful research on plant-physiology, Hugo de Vries gradually extended the scope of his experimental research to problems in the fields of variability, heredity and the origin of species, about 1880.

Studying growth-anomalies: fasciation, biastrepsis e.a., he found that monstrosities of rare occurrence in nature, could be accumulated by progressive selection from cultures. In several cases the proportion of the anomalous individuals could be raised in successive generations; but never a gradual transformation from the normal to the abnormal state was observed.

This sustained the critical doubts concerning the fundamental basis on which Darwin's theory of natural selection originally was established, the heritability of individual variations, influenced and selected by environmental conditions of life.

Like August Weismann, Hugo de Vries came to the conclusion that the continuous gradual variations, caused by the influence of environment during the individual life, could not be inherited, nor become heritable by sustained selection.

This negative statement led de Vries to the conception of a new basic principle for studying the diversity of the organic world and for explaining hereditary relations.

He called it "the theory of elementary characters".

For experimental evaluation and for the emendation of Darwin's theory of selection, de Vries wanted to bring his principle of elementary characters in the form of a more concrete working-hypothesis, a "physiological" theory of inheritance. Therefore he accepted from Ch. Darwin's "Provisional Hypothesis of Pangenesis" (1868) the idea of material carriers of hereditary characters, which particles Darwin called "gemmules".

De Vries rejected major parts of Darwin's hypothesis, i.e. Darwin's explanation of the alleged heritability of acquired characters by the transport of gemmules from the peripheral organs to the reproductive cells and into buds. To accentuate the difference of his new hypothesis from Darwin's, de Vries changed the name of the carriers from "gemmules" to "pangenes" (after Darwin's title Pangenesis), and supposed a specific pangene for every unit-character.

INTRACELLULAR PANGENESIS

In reality Hugo de Vries' book "Intracellular Pangenesis" (1889), (2) introduced an entirely new mode of explaining the way in which the properties of the parents are transmitted to the offspring. It is no less than the new basic constitutional law for modern genetics.

The most important aspects of the new theory may be summarised as follows:

Every animal and plant species is characterised by a large – but not indefinite – number of separate, independently and unalterably transmitted characters.

Each of these 'unit-characters' or 'elementary characters' is represented by a separate distinct particle of the living protoplasm, as a carrier, which de Vries called a 'pangene'.

The pangenes of all unit characters of the species, are situated, in a dormant state, in the nucleus of every cell of the organism.

For cell-division they produce, by selfdivision a replica of themselves, to furnish the daughtercells with identical copies.

The whole set of pangenes is transmitted to all cells of the organism, including the gametes, and thereby to the offspring.

De Vries here gives a detailed description of the mitotic division in its successive stages ('phases'), cited from E. Strasburger (1888), and he concluded that the pangenes must be sited in the chromosomes in one line, and are divided by the splitting of the chromosomes during the prophase of mitosis.

Single steps in evolution are marked by the acquisition of a new type of pangene, or by the loss of an existing one.

All living species are to be regarded as different combinations out of the total of existing pangenes.

470

Related species have a considerable number of their pangenes in common. The degree of relationship can be numerically expressed by the percentage of the pangenes in common.

In a chapter called "Current ideas on the carriers of hereditary characters" de Vries gives a critical survey of contemporary theories, comparing them with his own intracellular pangenesis. Aug. Weismann's renowned theory of the Germplasm is criticised because the hereditary units in the cells, the 'ancestor plasms' ('Ahnenplasmen') represent, everyone of them, the complete set of characters of the species. In Weismann's theory the cells of the germ tract bear the complete ancestorplasms; in the somatic sidelines only the characters which will be developed in the separate organs are present, the others being removed by a 'reduction-division'. Hugo de Vries' pangenes, being present in a latent (inactive) state, in the nucleus of every cell of the organism can penetrate into the cytoplasm. In somatic cells the distinct characteristics of the specific organ are expressed by the respective pangenes recovering their specific activity.

The occurence of a meiotic (reduction) division of the hereditary material, on the contrary, was denied by de Vries and by E. Strasburger, even after 1900 when the cytological figures of nuclear division in insect-spermatogenesis were tentatively explained as pairing and separating of the haploid sets. In order to obtain really independent assortment of all mutual characters of the parents, the paired chromosomes, de Vries thought, must fuse into a single body, mixing up their pangenes. A mitotic division is sufficient to restore the diploid number of chromosomes. H. de Vries devised an experimental proof of the independent assortment of more characters than the number of chromosomes with a study of heritable traits in a human population.

In the concluding chapters of his 1889-book Hugo de Vries recapitulates the differences between his theory of pangenes and Darwin's provisional hypothesis more consistently than in the foregoing parts. Then abruptly, as a final apotheosis, he differentiates it by a sharp, concise definition:

> "I call 'Intracellular Pangenesis' the hypothesis that the entire living protoplast consists of pangenes; ... there is no other living foundation to it."(4)

This aggressive epitome of his ideas, as a challenging device, greatly hampered the understanding and acceptance of de Vries' theory, for it was just at that time that the contrasting constitution and functioning of the nucleus as opposite to that of the cytoplasm was emphasised by the admirable achievements of E. Strasburger, O. Hertwig and others. The idea seemed unacceptable that living unit-particles out of the nucleus (the chromosomes) would pass into the cytoplasm to resume their activities. The cytoplast has its own specialised organelles, their activities being only indirectly regulated from the nucleus, perhaps by intervention of enzymatic processes.

De Vries' thesis seemed to be in contradiction even with his own principle of the 'panmeristic origin' of organelles (by selfdivision only), so energetically defended in a foregoing chapter, against the supposed 'neogenetic' mode of origin.

In his epitome de Vries argues that, if a character is changed, e.g. a colour-factor after hybridization, the constitution of the colour-producing organelle itself must be changed, not only the enzyme which activates it. In cases where the dominant factor comes from the male parent, which only contributes a haploid nucleus, the change must be brought about by the penetration into the organelle of the pangene in question. The newly introduced pangene in the zygote-nucleus after selfdivision sends its replica, or descendants, into the cytoplasm to build, or to rebuild, the organelle.

In terms of evolution: for every new character a new type of pangene must be settled in the cytoplasm-organelle before it can produce the new character. The new pangene has possibly been present in an inactive state for generations. By this formulation de Vries evaded the difficulty, inherent in the enzyme-theory, of dual evolution, i.e. always in parallel corresponding steps, in the nucleus and in the cytoplast-organelles, as the executing system. Nevertheless the enzyme theory held the upper hand: the hereditary units being located in the chromosomes, could only exert an indirect regulating or inhibiting influence on the activities of the plasmatic organelles.

THE LAW OF SEGREGATION

In a comprehensive paper Lindley Darden (6) gives an excellent evaluation of Hugo de Vries' Intracellular Pangenesis as compared to Darwin's Provisional Hypothesis of Pangenesis, and of the way in which de Vries "ultimately expressed" his thoughts in the "law of segregation", which came to so high credit as the principal tool for experimental research in "mendelian" genetics.

Darden's study is very welcome, because it clearly expounds Hugo de Vries' deduction of the 'Segregation-law' from his own original initial principle of independent 'unit-characters'. This deduction is very often misunderstood because, in his profound admiration for Darwin's person, de Vries intentionally imitated Darwin's expressions, while in reality he only makes use of some secondary points out of Darwin's provisional hypothesis, which as a whole, already had been rejected by Galton and others.

For the verification of pangenesis de Vries studied the percentage frequency of variant forms in selfed populations from monohybrid crosses.

The axiomatic thesis of separate independent and constant elementary characters supposes a random distribution of the pangenes carrying the different unit-characters. But at that time (1889) de Vries thought that an unlimited multiplication of the pangenes was demanded for their activity in the cytoplasm. This idea of the 'quantification' of the characters, inferred from his earlier pedigree-cultures of monstrosities, impeded de Vries in his analysis of the segregation-ratios. In the following years this quantitative evaluation of the fluctuating variability, which is not really compatible with his own principle of invariably constant unit-characters, was gradually altered and, by applying Quetelet's law and Galton's curves, the now usual image of fluctuation was established of a pendulum swinging to both sides of the mean value.

Formal proof that de Vries at least in the year 1896 was aware of the $\frac{3}{4}:\frac{1}{4}$ ratio is found in the collection of lecture plates of the Hugo de Vries laboratory in Amsterdam. One of these plates concerns the crossing experiments of the years 1893-1896, with two races of Papaver somniferum. The dark coloured type Mephisto (with black heart) is here indicated with A (active pangene), the lighter coloured (with white cross in the centre) Danebrog (Danish flag) designed with L (latent pangene). (7,8) The notes and dates on this plate, indicating the ratios in the first, second and third generations, are in de Vries' handwriting. Whenever de Vries mentioned this beautiful example of hybrid-segregation after 1900, the letters A and L were replaced by the indications: dominant and recessive: - for example, already in the German rediscovery paper "Das Speltungsgesetz der Bastarde", where also is immediately stated that these terms were borrowed from Mendel.

In "Die Mutationstheorie" the Papaver-crosses are amply discussed as the finest example for segregation. Summarising, the result in the second generation are given in round numbers:

25% Const. Dom.   50% Hybrid   25% Const. Rec.   (9)

On a photograph taken in the Oenothera-garden at Lunteren, dated 20 July 1932, de Vries, at the age of 84, noted, in Dutch and French: "Oenethera Lamarckiana brevistylis. This variety has led to the discovery of the law of segregation of hybrids." There is no real contradiction between these two statements: brevistylis was an exception among the Oenothera-mutations which de Vries studied, because the cross with the mother-species Oenothera lamarckiana was like the mother, and in the next generation always a minor part of brevistylis was thrown off. In the year 1892 de Vries noted 17-26% of brevistylis in the F_2.

The cross Oe. lamarckiana x brevistylis was repeated in 1899, without doubt because de Vries was eager to control the $\frac{1}{4}$-ratio. This time it turned out to be 22%. That this percentage of recessive homozygotes again is less than 25%, may be explained by the recessive type being weaker and more apt to die off as seedling. This normal segregation of Oe. brevistylis after crossing with Oe. lamarckiana is exceptional in the whole group of species and mutants, which was the object of Hugo de Vries' extensive research on mutations. He supposed that in this case the difference consisted in one pangene, active in Oe. lamarckiana being latent in brevistylis. Mendelian segregation only occurs, he suggested, when the parents differ in this way, an active pangene from one parent being brought together with the same pangene in a latent state.

A single progressive mutation, says de Vries, adding an elementary character to the complete species type, at once produces a new constant species. With Alexis Jordan's investigations on minor species in Erophila and Viola in mind, Hugo de Vries adopts the concept of the "elementary species" as the smallest possible, absolutely constant taxonomic unity.

Of course, applied taxonomic studies could not accept the use of so concise a standard unit.

In contradistinction to this minimised concept of the elementary species, the ultimate picture of a real plant species, such as Œnothera lamarckiana, the favourite object of de Vries' experiments, finally turned out to be an extremely complex and intricate one. This image was gradually built up in the course of half a century of Œnothera-research by Hugo de Vries, later on linking up with the studies by Blakeslee, Renner, Cleland and their schools. (10)

The apparent single steps in mutation showed to be complicated by hitherto unknown cytogenetical peculiarities. The "catenation of meiotic chromosomes", i.e. end-to-end linkage of the chromosomes of each haploid set, made them behave as one single chromosome in meiosis and in the resulting segregation of characters.

Other similar terms: "twin hybrids", "balanced lethals", "complex heterozygotes", designate separate lines of research, contributing to the final apotheosis in visualising the course of evolution.

The result of the common efforts in Œnothera-research is an analysis of the highly complicated process of "speciation" (species-building) in the subgenus Onagra.

Hugo de Vries' important contribution to this final success is fully accounted for in Cleland's book on Œnothera. Here, as well as in nearly all other contemporary and later research on heredity and evolution, de Vries' basic theory of elementary unit-characters, under the more efficient name of Mutation-theory, has been universally accepted as the leading principle for experimental research and for analysing problems of relationship and descent.

DE VRIES' EXPERIMENTAL PROGRAMME

De Vries, like Mendel, based his experimental programme upon the observation of the transmission of characters considered as independent entities. Unlike Mendel, however, de Vries was seeking to reveal the behaviour of the underlying hereditary elements, the pangenes, from his study of the behaviour of the characters. Mendel's aim was to show how new constant types can be obtained by crossing. His cardinal object was, therefore, the distinction between 'constant' and non-constant characters, the latter always being called 'varying characters', never segregating or splitting. The series A + 2Aa + a denotes that, by random combination of the two types of both pollen and germcells, out of four recombinants, two reproduce the hybrid-character, and one restores the maternal and one the paternal character. (11)

In the published text of the "Versuche", being a transcription of

Mendel's concept for the lectures, these points are given in an extremely condensed form, but in the 1865 meetings the lectures were preceded by a historical introduction, and elucidated with demonstrations and discussions. Nevertheless the auditory and other contemporary scientists could not grasp the importance of Mendel's achievement, because his research concerned separate characters, not the species-type as a whole, while, at that time, the thoughts on heredity and evolution were concentrated on Darwin's theory of natural selection (1859), which is based on a 'holistic' species-concept as a concrete indivisible, though variable, even plastic, entity.

Mendel himself, though adopting the general idea of evolution, was not at all biased by Darwin's 1859 book, because already before that date, experimental research had convinced him of the independent transmission of separate constant characters. In our hindsight view we might claim, that it is exactly by this original new concept of independent and unaltered hereditary characters, that Mendel laid the foundation of the science of genetics. But in his strictly maintained empirical custom, Mendel himself deliberately refrained from elaborating this concept in the form of a general theory or dogma, as de Vries did.

Hugo de Vries' ideas exposed in the 1889-treatise on Intracellular Pangenesis not generally being accepted, he decided to substantiate his views by ample experimental support, and, at the same time, by a general survey of the horticultural and agricultural literature in search of reports on the originating of new races and aberrations. From these two sides he hoped to accumulate sufficient essential facts for a comprehensive book, which was to be entitled: "Experimental Pangenesis". For the experimental control of his basic principle of independent unit-characters, de Vries devised crossings between individuals differing in only one such unit character.

Hybridising experiments on a large scale were undertaken to this purpose with a considerable number of genera. In the majority of these genera the F_1 was uniform and in the F_2 both parental types reappeared as well as the hybrid type. In 1896 with the Papaver-crosses, the fixed ratio of 1 : 2 : 1 was established; de Vries concluded that the different (colour-) character must be caused by one of the pangenes being active in the dominant parent and inactive (latent) in the other. One of the exceptional cases, the endosperm-hybrids in Maize being explained by the discovery of double fertilisation, de Vries hastened to publish his "Law of Segregation" (1899), as the rule, characteristic for 'true' hybrids.

It was not with the study of hybrids, however, that de Vries' was chiefly concerned but to initiate an experimental approach to a better understanding of the origin of species, in the light of his pangene-theory.

Immediately following the rediscovery of Mendel's work, experimental research was stimulated and pursued on pure scientific and on applied schemes all over the world. Complicated cases of segregation could be explained by secondary hypotheses like 'gene-interaction'. This rapidly expanding field of research since then is covered under the name of 'Mendelian genetics'. Hugo de Vries, more or less involuntarily, became isolated from these developments because he concentrated his experimental research on processes other than normal segregation, his purpose always being to elucidate the origin of new species by mutation. His idea of constant unit-characters involved the supposition that new characters originate at once by a single detectable step (by a progressive mutation), not by the accumulation of small gradual variations, caused and selected by the environmental forces.

In a waste field near Hilversum in the neighbourhood of Amsterdam Hugo de Vries detected a colony of the evening primrose, Oenothera lamarckiana, originating from a garden-escape.

Among the numerous quite uniform flowering plants and rosettes a small number of differently aberrant individuals struck his experienced attention.

From seeds and rosettes taken at the Hilversum-field he started pedigree cultures on a large scale in his experimental garden in Amsterdam.

The aberrant forms breeding true after selfing were accepted to be incipient new species and called 'mutants'.

These aberrating new constant forms must have originated by a newly formed pangene or by the transmutation of an existing pangene.

The terms 'transmutations' and 'sports' were already in use by Darwin, but these rare events were, in Darwin's thoughts, not of significance for the evolution of species.

The Oenothera-cultures provided concluding experimental support to verify de Vries' hypothesis of intracellular pangenesis and largely contributed to the transformation of that hypothesis into the 'Mutation-theory of Evolution'.

Since then Hugo de Vries is duly credited for initiating an experimental approach to the study of the course and causes of Evolution, which, until then, was largely based on theoretical suppositions and extrapolations.

Trying to elucidate the process of mutation, Hugo de Vries grew families of Oenothera lamarckana in thousands of individuals every year, during half a century.

After having retired from the Amsterdam University chair in 1918, he continued this research in his private experimental garden at Lunteren (prov. Gelderland).

In the lamarckiana-cultures some of the aberrant forms, not all of them, reappeared in a small percentage of the seedlings. In addition other new mutations were won.

During the season of 1895, when de Vries had succeeded in accelerating the growth so as to complete the life-cycle within one year, out of

14,000 seedlings, 7 different mutations were obtained, some of them in only one individual, others in the number of 8, 15 or even more than 70 individuals.

The new types, being selected as seedlings, were tested for true-breed after selfing, and were controlled by crossing with the mother-species Oe. lamarckiana.

DIE MUTATIONSTHEORIE

The unconventional explanations de Vries proposed for the aberrant phenomena, observed in his Oenothera-cultures, were mistrusted; the increasing number of mutants raised the suspicion of some complicated case of hybridisation. De Vries' files of citations from literature, compiled for being edited in the planned comprehensive book, were growing excessively. In the meantime the Oenotheras, cultured over more than ten years yet, though raising problems which asked for new specula- tive explanations, seemed to confirm his ideas on the origin of species, in accordance with the principle of separate unit-characters. Therefore de Vries decided no longer to pursue the accumulation of data for his comprehensive book, but to give what could serve as a preliminary edition of the planned conclusive "Experimental Pangenesis". The first volume (ed. Aug. 1901) is composed of the transcription and coordination of separate chapters from draft manuscripts intended for the ultimate project. It consists of three parts. The first is a historical analysis of theoretical views of selection and evolution.

The incriminated term 'intracellular pangenesis' seems deliberately to have been eliminated and replaced by the pre-Darwinian word "mutation". The intended title 'Experimentelle Pangenesis' was also not used; the subtitle of the first volume: "Die Entstehung der Arten durch Mutation" furnished the new well-chosen name for the two volumes together: "Die Mutationstheorie".

Hugo de Vries' Amsterdam-students know that this change of names was made at the very last moment. In the preface an allusion to the theory of unit-characters is only made at the end as an afterthought, recalling the original principle. (12) But this preface was written, as de Vries told us, after the complete manuscript of the volume was ready for print.

The first introductory sentence of the text, on the contrary, reads, under the heading "Einleitung":

> "Als Mutationstheory bezeichne ich den Satz, dass die
> Eigenschaften der Organismen aus scharf von einander
> unterschiedenen Einheiten aufgebaut sind".

In our hindsight-comprehension we notice that here the definition of de Vries' pangenes-hypothesis is assigned to the theory of mutation, as if, simply, the work 'Pangenesis' had been replaced by "Mutation-theory". Indeed this is the case.

In Professor de Vries' private library, to which his students had

access, the copy of "Die Mutationstheorie" was bound from printers-proof
sheets, marked, on every 16th page with the indication "bitte Revision"
for the German printer. In this copy the initial sentence: "Als Mutationstheorie bezeichne ich ..." etc. is not in print, but it is written by hand
at the top of the page to be added in the second proof ("Revision"). So
I dare guess that in the original manuscript, de Vries started with the
definition of his pangenesis-theory; as he did in the rediscovery-papers
of 1900. By the time the manuscript of the book was sent to the press,
the general interest was fixed on Mendelism. De Vries, well aware that
his original starting point, the separate unit-characters, had been
anticipated by Mendel, he preferred to shift the attention from that
common basic principle, to its application on the origin of species.
The proof-sheet leaving no room and not enough time for a deliberate
exposition, he solved the problem by adding the somewhat illogically
contracted sentence.

Certainly it is the title 'The Mutation-Theory' that made Hugo de Vries'
name famous, more than the comprehensive text of the book could have done.

In the first introductory chapters the terms "pangenesis" and "elementary
characters" are eliminated, and replaced by "variability", "selection",
"mutability", "mutations".

The second part of the volume gives the comprehensive description of the
pedigree-cultures of Oenothera lamarckiana and the mutants, found in
the field and detected in the cultures, until 1899.

In the preface of the second volume (1903) the development of de Vries'
ideas is recollected in more detail, from the original basic principle
of independent unit-characters and the theory of intracellular pangenesis,
this time without mentioning Darwin.

The first part of the volume, in accordance with the subtitle "Elementare
Bastardlehre", contains an exhaustive discussion on different kinds of
hybrids, their variability and development.

The chapter on segregating hybrids is entitled "Die Mendelschen
Spaltungsgesetze".

Of more principal importance is the second half of the volume, dealing
with the process of mutation and the behaviour of observed mutants.
The description of the bizarre phenomena, occurring after crossing the
Oenothera-mutants, raised the suspicion that perhaps some complicated
kind of hybrid-segregation was taken by de Vries as his leading
instance of mutation. For de Vries, reversely, these peculiar
processes supported the conviction that exactly these peculiar
mutations really contribute to progressive evolution of new species.

At the end of Hugo de Vries' career the number of his Oenothera-mutants
amounted to some 60. One of the 1895-mutants, appearing in one
individual only, was called Oenothera gigas, because of its robust
growth, broad leaves, thick buds, a dense spike of great brilliant
flowers and thick fruits. This Oe. gigas was de Vries' favoured instance
of a 'progressive mutation'. If originated in a natural habitat, it would

certainly be superior in the struggle for life and would conquer an area
as a true new species. When, later on, Oe. gigas was shown to be a
tetraploid (doubled genome) of Oe. lamarckiana, this fact for de Vries did
not in the least diminish the importance of Oe. gigas. On the contrary,
the dominance in crossing with Oe. lamarckiana, and the high sterility of
the crosses (being triploid) give additional support for the possibility
of its successful expansion as a progressive new species.

In the Hilversum field, one of the mutants, called Oenothera brevistylis,
was sterile, the pistil being deformed. It can only be propagated by
fertilising a lamarckiana-flower with brevistylis pollen. The F_1 is
uniform and of lamarckiana type, the F_2 shows normal segregation, like
true hybrids in other genera. None of the other mutants showed this
normal segregation.

The differentiating character of Oe. brevistylis only being a deficiency
(i.e. the loss of female fertility), de Vries concluded that normal
segregation (which later on was called 'mendelian'), occurs when an
active pangene is confronted with the same pangene in the inactive
(latent) state.

The shift of a character to the latent state, which he called a 'retro-
gressive mutation' could not significantly contribute to the evolution of
new species. So he concentrated his efforts on the cases with other
behaviour after crossing, in the idea, that there, by mutation a new
character was added to the stock.

For many years de Vries tried to maintain the differentiation between
varietal characters which show mendelian segregation and species-charac-
ters that do not. Mendelian segregation, de Vries thought, only can
occur where an active pangene from one parent is confronted with the
same pangene in inactive state. (12)

The resolution of the non-Mendelian behaviour of most of de Vries'
Oenothera material in terms of cytological processes which modify
hereditary transmission involved a cytogenetic approach. Cytology, which
had furnished de Vries with the basis for a physiological theory of
inheritance and variation in 1889, three decades later was equipping
Otto Renner, Ralph Cleland and others to solve the idiosyncratic
behaviour of the Oenotheras.

W. Johannsen (Copenhagen), in his textbook "Elemente der exakten
Erblichkeitslehre", 1909, rejected de Vries' concept of the pangenes as
not 'exact', because it is based on a hypothetical material carrier of
the characters. Therefore Johannsen abbreviated the term 'pangene' to
'gene', in order, as he expressed it, to free this concept from the
association with definite material carriers of heritable characters,
and to adapt it to the vague terms: 'Anlagen', 'Elemente', 'innere
Beschaffenheit' in Mendel's treatise. (13)

The easy word 'gene' has come in general use, but, against Johannsen's
intention, with the signification of a material carrier for a separate
hereditary character, exactly in the sense of Hugo de Vries' 'pangene'.

Although Mendel in his formulas and deductions never uses material
carriers of characters, but always the discernible characters themselves,
the current gene-theory of heredity is generally attributed to Mendel.
In reality it is Hugo de Vries' pangene-theory, which constantly is
employed, even, with suitably refined expression, in recent experimental
research in molecular genetics.

Hugo de Vries did not openly object to this alienation, because he -
better than most of the mendelists of that time - was fully aware of
Mendel's priority concerning the underlying basic idea of separate
independent constant characters. He reclaimed his rights by vindicating
the applicability of his principle on the study of descent and
evolution under the name Mutation-theory.

FOOTNOTES

(1) de Vries, Opera e periodicis collata, vol. iv, Utrecht, 1920.

(2) de Vries, Intracellulare Pangenesis, Jena, 1889 (Opera, vol. v).

(3) de Vries, Opera VI, p. 339.

(4) de Vries, Opera, vol. v. pp. 148, 149.

(5) Ibid, p. 86.

(6) Darden, "Reasoning in scientific Change: Charles Darwin, Hugo de
Vries, and the Discovery of Segregation", Studies in the
History and Philosophy of Science, 7 (1976), 127-169.

(7) de Vries, Mutationstheorie, II, p. 168.

(8) van der Pas, "Hugo de Vries and Gregor Mendel", Folia Mendiana
11 (1976), p. 4.

(9) de Vries, Die Mutationstheorie, Jena, 1903, vol. ii, p.
In a footnote the remark is added: "Aus diesen Zahlen habe
ich dann das Spaltungsgesetz zuerst abgeleitet, indem ich
damals die Arbeit Mendel's noch nicht kannte."

(10) An excellent survey of de Vries' contributions to this final
result is given in: R.E. Cleland, Oenothera, Cytogenetics and
Evolution, London and New York, 1972.

(11) See Heimans, "Notizblatt aus dem Nachlass Gregor Mendels mit
Analysen eines seiner Kreuzungaversuche", Folia Mendeliana, No. 4
(1968), 5-36.
and Olby, "Mendel no Mendelian?", History of Science, in press.

(12) de Vries, The Mutation Theory. Experiments and Observations on the
Origin of Species in the Vegetable Kingdom, translated
by Farmer and Darbishire, 2 vols., London, 1910.

(13) Heimans, "Exchange of ideas between Hugo de Vries and William
Hohannsen", Folia Mendeliana, 11 (1976), p. 295.

B. NORTON

FISHER AND THE NEO-DARWINIAN SYNTHESIS

INTRODUCTION

Historians of science are moving away from further analyses of
Darwin and his development of the theory of evolution by natural
selection. Increasingly they are beginning to consider the fate and
development of the theory of evolution once it had passed from Darwin's
hands. In particular, a handful of historians have begun to concern
themselves with the fate of Darwin's theory in the period after the
rediscovery of Mendel's ideas in 1900. Notable work has been done
by Garland Allen, Jonathan Hodge, William Provine, Lyndsay Farrall,
Allan Cock, Don MacKenzie and others.

There exists already, I think, what might be called the 'standard'
account of developments, perhaps most clearly laid out in Provine's
seminal Origin of theoretical population genetics. This suggests that
we may usefully see the period from Darwin to the 1930's as occupied
with the rise and consequent resolution of a 'biometric-Mendelian
controversy'. The resolution took the form of a neo-Darwinian
synthesis due to R.A. Fisher and others. Provine and many others have
noted that there arose in London, in the 1890's, a 'biometric' school
of mathematical biologists led by Karl Pearson and W.F.R. Weldon.
These men, it is generally agreed, believed in a Darwinian view of
evolution - that is to say, a view which stressed the continuity of
evolutionary change, a change which, by turn, was seen as due to the
moulding force of selection on the continuous variations displayed by
all members of large populations.

Being concerned with gradual and continuous change, the biometricians
did a great deal of work on normally distributed characters, like human
height. In particular, the biometric school laid great stress upon
the discovery of the correlation coefficients connecting ancestry of
different degrees. Thus, for example, Pearson found that the correlation between fathers and sons in respect of height was about 0.5, by
which he meant that we could expect that on the average, sons of
fathers deviating from the paternal mean height by z inches would
themselves deviate by only $\frac{1}{2}z$ inches.

Opposed to the biometricians, was another school of workers led by
the naturalist William Bateson, who worked in Cambridge. In the nineties this school took up the view that evolution was a discontinuous

process, taking place when large, discontinuous variations were formed.
Natural selection, in this perspective, did not mould or shape a
plastic nature into closer adaptation to the environment.Rather, it
allowed those variations which were viable to survive and disposed of
the rest.Selection in this view was a 'negative' force.

The clash that made the two schools famous was under way when
Mendel's ideas were rediscovered in 1900, but it was the advent of
Mendelism in that year which made the clash into a cause celebre.On the
one hand, it is said, there were the biometricians who used complex
statistical methods, who were Darwinians, and who opposed Mendelism.
On the other hand we find the discontinuist Bateson linking Mendelism
to his position by suggesting that large, discontinuous variations
were Mendelian 'unit characters' which would be integrally inherited.
Thus, it is said, there came to pass an antithesis between Mendelism
and statistics, and another between Mendelism and Darwinism. This
state of affairs is seen as holding outside of England as well as
within it. In Holland, after all, there w_as De Vries with his
'mutation' theory,and, in America there were several notable biologists
who felt unable to reconcile Mendelism with Darwinian continuity and
selection.

The clash is often seen as having come to a head in about 1906, the
year of Weldon's premature demise. But, the effects of the debate
are seen as having a long life, as having the power to prevent an
effective reconciliation between Mendelism and Darwinism until the
twenties and thirties. This reconciliation is seen as having been
produced centrally in the work of men like J.B.S.Haldane, Sir Ronald
Fisher and Sewall Wright, who were able to effect a synthesis of
Mendelism, statistical methods and Darwinism because they were
mathematically competent and in a position to to show or to realise
that Mendelian factors could account for continuous variations, which,
consequently, became seen as subject to Darwinian selection, and for
all sorts of small, non-discontinuous phenotypic effects - here one
thinks of ideas like those of gene interaction, position effect and
so on.[1] These three men produced work in Mendelian population
genetics, containing such results as the compatibility of Mendelism
with continuous variation, and analysing the effect of selection
upon Mendelian populations.As a consequence of these investigations
it became possible to describe the effects of selection acting in
different ways in different populations. Total agreement,however,
was not reached, Haldane and Fisher stressed the rapidity of
evolution in large, intercrossing populations; Wright stressed the
evolutionary potential of small populations showing varying degrees
of 'genetic drift'.

This analysis of the path to synthesis has faced criticism. Hodge
and Allen have stressed the role of the Russian school under
Tchetverikoff, with Hodge[2] emphasising the importance of Dobzhansky's
work as offering a first full neo-Darwinian synthesis. Dobzhansky,
he says, must be understood via the Russian connection.At the same
time, Hodge takes pains to stress that Fisher, Wright and Haldane
perceived themselves as tackling rather different problems, and,

consequently, that we must not too easily consign them to the historiographical mould marked 'resolution of the biometric-Mendelian controversy'. Quite generally, extant literature has little to say about the ways in which these men became involved in their work. But, these provisos aside, it must be said that Fisher's work does seem to conform to the aforementioned mould - better perhaps than that of his[3] colleagues. His synthesis is said to have begun with his great paper of 1918 on 'The correlation between relatives on the supposition of Mendelian inheritance.' This showed that normally distributed variations and observed correlations between relatives could be explained in Mendelian terms - thus reconciling Mendelism and continuous variations. The synthesis continued with a series of post-1918 papers, eventuating in his <u>Genetical theory of natural selection</u> of 1930, with its much discussed 'fundamental theorem of natural selection', describing the rate at which a population's fitness could change over time.[4]

In its most general outlines, at any rate, the aforementioned 'standard' account appears to have the support of notable evolutionary biologists. Julian Huxley, for example, in his <u>Evolution: the modern synthesis</u> argued that the early Mendelians 'refused to acknowledge[5] that continuous variation could be genetic', and that a synthesis occurred when the Mendelians found that 'mutations could be of any extent and accordingly that apparently continuous as well as obviously discontinuous variation had to be taken into account in discussing heredity and evolution.' Again, Gavin De Beer, in his paper on 'Darwin, Mendel and Fisher' argued that early Mendelians 'were wrong to pin faith on wide discrete steps as the invariable effects of genes because ..... Mendelian inheritance is perfectly compatible with the production of gradual variation by small and almost imperceptible steps'. In De Beer's view, Fisher's work was crucial in bringing about the integration of Darwinian selection and the Mendelian geneticist standpoints[6].

In this paper, I wish to focus upon Fisher and his putative role as biometrician-Mendelian synthesiser. In the first and longer part of my discourse I shall attack a question which, as Garland Allen and one or two others have noted[7] has been left curiously untouched-namely the question of why and how Fisher became involved in evolutionary biology, and of why he moved in the direction that he did. I shall suggest that his 1918 paper, for long regarded as a classic in the history of genetics, must be seen not simply as an attempt to resolve certain 'internal' problems within genetics, but also, and predominantly, as a contribution to the hereditarian social ideology of eugenics. I shall also wish to suggest that these early concerns and interests should <u>not</u> be kept out of sight when interpreting the remainder of Fisher's evolutionary enterprise. In the second part, I shall suggest that, while there <u>is</u> a strong sense in which Fisher's work <u>did</u> synthesise Darwinism, biometry and Mendelism, we should not easily suppose a too strong connection between his answers and the questions of the early, 'pre-synthetic' British Mendelians. We shall see, I hope, that the biometricians' objections to Darwinism included, quite centrally, issues and problems that arose <u>before</u> the advent of Mendelism in 1900 - issues and problems which led many late-Victorian commentators to remark that while [8] evolution lived, natural selection had foundered - and that these issues and problems, focused upon ideas of utility, adaptation, and phenotype-environment interactions, were not ones resolvable by demonstrations of the power of Mendelism to encompass continuous variation or by mathematical analyses of the effects of selection on arrays of genotypes. I shall offer, I hope, a reminder of the circumstance that a synthesis of lines A and B need not resolve the problems of both or either.

'On the correlation between relatives'

To understand Fisher's 1918 paper historically, we must surely attempt to uncover the problems that generated it. One wonders what were the problems that Fisher was trying to resolve in the paper, why he perceived these issues as problems, and what were the forces that gave his theorising its particular form

The traditional view is that Fisher was simply (or, perhaps, complicatedly) trying to resolve an outstanding problem in genetics. The biometricians had studied normally distributed variations in man and had found the correlations between relatives in respect of such characters - characters, that is, such as stature. Pearson and Lee,[9] for example, had found a value of about 0.5 for father and son, and of 0.54 for siblings. They had shown too that arrays of sons due to fathers of fixed stature were normally distributed about their means[10] with a uniform variance of $\sigma^2(1 - r^2)$. All of this, of course, referred to phenotypic variance - for, as stated, the talented Pearson would have no dealing with Mendelism, neither in 1900 nor at the hour of his death in 1936. This being the case, it might be thought that a notable problem was that of whether such distributions, correlations and variances could be accounted for within a Mendelian model. Pearson in fact investigated these issues in 1903, finding, he wrote, that a normal distribution could be accounted for on the hypothesis that characters such as height were controlled by many independent[11] Mendelian loci. He found also that linear correlation between relatives was a consequence of the model - but that the model gave the wrong values. It gave, for example, a value of 1/3 rather than the observed 1/2 for the correlation between fathers and sons, and, in Pearson's view, this was too bad for Mendelism, which he held to have been refuted by the facts.

But, in 1906, Pearson's one-time student and friend, G.U.Yule,[12] in a move that Duhem and Quine would have applauded, pointed out that the value of 1/3 was the consequent of two special assumptions - first that the fractions of a and A alleles at each locus were equal, at 1/2: and, second, that complete dominance prevailed. Relax these assumptions, Yule asserted, and the Mendelian model could after all be seen to explain the facts which Pearson had depicted as refuting it. Pearson, in effect, acknowledged and developed this point in 1909,[13] but nevertheless refused to accept Mendelism, claiming, perfectly contrary to fact, that Mendelians could have no truck with intermediate dominance. But, Pearson or no, by 1909, it was knowable for those with eyes to see that Mendelism and biometric results were quite compatible.Naturally, Pearson's and Yule's models were rather simple, not allowing for assortative mating for example, but the basic message was quite clear.

Fisher's reconsideration of the issue was presented to the Royal Society in 1916 by W.C.D.Whetham, already a fellow of the society.[14] Fisher's paper, one supposes, took care of the intricacies ignored by Yule and Pearson, but this seems to have done it little good, for it was withdrawn after being indifferently reviewed by Pearson and R.C. Punnett. Pearson, judging from his report, may not have read the paper at all thoroughly, and Punnett seems to have lost his nerve when faced with the prospect of n unidentifiable factors - confessing himself faced with something 'too much of the order of problem that deals with weightless elephants upon frictionless surfaces, where at the same time we are ignorant of the other properties of the said elephants and surfaces'. Fisher's work, consequently, did not see the light of day until its publication in Edinburgh in 1918 - and, if we are to go by Fisher's notes of thanks, we must suppose that he was encouraged to persevere in the quest for publication by Darwin's son, Major Leonard Darwin and by W.C.D.Whetham. Both were noted eugenists: both were to feel the lash of Lancelot Hogben, the anti-eugenist.

Until recently, Fisher's decision to become involved in this sort of work has remained somewhat mysterious. He was, after all, a mathematics graduate, who took his degree from Cambridge in 1912.Thus an understanding of his entering this line of country has been hard to acquire. But, recently uncovered documents show that Fisher's problems were ideological rather than biological, and that the ideology in question was one perfectly capable of propelling mathematicians into biological studies, or, more precisely, into sociobiological studies. The documents in question are the records of the Cambridge University Eugenics Society, founded in 1911 under the patronage of illuminati such as John Maynard Keynes and Lord Rayleigh. There was nothing odd in Cambridge having such a society, for eugenics achieved a considerable prominence in Edwardian England.[15]

Fisher took a hard eugenic line from an early stage. In 1911, for example, he was to be found assuring the members of the society that they were the 'agents of a new phase of evolution', whose task was to spread 'not by precept only, but by example, the doctrine of a new natural ability of worth and blood'. In these phrases, Fisher did [16] little more than prefigure later eugenic utterances which show him as an advocate of income-related family allowances, the restriction of professional opportunities to the scions of professionals and the view that eugenic factors could explain (and, if necessary, prevent, the fall of great empires and civilisations.) No fuller testimony to this may be found that the final five chapters of his Genetical theory of natural selection, dealing as it does with topic such as 'Reproduction in relation to social class', 'Conditions of permanent civilisation', and so on. Everywhere we find the message that inherited ability is a key to historical fate, and that its breeding must be promoted.[17]

From this one might suppose that the motivation for Fisher's 1918 paper was an eugenic one - and this indeed seems to be the case, for, in the records there is a paper by him on 'Heredity', given to the Cambridge Eugenics Society in 1911. In the paper Fisher discussed the [18] possibility and the utility to the eugenist of a biometric-Mendelian synthesis of the sort just described - i.e., that of bringing biometric results under the Mendelian explanatory schemata. The notion that a strong eugenic motivation underlay Fisher's work is reinforced, as noted, by the circumstance that Whetham and Darwin, his encouragers, were both authors of strongly written eugenic texts. Certainly such a motivation seems entirely possible, as recent work seems to show conclusively that Pearson's main reason for developing biometric science in the first instance was an eugenic one - that, for example, his development of the law of ancestral heredity arose out of a concern to show, contra Galton, that the offspring of a long-selected line would not show a reversal to some original population mean when selection was relaxed.[19]

Now, if all that could be said about Fisher's paper in this respect was that it arose out of eugenic concerns, we would, I think, have a significant result - a really good case of internal developments being due to external factors. But, I consider, it is possible to make a stronger connection than this between Fisher's ideology and his work represented in the 1918 paper. To appreciate this, it is first of all necessary to understand the key problems facing eugenists in the early 20th. century. They were in the position of wishing to substantiate the claim that the only effective forms of social policy were ones that were based in selective breeding, and this, by turn, was a consequent of a faith that nature vastly predominated over nurture. Clearly, they needed to establish this central dogma - for otherwise there was little point in eugenists showing that the lower classes were disproportionately fertile. If the supposed stupidity of the lower orders was thought to be environmentally produced, then eugenic policies would have little chance of gaining the upper hand in the political arena. Now, we know Fisher to have been a keen student of attempts made to establish the dogma, and we know too that none was more active in this field than Pearson, whose career was to lead him to become first incumbent of the Galton chair of eugenics. And, from 1906 onwards, Pearson found himself in command of a 'Galton Laboratory for National Eugenics', whose energies he channelled into the nature-nurture dispute - on the side of nature, so to speak.

Now, we know Fisher to have been aware of Pearson's work.[20] But, judging from a paper delivered to the Eugenics Education Society at about the same time as the publication of his 1918 paper, he was highly critical of Pearson's strategy. This, indeed, was a rather unsophisticated one, generally taking the form of a presentation of two columns of observed correlations - one said to give 'Strength of nature' and the other said to give 'Strength of nurture'. The first column would include correlations like that between filial and paternal statures, while the second included correlations like those observed, in restricted areas of data, between ,say, 'keeness of vision and home environment', or, to take another example, between 'moral state of parents and refraction of offspring'. (Further examples are given in the table below) The average value for the 'nature' correlations was about 0.5, whereas that for the 'nurture' correlations was about 0.05, and the conclusion to be drawn,

Appendix. From the Galton Laboratory's 1910 demonstration of the relative powers of nature and nurture.

TABLE II. STRENGTH OF NATURE.

PARENTAL RESEMBLANCES.

Physical Characters.

Pair	Organ	Correlation
Father and Son	Stature	·51
" "	Span	·45
" "	Forearm	·42
" "	Eye Colour	·55
Father and Daughter	Stature	·51
" "	Span	·45
" "	Forearm	·42
" "	Eye Colour	·44
Mother and Son	Stature	·49
" "	Span	·40
" ? "	Forearm	·41
" "	Eye Colour	·48
Mother and Daughter	Stature	·51
" "	Span	·45
" "	Forearm	·42
" "	Eye Colour	·51

Pathological Characters.

Parent and Offspring	Pulmonary Tuberculosis (Pearson)	·40 to ·60
"	Pulmonary Tuberculosis (Goring)	·43 to ·62
"	Insanity (Heron)	·53
"	(Goring)	·47
"	Deaf-mutism (Schuster)	·54
"	Corneal Refraction (Barrington)	·60

Mental Characters.

Father and Son	Ability (Oxford Class Lists, Schuster)	·49
" "	Intelligence (Family Records, Pearson)	·58
Mean Parental Correlation		**·49**

TABLE III. STRENGTH OF NURTURE.

Characters Dealt With.	Correlation.
Keenness of vision and home environment as measured by cleanliness of body and clothing	+·07
Eye disease and overcrowding	+·03
" " economic condition of home	+·06
" " physical " parents	+·02
" " moral " "	+·03
Myopia and age at which child begins to read	+·02
Liability to phthisis and destitution	—·10
Keenness of vision and number of persons per room	—·07
Myopia	—·09
Moral state of parents and refraction of offspring	—·07
Physical	·00
Economic condition of home and refraction of offspring	—·05
Moral state of parents and keenness of vision of offspring	—·02
Physical	·00
Economic condition of home and keenness of vision of offspring	—·01
Weight of child and mental capacity	+·04
Stature " "	+·08
Condition of teeth and mental capacity	+·09
Condition of clothing " "	+·04
State of nutrition " (Boys)	—·01
" " (Girls)	+·24
Cleanliness " (Boys)	+·08
" " (Girls)	+·14
Glands " (Boys)	+·07
Tonsils " (Girls)	+·08
Acuity of vision and time out of doors (Boys)	+·01
Shortsight " (Girls)	+·11
'Unhealthy' trade of father and weight of child	·00?
" " " height	+·04
Employment of mother and weight of son	+·07
" " stature of son	+·07
" " daughter	+·11
" " stature of daughter	+·14
" " intelligence of son	+·11
" " daughter	—·16
Wages of father and weight of child	+·12
" " health of child	+·08
Number of rooms and weight of child	+·10
" " stature	+·09
" " stature	+·11
"Mean nurture value, +·03	

Pearson averred, was that it was 'quite safe to say that the influence
of the environment is not one fifth that of heredity and quite
possibly not one tenth of it'. And, moreover, what we might see as
typically Fisherian consequences were seen to flow from this.[21]

> Hard environment may be the salvation of a race, easy
> environment its destruction.If you will think this point
> out in detail,I believe you will see the explanation of
> many great historical movements. Barbarism has often
> triumphed over covilisation, because a hard environment
> has maintained,an easy environment suspended, the force of
> natural selection - the power of the nature factor.

Though Fisher endorsed Pearson's conclusions, he was unhappy with
the arguments adduced in their support by Pearson, and, I wish to
suggest, there is a very strong connection between his dissatisfaction
with these arguments and the very shape taken by his 1918 paper.Here
it becomes necessary to mention the nature of the paper, for its
novelty did not lie in its having shown the consistency of Mendelism
and observed biometric results. Rather, it lay in Fisher's analysis of
phenotypic variance into a number of fractions - into environmentally
caused variance, into variance due to additive genetic effects and into
variance due to dominance effects. His analysis offered the prospect of
carrying out resolutions of observed variance into these different
fractions, because he was able to construct a series of equations
expressing these fractions as functions of observed correlations. Thus,
for example, he was able to show that when there was no assortative
mating, the difference between the fraternal and the paternal
correlation offered a quick way of estimating the different fractions.
When assortative mating was allowed for, things became more complex, but
the same general pattern held.[22]

Application of the new Mendelian calculus to Pearson's data suggested
that variance within the data for stature he had collected was due
almost entirely to genetic effects. For, as Fisher put it[23]

> An examination of the best available figures for human
> measurements shows that there is little or no indication
> of non-genetic causes. The closest scrutiny is invited on
> this point, not only on account of the practical importance
> of the predominant influence of natural inheritance, but
> because the significance of the fraternal correlation in
> this connection has not previously been realised.

Clearly, Fisher felt that he had shown the eugenic point to be correct:
nature did dominate over nurture. And, from an accompanying paper, it
would seem that he saw this new demonstration as filling the holes which
he had discerned within Pearson's arguments. For, it will be recalled,
Pearson had operated by comparing the magnitudes of two sorts of
correlation coefficients. The correlations measuring the influence of
'Nature', however, had a value of only about o.5. And, since the
variance of the array of sons due to fathers with a given value was
still $(1 - r^2)$ that of offspring in general, Pearson's 'proof' seemed
to leave some 75% of the observed phenotypic variance unaccounted
for. Thus, said Fisher, the road was left wide open for the intelligent
anti - eugenist to,[24]

point to a dozen causes to which height or shortness is
commonly ascribed, such as regular athletic exercise, or
accidental illness in childhood, and it would be difficult
to prove without a specially designed investigation for each
alleged cause that these do not contribute the important
proportions of the total. The task of ascertaining the
importance of the environment in this way is an endless one,
since new environmental causes could be suggested, each more
difficult than the last to define, measure and investigate.

But, using Fisher's methods, these objections could be overcome in
one step.

Here, it seems to me, we have it. Fisher, whose work was all along
stimulated by his connections with leading eugenists like Darwin and
Whetham, saw his work as an enterprise which could do away with the
need for this endless series of analyses of particular environmental
features - for, had he not offered a way of evaluating the total
effect of all environmental factors? His work was little short of an
eugenic triumph. It was so, not because it showed that Mendelism and
biometric results were compatible - which had been done before, even
if not at such a level of sophistication. Rather, its merit lay in
the way that it showed that a Mendelian approach could resolve the
longest-standing difficulty for the eugenist.

The consequence of this, I wish to suggest, is that we should desist
from seeing Fisher's early work as primarily a contribution to 'pure'
genetics, as a major staging post on the path to a neo-Darwinian
synthesis. We should see it rather, as a stunning contribution to
eugenics. That is all that I wish to suggest for the moment, though
were more space available, the tracing of connections between the
perceived eugenic problem situation and the construction of the theory
would be fruitful. Another point worthy of being followed up is
Fisher's use of his method - for he was shortly to be involved with
Lancelot Hogben in a prototypic version of the Jensen - Lewontin
debate over heritability. Here too, there is space for further
discussion of the relation between science and ideology. For the
moment though, I want simply to suggest this significant restructuring
of the historical perspective in which this landmark of genetics should
be seen. Fisher was bringing Mendelism to the aid of eugenics - a
point, I suggest, that is well worth documenting, not simply for the light it
throws upon his 1918 paper, but also for the guidance it may give to
historians of Fisher's later works.[25]

AFTER 1918
 I now pass to the remainder of Fisher's work and its historical
significance, inquiring further into the extent to which his labours
in the field of the mathematical analysis of selection, the evolution
of dominance and the fundamental theorem of natural selection may be
interpreted as representing a synthesis of the biometrician- Mendelian
controversy. And, to 'place' Fisher's work, we should perhaps look
at the ideas of early English Mendelians - the 'pre-synthetic'
Mendelians, so to speak. A standard view of these, as earlier
citations from the works of Huxley and De Beer seem to indicate, is

that, evolutionarily speaking, they believed in non-Darwinian
discontinuity,thinking of Mendelian mutations as being typically
discontinuous.In such a context, Fisher may be seen as having
overcome a nutationist school.

In practice, however, it seems that these Mendelians were more
sophisticated than generally presumed, and, I think, a knowledge of
this sophistication may materially aid the historian attempting to
place Fisher. For it seems that we may discern at least two distinct
senses of discontinuity. One doctrine,call it Discontinuity$_1$, was
that all observable forms of variation, bar certain environmental
effects, would eventually be shown as due to underlying atomic, genetic
factors, which did not blend in transmission. Thus, from an early stage,
Bateson suggested that continuous variation would succumb to Mendelian
analysis, and the general tenor of agreement on this point is brought
out by Lock's and Doncaster's receptivity to Yule's work, and by a
typically British Mendelian insistence that [26]

> Our present knowledge renders it easy for us to conceive
> of the existence of segregation without there being any
> obvious manifestation of its existence.

And, with this view, went the proposition that without 'the existence
of selection and elimination', 'there could be no evolution'.Fisher,
doubtless, agreed with both.

Where Fisher does appear to have differed substantially from the
Mendelians was over another doctrine, best called Discontinuity$_2$, which
found outlets in suggestions that evolution, even the evolution of new
species, might generally or frequently be due to large discontinuous
variations. And it seems that when they held this view, it was not
(as frequently suggested) because they supposed Mendelian variations to
be inherently discontinuous and large, but for several more
sophisticated reasons, notably (though not exclusively) the view that
the differences between species were not such as to make a Darwinian
scenario of cumulative, gradual, adaptive change of large populations
    plausible - not, at any rate, if one abandoned the possibility of
the inheritance of acquired characters which Darwin often used. This
style of plausibility argument had several variants, some more
compelling than others. But, so far as I can see,a crucial form of the
argument was not one easily answered or disposed of by the style of
work done by Fisher. This is the argument, derived from a line of
distinguished nineteenth century evolutionists, to the effect that
it was not plausible to suppose that all the differences between
predecessor and successor species in many evolutionary lines had been
brought about by continuous change with all intermediate forms between
predecessor and successor represented, and with each successive member of
the series enjoying some marginally greater 'adaptation' or 'fitness'
over its predecessor. Given these ideas, the Mendelians felt it more
plausible to suppose that important, species-distinguishing
characters had frequently arisen suddenly, and being possibly of
neutral 'fitness' or 'adaptive' potential, had merely survived and
persisted. A variant of this position was the claim that some
new characters, when wholly formed, conferred fitness (in the sense of
an enhanced ability to deal with the environment) but that the

490

imperfect forms through which Darwinian theory required them to have passed would have lacked such fitness:hence the need for discontinuity of some sort. Bateson was expressing such thoughts in [27] his Materials for the study of variation when he wrote of the colours of two species of ladybirds, existing in the same geographical area but showing very different levels of variability, that

> to be asked to believe that the colour of C.septempunctata is variable because it matters to the species, and that of C.decempunctata is variable because it does not matter, is to be asked to abrogate reason.

And, noting that the variation of C.decempunctata grouped around eight or ten clear 'types', he expressed essentially the same point when he wrote that,

> an expectation is created in the mind that the distinctness of these forms of varieties,all living (and probably breeding) together, may be of the same nature as the distinctness of species; and since it is clear that the distinctness of varieties is not the work of separate selection we cannot avoid the suspicion that the same may be true of the specific differences too.

Bateson's colleague, Punnett,[28] also put matters well when he stated that

> The vital point, which is whether natural selection does offer a satisfactory explanation of the living world, is too frequently lost sight of. Whether we are bound or not to interpret all the phenomena of life in terms of natural selection touches the basis of modern philosophy. It is for the biologist to attempt to find an answer, and there are few more profitable lines of attack than a critical examination of the facts of adaptation.

And, in another work, Punnett showed again the philosophical basis of the anti-Mendelian biometric position when he argued that the doctrine of Discontinuity, could release the biologist from that perennial late 19th.century problem, namely,[29]

> the burden of discovering a utilitarian motive behind all the multitudinous characters of living organisms.

Very similar remarks are also to be found in Bateson's work,often going hand in hand with ideas reminiscent of some thoughts of Koestler and Waddington in recent years, thoughts to the effect that understanding evolution required an understanding of how it was that organisms could vary in ways preserving an overall harmony. Bateson, seeing the vision of the monkey and the typewriter in conventional Darwinism, attacked always the view that living beings,[30]

> are plastic conglomerates of miscellaneous attributes... and that by variation any of these attributes may be subtracted or other any other attribute added in indefinite proportion.

The point of all this, I suggest, is as follows. We must not too easily assume that the main issue outstanding between Fisher and the early Mendelians in the matter of 'Darwinism' was that he asserted and that they denied the compatibility of continuous variation subject to selection and the Mendelian biological picture of the world. Rather, it was at the level of the doctrine of Discontinuity, that they differed - a vague and philosophical level developed in the nineteenth century, and one hinging crucially upon ideas about 'fitness' very different to those developed in, say, Fisher's fundamental theorem of natural selection. And, though the Mendelians' formulation of their view did remain obscure, and while they seem not to have followed it through in any detail, we can at least see that it comprised a line of argument against which formal Fisherian genetic analysis may have made very little progress: a case, perhaps, of scientific incommensurability. The upshot of this is the suggestion that, in so far as Fisher carried through a biometric-Mendelian synthesis, it was not the sort of synthesis that, per se, resolved the anti-Darwinian objections of the British 'pre-synthetic' Mendelians.

CONCLUSION

I want to propose that we view Fisher's early genetical work as esentially a response to an ideological, to an eugenic problem-situation, and to suggest that this eugenic perspective may prove useful in future historical analyses of his later work. At the same time, I wish to draw attention to an interesting incommensurability between his work in formal genetics and the problems of the 'pre-synthetic' Mendelians. This again has obvious implications for future historical work in the area of twentieth century evolutionary theory.

NOTES

1   Works mentioning or advancing this perspective include: W. Provine, Origins of theoretical population genetics, Chicago (1971) 177; G. Allen, 'Genetics, eugenics and society: internalists and externalists in contemporary history of science', Social studies of science, 6 (1976) 105 -122; G. Allen, Life science in the 20th century, New York (1975) 51; Open University course S 299, The history and social relations of genetics, Walton Hall (1976) 23 - 27

2   J. Hodge, unpublished typescript on 'Natural selection in mid-twentieth century biology: acontextual and conceptual analysis.'

3   R. A. Fisher, 'The correlation between relatives on the supposition of Mendelian inheritance', Trans. Roy. Soc. Edinburgh , 52 (1918) 399 - 433.

4   R. A. Fisher, The genetical theory of natural selection , 2nd. edn, Dover, New York (1958).

5   J. Huxley, Evolution: The modern synthesis, London (1942) 25.

6   G.De Beer,'Mendel, Darwin and Fisher', Notes and records of the
    Royal Society of London,19 (1964) 192 - 226.

7   G.Allen, op.cit (footnote 1)

8   See V.L.Kellogg,Darwinism today, New York (1907).

9   K.Pearson and A.Lee,'On the laws of inheritance in man',
    Biometrika 2 (1903) 357 - 462.

10 The sigma-squared refers here to the variance of sons en masse.

11 K.Pearson, 'On a generalised theory of alternate inheritance, with
    special reference to Mendel's laws', Phil.trans.Roy.Soc. , 203A
    (1904) 53 - 86.

12 G.U.Yule, 'On the theory of inheritance of quantitative compound
    characters on the basis of Mendel's laws - a preliminary note'
    Report of the third international conference on genetics, London
    (1907) 140 - 142.

13 K.Pearson, 'The ancestral genetic correlations of a Mendelian
    population mating at random', Proc. Roy.Soc., 81B (1909) 225 - 229.

14 B.Norton and E.S.Pearson, 'A note on the background to, and
    refereeing of R.A.Fisher's 1918 paper "On the correlation between
    relatives on the supposition of Mendelian inheritance", Notes and
    records of the Royal Society, 31 (1976) 151 - 162

15 These records kept at the offices of the Eugenics Education
    Society. For an account of British eugenics, see G.R.Searle,
    Eugenics and politics in Britain, 1900 - 1914, Leyden (1976).

16 See Searle, op.cit., p. 83. See also, R.A.Fisher, 'Positive
    eugenics', Eugenics review, 9 (1917) 206 - 212.

17 Op.cit. (note 4)

18 See Norton and Pearson, op.cit. (note 14).

19 See B.Norton, 'The biometric-Mendelian debate', Proc. 9th. internat.
    biometric conference (1976) 357 - 376.

20 See Norton and Pearson, op.cit. (note 14).

21 K.Pearson, 'Nature and nurture, the problem of the future'
    Eugenics Laboratory lecture series , 9 (1910) 27.

22 L.L.Cavalli-Sforza and W.Bodmer,The genetics of human populations
    San Francisco (1971).See chapter 9.

23 Fisher, op.cit. (note 3) 435.

24 R.Fisher,'The causes of human variability',Eugenics review,10,(1918)

213 - 220

25 J.Hodge makes some suggestive points in his paper, op.cit. (note 2)

26 'Ardent Mendelian','Methods and results', Mendel journal No 1
    (1909), 159 - 194. See p. 162. See also, for notes on Bateson, A
    Cock, 'William Bateson, Mendelism and biometry',Jnl.hist.biol., 6
    (1973) 1 - 36. See especially p. 11.

27 See W.Bateson, Materials for the study of variation, London (1894)
    572.

28 R.C.Punnett, Mimicry in butterflies, Cambridge (1915): 7.

29 R.C.Punnett, Mendelism, Cambridge (1912):137

30 Bateson, op.cit. (note 27) 80

Symposium 9.   CLASSIFICATION AND SYSTEMATIZATION IN THE SCIENCES

N. W. Fisher

INTRODUCTION TO SYMPOSIUM 9

Without order, there is no science. All scientific
disciplines share the need for classification and system,
though the extent to which these are major preoccupations of
scientists (or are taken very much for granted) varies from
subject to subject, and within a subject changes over time.
The aim of this symposium was to examine the extent to which
problems of classification and systematization are common to
different disciplines, and whether genuinely interdisciplinary historical analysis of these problems is possible.

Since classification was for so long a major preoccupation
of the life sciences, historical discussion of taxonomy has
inevitably been dominated by the problems of biological
classification - definitions of species, questions of the
reality of higher categories and the naturalness of classificatory systems - problems which do not necessarily crop up
in the classification of chemical elements or compounds,
rocks, or books. By sheer weight of interest and comment,
biological concerns were dominant again in Edinburgh, but it
is clear that many of the aims of classification are common
to different disciplines, even if the details of taxonomic
practice, the everyday work of the classifier, differ widely.

In his closing remarks at the end of the symposium, Ernst
Mayr suggested that the aims of classification are as follows:
(1) (and most obvious) Classification answers the need for
easy information retrieval; (2) Every classification embodies
a theory of affinity (in the 18th-century sense) - that is,
it brings together reasonably homogeneous classes of things
about which explanatory questions can be asked; and (3)
Classification is the foundation of comparative methods,
which in many sciences rank together with, if not above,
experimental methods. In all of those disciplines where
classification is of paramount concern, these aims are held
in common, and serve as the link between the sciences. The
emphases may be different: evolutionary taxonomy, unlike
pre-Darwinian taxonomy, is concerned far more with explanation of affinities than with information retrieval; conversely,
a library classification is, so far as the library user is
concerned, solely a means to locating information, and it is
only the bibliographer who will go beyond this to ask
theoretical questions. All classification has its formal
origin in the work of Plato and Aristotle, another link
between the sciences which helped to keep classification
fairly homogeneous until the late 18th century. In the last
two hundred years there has been a divergence accompanying
the differentiation of the various scientific disciplines,
and any historical discussion of classification must reflect
this. It is appropriate, then, that the main focus of this
interdisciplinary symposium was the late 18th century, when

classification still served as a link between the sciences, but when this link was beginning to break down. One of the more interesting aspects of the comments following the papers was the number of references to the abandonment of concern with classification, suggesting that this concern marks a stage which sciences go through on their way to maturity, but which, like a snake's too-confining skin, is sloughed at a certain stage in their development. At first glance, it might appear to be a paradox that when classification is at its most successful, scientists turn to other things. Once a system is established which answers the needs of scientists, it becomes unnoticeable and taken for granted. The successful classifier is one who makes himself redundant.

In medicine today, classification is implicit in the bedside diagnosis of the doctor, but the principles of nosology are certainly taken for granted. The classification of diseases is well established, and any alteration of the details of the classification is carried out within the established framework. This was not at all the case during the 18th century, when there were many competing nosologies, and medics believed (following the contemporary successes of natural history and chemistry) that the establishment of a correct rational nosology was a prerequisite for scientific medicine, and particularly for the teaching of medicine in the reinvigorated medical schools.

Randall Albury in his paper discusses three different nosologies of the turn of the century associated with the three French écoles de médicine established after the Revolution, and the extent to which these were inspired and informed by the recent successes of the new French chemistry of Lavoisier and his collaborators. Despite the fact that Baumes of Montpellier went the furthest in his adoption of the new information coming from the chemists, and their new language, his system was the most archaic of the three, and Albury shows that it was based ultimately on traditional humoral medicine; though they used more traditional terminology, Pinel of Paris and Tourdes of Strasbourg were more radical in their approaches, and ultimately much more successful.

The reasons for this were examined by John Pickstone and Ramunas Kondratas in their comments on the paper. Looking towards natural history and chemistry was nothing new in nosology; this had been the traditional approach throughout the century. The identification of disease depended on symptoms, and it was these that were arranged in nosological systems; as Linnaean classification in botany and in zoology depended on the identification of diagnostic surface features, so 18th-century medical theory and practice was based on the systematic arrangement of symptoms. The revolution in French medicine at the end of the century involved the abandonment of traditional nosologies, as medics came to realize that - unfortunately for patients in the 18th century - the ties between symptoms and the diseases themselves might be remote, and attempts to build systematic therapeutics on symptoms alone radically ill-conceived. The nosologies of Pinel and Tourdes represent steps on the road to revolution, as they tied nosology to empirical evidence from pathological anatomy, opening the possibility of vastly more sophisticated and secure diagnosis. It was Bichat who took the final step of abandoning rational nosology as the foundation of medicine, and this represents the real revolution: by basing their systems on ever deeper criteria, the nosologists had made themselves redundant.

For the user of a library, the successful classificatory scheme is the one which leads him to what he wants with the minimum of effort on his part. He will notice only the unsuccessful arrangement which frustrates him. It is in bibliographical classification above all that the most successful scheme is the one most easily taken for granted; the task of the library classifier is thankless indeed.

In the history of science scholars are particularly fortunate in the ease of information retrieval from the Isis critical bibliographies, begun by George Sarton and continuing under John Neu. With the recent publication of the first three volumes of the Isis cumulative bibliography, one can find what one wants without having to wade through ninety bibliographies. And yet this cumulation is far more than an index to back issues, as Magda Whitrow's account of her classificatory scheme makes clear. The aim is an arrangement which leads the scholar automatically to what he wants to know; the success of the venture is self-evident to anyone who has used the Cumulative bibliography. Information retrieval does indeed seem automatic, but this has been achieved only by great care in the arrangement to make the complex seem simple. Again, the successful classification is unnoticeable, but we are fortunate to get a rare glimpse behind the scenes through Mrs Whitrow's paper.

If biological classification today has achieved some unnoticeability, and taxonomy is no longer so heavily stressed in textbooks or courses of biology, this is a comparatively recent phenomenon. Arguments about the details, and indeed about the principles, of classification in biology were loud and long - they still continue, muted, today - and they were never of course louder than in the 18th century, when the very success of Linnaeus in getting botany into some sort of order called into question the foundations of biological systematics. Phillip Sloan discusses one such fundamental problem, the changing conception of biological species, and the implications of this for systems of classification in biology.

The major change that Sloan identifies is the introduction into the conception of species of the dimension of time, which he ascribes to Buffon. Leibniz and his disciple Wolff distinguished between absolute time, an abstraction, and real time which was guaranteed by the historical continuity of created beings. Buffon based his own 'natural histories' of man and animals on this distinction, seeking the real order of things in immanent time and space, as opposed to the

abstract ordering of nature of Linnaeus. His French successors paid little attention to this aspect of Buffon, taking
from him only a nominalist critique of Linnaean systematization. But Buffon's ideas had more impact in Germany, where
the Leibniz-Wolff tradition was still strong. Kant developed
Buffon's contrast between the two ways of looking at the
world into a distinction between abstract Naturbeschreibung,
and Naturgeschichte, which considered the real. Through his
disciple Girtanner, this distinction ultimately became basic
to Illiger, the founder of the German tradition in biological
systematics. When Darwin's work was published, there already
existed in Germany (but not in France or Britain) classificatory systems which involved a historical dimension, and which
could readily be developed into the foundation of modern
phylogenetic classification.

While applauding Sloan's concentration on the historical
in Buffon, and the attempt to see him as a consistent, rather
than a fickle, theorist, Jonathan Hodge questioned some of
Sloan's emphases. Historical thinking was more mature and
coherent in the 17th century than is often supposed; it made
sense after Descartes to reconstruct the creation of the
world and the development of its inhabitants, as would not
have been possible for Aristotelians. Though the Leibnizian
tradition was obviously important to Kant and the other
Germans mentioned, was it necessary to Buffon? He had other
problems to deal with, which had little or no connexion with
Leibniz's metaphysics of time. In particular, Buffon's
explanation of resemblances between creatures was in terms
of degrees of heat, internal moulds, organic molecules. In
inferring the nature of the internal mould (a conception
that is constant through all Buffon's writings), observation
over time is essential. It is the constancy of succession
over a number of generations that guarantees the inference
and shows that species are real, not abstractions. This is
the key to Buffon's preoccupation with time; he did not need
Leibniz's metaphysics of time to teach him the importance of
history.

In general discussion, Jacques Roger took up a related
point: that Buffon was not really historically inclined.
Even his cosmology was not historical, and should be characterized more as description of nature rather than history of
nature. Only late in life did he really turn to the latter,
and this coincided with a general renewal of historical
thinking late in the century - which may indeed have owed
something to the influence of Leibniz.

Camille Limoges turned to the legacy of Buffon, and asked
in his commentary whether the distinction between the history
and the description of nature is in fact of much help in
understanding the development of biology in the 19th century.
First, in the early part of the century there was no way in
which Girtanner's programme for a true history of species
could be fulfilled. There were no data on geographical
distribution of species, and no consistent picture from
palaeontology of distribution through time. Though Girtanner's bio-geographical programme was not forgotten, and

others such as Flourens in 1837 proposed similar schemes,
there were too many species flooding into Europe for
naturalists to cope with thus; morphology remained the only
practicable basis for classification. And then again, the
revival of Buffonian ideas in France in the 1860s came about
not through a tradition of Naturgeschichte, whether native
or German, but through a realization that the limits of the
traditional morphological approach had been reached. Was
the distinction between the two approaches really so radical?

From the chair, Ernst Mayr was more impressed than Hodge
with Sloan's attention to Leibniz; we need more such work to
differentiate Newton's influence on 18th-century thought from
that of Leibniz and Wolff. And wherever it came from, the
distinction between the ideal-abstract and the real-concrete
in taxonomy has plagued the whole field until recently.
Leibniz, and after him Buffon, were clearly central in the
18th-century rekindling of the nominalist-realist controversy:
does (and should) the taxonomist work in the real or the ideal?
This has only recently been settled with the introduction of
a terminological distinction between taxon, the concrete
object (a group having the logical status of an individual)
and category, the abstract rank in the Linnaean hierarchy.
The taxonomist in fact works with both real and ideal units.

Mayr also pointed to two further distinctions essential
for understanding the history of biological classification -
distinctions which Sloan did not always keep in mind. First,
there is the difference between micro- and macrotaxonomy:
the former deals with the species problem - how to distinguish
species, what to count as species, etc.; macrotaxonomy is
concerned with the higher taxa, their ranking in higher
categories, and the elaboration of classifications. Taxonomists have almost always been involved in one to the exclusion of the other. Buffon, for instance, contributed to the
species concept, but not to classification. As a micro-
taxonomist he may have had an impact on Kant (similarly
interested in the species problem), but his influence on
classifiers such as Illiger was very limited.

Second, there is the distinction between downward and
upward classification. Linnaeus started with the whole of
nature, and divided it dichotomously into smaller and smaller
pieces; since his differentiae were single characters, the
resulting classes were artificial. Buffon vigorously attacked
the whole enterprise. The next generation of French taxonomists took the inverse path of upward classification - the
grouping of organisms with apparent 'affinity'. This affinity
was judged not by single characters, but by 'all relations
of resemblance', as was continually stressed. Again, Buffon's
influence was limited because he was not much interested in
grouping above the species level.

If the lively discussion in the afternoon, with all the
different emphases adopted, did not have many lessons for
those interested in the history of non-biological classification, it at least served to demonstrate the special problems
of biological classification that set it in a class apart.

FRENCH NOSOLOGIES AROUND 1800 AND THEIR RELATIONSHIP WITH CHEMISTRY

Concomitant with the political revolution in France at the end of the eighteenth century was a medical revolution linked institutionally to the political sphere by governmental intervention in the reform of hospital organisation and medical teaching.[1]   Under the system of instruction established by the Convention in 1794, when it founded the three *écoles de santé* of Paris, Montpellier and Strasbourg to replace the previously-suppressed faculties of medicine, the hospital itself became the chief site of medical training.   Consequently the demand for rational order, which played so important a role in the political and economic spheres at this time, served also as a focal point for the revolution in medicine - a point where interests relating to the administration of hospitals, the instruction of medical students, and the advancement of research all overlapped and coalesced.   By instituting a rational distribution of patients into different spatial localities according to their different afflictions - physically separating those with unlike maladies and bringing those with like ones together in a common area - the well-administered hospital would first of all cease to be a place of contagion, a place where patients admitted with one disease soon acquired several more from those with whom they were thrown into proximity.   But this same spatial distribution was to have a pedagogical value as well - it would convert the hospital into an ordered museum of diseases in which each locale displayed to students a particular malady in isolation from the others, with all its stages of development revealed at the same moment: a temporal condensation brought about by a spatial displacement. Finally, this rational ordering of diseases would also serve the interests of medical research by making the hospital a field of unobstructed observation, where all maladies were brought together for study yet kept distinct, where the course of each illness could be scrupulously followed day after day without the perturbations introduced by new affections contracted in the hospital or by blind therapeutic interventions.   Unless diseases could first be thus observed in their simplicity and purity - uncomplicated and undisturbed - there was no hope of establishing either a rational pathology, based upon systematic understanding of proximate causes, or a reliable therapeutics, based upon systematic experimentation with possible remedies.

The success of this spatial distribution of patients in jointly serving the interests of administration, teaching and research, clearly depended upon a correct classification of diseases as a necessary precondition.   Thus it was that nosology could appear as the key to the medical revolution in France.   But nosology itself would have to be put on a new basis: the older systems of classification founded upon resemblances between symptoms and upon differences between the parts of the body affected would have to yield to a more profound analysis and a more 'natural' order.   In this connection, recent developments in natural history and chemistry offered themselves to the nosologist as models.

Through the works of A.L. de Jussieu in botany and F. Vicq-d'Azyr in zoology, a fundamental restructuring of the taxonomy of living things was begun in the late 1770s, with the focus shifting from the external properties of plants and animals to their internal organisation.[2]   Important though it was, however, this restructuring initially appeared only as a method-ological displacement within the confines of natural history; whereas the new system of chemistry introduced in the same period by Lavoisier, Berthollet, Fourcroy and other collaborators, appeared as a veritable chemical revolution: it seemed, as Bichat put it, "to have issued suddenly from nothing".[3]

What, then, was the relationship between nosology, the apparent key to the medical revolution then taking place, and the theoretical system underlying the chemical revolution recently accomplished? In order to investigate this question I propose to examine three French nosological systems dating from around 1800, one from each of the three *écoles de santé:*   the *Philosophical Nosography* by Philippe Pinel of the Paris school, first published in 1798;   the *Essay on a Chemical System of the Science of Man* by Jean Baptiste Baumes of the Montpellier school, also first appearing in 1798; and the *Sketch of a System of Nosology, Founded upon Physiology and Therapeutics* by Joseph Tourdes of the Strasbourg school, published in 1802.   Of these three systems, Pinel's was to achieve the most renown;   but it is important to remember that all three were equally 'institutionalised' in the first years of the nineteenth century, each one being taught by its author in his respective school as part of the established course on internal pathology.

The nosology which *prima facie* shows the closest relationship with chemistry, as appears even from its title, is the *Chemical System* of Baumes; and for this reason we shall consider it first. Baumes' system was outlined in a small volume of about 100 pages published in 1798 with the title, *Essay on a Chemical System of the Science of Man.*   In a slightly revised version this system formed the basis of Baumes' four-volume work, *Foundations of the Methodical Science of Disease,* published in 1801 and 1802, and reprinted in 1806 as an *Elementary Treatise on Nosology.*[4]

Baumes described himself in 1798 as having been "struck by the astonishing progress of pneumatic chemistry, by the mass of its facts and by the certainty of its results", and he said that his aim was

> to apply the doctrine of this science to
> that of medicine, to unite the chemical
> system with the medical system, to guide
> the observing spirit of the latter by the ex-
> perimental course of the former, and thus to
> bring medicine rapidly forward to the level
> of the other sciences behind which it has
> remained so long only because it neglected
> their accessory aid.[5]

Similarly in 1801 Baumes wrote:

> I find myself in the most brilliant epoch of this
> entirely regenerated science [of pneumatic
> chemistry]. ... And in view of the useful and
> happy revolution which this science has undergone,
> how could I not desire that it should become the
> cause and the instrument of the revolution which
> it seems to me that medicine requires?[6]

Let us see, then, how he attempted to bring about this medical
revolution.

Baumes' chemical system of nosology consisted of five
classes of disease, each one related to a particular element -
or simple substance - from the new chemistry.   In 1798 these
classes were arranged as follows: I. Oxygénèses, or disorders
of oxygenation;   II. Calorinèses, or disorders of calorification;
III. Hydrogenèses, or disorders of hydrogenisation;   IV. Azoténèses,
or disorders of azotisation [nitrogenation];   and V. Phosphorenèses,
or disorders of phosphorisation.   In principle each of these
classes could be divided into two sub-classes, the first designating
maladies caused by an excess of the chemical substance in question
(suroxygénèses, surcalorinèses, etc.) and the second designating
maladies caused by a deficiency of the same substance
(désoxygénèses, descalorinèses, etc.).   In practice, however, the
chemical balance of the body is such that an excess of oxygen
involves a deficiency of hydrogen, azote or phosphorous, and an
excess of any of these three substances involves a deficiency
of oxygen.   For this reason the last three classes of Baumes'
nosology were not formally sub-divided and referred only to
diseases of excess, while the first two classes referred both
to diseases of excess and to diseases of deficiency.

The distribution of diseases into these five classes was
carried out by Baumes on the basis of the results of experiments
in animal chemistry reported by the leading chemical investigators
of the day.   In the short text of 1798 his most frequently
invoked authorities were Fourcroy and Berthollet, who were cited
eleven and nine times respectively.   Other exponents of the
new chemistry whose findings were used by Baumes included
Lavoisier and Vauquelin (three references each);   Jurine,
Humboldt and Seguin (two references each);   and Chaptal,
Crawford and Proust (one reference each);   as well as a number
of lesser-known figures.

Applying the researches of these men, Baumes regarded oxygen,
first of all, as the principle of both combustion and irritability
in the living organism;   so he grouped as oxygénèses such
disorders as inflammation, spasms, and the loss of tone in
localised areas of the body.   He also considered syphilis to
belong to this class on the following grounds: "Since the
agent which heals venereal diseases is the oxygen released from
the oxides of mercury used in treating them, the pox must be
classified with the désoxygénèses."[8]

The second key element for Baumes' system, caloric, was closely related in its effects to oxygen. Baumes regarded it as "the true principle of animal heat; and," he added, "it genuinely influences irritability."[9] Among the surcalorinèses he included active haemorrhages and diseases of over-heating; while the descalorinèses were comprised chiefly of diseases characterised by weakness and languor. In his text of 1801-1802 the relation between oxygénèses and calorinèses was reorganised somewhat, with the oxygénèses (now Class II) becoming chiefly disorders of irritability characterised by general excitement or weakness,[10] and the calorinèses (now Class I) becoming chiefly disorders in the quantity of blood and in the production of body heat.[11]

Hydrogen, the basis of the third class of diseases, was considered medically important by Baumes as "the principle of water [in combination with oxygen] and a constituent part of oils [in combination with oxygen and carbon]."[12] Thus diseases marked by the production of oily or watery morbific materials, such as the bilious fevers or dropsy, were classified as hydrogenèses.

The fourth nosological element, azote, was significant for Baumes as the gas which, together with oxygen, "constitutes atmospheric air.... All known animal materials contain a more or less considerable quantity of it."[13] Azote is thus a characteristic component of the gases liberated by the putre-faction of animal matter; and the azoténèses, therefore, were maladies giving rise to putrid excretions and the decomposition of solid parts, such as cancers, ulcers, scurvy and the putrid fevers.

Finally phosphorus appeared in Baumes' system as the principle of solidity.[14] Diseases caused by "an over-abundant production of this earth,"[15] such as disorders involving excessive hardening and brittleness of parts, or the formation of solid deposits - for example, brittle bones, arterial aneurisms and gout - were placed by Baumes in the class of phosphorèneses.

Now on the basis of an 'empiricist' reading of Baumes' texts, such as we have thus far presented, the relationship between his nosology and the chemistry of Lavoisier was of the first order. We have seen Baumes' explicit statements concerning his intention to apply the new chemistry to medicine. We have seen further that he used five Lavoisian 'elements' as the basis for his five classes of diseases and that his distribution of maladies into these classes was governed by contemporary researches showing the effects, both pathogenic and therapeutic, which the five chemical substances were found to have upon the animal economy. We have even seen, by a primitive form of citation-analysis, his particular dependence upon the work of Fourcroy and

Berthollet, the two principal representatives of the chemical
revolution following the death of Lavoisier.

Nevertheless our 'empiricist' reading of Baumes' texts
presents us with a serious contradiction between the new
chemical theory and the traditional pathological conceptions
to which it was assimilated; for as Baumes said of his system:

> this application [of chemistry] serves to revive the
> humoural pathology of which Boerhaave was the
> great defender, combined to a certain point with
> the pathology of irritable fibre introduced by
> Hoffman, Cullen, Brown and Darwin, and brilliantly
> upheld by the great physicians of the Montpellier
> school.[16]

Nor was this link between the chemical revolution and medical
tradition fortuitous; it was Baumes' view that the two were
inextricably related: "chemistry," he wrote, "necessarily
causes [humoural pathology] to prevail."[17] We have no grounds,
then, for arguing that somehow revolutionary and traditional
theoretical elements merely co-existed, in a kind of contingent
contradictory juxtaposition, in Baumes' system; for that would
simply be to deny that it was a system at all. We are left with
only two possibilities for the understanding of Baumes' system:
either the new chemistry revolutionised medical tradition or
the old pathology 'traditionalised' the chemical revolution.

Now as the following chart illustrates, four of the five
chemical elements which Baumes selected as the basis of his
nosology served to replicate the four humours of classical
medical theory as mediated by the four elements of classical
matter theory, with oxygen - the privileged substance of the new
chemistry - forming the link between this humoural pathology
and the eighteenth-century theories based on irritability, such
as the Brunonian system.

bile ——— fire ——— caloric +⎫
blood ——— air ——— azote +⎪
phlegm ——— water ——— hydrogen +⎬ oxygen ——— irritability
black bile — earth ——— phosphorus +⎭

In addition to this structural equivalence Baumes' system also
revealed a functional equivalence to humoural medicine, with
the healthy state arising from a balance of the humours or elements
and diseases arising from the excess or deficiency of any of
these substances. Restoration of health in both systems, and
also in the pathologies based on irritability, depended upon
counteracting any disturbance to the balanced state by the
removal of excesses or the supplementation of deficiencies.
Furthermore the healthy balance of elements, according to Baumes'
system, still allowed for the constitutional predominance of
one element over the others, so that the 'temperaments' of
classical medicine were retained, although displaced onto new
elements. For example, the 'bilious temperament' was previously
associated with the element fire; however Baumes associated
caloric with the 'sanguine temperament,'[18] displacing the 'bilious
temperament' onto the element hydrogen.[19]

Baumes' system, then, reproduced the characteristic relations, both structural and functional, of traditional humoural pathology. Baumes 'applied' the new chemistry to classical medical theory in the way that one applies a new coat of paint to an old cabinet. Despite his claim that "medicine must naturally be tributary to chemistry,"[20] Baumes established precisely the opposite relationship. "Clinical observation," he wrote, "has accumulated the facts [of pathology], without making us any richer in our means [of explaining them]; difficulties remain, and we have grounds for saying that only chemistry can combat and remove them."[21] The findings of the chemists allowed Baumes to correlate existing clinical pictures with specific chemical substances and processes, but this in no way challenged fundamental conceptions of disease and therapeutics. If the appearance of the Lavoisian chemical system constituted an 'epistemological rupture' with the theoretical past of chemistry, then Baumes' chemical pathology constituted an 'epistemological suture' - an attempt to close the gap between traditional medical thought and the new theoretical system of chemistry in a way that was wholly dominated by the former. In the system of Baumes the new chemistry stood not as a break with traditional medical knowledge but as a vindication of it.

Nevertheless this domination of chemistry by medical tradition in the system of Baumes went largely unperceived by its author and his contemporaries, who regarded the system as a subordination of medicine to chemistry. We have already cited Baumes' views on this point, so we may now compare the assessment of one of his chief opponents, Philippe Pinel of the Paris medical school.

Pinel's *Philosophical Nosography*[22] was published shortly after Baumes' *Essay on a Chemical System of the Science of Man* appeared, and at the end of his treatise Pinel offered critical observations on the system of Baumes, appealing, in the process, to Baumes' chief authority, Fourcroy.[23]

What then of Pinel's own hopes for a medical revolution? If chemistry was only one of many "accessory sciences"[24] and not "the true basis of medical science" as Baumes claimed,[25] then what principles would serve to renovate medicine and to produce "the profound knowledge of diseases"[26] for which Pinel called?

For Pinel the correct approach to medicine was to regard it as a branch of natural history. His prescription for the reform of medicine was

> to introduce into its procedures the method
> currently followed in all the other parts of
> natural history; that is, severe exactitude in
> descriptions, correctness and uniformity in
> denominations, a sage reserve in ascending to
> general views without attributing reality to
> abstract terms, and a simple and regular
> distribution which is invariably founded upon
> relations of structure or the organic functions
> of the parts.[27]

It was the *method* of natural history, then, rather than the
results obtained in any particular branch of research, that
Pinel looked to for guidance - a method which was crystallised
in the single word, Analysis.   This method of analysis, as
systematised in the epistemological writings of the philosopher
Condillac, was to give Pinel's work its character as a
Philosophical Nosography - it was, as its subtitle announced,
The Method of Analysis Applied to Medicine.

    The method of analysis advocated by Condillac, whose work
Pinel frequently cited,[28] consisted of beginning with a large
and more-or-less undifferentiated grouping of phenomena, among
which the most prominent objects or general features are first
identified.   These most prominent items serve to establish a
primary division of the field of phenomena, with the less
prominent items clustering around one or another of the more
prominent ones.   Then each primary division of the field is
subjected to the same process to form secondary divisions, each
of which is again sub-divided, and so on.   This process of
analysis may continue until the least differences perceptible
to the senses have been reached, at which point the process
is reversed and the parts re-assembled mentally with great care
given to establishing the relations between these parts at
each level of complexity and to expressing the results of the
enquiry in an accurate terminology.   This method, according to
Condillac, is the one which the mind spontaneously follows in
forming correct ideas of species and genera of all kinds; but
owing to prejudice, bad education and self-conceit, it is
neither universally recognised nor universally followed.

    For Pinel, the epistemology of analysis entailed a new divi-
sion of diseases in a search for their most fundamental, primitive,
elemental manifestations.   The principle of continual subdivision led him to separate, in any collection of symptoms,
those which always occurred together from those which could
occur alone.   It was for this reason that in his primary division
of diseases into classes, Pinel distinguished general fevers
from local inflammations.   Is it not contrary to the natural
method of analysis, asked Pinel, to group

> the phlegmasies or local inflammations together
> with the primitive fevers, to make only one single
> class out of them, when nature has so to speak
> separated them, since they can exist independently
> of one another and since, when they are complicated
> [with one another] one can only be guided by
> obscure notions if one has not successively
> analysed them?   These general laws of methodical
> distribution, which are now followed in all parts
> of natural history, ought to preside over the great
> divisions of internal pathology; and it is from
> this point of view that I have made a distinct
> class of the phlegmasies...[29]

    On such grounds, then, Pinel established five classes of
disease: Fevers, Phlegmasies, Haemorrhages, Neuroses, and
Lymphatic Lesions (this last class being expanded in later

editions to Organic Lesions).   Further analysis of each class
allowed Pinel to distinguish a   number of orders, usually
based upon the principal seat of the disease.   Under the fevers,
for example, he separated those particularly affecting the
circulatory system, which he called angio-tenic fevers, from
those particularly affecting the stomach, duodenum and adjacent
parts, which he called meningo-gastric fevers; and so on with
the adeno-meningeal, adynamic, ataxic and adeno-nervous febrile
orders.

For his analysis of the phlegmasies, Pinel pointed out that
while the traditional anatomical division according to regions
of the body (head, chest, abdomen, etc.) was suitable for
specialised memoirs, a philosophical nosography required a
general perspective.[30] His division of the phlegmasies into
orders according to the nature of the tissue affected was
subsequently cited by Bichat as the inspiration for his general
anatomy of the tissue systems; and in later editions of his
Nosography Pinel revised his division of this class to take
account of Bichat's findings.

A similar principle of division was also applied to the
class of haemorrhages, with later editions showing four orders
related to the mucous, serous, cellular and cutaneous tissues;
although the first edition recognised only two orders, those
common to both sexes and those affecting the uterus in females.

For the last two classes of his Nosography, the neuroses and
the lymphatic lesions, Pinel noted that these diseases were not
sufficiently well known to allow for more than provisional
classification, even with the method of analysis.   For the
orders of the neuroses, therefore, he offered only an "artificial
disposition" which he established "on the most stable and least
variable basis, the properties of sensibility and irritability
and the organic functions of the parts."[31]   And concerning his
division of the lymphatic lesions he wrote, "I do not conceal the
fact that innumerable researches remain to be done on the diseases
of these various orders, and that this part of the Nosography
must still be in a state of imperfection."[32]

In establishing his genera, Pinel used the temporal criterion
of duration for the fevers - dividing them into continuous,
remittent and intermittent forms - and the spatial criterion of
location in the body for the other orders as far as possible.
When localisation was inadequate to distinguish between two
forms of disease under the same order - for example between
smallpox and measles under the order of cutaneous phlegmasies -
then characteristic symptoms were made the basis of the distinction.

For each genus thus established there was ordinarily only
one species which Pinel called 'simple' and which had no specific
characters apart from those of its genus.   Together with this
simple species there could be one or more complicated species,
according as this simple species was subject to complication
with simple species of other genera: for example, under the
genus Measles was listed one simple species and then a series
of complicated species resulting from the complication of

measles with meningo-gastric, adynamic and ataxic fevers, and with pneumonia.[33]   Finally below the species Pinel placed varieties, based on small differences attributed to the age, sex, temperament or external circumstances of the patient.

An 'empiricist' reading of Pinel's system, then, yields the following points: first, that Pinel attacked the approach of Baumes for its attempt to use chemistry as an ordering principle for pathology; and second, that Pinel explicitly sought to treat medicine as a branch of natural history by applying to it the general method followed in other branches of natural history, the method of analysis.   We have seen the characterisation of this method offered by Condillac, whose approach Charles Gillispie has described as producing "the simple pigeon-holing of the naturalist" on the model of botany,[34] and we have seen how Pinel attempted to follow Condillac's precepts in establishing his nosological division.   In the analysis of fevers, phlegmasies and haemorrhages, Pinel followed what taxonomists called 'the natural method' in which all the sensible properties of the phenomena to be ordered were taken into account, although given different weight.   For this reason François Dagognet has characterised him as "the physician-botanist Pinel, the Michel Adanson of medicine"[35] who "had principally to employ himself in victoriously transposing into medicine the heroic conquests of botany."[36]   In his analysis of neuroses and organic lesions, on the other hand, Pinel constructed what taxonomists called an 'artificial system' in which certain easily determined properties were taken as the key to classification, as in Linnaeus' sexual system of plant taxonomy.

On the evidence of our 'empiricist' reading of Pinel's *Nosography*, then, we are led to conclude that it was not chemistry but natural history, and particularly botany, which furnished the methodological model for his classification of diseases. Once again, however, we are faced with a serious contradiction, involving in this case, the logical status of 'complicated species' in a taxonomic system.   For if measles is a simple species under the cutaneous phlegmasies (Order I), and pneumonia is a simple species under the phlegmasies of cellular tissue (Order IV), to what order of the phlegmasies does the complicated species formed by measles together with pneumonia logically belong?   And when measles is complicated with a simple species from a different class entirely, such as one of the 'primitive fevers', to what class does the resulting complicated species logically belong?   There are only two possible solutions to this problem: either the complicated species belongs to the same genus, order and class as one or another of its component simple species (the solution adopted by Pinel), or else a special class of complicated diseases must be established in parallel to the other classes of simple diseases.   Neither of these solutions is adequate within the framework of a classical natural history taxonomy, however, as we shall show after our consideration of the nosological system of Tourdes, who adopted the second of these solutions to the problem of complicated diseases.

The nosological system of Joseph Tourdes was published in pamphlet form in 1802 under the title, *Sketch of a System of*

Nosology, Founded upon Physiology and Therapeutics, after having
been outlined in a contribution by Tourdes to the *Memoirs of
the Medical Society of Emulation* of Paris.[37]   The basis of
Tourdes' classification was the existence of three elementary or
primitive tissues from which, he held, all the organs of the
body are formed.   These three tissues - the nervous, fibrous
and cellular or lymphatic tissues - differ, he wrote, "in their
organisation, their chemical nature, their vital forces, their
action, their uses and their functions."[38]   The importance of
these three tissues for Tourdes' system was that despite the
multiplicity of pathogenic causes acting upon them, their
differences of structure and function were such

> that they are affected in isolation, independently
> of one another: from which come *three classes of
> disease* having their own proper symptoms, course,
> period, type, onset, and termination.[39]

For the division of each class into orders Tourdes relied, as
his title announced, upon physiology and therapeutics.   The
pathological affections of each tissue system, he maintained,
could be reduced to a small number of "primitive alterations, or
specific lesions" in the particular functions of each tissue.

> These lesions form in each class a natural
> series of *simple orders* characterised by particular
> phenomena, bringing together diseases dependent
> upon a similar alteration and submitted to the same
> methods of treatment.[40]

Under diseases of fibrous or irritable tissue, his first
nosological class, Tourdes listed five orders: phlegmasies,
haemorrhages, adynamies, dyskineses, and organic lesions.   For
the second class, diseases of nervous or sensitive tissue, he
set down six orders of neuroses. acute, malignant, convulsive,
comatose, irregular and mental; together with a seventh order
of organic lesions.   And finally, for the class of diseases
of cellular or lymphatic tissue, Tourdes again established
seven orders: catarrhs, gastroses, dropsies, acute eruptions,
chronic lymphatic affections, poisonous bites, and organic lesions.[41]

Within each order Tourdes applied the principle of localisation
to form his genera and species.   Since each order of disease
could affect its corresponding tissue system as a whole or in
some part only, "this double state gives two *simple* **genera** of
diseases, universal or local" for each order.[42]   Thus under the
order of convulsive neuroses, for example, were the genera
epilepsy and tetanus.[43]   The first of these genera was not
further subdivided and produced only one species, however the
second or local genus under each order was "subdivided into as
many species as there are local lesions or affected parts."[44]
Finally varieties were introduced by Tourdes to accommodate
the modification of species by climate, season, temperament and
other circumstances.

In none of the categories discussed thus far did Tourdes
include complicated diseases.   For these, as we have previously
indicated, he established a fourth nosological class "composed
of as many *orders* as there are simple lesions capable of
combining with each other."[45]   These orders could then be
divided into genera and species according to the same rules

governing the formation of simple genera and species in the first three classes.[46]   Thus for Tourdes the fourth class of his nosology was intended to parallel in a logically comparable way the three classes of simple diseases.

In spite of this difference in the approach to the problem of complicated diseases, the nosology of Tourdes was organised according to the same underlying principles as the nosography of Pinel.   This is clear if we regard Tourdes' classification as a three-dimensional matrix with elemental structural/functional systems (i.e. fibrous/irritable tissue, nervous/sensible tissue, and cellular/assimilative tissue) aligned horizontally to determine classes, and with elemental 'clinical pictures' of diseases (typical groupings of symptoms, with a characteristic onset, course, and termination) aligned vertically to determine orders, and finally with elements of anatomical localisation (specifying a particular position in the body) aligned in the dimension of depth to determine genera and species.   Now to transform this matrix into the classification of Pinel, we need only rotate it 90° about its axis of depth, or to put it another way, reverse the taxonomic significance of its vertical and horizontal dimensions so that the alignment of clinical pictures (fevers, phlegmasies, etc.) serves to determine classes while that of elemental structural/functional systems (mucous tissue, cellular tissue, etc.) determines orders.   The orientation, or taxonomic significance, of anatomical localisation remains the same for Pinel's system as it was for that of Tourdes, determining genera and species.   The only exception to this last point (leaving aside some anomalies in the neuroses and organic lesions, which Pinel admitted were incomplete and unsatisfactory) was the division of the essential fevers into genera according to their period.   But this division by a kind of temporal localisation was strictly parallel to the other divisions by spatial localisation: Tourdes' generic distinction between lesions affecting a tissue system universally or only in part matched Pinel's generic distinction between continuous and non-continuous (intermittent, remittent) fevers.

The only aspects of the two nosological systems which are not interconvertible by a simple transformation are the positions of complicated diseases.   In Pinel's nosography there are complicated diseases only at the level of species, while in the system of Tourdes the complicated diseases are represented at all taxonomic levels.   Unlike the other differences between the two systems, this one arises not from two variations on a single organisational plan but from two different responses to a single problem.   We have already pointed out that these two responses are the only ones the problem admits of, and we must now explain why neither is satisfactory within a classical taxonomic system.

Pinel's solution, we noted, was to locate the complication formed by two simple species under the genus of one of its constituent species.   But this approach gave no logical grounds for choosing between the two genera in question, so that Pinel was forced to resort to the pedagogical expedient of listing under each genus only those complicated species formed by the simple species of that genus together with other simple species dealt with earlier in the order of exposition.[47]   Thus all complications between phlegmasies and fevers would be listed under the genera of the phlegmasies, since the class of fevers was the first one to be described in the *Nosography*. It is clear, however, that so far as the 'nature' of such a complicated disease was concerned, it could just as well have been listed under the genera of the fevers; or, to put it another way, the taxonomic position of Pinel's complicated species was in principle ambiguous.

The alternative solution adopted by Tourdes for complicated diseases overcame this inherent ambiguity but did so at the cost of establishing a wholly anomalous class.   In the first three classes of Tourdes' system the unity of each class was founded upon the structural/functional unity of one of the elemental tissues, while the internal division of each class into orders was dependent upon the different pathological modifications of function or structure which the elemental tissue in question could undergo.   In the fourth class, however, the relations of dependence were reversed.   The complicated orders were founded upon modifications of the simple orders from the first three classes, namely their binary and ternary combinations, while the unity of the class as a whole was dependent upon the composite nature of these orders.   Or, to put the contrast another way, in each of the first three classes the orders were unified as any modification whatever affecting one particular tissue, while in the fourth class they were unified as one particular modification (combination) affecting any tissues whatever.

Neither Pinel nor Tourdes, then, could find a logically-consistent way of accommodating complicated diseases within the framework of their nosological classifications.   This situation cannot, however, be taken as a sign that the subject of complicated diseases was regarded as one of small nosological importance.   On the contrary, Pinel announced it as one of the principal advantages of his analytic method that it promised to bring order to the nosological chaos of complicated diseases.[48]   And Tourdes, for his part, held similar views.[49] Given, then, the degree of importance which both Pinel and Tourdes clearly accorded the problem of complicated diseases, it is not possible to maintain that the logical difficulties raised by the position of these diseases in each system resulted from their theoretical insignificance.

Indeed, it is precisely from the theoretical significance of complicated diseases that the difficulties surrounding them arose; for beneath the traditional, hierarchical structure of classes, orders, genera and species into which both nosological systems were cast, there operated a conception of simple diseases as combinatorial units, capable of existing independently or of combining and recombining to form various compound entities without ever losing their identity - a conception, in other words, of simple diseases which behave exactly like the simple substances of chemistry.   An adequate representation of such a combinatorial system - whether in pathology by Pinel and Tourdes, in chemistry by Lavoisier, or in epistemology by Condillac - could never be constructed using the categories of classical logic on which natural history taxonomies were based. Only a mathematically-founded logic of combinations could offer the means for such a representation; yet such a tool was scarcely developed at the time, and in any case could not have been applied to the fields in question so long as knowledge continued to be defined in essentialist rather than relational terms.[50] Thus Condillac's epistemology of simple sensations, and their combinations, Lavoisier's chemistry of simple substances, and their combinations, and Pinel's and Tourdes' pathologies of simple diseases, and their combinations, were all depicted through the language and categories of classical logic, with due concern for the determination of species, genera and so on.[51]

We are not arguing here, any more than we did in considering the nosology of Baumes, that we are dealing with systems in which novel and traditional theoretical elements merely co-existed. Rather we are suggesting that the fundamental functional principle of the systems of Condillac, Lavoisier, Pinel and Tourdes was that of a combinatorial logic, but that the structural representations of these systems were governed to a greater or lesser extent by a logic of class-inclusion. The resultant lack of correspondence between the functional basis of these systems and their structural representations produced certain contradictory effects in these representations, which in the cases of Pinel and Tourdes we have highlighted.[52]

The nosologies of Pinel and Tourdes, then, shared a common systematic principle with the new chemistry of Lavoisier, and like the new chemistry - but to an even greater extent - the combinatorial basis of this principle was masked by the language and categories of traditional logic. In the nosology of Baumes, on the other hand, the language and empirical results of the new chemistry served to mask a traditional pathological system. For these reasons we must conclude that the relationship of the new chemistry with the nosology of Baumes was apparent but superficial, while its relationship with the nosologies of Pinel and Tourdes was concealed but fundamental.

514

NOTES

1. These introductory remarks are based on Michel Foucault *The Birth of the Clinic* [1963] (New York: Vintage, 1975), chapters 2-7; François Dagognet, *Le catalogue de la vie* (Paris: P.U.F., 1970), chapter 3; and E.H. Ackerknecht, *Medicine at the Paris Hospital, 1794-1848* (Baltimore: Johns Hopkins, 1967), chapters 1-4.

2. Michel Foucault, *The Order of Things* [1966] (New York: Pantheon, 1970), pp.226-232.

3. Xavier Bichat, 'Discourse on the Study of Physiology' [1798], in W.R. Albury, 'Experiment and Explanation in the Physiology of Bichat and Magendie', *Studies in History of Biology*, 1977, 1:102.

4. Jean Baptiste Baumes, *Essai d'un système chimique de la science de l'homme* (Nismes, 1798); *Fondemens de la science méthodique des maladies* (4 vols.; Montpellier, 1801-1802); and *Traité élémentaire de nosologie* (4 vols.; Paris, 1806).

5. Baumes, *Essai*, p.6.

6. Baumes, *Fondemens*, I, p.12.

7. Baumes, *Essai*, p.48.

8. *Ibid.*, p.96.

9. *Ibid.*, p.62.

10. Baumes, *Fondemens*, I, pp.175-6.

11. *Ibid.*, pp.174-5.

12. Baumes, *Essai*, p.65.

13. *Ibid.*, pp.72-3.

14. *Ibid.*, pp.80-1.

15. *Ibid.*, p.82.

16. *Ibid.*, p.89.

17. Baumes, *Fondemens*, I. p.13.

18. *Ibid.*, II, p.217n.

19. *Ibid.*, III, p.525.

20. Baumes, *Essai*, p.7.

21. *Ibid.*, p.47.

22. Philippe Pinel, *Nosographie philosophique, ou la méthode de l'analyse appliquée à la médecine* (2 vols.; Paris, 1798). Pinel's *Nosography* dominated French medical thought for twenty years, going through six editions in that period. Except where otherwise indicated, all subsequent references are to the first edition of this work.

23. Pinel, *Nosographie*, II, pp.368-9.

24. *Ibid.*, I, p.iii.

25. Baumes, *Fondemens*, I, p.viii.

26. Pinel, *Nosographie*, I, p.ii.

27. *Ibid.*, pp. iii-iv.

28. Cf. *Ibid.*, pp.xi, xvi, xxii-iii, and II, p.322. The principal methodological writing of Condillac's cited by Pinel was *La logique, ou les premiers développemens de l'art de penser* (Paris, 1780), on which the following outline of his method is based.

29. Pinel, *Nosographie*, I, p.xxiii.

30. *Ibid.*, p.2 xxv.

31. *Ibid.*, p. xxxii.

32. *Ibid.*, p. xxxiv.

33. Pinel, *Nosographie* (5th edition, 3 vols.; Paris, 1813), III, p.584.

34. C.C. Gillispie, *The Edge of Objectivity* (Princeton: Princeton Univ. Press, 1960), p.170.

35. Dagognet, *Catalogue*, p.149.

36. *Ibid.*, p.147.

37. Joseph Tourdes, *Esquisse d'un système de nosologie, fondé sur la physiologie et la thérapeutique* (Strasbourg, 1802);'Esquisse d'un système de Nosologie, fondé sur la Physiologie et la Thérapeutique', *Mémoires de la Société Médicale d'Emulation (1797)* [2nd edition, 1802], 1:85-90. I have not been able to examine Tourdes' Strasbourg pamphlet of 1802; however the principles of his classification are described in the memoir version of his 'Esquisse' and his classes, orders and genera are set out in Philippe Pinel and Isidore Bricheteau, 'Nosographie', *Dictionnaire des sciences médicales*, v.XXXVI (1819), pp.228-30.

38. Tourdes, 'Esquisse', p.85.

39. *Ibid.*, p.87.

40. *Ibid.*, p.88.

41. Pinel and Bricheteau, 'Nosographie', pp.288-91.

42. Tourdes, 'Esquisse', p.88.

43. Pinel and Bricheteau, 'Nosographie', p.229.

44. Tourdes, 'Esquisse', p.88.

45. *Ibid.*

46. *Ibid.*, p.89.

47. Pinel, *Nosographie*, I, pp. xvii n., xxvi-ii n.

48. *Ibid.*, pp.x-xi.

49. Tourdes, 'Esquisse', p.89.

50. The mathematician Condorcet seems to have been the
     only one from this period to attempt a combinatorial
     system of taxonomy.   See K.M. Baker, 'An Unpublished
     Essay of Condorcet on Technical Methods of Classification',
     Annals of Science, 1962, 18:99-123.

51. Cf. Condillac, *Logique*, part I, chapter 4; and A.L.
     Lavoisier, *Elements of Chemistry* [1789] (New York:
     Dover, 1965), pp.xxvi-xxx.

52. For a discussion of some of these effects in the work of
     Condillac and Lavoisier, see, respectively: Foucault,
     The Birth of the Clinic, p.116; and Gillispie, *The Edge
     of Objectivity*, pp.241-50, and François Dagognet,
     Tableaux et langages de la chimie (Paris: Seuil, 1969),
     chapter 1.

Magda Whitrow

A CLASSIFICATION SCHEME FOR THE HISTORY OF SCIENCE, MEDICINE AND
TECHNOLOGY

Although the title of my paper is "A Classification Scheme for the
History of Science, Medicine and Technology", I do not propose to
devote all the time at my disposal to it.   I have available an outline
of my scheme together with the classification schedules and an alphabetical index to the schedules, which I shall be pleased to give to
anyone interested.   It is an extended version of my essay at the end of
Volume 3 of the *ISIS Cumulative Bibliography* except that the index at
the end is, of course, an index to the schedules and not to the Bibliography and that the schedules also include the section on 'Civilizations
and periods'.   I want to discuss some of the problems that I had to
tackle in devising the scheme, but before I do so I should like to say
something about classification of the sciences, which is the theme of
this Symposium.   I need not stress here that the classification of
objects according to their resemblances is a fundamental process of
thought.   Jevons claimed that classification is essential to any true
science and may almost be said to constitute its framework.[1]   Indeed
its importance in the sciences increases according to the number of
objects which need to be arranged in some systematic fashion, so that it
is greatest in the natural sciences.   The advancement in the state of a
scientific subject field seems to go hand in hand with the degree to
which the objects with which the science is concerned can be classified.
Although the classification of objects according to chosen characteristics of division has its most important application in biological
taxonomy, the subject of the other two papers in this Symposium, it is
of importance in all the sciences.   The classification of the elements
contributed in no small measure to the development of chemistry, the
classification of stars helped to advance stellar astronomy and the
classification of subatomic particles has played its part in the more
recently established science of particle physics.

The classification of objects according to chosen characteristics of
division is only one aspect of the classification of the sciences.
This is a quasi-inductive approach, if I may call it that, involving the
arrangement of different species into families and genera.   The other
aspect is the construction of systems of knowledge which follows a quasi
-deductive procedure.   It is, of course, also based on some characteristic of division, although often this is not explicitly stated.   It
presents an attempt to uncover the relations between the different
branches of knowledge, a kind of epistemological exercise, usually performed by philosophers, particularly philosophers of science.   Indeed,
some claim that without a satisfactory classification of knowledge no
full understanding of the universe is possible.   Often systems of
knowledge have been constructed to serve as the framework for an en-
cyclopaedia, and they have of course been applied to the organization of
libraries and bibliographical material and the planning of an educational syllabus.   "The original purpose of the classification may be
logical, or theological, or encyclopaedic; the fruits are similar" says

Sarton in his *Introduction*, and he pleads for a detailed comparison to be made of systems of knowledge.[2]

I do not intend here to make a survey of classification schemes from Plato and Aristotle through Hugues de Saint Victor and Comte to the present day. There are many brief informative surveys of the subject, although there are very few large-scale works, and none of them really comprehensive. "From the time of Plato to the present day there has been a continuous series of attempts to classify the sciences", says Flint in the introduction to his *History of the classification of the sciences*, published in book form in 1904. Parts of it appeared, however, as early as 1885.[3] His account concentrates on modern systems of knowledge from the seventeenth to the nineteenth century, although his coverage is restricted. Flint devotes some space to systems developed in antiquity from Plato to Capella, but he disposes of the Middle Ages in eight pages and completely ignores non-Western systems. Sarton refers *passim* to many schemes for the classification of the sciences, mostly by Islamic philosophers and scientists, but his account stops before the modern period. E.I. Samurin, in his history of bibliographical and library classification[4], published in Moscow in 1959 and translated into German in 1967 - unfortunately, there is no English translation available - devotes many pages to schemes evolved by philosophers with no practical application in mind, but again pays little attention to mediaeval systems or non-Western systems, and his book is written from the angle of dialectical materialism.

Although I doubt whether the construction of systems of knowledge has had any influence on the development of science, their study is important to the history of science because systems of knowledge mirror the state of knowledge at any given time and throw considerable light on the state of education. Perhaps then it should not be a matter for astonishment that in this century, when the growth of science has been so much more rapid and extensive than ever before, there has been such a lack of interest in general systems of knowledge on the part of scientists and philosophers. Or is this due, I wonder, to the divorce between science and philosophy during this century which Collingwood so deplored?[5] Most of the outstanding contributions to the subject have been made by bibliographers and librarians, such as Bliss, Dewey and Ranganathan.

Flint believed that the classification of the sciences and the classification of books were essentially distinct. "The classification of the sciences", according to him, "is a fundamental problem of philosophy, the first step toward the positive philosophy of the sciences; the classification of books is merely a practical problem of very limited interest, the convenience of bookish people".[6] Bliss thought this distinction invalid. He believed that when books and documents are classified by subject matter, regarded as knowledge or as sources of knowledge, such classifications are virtually classifications of knowledge.[7] Recently, there have been attempts at a rapprochement between philosophers of science, classificationists and taxonomists, thus bringing together the three important aspects of classification: logical, bibliographical and taxonomic. I believe that the gap between taxonomists and builders of classification systems has narrowed, because on the one hand modern taxonomists consider that their science should no longer be based on descriptive morphology alone but should be regarded as multi-dimensional, and because on the other hand the tremendous growth and increased specialization of scientific literature has led to a more inductive approach to the organization of knowledge. Although at present it could not be said that there exists a proper discipline of classification science, there are signs that such an interdisciplinary subject may well be in course of formation. A Classification Society was founded in Great Britain in 1964. It has recently been organized into two branches, a European and a North American Branch. Its main purpose is the promotion of cooperation and interchange of views and information among those interested in the principles and practice of pattern recognition and classification in any discipline that uses them. As a result, its membership not only includes biologists, anthropologists and geologists, but also computer and information specialists, librarians, linguists and others. In 1971 an international conference was held in Ottawa on the 'Conceptual basis of the classification of knowledge' with the aim of bringing together philosophers and specialists in classification. More recently a journal entitled *International Classification*, a journal on theory and practice of universal and special classification systems and thesauri, was started in 1974 to be devoted to theoretical and practical problems of concept systems.

Although many modern methods of information retrieval do not depend directly on classificatory principles, it has become increasingly clear that classification can only be neglected at one's peril and is as necessary to the construction of a thesaurus as it is to that of a hierarchical system of knowledge. In fact, the increased interest in the theory and practice of classification is largely due to the development of special techniques to deal with scientific literature, rather than a sudden interest on the part of philosophers of science. One of the techniques developed early in this century by the eminent Indian scholar S.R. Ranganathan, 'facet analysis', was used by him in conjunction with a more or less traditional scheme of main classes in his 'Colon classification'[8], published in 1933. As I have made use of this technique, a word of explanation may not be out of place here. 'Facet analysis' is, in essence, somewhat similar to systematic classification carried out on the usual logical principles. The subject field is analysed into mutually exclusive facets, or categories, derived by using a single characteristic of division. However, unlike the traditional systems of classification in which every term is enumerated, in a faceted scheme the terms in each category are listed and then when it is necessary to represent a multi-faceted subject the terms can be combined in a predetermined order to form the appropriate code or classmark. The fact that the order is predetermined helps to solve the classifier's difficulty in deciding where to class a composite subject that might well fit into more than one category.

The Classification Research Group of London, founded in 1952, applied and developed the faceting technique originated by Ranganathan. Members of the Group successfully devised a number of classification schemes for special fields based on this technique which have been found very satisfactory in practice. However, these were schemes devised for comparatively narrow fields of knowledge and the lack of a satisfactory general scheme to provide a framework for marginal fields has become increasingly felt.

The need for such a scheme was stressed in the first Newsletter issued by the Organizing Committee of the Ottawa Conference.[9] "For two thousand years", the Newsletter said, "the Classification of Knowledge has been a major philosophical problem. Lately, it has evolved from a philosophical preoccupation into a specialized technique, a species of the general techniques of classification, far removed from philosophical preoccupations and equally ignored by philosophers. At the present time there exist powerful technical models of classification and highly developed classifications of various branches of knowledge. What is lacking is a satisfactory general classification of knowledge and an adequate clarification of the philosophical problems underlying such classification."

The search for a general classification of the sciences continues and as far as I am aware no satisfactory scheme has so far been developed. The Library Association, on behalf of the Classification Research Group, received a grant to conduct research into the principles which might lead to the construction of a general classification system, this research being carried out during the period 1964-69. The Group spent much time discussing these principles. They favoured a scheme based on the theory of integrative levels, as developed by Joseph Needham[10] and J.K. Feibleman[11], but without the framework of the accepted boundaries of the traditional disciplines. The results of the research were somewhat disappointing. We are still a long way from a satisfactory general scheme, but some of the problems involved in the construction of a scheme were clarified. Derek Austin, who carried out the research in its final stages and who has reported on its result[12], was led to believe that a systematic arrangement of 'organized fields of knowledge' was incompatible with the categorization of terms. This entails a direct analytical approach to the universe of individual concepts that is essential for the retrieval of specialized information, particularly by machine. Out of this research into the principles for a new general classification, Austin was led to develop the 'Precis' system, or the Preserved Context Indexing System, which is a machine-produced alphabetical indexing system, based on a syntax derived from a study of natural language.

However, it seems that the desire for constructing a general system of classification of the universe of knowledge cannot easily be suppressed. Austin is now trying to discover, by an inductive process, whether a general classificatory framework could be established by listing a hierarchy of increasingly general terms for each term indexed. I personally do not believe that a general system of classification can be constructed without the framework of subject disciplines, although the boundaries of the special sciences are ephemeral. In particular, in the history of science and the sciences it would be difficult to order concepts and phenomena without reference to their subject fields, however vaguely these may be defined. Even a cursory glance at the literature makes it evident that historical writings, particularly large-scale works, deal with the history of the traditional disciplines and it would hardly be convenient for the user to find these widely separated from the histories of specific topics.

A scheme for the classification of the history of science, medicine and technology differs from a scheme for a specialized field of knowledge in that it has as one of its facets the whole of science, medicine and technology and thus is in some ways similar to a general scheme of classification; on the other hand, it is special in that it considers so much of knowledge from a particular angle. The scheme that I have devised for use in the *ISIS Cumulative Bibliography* is based on the arrangement by which entries are classed in the annual *Isis* Critical

Bibliographies, which is a simplified version of that adopted by Sarton before he retired as Editor of *Isis*. During the time Sarton edited the Critical Bibliographies - a time-span of about forty years - he evolved many classification schemes. Nearly every one of the early bibliographies shows some change in arrangement on the previous one, and in the seventh he adopted a partly alphabetical system. Not until 1926 (the eighteenth Critical Bibliography) did Sarton abandon the alphabetical arrangement and revert to an improved systematic scheme. In his apologia for abandoning his original scheme[13] he blamed the war for his demoralization. "I now realize", he said, "that this abandonment was a lack of courage and faith, which was especially shocking in a journal like *Isis*, the very purpose of which is to illustrate the unity of knowledge and the unity of mankind."

What makes Sarton's scheme so unusual was its faceted structure, which enabled me to construct a scheme based on this technique without materially changing Sarton's analysis of the subject. The history of science fits naturally into a three-dimensional matrix in which period/civilization is on the first coordinate axis, the subject on the second, and the aspect from which the subject is considered on the third.

The question which Sarton discussed more than once, and to which the Committee that devised the modified scheme still in use in the *Isis* Critical Bibliographies applied themselves, is whether period/civilization or subject should be what we would now call the 'preferred facet'. A historian of a particular science who is interested in the history of his subject irrespective of period might well prefer the subject to be the first facet, so that all the material he requires would be collected together. However, most historians of science, even if their main interest lies in a special subject field, are usually concerned with a particular period, and it was therefore decided to retain the period/civilization facet as the main one. All titles that relate to the history of science or the sciences in one particular civilization or period and to not more than two centuries during the modern period are first of all classified by this characteristic. Titles that refer to two centuries are classified in the later century. The only exceptions are titles that deal with the 19th and 20th century combined or with the 20th century alone. They are treated as if they were general. Like all material which deals with the history of the subject generally, without special reference to a particular period or civilization, or which follows the subject into the modern period, they are classified by subject first, the period/civilization code not being applicable.

Although in the scheme currently used in the annual *Isis* Critical Bibliographies there is a strict rule that all material that fits into the chronological sections is classed there--a rule which, as I have explained, I have retained--no consistent citation order is adhered to regarding aspect or point of view on the one hand and subject field on the other. For example, some titles dealing with the philosophy of biology have been classed under philosophy of science, others under biology; similarly, the section on Institutions has sometimes contained references to botanical gardens, at other times these have been classed under botany. In my scheme however, the following rule has been adopted consistently: first classify by period/civilization, then by subject, and finally by aspect (or point of view) and bibliographical form. This means that all aspects of a subject, however narrow, are classed under that subject, so that, for example, material on the philosophy of science and the sciences or on the social relations of science and the sciences is distributed among the subject fields. The emphasis in this scheme is on the history of each subject, and so, however narrow the subject, all its aspects are gathered together. However, anyone wishing to adopt a different order and yet follow the scheme can do so by changing the order in which the code is formed. For example, a philosopher of science who wishes to have all references to philosophical aspects collected together and subdivided by subject, and similarly a sociologist who prefers to find material on the social relations of all the subject fields brought together, can modify the scheme to meet these requirements.

The notation has been chosen to emphasize the three main facets: numerals denote civilizations and periods, capital letters subject fields, and lower case letters aspects and forms. Lower case letters are also used to form compound subjects and to extend subjects in a non-systematic order. Although purists may object to the use of lower case letters for this purpose since this spoils the idea that the three different types of symbol correspond respectively to the three axes which form the framework of the scheme, in practice this device obviates the use of punctuation marks and the consequent need to assign to them a filing order.

The numerals are quoted as the first element of the code, followed by capitals and lower case letters or, in the case of some general historical material, by lower case letters. Much of the notation has been devised with the aim of retaining, at least in broad outline, the arrangement now in use in the *Isis* Critical Bibliographies, which has been generally accepted by historians of science. This accounts for some awkwardness in the notation, particularly in the bibliographical section, since the aim has been to keep all the bibliographical material concerning the general history of the subject together; also it was important that each section should begin with the general history of the subject.

It must be emphasized that the notation does *not* reflect the hierarchy of the classification – it is ordinal, with a decimal point imagined before the first digit. For example, the main section on the Ancient World is 15, but the subsection on Judaea is 2. Similarly, although a single capital letter frequently represents a main subject, it may indicate a subdivision of one described by a two-letter code: for example, meteorology G is one of the earth sciences FZ.

One of the most important features of a faceted scheme is its predictability: just as a rule is laid down that the period/civilization facet is cited before the subject and the subject before the aspect and form, so within the individual schedules an order is prescribed in which the different elements are quoted. This order, unless specially stated, is usually the reverse of the schedule order so that terms later in the schedules are quoted before the earlier ones.

As this Symposium is concerned with the classification of the sciences I want to devote most of my remarks to the classification of the scientific subject fields, but I would like to mention one or two of the problems concerned with the chronological classification. The division of history into ancient, mediaeval and modern only applies, strictly speaking, to European and Near Eastern civilizations and has little meaning for the rest of the world. My scheme makes the division into Western, that is European and Near Eastern, civilizations, on the one hand and Eastern and American civilizations on the other hand more explicit. The schedule begins with a general section on prehistory and primitive societies, the prehistory of specific areas being classed under those; for example, the prehistory of China is classed under China and that of Egypt under Ancient Egypt. It is very difficult to draw the line between prehistory and early civilizations and similarly between primitive societies and early civilizations. The general section on prehistory is followed by the prehistory and early civilizations of Europe and the prehistory and ancient civilizations of the Near East and Mediterranean area. The next main section covers the prehistory and indigenous civilizations of Asia (except the Near East), subdivided into large geographical areas representing the different cultures, followed by those of Africa (not including North Africa) and America. When these have become fused with Western cultures the material is treated like the rest of modern knowledge. My main problem here concerned the countries of Western Asia. Their ancient history is included among the ancient civilizations in the Near East, and in mediaeval times most of them are Islamic and are classed under Islam. However, I had to introduce a section on Western Asia to take care of general histories of these countries.

I have already emphasized that in some ways a classification scheme for the history of science, medicine and technology is similar to a general scheme in that it covers such a large section of knowledge. I could, of course, have adopted one of the modern schemes, but the only one that might have been satisfactory - Bliss's Bibliographic Classification[14] - is in the process of complete revision on faceted lines and has not yet been completed. Also, it was clear that nothing like the detailed schedules now being developed for Bliss were needed, and since only part of the schedules were required the notation would have been uneconomical. However, the main reasons that led me to devise my own scheme, which although based on that used in the *Isis* Critical Bibliographies needed to be far more detailed, are first that all modern schemes aim at providing a framework for the classification of contemporary literature and secondly, that historical writings cover the scientific, medical and technological fields quite unevenly. Only a small proportion of the references deal with very specific topics. Although there are fashions in science and there will always be some fields that are more written about than others, in our field spotlights are turned on here and there in a manner that is probably accounted for by historical rather than scientific reasons. For example, a scheme for organizing writings on 'Medical bacteriology' would provide for the systematic arrangement of all the causal agents; historical writings, on the other hand, spotlight certain diseases, very often just because they were historically far more important and had more serious social consequences.

In the *Isis* Critical Bibliographies somewhat different subject schemes are used for ancient and modern periods. Lund and Taube[15] some time ago proposed that literature should first be divided by the period in which it was written and that a different scheme of classification should be used for each of these periods. Of course, they must have had mainly primary literature in mind. For a number of reasons I have applied the same overall scheme to all periods. First, the scheme would have had to be enumerative rather than faceted; secondly, using the same subject scheme makes it possible to change the citation order as between subjects

and periods, if desired; thirdly, the task of making a comprehensive
index for different schemes would not be easy. However, the way the
scheme is constructed allows considerable latitude both in the choice of
more general headings when the borderlines between subjects are vaguer,
and in the use of non-systematic subject extensions. Also, even if a
subject was not known in earlier periods under its present name, there
is no reason why it cannot feature as a term in the classification.
For example, although Lamarck was the first to use the term 'biology',
this is no reason why a biological topic discussed by a Renaissance
scholar may not be classed under biology.

The order of subjects which I have chosen is based on the arrangement
of subjects used in the *Isis* Critical Bibliographies for the modern
periods, which leans heavily on some nineteenth century classifications
of knowledge. It begins with science in general, followed by mathematics and the physical sciences, the earth sciences (including geography and cartography), biology, the sciences of man, medicine, agriculture, technology and finally some ancillary disciplines which, when
applied to any of the foregoing subject fields, appear in the aspect
subdivisions which I will mention later. I have made a few changes in
the order of subjects used at present in the *Isis* Critical Bibliographies.
I have placed astronomy after physics and chemistry, although in the
earlier periods it may be more useful to have astronomy follow mathematics. However, since the inclusive headings 'Mathematics and the
physical sciences' may be chosen for studies that cover both mathematics
and astronomy, there should be no difficulty here. I have changed the
order within the Earth sciences and the Biological sciences, and transferred human anatomy and physiology to the medical section. I have
collected together the different branches of agriculture, instead of
separating agricultural botany, zoology and technology. I have retained the divisions between science and technology.

I claim no originality in the schemes which I have devised for the
different subject fields; they are eclectic and lean heavily on
schedules prepared by others, notably the draft revisions of the Bliss
classification[16], the draft of the scheme for classifying medical
literature prepared by Miss Ruth Daniel for the Wessex Regional Hospital
Board[17] and Barnard's *Classification for Medical and Veterinary
Libraries*[18]. Dr. Norman Smith helped me to construct the schedules
for Technology in which the scheme used by *Technology and Culture* [19]
proved helpful. Although no detailed classifications of subjects were
required, I was anxious to introduce a faceted structure in all the
schemes used and to allow for expansion if needed at a later date. For
simple subjects I have devised a one or two letter notation (capitals),
a third letter being used only in very few instances and always where
the further breakdown is of a systematic nature.

I have already emphasized the way in which quite specific topics
tend to be spotlighted in historical writings. To avoid complicated
systematic schedules to accommodate these topics I have used the device
of non-systematic extensions introduced by the lower case letter z.
(Lower case letters from a through x represent aspects and forms, the
letter y being used to combine two facets.) For example, in the field
of general and solid mechanics, which has the classmark CD, the topics
of gravitation and inertia occur much more frequently than others.
Consequently, instead of devising a scheme that would involve a number
of facets, I have chosen the codes CDzg and CDzi to represent gravitation

and inertia, respectively. Had I chosen a third letter to form CDG and CDI respectively, I would have blocked the way to a systematic sub-division of the field of mechanics which may become necessary when a larger number of references is involved and it would not have warned the user that here the systematic arrangement ceases.

I cannot describe here the way in which each subject field has been classified - you will find more details in the publication which I have prepared and which is available for those interested - but I should like to outline the faceting principle as it is applied to the field of 'Medicine', in which it can be very clearly demonstrated. The main facets that I have taken into account are: systems of the body; structure (anatomy), function (physiology) and malfunction (pathology); causal agents of disease; specific therapeutic methods, including surgery, drugs and medicaments; environmental factors; public health and safety. To simplify the scheme and the notation I have omitted the organ and part of the body facet which is subsumed in the system of the body concerned, for example, heart under cardiology and the liver under gastroenterology. The order of the schedule is: medicine, general, public health and social medicine, including hospitals and nursing, anatomy and physiology, general, special systems of the body (the medical specialties), pathology and clinical medicine, environmental medicine, therapeutics, including surgery, dental surgery, and pharmacy and materia medica. Unlike the rules governing most other subject fields, the citation order, that is the order in which the facets are joined up to form the classmark, which is generally in reverse of schedule order, is sometimes in a forward direction. The central part of the medical schedules is the section on the medical specialties, covering the anatomy, physiology and pathology of the systems of the body. Disorders and diseases, their diagnosis and treatment, are classed under the part of the body affected. For example, a title on lung cancer is classified as: respiratory diseases - cancer, although the pathology section follows that on respiratory diseases in the schedule. However, when the disease is caused by a live organism it is classed under medical bacteriology, entomology or mycology, and if the treatment is specified, the treatment concerned is cited first.

I have spoken of two of the coordinates which form the framework of the scheme, Civilizations and periods, and Subject fields, but I have so far not mentioned the third on which the aspects and form lie. Each subject, however narrow, is subdivided in the same way into what may broadly be described as: its general history, including bibliographical, biographical and historiographical aspects; its psychology and philos-ophy; communication in, and teaching of, the subject; its sociology, including organization, institutions, the profession and its relation with society as a whole; its relations with the humanities; its popular aspects; its history in different linguistic and ethnic groups and in different countries; and finally equipment, techniques and instruments used. Nearly all these aspects are also considered as disciplines, the history of which is the subject of study. Biblio-graphical, historiographical, linguistic and educational aspects re-appear as subjects under the ancillary disciplines. Philosophy, in-cluding the philosophy of values, is classed between science, in general, and mathematics. Psychology and sociology are part of the 'Sciences of man', but they also feature as aspects of the history of science. It may be thought that it is difficult to distinguish between subjects when they form the core of the study and when they appear as aspects of the history of science, but in classifying hundreds of entries I have found no difficulty in making this decision. The criterion I have adopted is this: when the subject is treated generally, without reference to science, medicine or technology, it is classed with the appropriate subject field. For example, a general work on the nature of genius is classed under psychology, but the psychology of discovery in a special field is an aspect of whatever subject is involved. Similarly, a history of sociology is classed under sociology, but the sociological aspect of a subject is classed under the latter with the aspect code added.

The treatment of the geographical and regional subdivisions is another source of difficulty. They appear in all parts of the schedules wherever a regional aspect occurs. First of all, there are the cultural regions in the Civilizations and periods schedule, mostly associated with particular periods of history. Then there are the geographical subdivisions towards the end of the aspect schedules under w which represent the national regions (largely based on the contemporary situation) as associated with modern history. When it is necessary to denote these regions purely as the location of sources or studies they are at the beginning of the aspect schedules under a, following exactly the same pattern. It must be remembered that different aspects of a subject are combined in reverse of schedule order with the effect that organizational, professional and general social aspects which are very relevant to the history of a subject in a particular country are brought together, so that for example, the medical profession in Great Britain will be analyzed as medicine - Great Britain - professional aspects. Important regional aspects also occur in the sciences, particularly in the earth sciences and in natural history. Instead of using a completely different notation for cases where the region is the subject rather than the aspect, I have used the second letter from the a and w aspect subdivisions, but in capitals, and added them to the code for the regional subject. I had to differentiate between, for example, the history of the geology of America and the history of American geology which may include, say, the geology of France or Great Britain; the fauna of New Zealand may have been studied by British zoologists and the code must express these different elements.

There are two more points I should like to make. One concerns the need for distinguishing between history as a record of events and history as a subject of study. The study of history is organized in a similar way to the arrangement of the subject as a whole. The construction of the section on the teaching of the history of the subject mirrors that of the teaching of the subject. The other point relates to the classification of instruments, always a cause of difficulty. They are classed in the last of the aspect subdivisions x, but they also appear under Technology. If the principle or theory of their construction is involved or their use for a special purpose, then they are classed with the latter, but their technology is included among the subject fields. This is one of the relatively few cases where the classifier must use his discretion and the user may have to look in more than one place. I regret this, since my aim in devising this scheme has been to lighten the task both of the classifier and user by making the position of a title predictable.

I think that with a little practice the scheme should be found easy

to apply and use.   I admit that it is rather more sophisticated than
that adopted more than two centuries ago by a director of the Russian
Academy of Sciences - a scheme criticized by Lomonosov - in which books
with beautiful bindings were most prominently displayed because the
Director thought that aesthetic considerations were the most important.
I also hope that you will find my scheme more logical than the arrangement in a well-known Oxford bookshop where an assistant, when asked by
an American friend of mine where to find a book on space and time, replied "The hardback is under philosophy and the paperback under science".

R E F E R E N C E S

(1)   JEVONS, W.S. The principles of science, vol. 2, p. 345. London:
            Macmillan, 1874.
(2)   SARTON, George. Introduction to the history of science, vol. 3,
            p. 76. Baltimore, Md.: Williams and Wilkins, 1948.
(3)   FLINT, Robert. History of the classifications of the sciences.
            Bound with Philosophy as scientia scientiarum. Edinburgh:
            Blackwood, 1904. Reprinted New York: Arno, 1975.
(4)   SAMURIN, E.I. Geschichte der bibliothekarisch-bibliographischen
            Klassifikation. Transl. from the Russian by Willi Hoepp.
            2 vol. Leipzig: VEB Bibliographisches Institut, 1964-67.
            (Russian original Moscow: 1955-59.)
(5)   COLLINGWOOD, R.G. The idea of nature, p. 3. Oxford University
            Press, 1945.
(6)   FLINT, Robert. Op.cit., p. 125.
(7)   BLISS, Henry E. Theoretic principles of bibliographic classifica-
            tion. 14th Conference of Federation Internationale de
            Documentation (Oxford-London, 1938), vol. 1, pp. C57-C65.
(8)   RANGANATHAN, S.R. Colon classification. 6th ed. London: Asia
            Publishing House, 1960.
(9)   CONCEPTUAL basis of the classification of knowledge. Proceedings
            of the Ottawa Conference, October 1971, ed. by J.A.
            Wojciechowski. Pullach/Munich: Verlag Dokumentation, 1974.
(10)   NEEDHAM, Joseph. Integrative levels: a revaluation of the idea of
            progress. (Herbert Spencer Lecture) In his: Time: the
            refreshing river, pp. 233-72. London: Allen and Unwin, 1943.
(11)   FEIBLEMAN, J.K. Theory of integrative levels. *British Journal
            for the Philosophy of Science*, 1954, vol. 5, pp. 59-66.
(12)   AUSTIN, D. Report to the Library Association Research Committee
            on the use of the NATO grant. In: Library Association.
            Classification and information control. London: Library
            Association, 1969.
(13)   SARTON, George. *Isis*, 1926, vol. 8, p. 526.
(14)   BLISS Bibliographic Classification. 2nd greatly enlarged and com-
            pletely revised edition by Jack Mills and Vanda Broughton.
            London: Butterworth, 1977- (in progress).
(15)   LUND, John J., and Mortimer Taube. A nonexpansive classification
            system: an introduction to period classification. *Library
            Quarterly*, 1937, vol. 7, pp. 373-94.
(16)   BLISS Bibliographic Classification. *Op.cit.*
(17)   Classification of the health sciences by Ruth Daniel (for the
            Wessex Regional Hospital Board). Private communication.

(18) BARNARD, Cyril C. A classification for medical and veterinary
          libraries. London: H.K. Lewis, 1955.
(19) *Technology and Culture*. The international quarterly of the Society
          for the History of Technology. Vol. 1- Detroit: Wayne
          State University Press, 1959- The first of the annual
          bibliographies was published in 1964, vol. 5, pp. 138-48, an
          outline of the classification scheme used being on page 139.

A P P E N D I X

S Y N O P S I S   O F   T H E   S C H E M E

CIVILIZATIONS AND PERIODS

1	Prehistory and primitive societies, in general
12	Europe (except Greece and Rome): prehistory and early civilizations
15	Near East, including Mediterranean: prehistory and ancient civilizations
25	Classical antiquity
35	Asia (except ancient Near East): prehistory, primitive societies and indigenous civilizations
45	Australia and New Zealand, including New Guinea: prehistory, primitive societies and indigenous civilizations
46	Africa (except ancient Egypt and North Africa): prehistory, primitive societies and indigenous civilizations
47	America, in general; North America: prehistory and primitive societies
48	Latin America: prehistory, primitive societies, and pre-Columbian civilizations
49	Ancient and mediaeval periods combined
5	Middle Ages
58	Middle Ages and Renaissance combined
6	Renaissance (Fifteenth and Sixteenth centuries)
7	Seventeenth century
8	Eighteenth century
9	Nineteenth century

SCIENTIFIC SUBJECT FIELDS

A	SCIENCE	GV	Geomorphology
AC	Natural magic; Pseudo-sciences	H	Geography
AF	Philosophy	HU	Palaeontology
AK	Logic	HY	Natural history
AZ	Mathematics and the physical sciences	HZ	Biology and the medical sciences
B	Mathematics	J	Biology
BZ	Physical sciences	JR	Microbiology
C	Physics	K	Botany
D	Chemistry	L	Zoology
F	Astronomy	M	Sciences of man
FZ	Earth sciences	MA	Anthropology
G	Meteorology	MG	Psychology
GG	Geology	MP	Social sciences
		N	MEDICINE

NA	Public health and social medicine	TC	Civil engineering
O	Anatomy and physiology	TN	Transport engineering
OH	Systems of the body; specialties of medicine	U	Mechanical engineering
		UT	Electrical engineering
Q	Pathology and clinical medicine	V	Materials and processes
R	Therapeutics	WJ	Town and country planning
RB	Surgery	WQ	Special mechanical technologies
RE	Dental surgery	WW	Special crafts
RF	Therapeutic practice	X	Communications
RM	Pharmacy; Materia medica	XJ	Electrical communications
S	Veterinary medicine	XR	Military technology
SA	AGRICULTURE	Y	ANCILLARY DISCIPLINES
SO	Forestry	YA	Historical sciences
SP	Animal husbandry	YJ	Linguistics
SV	Food technology	YR	Education
T	TECHNOLOGY	YS	Bibliography

FORM AND ASPECT SUBDIVISIONS

a	geographical subdivisions (for location only)
b	common form subdivisions
bu	sources of information; documentation
bz	prehistory
c	general history
cb	libraries and archives
cc	historical sources
ce	literature
cp	non-bibliographical sources
d	biography
da	collective biography
db/ds	biography of individual personalities
f	the history of the subject as a study
g	teaching the history of the subject
h	history of the subject: historiography; philosophical and evolutionary aspects
j	psychological aspects
k	philosophy and methodology
m	fundamental concepts
mz	relation with other scientific subject fields
n	communication
o	the study and teaching of the subject
p	organization of the subject
q	societies and institutions
r	the profession: the person in the subject
s	the subject in relation with society as a whole
t	the subject in relation to culture and the humanities
u	popular aspects of the subject
v	ethnic and linguistic subdivisions
w	national subdivisions
x	techniques and instruments
xw	museology

---

y	facet indicator
z	subject extension symbol

Dr. Phillip R. Sloan

THE IMPACT OF BUFFON'S TAXONOMIC PHILOSOPHY IN GERMAN BIOLOGY:
THE ESTABLISHMENT OF THE BIOLOGICAL SPECIES CONCEPT

In the literature on the French naturalist George Louis LeClerc, Comte de Buffon (1707-1788), his statements on the meaning and value of classi-fication in biology have been seen either as negative and nominalistic, or fluctuating and incoherent. Buffon's intent in his polemic against sys-tematic biology, however, takes on a new dimension when seen against the background of the Leibniz-Clarke controversy. The insight this gives in-to Buffon's understanding of classification, and the consequences of this for a further elaboration of a historical and genetic understanding of biological systematics, form the subject of this paper.

One central issue in this controversy between Leibniz and Newton con-cerned Newton's conception of time and space as metaphysically real and subsistent entities, identified with the very duration and omnipresence of God, and existing independently of all material and created bodies.

To this claim Leibniz had replied with a battery of metaphysical and logical arguments, deriving from his two foundational principles--suffi-cient reason and the identity of indiscernibles. Since both time and space, apart from existent bodies, would lack any internal distinctions, God could not choose one order of time and space over any other.

Leibniz' answer was that both time and space, apart from created beings, have no reality, and that a sharp distinction must be inserted between two metaphysical and conceptual orders. The first, an abstract and ideal order, would be the locus of the concepts of absolute and independent time and space, conceived in abstraction from all bodies or creatures.

The second order, a real and concrete domain, is immanent in created nature, and extends as far as the real relations of coexistence and suc-cession of bodies themselves:

> The parts of time or place, considered in themselves, are ideal
> things; and therefore they perfectly resemble one another like two
> abstract units. But it is not so with two concrete ones, or with
> two real times, or two spaces filled up, that is, truly actual.
> ...Space is nothing else but an order of the existence of
> things, observed as existing together. . . .There is no real
> space out of the material universe.

And concerning time:

> [Newton makes] time a thing absolute, independent upon God;
> whereas time does only coexist with creatures, and is only con-
> ceived by the order and quantity of their changes.[1]

Leibniz' views, left somewhat unsystematic and undeveloped by his death, but subsequently amplified and synthesized in influential ways by his disciple, Christian Wolff, implied a profound reorientation of the rela-tions of time and space to the natural world. The world, for Leibniz and Wolff, no longer stood in a Newtonian four-dimensional box of an indepen-dent and infinite time and space, in which its history could remain delim-ited by the almost a-historical framework of the mechanical philosophy and literalist Mosaic cosmology.

The incursion of the Leibniz-Wolff philosophy into France in the 1730's, particularly through the circle of figures associated with Madame du

Châtelet's residence at Cirey, has now been well described by several commentators.[2] In 1740, her Institutions of Physics appeared, in which the first eight chapters prefacing her discussion of mechanics gave a general summary and exposition of Wolff's philosophy.[3]

In this curiously ecletic work, du Châtelet openly sides with Leibniz and Wolff on the nature of time and space, distinguishing between the "abstract" and "ideal" nature of Newtonian time and space, and the concrete meaning of it in the Leibniz-Wolff philosophy. As a consequence of this, she warns of the danger in science generally of confusing the "abstract" with the "real and concrete" dimensions of scientific understanding. As she writes concerning time:

> Time is in reality nothing else than the order of successive beings.... Thus there is no time without true beings, arranged successively in a continuous series.[4]

Buffon had apparently visited du Châtelet at Cirey in the critical period of 1738,[5] and by late 1740 had both read and highly praised her Institutions of Physics.[6] More significantly, it is precisely the central Leibniz-Wolff epistemological thesis--that there are two distinct conceptual orders, that of arbitrary and abstract concepts, and that of real relations-- that becomes virtually axiomatic in all of Buffon's subsequent writings after 1740.[7]

The evidence that this same line or argument underlies Buffon's basic philosophy of taxonomy is encountered in the paradoxical "First Discourse on the Manner of Studying and Treating Natural History," which appeared as the opening discurse of the first volume of his Natural History in 1749. The key to Buffon's argument in this text lies in his apparent endorsement of the Leibniz-Wolff distinction of the two conceptual orders. One, that of abstract concepts, stands divorced from the reality of the world order in a manner akin to the Newtonian concepts of time and space. This is where he places Linnean taxonomy:

> [The Linnean system] is not a science, and at most is only a convention, an arbitrary language, a means of understanding, but from it results no real knowledge.[8]

The second order, a "real and physical" order, as he terms it, is not an abstract order of concepts, but the actual interconnection of objects and processes that constitute the physical world.

This takes on its concrete significance for the interpretation of the concepts of taxonomy when Buffon redefines the meaning of a "species" in natural history in the second volume of his Natural History:

> This power of producing its like, the chain of successive existence of individuals, constitutes the real existence of the species.[9]

The significance of this for Buffon's general approach to taxonomic systems emerges in the diffuse discussion of taxonomy that can be followed through the Natural History of the Quadrupeds. These discussions consistently reveal an attempt to reinterpret the familiar categories of the Linnean hierarchy at varying levels, not as classes defined in terms of common characters, but as physical networks of organisms standing in historical and material connection.[10]

Buffon's novel insights on the meaning of taxonomic groups was misunderstood by his French successors. In the works of French natural historians, who at least showed a superficially direct impact of Buffon's taxonomic philosophy--Daubenton, Adanson, and Lamarck--there is an absence of any comprehension of the Leibnizian foundation of this polemic against

532

"abstract" systems, and an attempt to assimilate Buffon's critique to a taxonomic nominalism that derives from Locke and Condillac. Closely related to this, French taxonomists generally thus sought to resolve the search for the "natural" system by a calculus of morphological resemblance, rather than in terms of historical and genealogical connection.[11]

The German reception of Buffon's views on taxonomy presents a much more complicated picture. Although the deep entrenchment of rational systematics in the Germanies delayed an appreciation of his arguments until the 1770's, from that date, Buffon's conception of the "physical" meaning of the taxonomic system was increasingly inserted into German biology.

The central issue producing this change in understanding concerned the status of man in growing systematization of nature. From the publication of the first edition of his Systema naturae in 1735, Linnaeus had triggered the problem with his inclusion of man among the Quadrupeda in the Order Anthropormopha, along with the apes and sloths. In this taxonomy, man had formed a separate genus, Homo, with four varieties, European, Asian, American and African, distinguished on the basis of color.

In 1774, the Scottish jurist and moral philosopher, Henry Home (Lord Kames) had further complicated the issue by his explicit revival of the polygenetic theory of the origin of these varieties in his Six Sketches on the History of Man, a work translated into German and published at Leipzig in the same year. In the first part of this influential work, Kames had raised what is essentially a taxonomic argument. The divisions that could be made in mankind showed an analogy, he argued, not to varieties of a single species, but a constancy of character and geographical localization that was fully analogous to distinct species recognized in animals and plants.

The entry of the philosopher Immanuel Kant into this discussion in the 1770's had profound implications for the establishment, clarification and amplification of the taxonomic concepts originally developed by Buffon.

Kant's concern with taxonomic subjects was connected with the course he had initiated as a Privat-docent in 1757 in physical geography, a course he taught regularly for the next forty years.[12] The year following the appearance of Kames' work, Kant scheduled for the topic of the summer half of this course a discussion of the topic of the races of man.

Kant's arguments on this issue were closely connected to the technical distinction he consistently makes in his scientific writings between an a-temporal description of nature (Naturbeschreibung), and a historical and genetic history of nature (Naturgeschichte). Among many other aspects of the Wolffian philosophy, which had served as Kant's original intellectual beginning point in both philosophy and science, Kant had embraced the distinction of the "real" from the "abstract" orders, following, as a consequence, Leibniz and Wolff against Newton on the nature of time and space. After 1770, Kant had rejected the Leibniz-Wolff understanding of time and space, replacing them with an anthropological grounding as the pure forms of sensible intuition, with space governing outer sense, and time inner sense.

The bearing of this on taxonomic questions emerges most clearly in the introduction to his lectures on physical geography. Judging from the late collations of these lectures, Kant indicates three kinds of inquiry into "natural history" in the broad sense.

The first of these approaches nature in terms of a logical ordering of concepts. And much as Buffon had done previously, this is where he places Linnean taxonomy and the Linnean understanding of the natural system:

The division of the objects of knowledge according to concepts is a logical division. That according to Time and Space, however, is a physical division. By means of the first we obtain a Natural System (Systema naturae), as for example, that of Linnaeus. Through the latter means, on the other hand, we obtain a geographical description of Nature.[13]

As Kant develops this point, it becomes clear that there is really a bifrication in the last concept. This "physical" division can proceed in two ways: the first analyzes organisms in terms of their spatial relations, treating them by locale, habitat, and geographical distribution. But just as the concept of time took precedence over the concept of space in the mature Kantian philosophy, so this a-temporal description could be subsumed under a more comprehensive historical and genetic explanation of nature:

The History of Nature contains the multiplicity of geographical objects, as it has been in different times, but not how it is now at the same time. Because in this case it would be a Description of Nature. But if the events of collected nature, as they have been constituted through all times, are brought forth, we are supplied for the first time with a Natural History properly called.[14]

As an example of the latter, Kant then gives as an example the conception of the different races of dogs and horses understood as having come from a single stock or Stamm.[15]

In light of these distinctions, the 1775 paper on the races of man clarifies how Kant was then thinking on concrete taxonomic issues. In the introduction to this paper, Kant indicated clearly that he understood Buffon's concept of a "physical" species to be one intimately connected with what he had distinguished as a History of Nature:

In the animal kingdom the division of Nature into Genera and Species is grounded on the general law of reproduction, and the unity of the genus is nothing else than the unity of the generative force, which is considered as generally active for a determined Manifold of animals. Thence, the Buffonian rule--that animals which can generate fertile young and which might show differences in form, belong to one and the same physical genus--properly can be applied only as the definition of a Natural Genus of animals generally, to differentiate it from all Logical Genera.[16]

As Kant continued his argument, the claim was again, with Buffon, that in terms of such a "physical" meaning of the group, the key to classification in the History of Nature was not character resemblance, but the recognition of groups united by the historical unity of the Stem:

The Logical Division [of taxonomists] proceeds by Classes according to similarities; the Natural Division considers them according to geneology and with reference to reproduction. One produces an arbitrary system for the memory, the other a natural system for the understanding. The first only has the intention to bring creation under titles. The second intends to bring it under laws.[17]

As Kant developed this essay, his argument was that understood in terms of a "natural-historical" point of view, the necessary unity and law-like character of the "natural" genus and species, could only be grasped if it was understood as having a unitary origin from a common historical stock. The polygenecists were thus refuted on two levels: first, they had only recognized "logical" divisions, based purely on morphological similarities and differences, in their recognition of several distinct species of men. Secondly, they had, by postulating several separate historical creations, failed to seek a unification of a plurality of phenomena under necessary unifying principles and an economy of explanatory causes.[18]

Kant's views on this issue were further developed, elaborated, and clarified in two papers appearing in 1785 and 1788, which served to bring him into a public controversy with the Linnean disciple and zoogeographer, Johann Reinhold Forster. In these writings, however, his concerns remained confined to specifically anthropological issues, and received no wider generalization at his own hands.[19]

In 1796, however, Christoph Girtanner's On the Kantian Principle for Natural History appeared at Göttingen which finally brought together in a coherent and synthetic way Kant's basic arguments, connected these with dimensions of reigning biological theory, and then generalized these into a comprehensive program for the reform of the classification of organisms generally in terms of genealogical relationship.

Following a period of close contact with two disciples of Kant's philosophy, most notably Kant's personal scribe Johann Jachmann, with whom Girtanner roomed during medical studies in Edinburgh in 1788-1789, Girtanner made an explicit attempt to apply Kant's principles on a broader scale:

In long reflections on the Kantian foundation, I have found that this had value not purely for the races of mankind, to which the famous philosopher has applied it, but this foundation is a general law which can be applied to all of organized Nature.[20]

Girtanner then generalized Kant's conclusions on the "physical" meaning of a species in Naturgeschichte:

All animals or plants, which generate fertile young with each other, belong to a physical species. This is the great law of nature on which the history of nature is grounded.

Organized bodies, which belong to one and the same natural species, are, through their capacity to reproduce, united to each other in a connected group, and have arisen from one stem.

Organized bodies, which belong to the same artificial species, are united purely through a comparison of common characters.[21]

In light of these explicit principles, Girtanner then proposed a new classification of the organic world to be made in terms of a subordination of groups based on relationships of reproduction and presumed common descent. Furthermore, such inquiries were to be undertaken on the experimental level, following the lead of Joseph Koelreuter.

Girtanner's appeal for a new and comprehensive research program in natural history served as a logical conclusion of the development of concepts, begun in a general way in the philosophy of Leibniz and Wolff, given a concrete, if elliptical application to natural history by Buffon, and subsequently clarified and reinterpreted by Kant. Understood correctly, it meant that a new way of looking at the systematic relationships of life was to be sought, not antagonistic to the rational systematization of Linnaeus, but now clearly separating traditional taxonomic and morphological approaches to taxonomy from a temporal and genetic inquiry into the relations of life.

Beyond a circle of devoted Kantians, however, it appears that this proposal for a complete reform of classification in terms of stem-relations and reproductive compatibility had little immediate endorsement by the authoritative zoological tradition, represented by Johann Blumenbach.[22]

But in more indirect ways, certain aspects of the Kant-Girtanner interpretation received a highly important, if ambiguous, endorsement in the
writings of the zoologist and rising codifier of German taxonomy, Johann
Karl Illiger, whose Attempt at a Systematically Complete Terminology, published in 1800, was to serve as an authoritative handbook for a new generation of working taxonomists.

Although the names of Kant, Girtanner, or Buffon never appear in the
treatise, key definitions in Illiger's text are either verbatim transcrip-
tions or close paraphrases of those found in Girtanner's work. In the
lengthy preface to the work, Illiger proceeds to examine the concepts of
species and genera as they are to be understood in Naturgeschichte, and
the debt to Girtanner is prominent: a "species" in the History of Nature
is thus "...the Totality of all individuals which generate fertile young
with one another," repeating very closely what Kant and Girtanner had defined as the "physical" species.[23]

Illiger then follows the Buffon-Kant-Girtanner tradition in distinguishing this concept of species from that belonging to logic and
Naturbeschreibung. The latter would constitute the species on the basis
of character similarity and difference.

If much of this can be shown to be directly derivative of Girtanner
and Kant, there is, however, a fundamental ambiguity in Illiger's discussion that represents a confusion of the two levels of inquiry carefully
distinguished by Buffon, Kant and Girtanner. Although the prefatory
discussion is oriented to a discussion of concepts in the domain of
Naturgeschichte, the work itself aims at giving a terminology in Naturbeschreibung. This ambiguity, noted by contemporaries, thus results in
Illiger's actual endorsement of definition of the categories of the
Linnean hierarchy not in terms of the "natural-historical" concepts advocated by Girtanner, which would have placed the emphasis on making taxonomic divisions according to relations of common descent and reproductive
compatibility, but rather in terms of purely morphological definitions,
in terms of characters held in common. Illiger thus defines the "natural"
system as that classification based "...on all the essential characters,
and thus on the whole habitus," in contrast to the "artificial" system,
based on similarities in a single part.[24]

Thus in spite of Illiger's clear link to the tradition of Buffon, Kant
and Girtanner, in the end it is the morphological perspective of Linnaeus
and Blumenbach that triumphs in his work. The "natural-historical" concepts thus established in the work are ultimately anomalous--simply definitions which have no ability to reform a taxonomy based purely on morphological affinity, and what I have argued is the main intent of Buffon's
critique of taxonomy is essentially obscured.[25]

Illiger's influential work typifies the ambiguous impact of the alternative meaning of the taxonomic system instituted by the Buffon-Kant-
Girtanner tradition. It is not difficult to find a persistence of these
"natural-historical" concepts in the writings of German natural historians
after 1800. It is, however, increasingly impossible to determine the
grounds upon which they are differentiating these taxonomic concepts from
the traditional morphological definitions of taxonomic groups and the
"natural" system, as employed by contemporaries not concerned with the
historical relations of life. Prior to the advent of the Darwinian period,
the two alternative approaches to the natural system stand, in German
natural history, in an incoherent relationship.

Unlike the situation in either French or English natural history, however, the German tradition continued to contain within it a set of concepts which could be readily developed by Haeckel and others in the post-Darwinian period as the foundations of phylogenetic classification, the "biological" species concept, and other important taxonomic concepts that[26] Darwin himself was conceptually unable to develop fully in his own work.

NOTES

[1] The Leibniz-Clarke Correspondence, ed. H.G.Alexander (New York: Philosophical Library, 1956), 5th letter, pp. 63, 75.

[2] See esp. W.H.Barber, "Mme, du Châtelet and Leibnizianism," in: The Age of Enlightenment, ed. W.H.Barber et. al. (Edinburgh: Oliver and Boyd, 1967); and C. Iltis, "Madame du Châtelet's Metaphysics and Mechanics," Stud. Hist. Phil. Sci. 8 (1977), p. 29-48.

[3] Du Châtelet's preface is more correctly seen as an exposition of Wolff's philosophy than that of Leibniz'.

[4] Gabrielle Émélie du Châtelet, Institutions de Physique (Paris: Prault, 1740), p. 119.

[5] See letter of 4 Nov., 1738 in L. Hanks, Buffon avant l'histoire naturelle (Paris: Presses universitaires de France, 1966), p. 259.

[6] Letter from Helvétius to du Châtelet, ca. Dec. 1740 in: Les lettres de la Marquise du Châtelet, ed. T.Besterman (Geneva: Institut et Musée Voltaire, 1958) II, 36.

[7] See Buffon's translation of Newton's Fluxions of 1740, where he argues that Newton's defense of the concept of the actual infinite is unintelligible since it is only an abstraction. See also his similar criticism against the "abstract" nature of time, space and number. Buffon, Oeuvres philosophiques, ed. J. Piveteau (Paris: Presses universitaires de France, 1954), pp. 448.

[8] Ibid., p. 11.

[9] Ibid., p. 236.

[10] See "Le Mouflon," in: Oeuvres complètes de Buffon, ed. de Lanessan (Paris: Pilon, 1884) I, 408.

[11] This point will be analyzed at length in the complete version of this paper, to be published subsequently elsewhere.

[12] Kant's 1757 prospectus for the course announced that Buffon's Natural History would be one of the three main texts for the course. See Immanuel Kant, Frühschriften, ed. G.Klaus (Berlin: Akademie-Verlag, 1961) I, 284.

(13)Kant, Physische Geographie in: Sämmtliche Werke, ed. K. Vörländer (Leipzig: Meiner, 1905) IX, 11.

(14)Ibid., p. 14.

(15)His examples here are found in Buffon's "Abartung der Thiere," which had appeared in German in 1772.

(16)Kant, "Von der Verschiedenheit der Racen uberhaupt," in: Kant's Werke (Berlin Akademie; Berlin: Reimer, 1912) II, 429. This originally appeared in print in Jacob Engel's Philosophie für die Welt in 1777. An ambiguity in the German word Gattung in the biological literature of the time complicates the translation of this. As Kant explains in 1785, however, in Naturgeschichte, there is no fundamental subordination of genera and species.

(17)Ibid.

(18)See the elaboration of this point in the Kritik der reinen Vernunft, A 658-662.

(19)Kant, "Bestimmung des Begriffs einer Menschenrasse," and "Ueber den Gebrauch teleologischer Prinzipien in der Philosophie," in: Kant's Werke, ed. E.Cassirer (Berlin: Cassirer, 1922) Vol. IV. Both appeared originally in the Teutscher Merkur.

(20)Ueber das Kantische Prinzip für die Naturgeschichte (Göttingen: Vandenhoek & Ruprecht, 1796), "Vorrede," p. ii.

(21)Ibid., p. 4.

(22)This point is developed at length in the full version of this paper.

(23)Versuch einer Systematischen vollständigen Terminologie für das Thierreich und Pflanzenreich (Helmstadt: Fleckeifen, 1800), p. xxvi.

(24)Ibid., pp. 107-108.

(25)This is clearly evidenced by Illiger's later influential reclassification of the Mammals and Birds (Prodromus systematis mammalium et avium, Berlin, 1811). While based heavily on comparative anatomy, and the morphological definitions of taxa, historical and genealogical issues play no role.

(26)E.Haeckel, Generalle Morphologie der Organismen (Berlin: Reimer, 1866) II, esp. chp. 26; Carl Nägeli, Entstehung und Begriff der Naturhistorischen Art (Munich: Königlichen Akademie, 1865).

Symposium 10.   COSMOLOGY SINCE NEWTON

M.A. Hoskin and G.J. Whitrow

INTRODUCTION TO SYMPOSIUM 10

This symposium, arranged by Dr M. A. Hoskin, Cambridge University, and Professor G. J. Whitrow, Imperial College London, took place on Thursday 18 August. The morning session, under the chairmanship of Dr Arthur Norberg of the Bancroft Library, University of California at Berkeley, was devoted to cosmology in the eighteenth and nineteenth centuries, and this was the title of the opening paper by Dr Hoskin. Instead of attempting a systematic exposition of the history of cosmology over the two hundred years, Dr Hoskin preferred to discuss some of the concepts and methodological principles which feature in cosmological theories of the time, arguing that in a science which of its nature studies the remotest regions of the observable universe, the observational evidence must always be woefully inadequate and the role of methodological principles correspondingly prominent. Dr Hoskin singled out for discussion simplicity, uniformity, change with time, the reconciliation of human and cosmic time-scales, probability, and the use of stellar statistics, and he then went on to illustrate the roles played by these concepts and principles. In particular, the career of William Herschel could be seen as a rear-guard action in defence of the concept of uniformity against advancing evidence to the contrary, and the recognition by the early nineteenth century that the stars are extraordinarily diversified helped to explain the poverty of cosmological thinking in that period; later in the century the rise of astrophysics and the revelations of photography opened new realms to the astronomer and diverted attention away from the problems of cosmology.

There followed two papers on important topics not discussed by Dr Hoskin. The first, by Dr J. A. Bennett of the National Maritime Museum, London, stemmed from his researches while Archivist of the Royal Astronomical Society of London, when he had care of the Herschel papers. Dr Bennett discussed the role of the great reflecting telescopes of the Herschels, Lassell and the Earls of Rosse, instruments which because of their great light-gathering power permitted the study of very distant and therefore faint objects. The second, by Professor Jacques Merleau-Ponty of the University of Paris, dealt with the cosmological implications of classical thermodynamics.

The afternoon session was devoted to cosmology in the twentieth century under the chairmanship of Professor D. Ivanenko of the University of Moscow. Two papers were read, one by Dr N. S. Hetherington, U.S.A. and the other by Professor Whitrow. Attention was confined mainly to developments in the first half of the century.

Dr Hetherington spoke on Observational Cosmology in this period. This is a subject that has begun to attract the attention of historians of astronomy. In particular, the famous Shapley-Curtis debate in 1920 before the National Academy of Sciences in Washington has been critically investigated by Professor Richard Berendzen and Dr M. A. Hoskin. Papers have also been published by Dr Hoskin and Dr Hetherington on some of the observational work in the early decades of this century on the properties of galaxies, including the controversial claims made by A. van Maanen on rotation in spiral nebulae, which conflicted with the hypothesis that these objects lie outside our stellar system because otherwise the linear velocities involved would be far too large. In his paper Dr Hetherington discussed the two main themes of observational cosmology in the period under review: the structure of the universe as either a stellar system or one composed of galaxies (the 'island universe' hypothesis), and the expansion of the universe as revealed by the extragalactic red-shifts.

A lively discussion on this paper was opened by Mr R. W. Smith of the University of Cambridge. He pointed out that, since our own stellar system (the 'Milky Way' system) is the yardstick which supposed external galaxies were measured against, it is essential to bear in mind contemporary ideas about the dimensions of our system. For example, in 1911 F. W. Very assigned a distance of 1,600 light-years to M31, the great nebula in Andromeda, which was regarded as one of the main candidates for the class of external galaxies, if such a class existed. Very assumed that our own stellar system is only about 100 light-years in diameter, and so the Andromeda nebula appeared to lie well beyond its boundaries. Consequently, Very concluded that this nebula must be an external stellar system. However, a few years later no astronomer would have accepted so small size as Very did for our own system. It was then thought by many that its diameter was roughly 10,000 light-years, and hence Very's result for M31 then seemed to refute the island universe theory rather than support it.

Mr Smith drew attention to the contribution of German astronomers in the early part of this century to what we now call observational cosmology. In particular, Max Wolf made important studies of the spectra and distribution of spiral nebulae. In 1899 another German, Julius Scheiner, became the first astronomer to announce that the continuous spectrum of M31 is crossed by dark absorption lines. These contributions are often overlooked.

As regards the controversial measurements made by van Maanen, Mr Smith pointed out that they soon became linked to the highly influential theoretical researches on the nature of spiral nebulae made by Jeans. The internal motions in these objects that van Maanen claimed to have observed were regarded by Jeans as supporting evidence for the theory that he had developed to explain the appearance of spiral nebulae. According to his theory, these should be a flow of material along the spiral arms from the centre outwards. After reading a brief report in <u>The Observatory Magazine</u> in which it was stated that van Maanen had found an outward flow of material, Jeans wrote a letter for publication in that periodical asserting that here was observational support for his views. Indeed, Jeans became so firmly attached to van Maanen's measurements that in 1923 he even went so far as to propose a new law of force to account for the large-scale motions in spiral nebulae. For, if van Maanen's measurements did not fit in with the Newtonian law, then it must be the latter which was in error.

The measurements made by van Maanen were, however, at variance with those made by other observers at the time. Seth Nicholson, a colleague of van Maanen's at Mount Wilson Observatory, repeated some of van Maanen's measurements as a check. Although van Maanen stated that Nicholson's results and his own were in good agreement, a careful reading of Nicholson's paper reveals that his measurements indicated an inward flow of matter and not an outward flow as van Maanen claimed. Jeans and van Maanen tended to ignore anomalies such as this.

Another point made by Mr Smith was that Hubble's famous discovery of Cepheid variables in M31 in 1924 was not the result of a systematic search for such stars. Instead, he was looking for novae. Since Hubble's discovery was decisive in the debate on the existence of 'island universes', why were Cepheids not looked for before 1924. Mr Smith considered three possible answers:

(1) Astronomers did not realise previously the importance of Cepheids as distance-indicators.

(2) Astronomers did not believe that Cepheids were to be found in spiral nebulae.

(3) Astronomers who believed that Cepheids are present in these nebulae thought that they would be too faint to be detected against the bright luminous background of the spiral arms.

The first answer is untenable because Cepheid variables were widely discussed for some years before 1924, and Harlow Shapley's theory of our galaxy that he proposed in 1918 was based on the use of these stars as distance-indicators. Nor can the second answer be correct, because some astronomers in the early nineteen-twenties, when van Maanen's measurements of internal motions were widely accepted, still believed that spiral nebulae were independent stellar systems like our own and therefore presumably must contain Cepheid variables. Only the third answer appears to be a likely explanation. We should note here that there was one astronomer, Knut Lundmark, who stated in print that a search should be made in spiral nebulae for Cepheids. He wrote this in the middle of 1924, but unknown to him Cepheids had by then been discovered in M31 by Hubble.

Although Hubble's observations were soon accepted by most astronomers, there was still some opposition to the island universe theory. For example, as late as 1928 H. H. Plaskett argued, in a colloquium at Harvard, that spirals are not external galaxies. On the other hand, Hubble's discovery of a relationship between red-shift and distance was more rapidly accepted. However, not all astronomers accepted the idea of an expanding universe, leading to the possibility that there was a time in the past when all the galaxies were crowded together. This raised the question of whether all the matter in the universe was created then.

In further discussion it was pointed out that van Maanen and Hubble had used different eyepieces in their telescopes, and this might have been a reason why they came to different conclusions about the status of spiral nebulae. Attention was also drawn to the recent, but much disputed, claim by Arp and others that the law of red-shifts is not isotropic.

Professor Whitrow then spoke on his paper on Theoretical Cosmology in the twentieth century, in which he mainly concentrated on developments up to about 1950. This is a subject on which two important books were published just over ten years ago, by Dr John North of the University of Oxford and Professor Merleau-Ponty. After Professor Whitrow had spoken there was a lively discussion led by the Chairman, who drew attention to some of the recent attempts by theoretical cosmologists to investigate the earliest stages in the evolution of the universe involving the creation of elementary particles, etc. and then of stars and galaxies. Also more complicated world-models than the homogeneous and isotropic type have been considered, including the introduction of the geometrical concept of torsion, to see whether the possibility of an initial singularity can be avoided.

Dr M A Hoskin

COSMOLOGY IN THE EIGHTEENTH AND NINETEENTH CENTURIES

The term 'cosmology' can be made as all-embracing as the cosmos itself. I shall take cosmology to mean "efforts to establish and to explain the large-scale structure of the universe and changes in that structure". I exclude all issues limited to the solar system even when, for example, the discovery of a new planet may confirm the existence of a universal law. The explanations I consider will be mainly in the natural order and the explanations of changes will mostly be limited to the action of gravitational attraction, established by Newton as a universal law at the very beginning of our period; the unquestionably cosmological issue of the amount of motion in the universe and its possible reduction through the retarding action of ether, a major interest of Newton's contemporaries, will be better understood by the time of our next Congress, while the allied problems of thermodynamics form the subject of a separate paper by Professor J. Merleau-Ponty. Lastly, I restrict myself to theory: the development of what we might term 'cosmological telescopes', the huge reflectors designed to collect light from remote and therefore faint objects, is likewise the subject of a separate paper, by Dr J. A. Bennett.

Even with these restrictions, two centuries of cosmology is too vast a topic to be systematically treated in a short survey. Instead I shall seek to highlight some of the major issues in the period, and begin with remarks on key concepts and techniques.

First, simplicity. Simplicity of course plays an important role in all scientific theorising, Ockham's razor being an obvious manifestation. To the observational cosmologist (of any period) simplicity will be of special significance, for he is by definition studying the most remote objects accessible with contemporary apparatus and the evidence at his disposal is inevitably scanty and ambiguous; evidence alone is therefore likely to be less effective in deciding between theories in cosmology than is the case with other sciences. The almost-ruthless search for simplicity lies behind the extraordinary swings of opinion, from one extreme to the other, over the nature of the nebulae, and the repeated suppression or disregard of evidence to the contrary. William Herschel, for example, found that some of these milky patches were star systems disguised by distance, and claimed that they all were -- concealing the changes which he believed he had himself observed in the Orion nebula, changes which could not have occurred so quickly if the nebula was indeed a vast star system. A century later, William Huggins used spectroscopy to establish beyond question that some nebulae are gaseous; but when the Andromeda nebula was found to have the spectrum to be expected of a galaxy of stars, Huggins devised a physical theory for the nebula which would enable it nevertheless to be classed with the gaseous nebulae. In fact some nebulae are gaseous and some are galaxies of stars and some lie between these extremes, yet astronomers resisted this easy but multiple solution, preferring to hold to a simple theory while suppressing or explaining away evidence to the contrary.

My second concept is uniformity, by which I mean the extent to which celestial bodies may be presumed to be of similar nature, to lie at similar distances, to move at similar speeds, and so forth, until proved to the contrary. It will often be the case that an answer, however provisional, can be made to a question only if gaps in the evidence can be filled by an assumption of uniformity. This is notably the case with measurements of very great distances. As we shall see, Newton established the distances of the nearest stars by assuming they are uniform in nature with the Sun and their faintness entirely the consequence of their great distance; for an illustration of the same concept and methods at work in modern astronomy, one only has to read the chapter on "The Measurement of Astronomical Distances" in Hoyle's Frontiers of Astronomy. In the absence of direct measurements of annual parallax, Newton's assumption permitted provisional answers to be given when otherwise the question would have gone unanswered. Similarly, Herschel derived an outline for our own star-system by assuming that within the system the stars are uniformly dense, so that the more stars there are in a given direction, the further the system extends in that direction. Later his own studies of star clusters forced him to recognise that the distribution of the stars is very far from uniformly dense, to such an extent that his outline of our system must be withdrawn; but students of nature abhor a vacuum and it is a striking indication of the pressure to supply an answer, any answer, to a scientific question, that his outline was still being reproduced decades after his death.

The assumption that the stars are uniform in physical properties was especially necessary to Herschel for it offered the only hope of establishing the distribution of the stars in three dimensions, the "construction of the heavens" that was his principal goal. Yet even in Newton's day there were warning signs to be seen, for example in the numerous new and variable stars, and by a century later a crisis had developed. Herschel personally observed binary stars, pairs of companion stars which in some examples differed strikingly from each other; and during his lifetime certain faint stars, notably 61 Cygni, were discovered to have rapid proper motions across the sky which implied that they must be near, notwithstanding their faintness. Herschel's career can be seen as a desperate rearguard action in defence of the assumption of uniformity among the stars, and the abandonment of the assumption by the next generation accounts in part for the poverty of their cosmological thinking. Not for them the astonishing claims of Herschel that he had seen light that had taken two million years on its journey to Earth, and that its source must have existed two million years ago, but might no longer exist.

My third concept is that of change with time, which becomes a central issue in our period. A universe unified by gravity is a universe in which changes may be expected as a result of the action of gravity. How the stars could be 'fixed' and motionless despite gravity, as seemed to be the case in Newton's day -- how the universe could be for practical purposes static -- was a problem brought forcibly to Newton's attention by Richard Bentley. Lambert, by contrast, described a universe in which, as with the solar system, the components move but the overall structure remains the same, the stars, star-systems and so forth orbiting stably around a centre. Herschel accepted an evolving universe in which gravity had been at work for a long but finite time to produce a unidirectional change towards ever-increasing condensation of matter into stars and of stars into clusters.

Related to this is my fourth concept, the <u>reconciliation of human and cosmic time-scales</u>. When human life is but a moment, how can we hope to observe changes of cosmic dimensions? Herschel's answer is to parade for our inspection similar objects at different stages of their development: "Is it not almost the same thing", he asks, "whether we live successively to witness the germination, blooming, foliage, fecundity, fading, withering, and corruption of a plant, or whether a vast number of specimens, selected from every stage through which the plant passes in the course of its existence, be brought at once to our view?"

My fifth concept is the employment of <u>probability theory</u>, notably by John Michell in his great <u>Philosophical Transactions</u> paper of 1767. Michell asks himself whether double stars and clusters like the Pleiades come about by chance as the optical effect of isolated stars which happen to lie in the same line of sight from us though at different distances, or whether they are true physical associations, companions in space. This was the first demonstration that gravity or similar forces operate among the stars -- and a powerful blow against those who would defend the uniformity of stars, for a cluster like the Pleiades contains stars of very different apparent brightnesses though all are at much the same distance from the observer, so that the differences among them are not merely apparent but real.

Sixthly, the employment of <u>stellar statistics</u>, that basic tool of the modern astronomer. In his investigations of the regularity or otherwise of the stars in our neighbourhood, Newton analysed the total numbers of stars of successive magnitudes. Only the barest bones of this analysis saw publication, but the technique of stellar statistics came into the possession of astronomers when it was employed in mature form by Herschel in his investigation of the outline of our star-system. Herschel would count the number of stars in ten neighbouring fields of view and average his results to derive the distance to the border in that direction.

I have dwelt on the concepts and methods developed in our period, not only because systematic accounts of particular theoretical issues are already available, but also because the concepts and methods may be said to form the true and enduring legacy to later astronomers. Let us now see how these concepts entered into the responses of our period to questions of cosmology, beginning with the distances that separate the stars one from another.

The failure of the Copernicans to detect annual parallax, the apparent movement of stars as we the observers on Earth orbit around the Sun, implied that even the nearest stars were too distant from us and their apparent movements therefore too slight for the accuracy of the instruments then in use: that is, it supplied a <u>minimum</u> distance from the solar system to the nearest stars, but said nothing positive about the <u>actual</u> distances to the nearest stars, nor of course about the distances of the further stars from us or of the stars from each other. On each of these questions Newton made fundamental advances, but of this little was known in his own lifetime and much of his analysis had to be recapitulated a century later by Herschel.

Newton based his investigation on a method published by James Gregory in 1668 but overlooked by their contemporaries. Gregory made the uniformity assumption that the Sun and the stars are similar in nature -- he termed the Sun <u>stella fixa vicina</u>, our local fixed star -- so that the Sun looks bright to us only because it is near. Since light falls off

with the square of the distance, the ratio of the distances of the Sun
and (say) the prominent star Sirius is the square root of the ratio of
their apparent brightnesses.

The same reasoning occurred to others in the late seventeenth century.
Gregory's master-stroke lay in proposing a workable technique for actually
comparing two such different apparent brightnesses and so arriving at a
distance for Sirius as a multiple of the distance of the Sun. He pointed
out that if we seize on one of the outer planets at the moment when it
equals Sirius in brightness, we can replace the comparison of Sirius and
the Sun with the equivalent comparison of the planet and the Sun; and this
latter comparison depended solely upon quantities within the solar system,
assumed to be known. Gregory's own calculations resulted in too low a
figure for the distance of Sirius, because in his day the Earth-Sun
distance (the 'astronomical unit') was poorly established, as he himself
recognised. Newton, in his System of the World and in manuscript drafts,
places Sirius at around one million astronomical units, a figure which is
of the correct order of magnitude -- a few light years, we might say --
and as it happens somewhat too large.

Newton, then, was the first to have an adequate grasp of the distances
separating the Sun from the nearest stars. For differing reasons his
attempts in his lifetime to share his insight with others came to little,
and in the first quarter of the eighteenth century the question of the
distances of the nearest stars was dominated by the opinion of Huygens,
who made the same assumptions but could contrive only the crudest of
techniques and so arrived at less than 28,000 astronomical units for the
distance of Sirius. When in 1728 Newton's System of the World at last
saw posthumous publication, its appearance virtually coincided with
Bradley's announcement of a revised minimum distance to the nearest stars
of some 400,000 astronomical units; the convergence of these two
techniques carried general conviction and so Gregory's method was
repeatedly applied until the uniformity assumption on which it relied had
to be abandoned early in the nineteenth century.

In theory the uniformity assumption could be used to derive the
distances of the fainter stars, but the necessary photometric techniques
were lacking until Herschel hit upon the idea of using twin telescopes,
each pointing to one of the stars under comparison, the aperture of one
telescope being masked so that both stars appeared equally bright;
comparison of the apertures then led to comparison of the apparent
brightnesses and so to comparison of the distances. Instead Newton
somewhat thoughtlessly assumed that an n-th magnitude star could be
assigned to n units of distance, but he found that a bizarre stellar
distribution resulted and further investigation led him to a good
understanding of the correct relationship between the traditional stellar
magnitudes and so to the distances of the fainter stars visible to the
naked eye. He found that the star density was then reasonably uniform in
the neighbourhood of the Sun; in other words, that the distances between
one star and another are similar to the distances separating the nearest
stars from the Sun. But all this remained in manuscript and the
identical path had to be beaten afresh by Herschel a century later.

Bradley's revelation of the minuteness of the angle of parallax and
of the multiplicity of factors complicating the delicate measurements led
to a failure of nerve on the part of astronomers of the second half of
the century. Yet, thanks to the Gregory method, the challenge was now
properly assessed and, as Michell pointed out, the situation was by no

means hopeless. And estimates based on the Gregory method were no
substitute for direct measurements, especially with the uniformity
assumption itself coming under increasing criticism: indeed, direct
measurements would allow astronomers to replace the assumption with
factual knowledge of the luminosity of stars, and if Michell proved right
and some double stars were indeed binaries with the companions in orbit
around each other under gravity, then knowledge of the distance of the
system could lead to knowledge of the combined masses of the component
stars.

The search for measurements of annual parallax was therefore resumed
by a number of astronomers early in the nineteenth century, most notably
by Bessel and Struve, who were fortunate to have at their disposal
instruments by Fraunhofer of a delicacy previously unattained. They also
considered very carefully the strategy of their campaign. Since the
angle sought falls off directly with increasing distance, it was essential
that the stars selected for scrutiny be the nearest: astronomers should
therefore concentrate on stars that were bright or, more especially, had
a large proper motion, and, if they were binaries, an unusually large
angle of separation. If possible the position of the star under scrutiny
should be measured relative to a distant star in the same line of sight,
as Galileo had recommended long ago, for the distant star would be almost
unaffected by parallax and would therefore act as a fixed reference point
in the sky -- fixed, yet mercifully affected by aberration, atmospheric
refraction and so forth equally with the star under scrutiny. And by
1840 this carefully considered strategy had led to the first successes.

The question of stellar distances apart, cosmological issues seemed in
the early eighteenth century unpromising by comparison with the
investigation of the operations of gravity within the solar system.
True, Newton's Principia claimed gravity as a universal law, but the
first visual evidence of gravity actually at work among the stars came
only at the beginning of the nineteenth century, when Herschel
re-examined some of the numerous double stars he had collected two
decades earlier. He had hoped, following Galileo, that the distant
member of each double might serve as a reference point for measuring the
parallactic movement of the nearer member, but Michell had warned him that
probability arguments showed that most of the doubles would be true binary
systems of stars in orbit around each other. And so it proved to be: in
several doubles the components had clearly moved around each other. There
was little doubt that the force at work was that of gravity, though proof
of this had to wait until after Herschel's death.

To understand the poverty of interest in cosmology before Herschel, we
must remind ourselves of the contrast between the rich pickings the solar
system offered to mathematician and observer alike, and the doubtful
prospect of worth-while progress in understanding the fixed stars. Until
Halley's announcement in 1718 of three proper motions the stars were
seemingly fixed for ever, but even after 1718 the advance in understanding
was slow. It is not always realised that Bradley's discoveries of
aberration and nutation threw doubt on what had been thought to be
reliable star catalogues and this postponed the time when present and past
positions of stars could be accurately compared and their proper motions
established. So it was that only in 1760 did Mayer significantly add to
the very limited announcement by Halley, and only in 1775 were Mayer's
data published. It was therefore more than half-a-century after Halley's
announcement that sufficient proper motions were known for astronomers to
investigate whether the movements showed a pattern which might reflect the

motion of the solar system through space, and Herschel was in fact the first to announce such a pattern.

If stellar positions showed no changes at the beginning of the eighteenth century, the opposite was the case with stellar brightnesses. From the astonishing novae of Tycho and Kepler, astronomers of the later seventeenth century had acquired a taste for variable stars and until Herschel later compiled catalogues of stars delicately ranged in order of diminishing brightness -- so that even a small change would reveal itself by disturbing the order -- it was all too easy for observers to convince themselves that they had discovered another variable star. Theoretically these variables, real or imaginary, posed a problem that was at once too difficult and too easy: too difficult, because the astonishing variety of novae and variables taxes the understanding of the modern astronomer; too easy, because an analogy with sunspots could explain all but the most dramatic of novae. The spots on the Sun were known to vary, so that to a small extent the Sun itself was a variable star, and these variations were partly cyclic (because of the rotation of the Sun) and partly non-cyclic; and so a hypothetical rotating star with variable dark patches could display almost any sequence of fluctuations in apparent brightness.

The individual stars apart, astronomy beyond the solar system was confined mostly to a description of the Milky Way, and to a listing and description of a handful of nebulae, which some held to be formed of a luminous fluid but others regarded as star clusters owing their milky effect, as did the Milky Way itself, to large numbers of stars lying at great distances. Even the Milky Way, which to us cries out for further explanation, was a static feature of the night sky adequately demythologised by the telescope. And so it was that the questions in stellar astronomy and cosmology were for a century posed, not by 'hard' scientists but by those on the fringes of physics: Richard Bentley the theologian demanding that Newton tell him whether the stars are finite or infinite in number; Thomas Wright the adult educationist seeking to reconcile the elementary astronomy he knew with his theological vision of a universe with the stars in orbit about the divine centre; Immanuel Kant the philosopher pondering the likely effects of gravity working on a universe of diffused matter; J. H. Lambert, in what Herschel termed a-work "full of the most fantastic imaginations", extending a thousand-fold the hierarchy of moon-planet-star-starcluster; John Michell, pioneering the study of star clusters in the very year he became rector of a parish in the north of England; William Herschel, the organist -- these were the men who for a hundred years sustained a flickering interest in cosmological issues. And, until Herschel, so little of what they wrote had the slightest impact. Newton in private developed a static and near-stable model of the universe in which each star, the Sun included, is surrounded by other stars which together approximate dynamically to a series of concentric shells, which shells, by Proposition LXX of Book I of the Principia, have zero gravitational influence on any matter within them; but his thoughts remained in manuscript.

The attempts of Wright, Kant and Lambert to explain the Milky Way as the optical effect of our immersion in a layer of stars make fascinating reading for the historian but may well have been totally without influence. Even Michell's brilliant insight into the nature of double stars and clusters was unknown to Herschel when he began his career - and disregarded by him when it was known, since it undermined the uniformity principle which offered to Herschel his only hope of understanding the construction of the heavens.

In a very real sense, then, modern cosmology begins with Herschel. His vast achievement can be readily summarised. First, he understood that distant objects are faint objects and call for telescopes with great light-gathering power, that is, with large mirrors to collect as much light as possible: the larger the mirror, the further the telescope would penetrate into space. Such telescopes he designed and built with his own hands and so won for himself an overwhelming advantage over all other observers in the study of distant objects; often enough the evidence would be accessible to him alone.

Second, although he was middle-aged before he could devote himself to astronomy, he observed with a persistence and fortitude never surpassed and seldom if ever equalled. He saw himself as a natural historian of the heavens, spending twenty years in collecting specimens of nebulae and increasing the number known from a hundred or so to two and a half thousand, providing long lists of double stars, and cataloguing the comparative brightness of stars as a record against which to check possible variations in stellar brightness. The influence of this natural history was immense. His nebulae, for example, were reobserved by his son John and the coverage extended to include the southern skies, and John's General Catalogue (1864) was later enlarged by Dreyer to form the New General Catalogue (1888) or NGC which is standard today.

Third, and very consciously, Herschel avoided the trap that snares most dedicated observers: he was not only prepared to theorise, he held it to be the greater fault to theorise too little rather than too much. In consequence he asked and offered provisional answers to many of the great questions of cosmology. What is the three-dimensional shape of our star system which we see as the Milky Way? - Herschel used stellar statistics and a uniformity assumption to attempt an answer. In which direction is the Sun and solar system moving through space? -- Herschel found a pattern in the published data on proper motions which he explained, in striking conformity with modern theory, as caused by the motion of the solar system towards Hercules. But, above all, the nebulae: were they star clusters or clouds of luminous fluid and how did they relate to the cosmos as a whole? - Here was a question the evidence for which was known to Herschel and to him alone.

Herschel had begun by examining the nebulae already known, and he found that his mirrors were powerful enough to resolve many of these into their component stars. He saw no reason to doubt the simple theory that all nebulae are star clusters. And while he swept the skies year in and year out in his search for more nebulae, he catalogued these specimens of star clusters (as he thought) and he classified them according to the degree of clustering they exhibited. Clustering implied a clustering power (no doubt that of gravity), and the continued action of that clustering power would surely lead in the future to a higher degree of clustering perhaps accompanied by fragmentation into sub-clusters; conversely, the present clusters must have developed out of more scattered systems in the past. And so he was naturally led to develop a true cosmogony, whereby the universe began with stars scattered throughout space, with here and there an above-average concentration of stars; and these concentrations would pull into themselves any isolated stars nearby, so that gradually gaps would form between one concentration and the next, and the subsequent history would be one of ever-increasing condensation and fragmentation.

Our Milky Way system was in the present time a layer or stratum of stars gradually breaking up into subordinate clusters. Looking towards Coma Berenices, Herschel saw nearby stars of the constellation and, beyond them, large numbers of nebulae, and all these he interpreted as fragments of another stratum. As to a large nebula such as that of Orion, in which even Herschel's telescopes could not actually discern individual stars, this must be at a very great distance indeed. But to our eyes the Orion nebula occupies a large area of sky, and to look so large and be so distant its actual size must be immense -- comparable to and perhaps larger than our Milky Way, and so a fully-fledged galaxy or (in the nineteenth century term) 'island universe'. Herschel, in short, believed in the 1780s in island universes.

In 1790, in his routine sweeping for nebulae, Herschel encountered a planetary nebula in the form of a central star apparently surrounded by a halo of light. It could have been a foreground star which chanced to be in line with a distant cluster, but this was against all likelihood. It could have been a large star surrounded by small ones, but the uniformity of stars was a fundamental and non-expendable concept. It must therefore be a central star physically surrounded by nebulous fluid: Herschel had been too hasty in deciding that all nebulae were simply star clusters.

What was the effect of this on Herschel's cosmogony? First, he envisaged the central star as condensing under gravity out of the nebulous fluid, and so he carried his cosmogony further back in time to a stage before stars were formed, when the universe was filled with luminous fluid, which gradually condenses under gravity and eventually formed into stars which developed into systems as before; indeed Herschel could (as it were) parade before us samples of nebulae at every stage of this development. But he had lost control of the spatial dimension, for a nebula in which he could discern no stars (like the Orion nebula) might still be a distant star system, but might also be nothing more than a nearby cloud of luminous fluid. So he could no longer with conviction point to island universes comparable with our own.

And there the matter rested until in the mid-nineteenth century history repeated itself. Lord Rosse built even bigger telescopes than had Herschel, these telescopes resolved (or so it seemed) the principal nebulae into stars, and Rosse's contemporaries once more equated nebulae with star clusters, abolished nebulous fluid, and believed in island universes -- only for the pendulum soon to swing violently once more, after the spectroscope demonstrated beyond question the existence of gaseous nebulae. The instinct for simplicity was reinforced by two striking observational discoveries: the extraordinary flare-up in the Andromeda nebula in 1885, in fact a true supernova explosion in another island universe but much more plausibly explained as the encounter of a star with a gas cloud inside our own Galaxy; and the absence of any possible island universes in the sky near the plane of the Milky Way, decisive evidence against island universes ranking with our Galaxy until it was realised in the second decade of the twentieth century that this might be nothing more than the visual effect of obscuring matter around us in the galactic plane. And so by the end of the century only a few brave souls dared to admit belief in island universes.

Until the advent of the spectroscope enriched and transformed the nature of the evidence available, many of the other cosmological questions were debated along the lines laid down by Herschel. Bessel used the legacy of Bradley's Greenwich observations to establish with great

accuracy the positions of the stars as they had been in Bradley's day and
so obtained greatly improved data on proper motions; these seemed to
Bessel to reveal no pattern, but a thorough analysis by Argelander in
1837 showed that there was indeed a pattern, and that the solar system
was moving through space, in much the direction Herschel had claimed.
Herschel's stellar statistics were re-examined by Struve to provide
support for a model of the universe in which space is filled with stars
whose density is always uniform in any plane parallel to the Milky Way
but diminishes as one moves vertically away from the Milky Way on either
side. As to the Milky Way itself, with belief destroyed in the
uniformity of stars, many observers took it to be what it appeared, a
ring of small stars surrounding the Sun and other large stars congregated
in the midst of the ring. Olbers drew attention to the theoretical
consequence to be expected if we live in the midst of an infinite and
regular star system, that we would see the whole sky ablaze with light;
but too many ways of avoiding the difficulty were known for the
argument to be of importance, and in any case with the rapid development
of high-quality telescopes in sophisticated mountings, the extension of
regular observations to the southern skies, and above all the introduction
of photography and spectroscopy, the later-nineteenth century was a time
for astronomers and astrophysicists rather than cosmologists.

SELECT BIBLIOGRAPHY

A more systematic survey of the stellar astronomy and cosmology of the
period, with extensive references, is my 'Sidereal Astronomy in
Adolescence', in Avant, avec, après Copernic (XXXIe Semaine de Synthèse:
Paris, 1975), 315-42.

Recent works dealing with particular themes include:
S L Jaki, The Paradox of Olbers' Paradox (New York, 1969)
S L Jaki, The Milky Way (New York, 1972)
M A Hoskin, 'Newton, Providence and the Universe of stars',
     J. Hist. Astron., viii (1977), 77-101
M A Hoskin, 'The Cosmology of Thomas Wright of Durham',
     J. Hist. Astron., i (1970), 44-52
M A Hoskin, William Herschel and the Construction of the Heavens
     (London, 1963)
M A Hoskin, 'Apparatus and Ideas in Mid-nineteenth-century Cosmology',
     Vistas in Astronomy, ix (1968), 79-85
M A Hoskin, 'The English Background to the Cosmology of Wright and
     Herschel', in W.Yourgrau and A.D.Breck (eds), Cosmology, History
     and Theology (New York, 1977), 219-32.

Dr J.A. Bennett

THE GIANT REFLECTOR, 1770-1870

It was no accident that William Herschel pioneered two lines of development in astronomy, and both on a grand scale. One was to extend cosmology beyond the solar system and into the starry heavens, for so long simply a convenient backdrop for lunar or planetary studies, at least as regards empirically-based research. The other involved his determination 'to carry improvements in telescopes' - and here he was thinking of large reflectors - 'to their utmost extent'. Both subjects had seen some movement during the eighteenth century, but to inaugurate the serious study of what Herschel called 'The Construction of the Heavens' required a spirit sufficiently self-confident, perhaps sufficiently naive and untrained, to take on two audaciously expansive programmes at once. Herschel demonstrated the link between the two in a paper presented to the Royal Society in 1799, 'On the Power of Penetrating into Space by Telescopes'. A telescope's range, its 'space-penetrating power' in Herschel's terms, was directly related to the width of the objective aperture. For about 60 years it seemed that the extent of the known universe would be a function of the size of the largest useful telescope.

Almost from the beginning cosmological questions centred around the nature of the nebulae. Were they all distant clusters of stars, or did there exist luminous areas of fluid or gaseous material? If all nebulae were clusters and some sizable nebulae could not be resolved by powerful telescopes, then they must be large enough to be classed as star-systems independent of our own, other galaxies or 'island universes'. If, on the other hand, true nebulosity did exist, how could nearby nebulae be distinguished from distant clusters or galaxies?

Our concern is the telescopes, but it is important to realise that a compelling theoretical debate motivated much of the instrumental development. To resolve distant clusters into stars was a spur to building ever larger telescopes and theoretical advance was very much dependent on their improvement. Though the larger telescopes did not really clarify the issues, they did provide material for valuable speculation and discussion.

I want first to give a scale to our subject and an impression of the difficulties involved by outlining the achievements of the three foremost builders of giant reflectors in the period, William Herschel, William Parsons, third Earl of Rosse, and William Lassell.

The story begins with Herschel's so-called small 20ft Newtonian. His smaller telescopes, such as the famous 7ft Newtonian, were not really part of this development, since their design was a success from an early stage and continued unchanged for many years. The more delicate skills of grinding and polishing specula by hand could be best employed on a relatively small objective, thus giving good definition for planetary studies. But for nebulae the prime requirement was more light, and a larger aperture generally meant a longer focal length.

The small 20ft had an aperture of 12ins and was built in 1776. For the mounting Herschel's first idea was to imitate the traditional arrangement for long refractors by using a single pole and pulley system.

With a Newtonian, of course, he needed a ladder to reach the eyepiece, but even with this cumbersome arrangement he was able to have fine vertical and horizontal motions worked from the observer's station on the ladder. The telescope was naturally very susceptible to wind and difficult to manage by an uncomfortable and often very cold observer.

Herschel tried to build a similar 30ft with a 3ft speculum in 1781, but the casting, which was attempted in the basement of his house, was a spectacular failure. As we move on to consider larger telescopes, the efforts of their builders centre on three areas of difficulty. First there is the problem of casting a large mirror of an alloy that will take a high and lasting polish. Secondly they required a method of grinding and polishing a large and heavy speculum to an approximately parabolic figure, and thirdly a manageable mounting for a heavy speculum and a large tube. These are great technical problems, which were confronted by men who were amateurs working in a new field and relying largely on their individual resources.

Herschel's large 20ft, aperture 18.7ins, followed in 1783 and represents an advance in mounting and management. It began simply with the idea of a triangular framework to replace the single pole, giving a more stable support for the tube and carrying an observing platform. But from this Herschel developed the design slowly and carefully over several years. He realised that the mounting for the base of the tube could be incorporated into the framework and the whole structure carried on wheels and rotated to any azimuth. He experimented a great deal with the motions of the telescope and the methods of recording right ascension and polar distance, and he changed from the Newtonian construction to a front view in 1786. All of these developments took place gradually, by a process of trial and error, and were tried in situ rather than being designed before construction. The result was a very fine telescope and certainly the most productive that Herschel ever built. It became a model for all his subsequent large reflectors.

It seems to be a rule, which applies to all three of our principle builders with their enthusiasm for penetrating into space, that once a successful telescope has been proved, the next is of twice its dimensions. However a technology that has been developed for a smaller telescope is not necessarily successful for one much larger, and so it proved with Herschel's 40ft. built from 1785 to 1789

It was built on the model of the large 20ft, with a speculum of 4ft diameter weighing 19cwt. Improvements in the design included the accommodation for the amanuensis and workman, the speaking pipe for communication, the apparatus for keeping the tube in the same azimuth while sweeping and the support for the base of the tube. These and other modifications were mainly to accommodate the greater size and weight. This was to have been Herschel's greatest achievement in telescope building. As he said himself, all the developments made with the 20ft 'were at once, and with great advantage, designed and executed in their most improved state'.

The problems of the 40ft cannot be detailed here, but one basic difficulty concerned its management - it simply took too long to set up the telescope for observing. Then the speculum tarnished quickly, because Herschel had tried a higher than usual proportion of copper to increase its strength and because the large mass of metal attracted so much condensation. For practical purposes the 40ft was a failure.

One positive development however was that polishing such a large speculum by hand proved impossible. Up to 22 men were required to move

it on the polisher and their collective movements could not be directed
with sufficient precision. Thus Herschel was forced to devise polishing
machinery, which he subsequently used for objectives of all sizes.
Polishing could now be reduced to a mechanical rule, rather than depending on individual skill.

Herschel did not lose interest in building large telescopes, but he
never again attempted anything on the scale of the 40ft. For the King
of Spain he built a 25ft of 2ft aperture, which he erected at Slough and
which performed very well. It was in fact his largest successful telescope. He built a 10ft, also of 2ft aperture, for his own use, with
ease of management as its prime virtue. Also the large 20ft was later
refurbished and adapted for John Herschel. The limits of Herschel's
methods had been established and future developments would have to take
a different direction. Sir James South recorded that towards the end of
his life Herschel had told him that the reflecting telescope was only in
its infancy, but had stressed at the same time that John must never
consider rebuilding the 40ft.

The challenge of following Herschel was taken up by the Earl of Rosse
and in many technological respects Rosse had to start afresh. From the
beginning he strikes a different note from Herschel. Herschel, for
whatever reason, had not published his methods of making specula, but
Rosse clearly wanted to devise and make known techniques that anyone
could master, 'so that the art might be no longer a mystery, known to
but few individuals, and not to be acquired, but after many years of
laborious apprenticeship'.

He began in 1827 by making compound specula - first in an attempt to
reduce spherical aberration, but later as a way of building large
specula from small castings. After a great many experiments he eventually made a compound speculum of 3ft diameter and 27ft focal length.
The mounting for this telescope was in fact largely derived from Herschel,
as Rosse himself pointed out. Shortly afterwards he succeeded in casting
a solid objective of the same size and thereafter abandoned compound
specula.

Rosse's greatest achievement was probably in the successful casting
and machine finishing of very large specula. We cannot go into detail
here, but he paid great attention to techniques of casting and uniform
cooling. He introduced an annealing oven where a large speculum might
remain for up to 16 weeks. He devised and built machines for grinding
and polishing which were effective even with very large mirrors. And
all this work was done on his own estate in Ireland with the help of
previously untrained labourers. His greatest success was a speculum of
6ft diameter that weighed no less than 4 tons.

The mounting for this speculum, built 1842-4, was like nothing before
or since. Two massive walls were built in the meridian. They were 72ft
long, 56ft tall and 24ft apart. Between them the speculum, with its box
and 56ft tube, was mounted on a cast iron universal joint set in a stone
pier. The support for the speculum included an ingenious idea of Thomas
Grubb's used with both the 3ft and 6ft, whereby its weight was equally
distributed on to a number of levers, so that the speculum could maintain
its figure when in use and when being polished. The tube was supported
and moved by chains and, with the help of a clever system of counter-
poises, could be moved by one man. Workmen on the ground controlled
coarse motions in right ascension and declination, and the observer had
two corresponding fine motions.

The eastern wall carried a cast iron arc of 40ft radius, used to keep

555

the tube in the meridian or to take one end of the rackwork for the right ascension motion. For lower elevations the observation galleries were moved on staging in front of the tube; for higher the galleries were carried by the western wall and had of course to move out over the centre to follow the tube in right ascension.

The Rosse 6ft was a brilliant achievement and under the best conditions was capable of very fine work. But although the original intention was a programme of sweeping the heavens, it was never used in so systematic a way as the instruments of Herschel and Lassell. Climatic conditions of course were a serious limiting factor. In 1848 when Airy, the Astronomer Royal, visited Rosse, he wrote to his wife on one of the better days, 'The weather is still vexatious: but not absolutely repulsive'.

Rosse issued an open invitation to astronomers to come and use his telescopes and examine his methods of making them. One of those who accepted was the Liverpool brewer William Lassell, who had been experimenting with casting and polishing mirrors of up to 12ins diameter. He was already collaborating with the engineer James Nasmyth, but so far his polishing had been done by hand. In his observatory Lassell had a good 9in. speculum of 9.3ft focus mounted in a Newtonian telescope on an equatorial stand.

He visited Rosse in 1844 and what he saw greatly affected his subsequent work. The following year he was at work on a 2ft speculum of 20ft focus, and at first he adopted many of Rosse's ideas relating to casting and polishing. His first polishing machine was copied from Rosse's, but when it was not a success, he devised a new one and gave the principle to Nasmyth, who actually designed and constructed the machine. It was an elegant and effective piece of engineering. The mounting for the telescope was a larger version of his small equatorial, and this was the first time that a very large reflector was given an equatorial mount. Grubb had built a 15in. equatorial for Armagh Observatory in 1835.

The telescope was used at first near Liverpool, but after 1852 at Valetta in Malta. Here improved observing conditions persuaded Lassell to build a telescope of twice its dimensions, and an equatorial of 4ft aperture with a 37ft tube was built near Liverpool in 1858-60. It was dismantled and re-erected in Malta in 1862, and used mainly for observing nebulae.

Lassell's achievements were in using an equatorial mounting for the large reflector and, more generally, in applying advanced precision engineering to its construction. Here the role of Nasmyth was of the first importance. Lassell owed much of his success to Nasmyth's experience as an engineer.

Our account of the early history of the large reflector has been confined to Great Britain and Ireland, but we have not in fact neglected any major developments elsewhere. One exception would be experiments concerning silver-on-glass mirrors, but these telescopes were not so large and involved different kinds of problems. When William Huggins came to write the Royal Astronomical Society's obituary for Lassell in 1881, he said that his name would 'rank with those of Herschel and the late Lord Rosse in connection with that essentially British instrument, the reflecting telescope'. What did Huggins mean by describing the reflector as 'essentially British'? Was he referring simply to the pioneering work of Gregory and Newton and the fact that the reflector had indeed been largely developed in Britain, or does 'essential' imply something more significant? At any rate it is interesting to ask why the giant reflector was developed in one social context rather than another.

One obvious answer is that our story has coincided with the period of British technological supremacy, and the new technology was soon applied to the telescopes' construction. Each of our builders was interested in technological advance in general. William Herschel made trips north to Coventry, Birmingham, Manchester, Liverpool, Sheffield, Preston, Glasgow, and to judge by his travel diaries, his purpose was to learn of new technological developments. They are crammed with reports and drawings of ironworks, furnaces, steam engines and machines of all kinds. He visited Watt and Boulton and the Soho works and Wilkinson's works at Coalbrookdale; everywhere he was fascinated by manufacturing processes and machinery. Rosse, at one time President of the Mechaninal Section of the British Association, was perhaps really more of an engineer than an astronomer. The bulk of the actual observing with the Rosse telescopes was done by a series of assistants. Lassell was, of course, geographically well placed to take advantage of engineering advance. He cast his 2ft specula, for example, at a Liverpool foundry, using a blast-furnace to melt the copper. In Nasmyth he had the help of a first class engineer, who was engaged in the machine-tool industry and famous for his invention of the steam-hammer. To a lesser extent Grubb played a similar role for Rosse.

Techniques applied to building large reflectors paralleled more widespread developments in technology. Herschel's polishing machines were hand driven. He had tried using a horse, but without success. Rosse was able to employ a steam engine and so, of course, was Lassell. Through Nasmyth, Lassell was able to take advantage of developments in precision engineering, as seen in both his polishing machine and his mountings. He could use more metal castings in his mountings than had been done before, and he likened the principle of the framework for the 4ft to the Britannia Tubular Bridge. There has been much discussion among historians over whether science influenced the course of the Industrial Revolution; here at least is an area of astronomy where the influence acted in reverse.

In addition to a ready technology, there was perhaps a second and related social factor at work. A common characteristic of our telescope builders is that they were scientific amateurs, individuals applying their own resources to grand projects in a determined way. Builders of lesser reflectors such as Ramage, Nasmyth, who in his own right had a 20in. Cassegrain, and De La Rue fall into the same pattern. When in 1848 Airy lectured to the Royal Astronomical Society after examining the telescopes of Rosse and Lassell, he said that these instruments should interest the Society on two counts: 'first, that the reflecting telescope is exclusively a British instrument in its invention and improvement, and almost exclusively so in its use; and secondly, that it has been almost exclusively the instrument of amateurs'. The same year John Herschel presented the Royal Astronomical Society's gold medal to Lassell, saying that he felt, as he put it, 'a sort of hereditary fellow-feeling with Mr. Lassell, seeing that he belongs to that class of observers who have created their own instrumental means – who have felt their own wants, and supplied them in their own way'.

To build a large reflector, at least at this stage in its development, required single-minded determination and ability to meet new challenges and overcome unpredictable difficulties in novel and resourceful ways. It was a field where an individual with drive and vision was more likely to succeed than a government or an academic institution. These were just the qualities that were apparently making a success of industrial and commercial development in Britain, and in this sense our telescope builders, applying their energies to impressive technological projects, were the entrepreneurs of science. It was only at a later stage that governmental or commercial concerns would take over from the individuals. The first example of this was Grubb's Melbourne 4ft Cassegrain equatorial of 1867.

Perhaps Huggins felt correctly that these builders of giant reflectors had tapped energies characteristic of contemporary British society and applied them to the current instrumental needs of cosmology. It is ironic that it was he who had demonstrated with his spectroscope that the next step in answering the problem of the nebulae lay not in building larger and more impressive machines, but in a new technique based on quite a different principle.

SELECT BIBLIOGRAPHY

G.B. Airy, 'Substance of a Lecture ...', Monthly notices of the Royal Astronomical Society, ix (1848-9), 110-23.
   Autobiography of Sir George Biddell Airy, ed. W. Airy (Cambridge, 1896)
J.A. Bennett, '"On the Power of Penetrating into Space": the Telescopes of William Herschel', Journal for the history of astronomy, vii (1976), 75-108.
J.F.W. Herschel, 'Address delivered by the President ...', Monthly notices ..., ix (1848-9), 87-92.
W. Herschel, MS travel diaries, copies at the National Maritime Museum, London.
M.A. Hoskin, 'Apparatus and Ideas in Mid-nineteenth-century Cosmology', Vistas in astronomy, ix (1967), 79-85.
W. Lassell, Papers in Memoirs of the Royal Astronomical Society, xii (1842), xviii (1850), xxxvi (1867); announcements in Monthly notices, and MSS at the Royal Astronomical Society.
The scientific papers of William Parsons, third Earl of Rosse, 1800-1867, ed. C. Parsons (London, 1926)
T. Woods, The monster telescopes erected by the Earl of Rosse ..., 3rd edn (Parsonstown, 1845)

The paper was illustrated with slides. Necessary illustrations can be found in Bennett, op. cit., and H.C. King, The history of the telescope (London, 1955), ch. 10.

Prof. Jacques Merleau-Ponty

THEMES COSMOLOGIQUES CHEZ LES FONDATEURS DE LA THERMODYNAMIQUE CLASSIQUE.

Les remarques qui suivent se rapportent à une recherche plus générale dont il faut d'abord présenter l'intention et le but; savoir: comment l'idée de l'univers et le problème cosmologique apparaissent-ils dans les oeuvres scientifiques, et même plus généralement, rationnelles, du XIX° siècle?

Ce siècle occupe en effet une situation assez particulière dans l'histoire de la cosmologie. Alors qu'il fut extraordinairement fécond en inventions et découvertes scientifiques de toute sorte, il fut, après W.Herschel, peu productif en systèmes cosmologiques. Il n'a pas vraiment construit de cosmologie.

C'est au point que certains ont cru, dans la ligne du positivisme d'Auguste Comte, qu'il ne pouvait pas y avoir de cosmologie scientifique, que l'idée d'une théorie de l'univers appartenait à un âge révolu,"théologique" ou "métaphysique",de la pensée rationnelle.

Or la science du XX° siècle a démenti avec éclat cette conception, puisque depuis les années 1920, il existe une cosmologie scientifique, cette existence étant empiriquement attestée par la pratique quotidienne des institutions scientifiques.C'est donc une question intéressante pour l'histoire des idées de se demander ce qu'est devenue la cosmologie dans la science et la philosophie du siècle dernier et quelle place y a tenue le concept d'univers; en admettant, bien entendu, le postulat discutable d'une sorte de conservation des idées - comme si un concept qui n'est plus utilisé, ni même bien défini à une certaine époque, continuait cependant à vivre et à garder certaines virtualités quelque part dans le milieu social qui l'a produit.

Il est certain, d'abord,que les astronomes du XIX° siècle ont, en général, été peu portés à la spéculation sur l'univers et pour des raisons qui n'ont rien de mystérieux; d'abord en raison de l'état de l'observation qui a engendré ce que l'on pourrait appeler la "docta ignorantia" des astronomes: alors que les méthodes et les instruments faisaient des progrès considérables, le premier résultat de ces progrès fut de faire apparaître combien hasardeuse était toute tentative de description systématique du monde stellaire, a fortiori de l'univers entier; les astronomes du XVIII° devaient se borner à estimer les distances stellaires, et les plus hardis tiraient de ces estimations des hypothèses sur le système du monde; ceux du XIX° siècle, bien contents de mesurer avec certitude quelques distances ne se risquaient plus, sachant désormais qu'il y avait très loin de la coupe aux lèvres; ils firent de grands catalogues d'

étoiles et peu de systèmes du monde.
Il y avait aussi des raisons théoriques à cette abstention;
La fin du siècle des Lumières avait été marquée par le discrédit de la méthode métaphysique a priori conformément
à laquelle Wolff avait inventé le mot même de cosmologie.
Il est vrai que d'un autre côté, les étonnants succès de
la mécanique céleste permettaient d'ériger les axiomes de
la mécanique et la loi de la gravitation universelle en un
"vrai système du monde" -selon l'expression favorite de La
place, par opposition aux vaines constructions édifiées sur
l'incertain principe des causes finales.
Mais, en fait, autant la théorie newtonienne se montrait efficace pour l'analyse des mouvements individuels des astres
dans le système solaire, autant elle laissait indéterminés
les problèmes d'ordre systématique et encore plus de genèse
et d'évolution. Laplace était sans doute allé aussi loin
qu'on le pouvait à cet égard en utilisant la mécanique newtonienne pour esquisser une histoire du système solaire.
D'ailleurs si son oeuvre en mécanique céleste a connu des
développements considérables au XIX° siècle, son hypothèse cosmogonique fut souvent mentionnée, parfois critiquée,
mais peu développée et jamais remplacée jusqu'au XX° siècle. Les quelques rares recherches qui, selon les idées
de notre siècle, auraient pu relancer le débat cosmologique, n'ont guère eu de retentissement; ainsi du mémoire d'
Olbers "sur la transparence de l'espace cosmique; dans l'
abondante correspondance entre Olbers et Bessel, il est
mentionné une fois par son auteur, sans que Bessel ait même pris la peine de répondre à ce sujet.
En revanche, si elle est remarquablement absente, ou rare
chez les astronomes, la spéculation cosmologique, ou une
certaine forme de spéculation cosmologique, se retrouve là
où on l'attendrait moins, chez les physiciens.
Car, si l'on y regarde de près on voit que le développement
des sciences physiques est très loin, au moins dans certaines branches, d'être conforme aux normes de Comte pour
qui la physique, science plus spéciale que la mécanique,
confinée aux phénomènes observables sur la terre, était
encore plus éloignée de toute perspective sur la cosmologie. Les physiciens du XIX° siècle ne se gênaient pas, à
l'occasion, pour discourir sur l'univers, comme sur l'atome
et même sur Dieu; discours parfois étrangers, mais parfois
aussi étroitement liés, à leurs recherches physiques. Cela
est, je crois, particulièrement vrai de la thermodynamique
et c'est sur ce point que je voudrais m'arrêter en me bornant à la période de fondation, c'est-à-dire, grosso modo
entre 1824 et 1865.
L'intérêt cosmologique des fondateurs de la thermodynamique, Carnot déjà, Mayer, Helmholtz, W.Thomson, à un moin
dre degré Joule et Clausius, peut être envisagé sous deux
aspects qui se répondent et se complètent:
A - Leurs recherches sont pour une part guidée par quelques principes et concepts qui ne sont ni exactement

expérimentaux ni strictement mathématiques et dont certains sont manifestement en rapport avec l'idée de l'univers physique pris dans toute son extension.

B - Plusieurs d'entre eux ont immédiatement pensé à appliquer les lois qu'ils étaient en train d'établir à l'examen de quelques problèmes que nous dirions "astrophysiques" mais qui, pris dans le contexte, avaient manifestement une portée cosmologique·et qui étaient inaccessibles par les méthodes de la mécanique céleste. C'est le cas de Mayer, de Joule, de Helmholtz, de Thomson.

A - Les thèmes conceptuels dominants activement impliqués dans la recherche thermodynamique à ses débuts ont déjà été plus ou moins mis en évidence par les historiens, voici ceux qui nous intéressent particulièrement.

I - L'idée de l'interconnexion et de la convertibilité réciproque de toutes les"forces physiques" était devenue banale vers I830 et T.S.Kuhn a montre (I959) son rôle dans l' invention du premier principe; toutefois l'influence de cette idée sur l'évolution de la pensée cosmologique n'est qu'indirecte et de trop longs détours seraient nécessaires pour la mettre en évidence.

2 - un second concept directeur est, lui, de portée plus directement cosmologique; c'est celui de l'universalité, de l'originalité, et, à la limite, de l'irréductibilité des phénomènes thermiques (qu'il faut soigneusement distinguer du principe suivant lequel la chaleur serait une substance) Avant même la thermodynamique, ce concept est déjà activement présent, comme l'a noté D.S.L.Cardwell, dans la pensée du premier grand théoricien moderne de la chaleur, Joseph Fourier; c'est l'un des point sur lesquels il se séparait nettement des laplaciens Biot et Poisson comme de Laplace lui-même; la chaleur, aussi universelle que la gravitation, présente comme elle dans tout l'univers, requé rait selon Fourier une théorie mathématique sui generis indépendante de la mécanique, dont la première application fut le refroi dissement de la terre phénomène cosmique indépendant de toutes les actions mécaniques connues.

Cette idée se retrouve de façon décisive chez les fondateurs de la thermodynamique, ceux au moins qui ont découvert le second principe; chez Carnot par exemple, les deux idées initiales des "Réflexions sur la puissance motrice du feu" sont l'une politique: c'est la machine à vapeur qui est désormais l'instrument de la puissance des nations; l'autre est cosmologique: la chaleur est à l'oeuvre dans tous les grands phénomènes cosmiques et elle constitue un immense réservoir pour l'industrie humaine. C'est d'autre part le concept original et spécifique de "déséquilibre dans le calorique" qui sert de base à toute la théorie. Carnot, ce - pendant, reste mécaniste et ne se dégage pas complètement de l'analogie originale et efficace mais fausse à la rigueur entre machine hydraulique et machine thermique.

En revanche, chez Clausius, comme chez W.Thomson, l'énoncé

du second principe comporte une référence directe et exclu
sive à une propriété spécifique de la chaleur pour laquelle aucune analogie n'est proposée; c'était une nouveauté
considérable, car depuis le XVII° siècle ,il était généralement admis que les principes fondamentaux de la connaissance de la nature devaient concerner les propriétés géométriques et mécaniques des corps; Clausius n'était pas
très spontanément porté à la spéculation cosmologique, mais
Thomson donna immédiatement au second principe une dimension cosmique en en faisant l'expression d'une orientation
fondamentale et universelle du cours des choses
3 - Le troisième thème ou concept directeur, lié à l'impos
sibilité du mouvement perpétuel, concerne l'insertion de la
vie et de l'action humaine dans l'univers, mais par là même tend à modifier profondément le concept d'univers lui-
même.
L'horloge fut, comme on sait l'analogie technique priviligiée
pour la représentation de l'univers à l'époque des Lumières
Or, dans une cosmologie mécaniste, de type newtonien, entre l'horloge cosmique et les horloges humaines, la différence est , et est seulement, celle du parfait à l'impar-
fait; d'autre part, cosmologiquement parlant, on pouvait
raisonnablement concevoir l'univers comme un emboîtement
spatio-temporel de systèmes périodiques; c'est ainsi que
Lambert imaginait le monde et sur le plan strictement scien
tifique, cela correspondait assez bien aux résultats de Laplace sur les mouvements dans le système solaire: que les
"perturbations" des mouvements périodiques simples des pla
nètes et des satellites sont elles-mêmes périodiques.
Mais les lois de la thermodynamique telles que les conçu-
rent dès le début leurs inventeurs rendaient très clairement illusoires les ambitions de toute mécanique humaine
au-delà de certaines limites que ces lois rendaient déter-
minables indépendamment de toute référence aux propriétés
particulières et contingentes de l'être humain. Ils impli-
quaient le déréglement inévitable de tout système périodique.
Particulièrement significatif est à cet égard le témoignage
de Helmholtz, parce que de tous c'était le plus proche, sur
le plan physico-théorique, du mécanisme newtonien, et qu'il
n'a d'autre part jamais montré un très grand intérêt pour
le second principe. Pourtant il avait très clairement compris qu'une théorie de l'univers, dans la mesure où elle
est énergétique,entraîne des conséquences très précises
sur l'insertion du vivant en général, et de l'homofaber
en particulier: dans une conférence où il expose à un public scientifique mais non spécialisé (1854), les éléments
de la nouvelle science, il commence son exposé en évoquant
la mémoire desdeux Droz, père et fils, Jean-Jacques et Hen-
ri-Louis, ces deux horlogers de la Chaux de Fonds, qui s'é-
taient rendus célèbres, le fils surtout, par leur habileté
à construire des automates; selon Helmholtz, les Droz se

sont laissés prendre (il les en soupçonne en tout cas) à la chimère du mouvement perpétuel dont l'impossibilité est un des axiomes de la nouvelle science. Que ce soit vrai ou faux, ce que Helmholtz néglige de souligner, c'est que le mouvement perpétuel n'avait pas à l'époque des Droz le sens qu'il lui donne et que c'est précisément ce changement de sens qui ouvre une nouvelle perspective dans la conception de l'univers. Car, dans le contexte du mécanisme des Lumières, il n'est pas absurde que l'univers lui-même soit un <u>perpetuum mobile</u> au premier sens; alors que dans le contexte thermodynamique, c'est le second sens qui importe et, une fois ce sens fixé, l'axiome d'impossibilité non seulement fixe des limites infranchissables à toute action technique locale, mais il interdit toute inférence du technologique au cosmologique: l'univers ne peut être, en aucune façon, l'analogue d'une machine.

B - Après avoir examiné dans quelle mesure le motif cosmologique a pu intervenir dans le processus de découverte de la thermodynamique, il reste à confirmer son importance en montrant comment les lois nouvellement établies ont fourni un instrument permettant au moins de poser, en des termes scientifiquement acceptables, des problèmes de portée cosmologique inaccessibles par d'autres voies.

La priorité sous ce rapport revient apparemment à R.J.Mayer qui soumit à l'Académie française des Sciences un mémoire "Sur la production de la lumière et de la chaleur du soleil", dans lequel l'origine de la chaleur solaire était attribuée à une chute incessante d'astéroïdes dont l'énergie cinétique était transformée en chaleur suivant la loi de l'équivalence et Mayer n'hésitait pas à généraliser cette explication au rayonnement des étoiles et de toutes les sources de lumière dans l'univers. L'académie ne publia pas ce mémoire, pas plus qu'un second (peut-être le même) dont il est question dans les compte-rendus de septembre, 1848, ce qui semble indiquer que la communauté scientifique française était encore peu accueillante aux idées énergétiques; c'est seulement deux ans plus tard que Mayer publia sa théorie, considérablement développée dans ses "Beiträge zur Dynamik des Himmels" (le premier mémoire ne fut publié qu'après sa mort). Mais entre temps, Joule avait publié dans le Philosophical magazine son article "On shooting stars"; cet aspect secondaire de la fameuse querelle Joule-Mayer prouve cependant que les deux grands inventeurs de la loi d'équivalence avaient immédiatement pensé l'un et l'autre à ses applications cosmologiques.

Au sujet de ce mémoire de Joule on notera seulement que de la part d'un praticien de la méthode expérimentale si justement célèbre pour son habileté et sa rigueur méthodologique, le mémoire est remarquablement déductif: Joule se donne un "modèle" plausible de météore moyen et en déduit l'ordre de la grandeur de la chaleur produite par le freinage atmosphérique; chaleur si énorme qu'un centième suf-

fit à plulvériser la météorite et Joule conclut en remerciant
l'auteur de la nature d'avoir protégé ses créatures terres-
tres d'un bombardement aussi redoutable; dès ce premier essai apparaissent déjà l'originalité et les avantages des rai
sonnements énergétiques.
Les Beiträge zur Dynamik des Himmels sont un ouvrage très
développé sur lequel je dois me borner à quelques indications; d'abord Mayer associe explicitement la question de
l'origine de la chaleur solaire à celle de la formation du
système planétaire; les corps qui tombent incessamment sur
le soleil sont des restes de cette matière originaire, de
cette Urmaterie dont ont été formées les planètes; de tous
les problèmes -finalement insolubles - que soulèvent son hy
pothèse Mayer en mentionne un: la constance de la masse solaire qui semble bien attestée par les résultats de la mécanique céleste; il y répond de façon intéressante du point
de vue de l'histoire des idées énergétiques en cosmologie:
Le soleil perd par rayonnement ce qu'il gagne par accrétion;
cela dit Mayer est vrai même dans une théorie ondulatoire de
la lumière, car il y a de toutes façons une "action centri-
fuge" du rayonnement; autrement dit, en termes modernes, le
rayonnement est porteur d'une énergie transformable en éner
gie cinétique. En reliant la question de la chaleur solaire
à celle de l'origine du système solaire Mayer ouvre une
voie dans laquelle Laplace n'avait pu s'engager; car s'il
parle du "refroidissement" de la nébuleuse originaire, ce
refroidissement n'est l'objet d'aucune estimation ni qualitative ni quantitative et il ne joue aucun rôle dans le processus cosmogonique qui reste chez Laplace purement mécanique.
Dans l'histoire des théories de la chaleur solaire, la contribution de Helmholtz est en général considérée comme plus
importante que celle de Mayer mais nous la laisserons de cô
té parce qu'elle est plus tardive: la première étude de
Helmholtz sur ce sujet date de 1871, et n'appartient plus
ni par sa date ni par son contenu à la période que nous considérons.
En revanche, dans ces démarches d'exploitation cosmologique
de la thermodynamique, l'intervention de W.Thomson fut à la
fois très précoce et très originale par rapport à toutes les
autres. A vrai dire l'intérêt de Thomson pour les questions
cosmologiques a commencé dès le début de sa carrière, selon
les deux récentes études de Crosby Smith, et n'a jamais
cessé; c'est donc un peu arbitrairement que nous avons iso-
lé quelques mémoires (1854, 1859, 1862), qui ne sont que la
phase initiale d'une suite presqu'ininterrompue de recherches et de réflexions. En voici les éléments les plus frap-
pants du point de vue de la présente étude:
Comme ses prédécesseurs et contemporains dans l'invention
de la thermodynamique, il voit dans la théorie de l'accrétion la seule solution plausible du problème de la chaleur

solaire, mais il en voit plus clairement les difficultés;
ce qui est particulièrement intéressant pour notre propos
c'est qu'il pose une question implicitement écartée par
Mayer qui n'en parle même pas   ', savoir s'
il faut considérer l'état thermique du soleil comme sta-
tionnaire; car dit Thomson, le soleil se réchauffe et se
refroidit; lequel des deux processus l'emporte sur l'autre? il n'en sait rien, mais - en fait sans donner de rai-
son- il opine pour le refroidissement et l'on voit par là
et par bien d'autres indices, que le problème central, vers
la solution duquel tend toute cette ébauche de physique du
soleil, c'est l'âge de l'astre, et l'ordre de grandeur temporel de son évolution, dans des limites compatibles avec
les conditions prévalant actuellement sur la terre. Son propos est donc cosmogonique; sa conclusion en 1862 (il ne la
modifiera d'ailleurs pas essentiellement par la suite) est
que le soleil n'a pas éclairé la terre pendant plus cinq
cent millions d'années et probablement pas plus de cent mil
lions, à moins, ajoute-t-il d'une façon qu'on peut considérer comme prophétique, "que des sources inconnues   soient
préparées dans le grand réservoir de la création". Et déjà
s'amorce le différend qui l'opposera pendant des années aux
géologues et aux darwiniens, car il note que ses calculs le
mènent très loin des trois cent millions d'années que Darwin avait estimé nécessaires à la dénudation du Weald.

De ces mémoires ressortent nettement en tout cas les idées
directrices qui guident Thomson dans ses considérations
cosmologiques et qui annoncent des orientations importantes
de la théorie de l'univers aux XIX° et XX° siècles.
W.Thomson est frappé par la dyssymétrie temporelle qu'intro-
duisent les phénomènes thermiques dans le cours de la nature et, parallèlement par la situation particulière dans
lesquelles se trouvent les inférences que l'on peut faire
vers le passé de l'univers, en raison de cette dyssymétrie.

Il faut se souvenir que dans l'une de ses premières publications à une époque où il ne connaissait probablement pas
le mémoire de Carnot, Thomson avait eu l'idée - qui semble-
t-il n'était jamais venue à Fourier lui-même, d'utiliser l'
équation de conduction de la chaleur pour reconstituer le
passé thermique de la terre; pour des raisons tenant à la
structure des équations, le problème est bien plus compli
qué que le problème inverse et peut conduire à une extrême variété de solutions, suivant la distribution initiale
(ou plutôt finale) que l'on se donne.
Cette notion de la dyssymétrie temporelle était insépara
ble de sa conception du second principe; très vite dès
1852 Thomson découvrit l'implication cosmologique du second principe et le décrivit comme une loi de dissipation
qui marque l'orientation du cours spontané de la nature;
à cette époque, Clausius était indifférent à cet aspect
temporel du second principe; c'est seulement dix ans plus
tard qu'à la suite de sa controverse avec Rankine sur l'
application du second principe au rayonnement   il de -

vait écrire ses célèbres phrases: "L'énergie de l'univers est
constante. L'entropie de l'univers tend vers un maximum."
Dans les mémoires cosmoogiques de Thomson auxquels nous
nous référons, le second principe intervient soit implicitement, soit explicitement (alors que Mayer et Joule l'i-
gnorent complètement, et que Helmholtz ne s'y réfère guère)
D'autre part la question de la possibilité et de la sécuri-
té des inférences vers le passé de l'univers est posée par
Thomson en des termes intéressants quoique parfois équivo-
ques parce que s'y mêlent trois éléments tous importants
mais imparfaitement conciliés et conciliables: a - un sché
ma de pensée mécaniste auquel Thomson, comme tous les savants de son époque est habitué et reste très attaché: les
lois de la nature fondent les inférences dans les deux sens
et rien ne peut arrêter l'inférence scientifique vers le
passé; b - la dyssymétrie proprement thermodynamique: l'é-
nergie de gravitation transformée en chaleur ne se restaure
pas; si l'on veut trouver là l'origine de l'énergie rayon-
nante dans l'univers il faut éloigner les corps les uns
des autres à l'infini à mesure qu'on recule dans le passé;
c - et enfin la conception théologique de Thomson pour qui
la présence d'un dessein divin dans les êtres vivants est
une sorte d'évidence qui renvoie à un acte créateur .

Norriss S. Hetherington

OBSERVATIONAL COSMOLOGY IN THE TWENTIETH CENTURY

Two fundamental developments in cosmology during the first half of the twentieth century were the determination of the distances of spiral nebulae, along with which came the proof that they are galaxies lying at tremendous distances from us, and the discovery that the universe is not static but rather is expanding. Both of these major advances in our knowledge of the structure and the evolution of the universe came about in large part because of new instruments and new observations.

THE STRUCTURE OF THE UNIVERSE

At the beginning of the twentieth century, the spiral nebulae had become the focus of attention of observational cosmology. Were they stellar systems comparable to our own galactic system and at tremendous distances beyond the confines of our own galaxy, or were they merely appendages of our galaxy? The eighteenth and nineteenth centuries had witnessed several swings of opinion on this question, focusing on whether nebulae were gaseous or stellar, with each change of ideas usually following from an innovation in observational equipment, as Michael Hoskin has shown.

William Herschel's initial successes with his large reflecting telescopes suggested that all nebulae were resolvable into stars, but in 1790 he encountered a planetary nebula which defied resolution. Astronomers began to doubt that all nebulae were composed of stars and were hence island universes. Even the Andromeda nebula (now known as the nearest galaxy) was pointed to as an example of a mass of nebulous matter definitely not a stellar system! Larger reflecting telescopes constructed by the Earl of Rosse around mid-century increased the number of nebulae resolved into stars and opinion once again swung toward the stellar composition of all nebulae. Even the Orion nebula was allegedly resolved into stars! The end of the nineteenth century, though, saw opinion once more set against the spiral nebulae as stellar systems when the spectroscopic observations of William Huggins revealed that about one-third of some seventy nebulae displayed a gaseous character.

Astronomical Spectroscopy and Radial Velocities of Spiral Nebulae

The final swing of opinion on the nature of the spiral nebulae came early in the twentieth century with one of the more significant observations of modern cosmology: the detection of very high radial velocities of spiral nebulae.

Huggins had applied the new science of spectroscopy not only to the constitution of celestial objects, but also to the measurement of motions of celestial objects. Christian Doppler and Hippolyte Fizeau had predicted a shift of the spectral lines from moving objects, and Huggins found the predicted shift for a few bright stars and gaseous nebulae. He was, however, unable to extend his results to the spiral nebulae because of instrumental limitations.

Astronomical entrepreneurship in America's gilded age led to the construction of new and larger instruments and a shift of the center of spectroscopic research from Britain to the United States at the end of the nineteenth century. For example, the Lick Observatory, constructed in California in the 1880's, possessed a 36-inch refractor (Huggins had used his 15-inch refractor and an 18-inch speculum reflector borrowed from the Royal Society) and a diffraction grating (giving a dispersion about 24 times greater than a prisim, though Huggins did obtain a grating in 1893). The Lick Observatory chose to concentrate its attention on the spectra of planetary nebulae which, being bright emission line spectra, were easily obtained in large numbers with relatively short exposure times.

The more laboriously won and consequently fewer observations of the dark absorption line spectra of spiral nebulae came from another of the new American observatories appearing at the end of the nineteenth century: the Lowell Observatory. Founded in 1894 by Percival Lowell, a wealthy Boston investor, the Lowell Observatory's major concern was an examination of the surface of Mars for signs of life. However, Lowell suspected that spectral lines as yet unidentified in the spectra of planets might also be found in the spectra of spiral nebulae, and in 1909 he asked Vesto M. Slipher to obtain spectra of spiral nebulae. Slipher initially saw little hope of obtaining spectra of the faint nebulae, but later he examined the question of what instrumental improvements were needed. He came to realize that for spiral nebulae with extended surfaces, the crucial instrumental factor was not the size of the telescope nor the power of dispersion, but simply the speed of the camera. He increased that, and in 1912 he obtained a spectrogram of the Andromeda nebula of sufficient quality that the velocity could be measured.

The radial velocities of the Andromeda nebula and of other spiral nebulae observed by Slipher were extraordinarily high. Since all galactic objects were then thought to have low radial velocities, the peculiarly high velocities suggested to many astronomers that the spiral nebulae were extragalactic objects. Edwin Hubble wrote in 1917 that "the great spirals, with their enormous radial velocities and insensible proper motions, apparently lie outside our system." The evidence of high radial velocities was not conclusive proof, though, that the spiral nebulae lay beyond the boundaries of our galaxy, especially when a few stars within our galaxy were found to have high radial velocities.

Astronomical Photography and Novae in Spiral Nebulae

Another observation which commanded a significant and also ambiguous role in cosmological debates of the early twentieth century was the discovery of novae in spiral nebulae. Novae and other special types of stars were potential distance indicators, their distance to be estimated from a comparison of their observed brightness with their expected luminosity.

In 1885, a new bright star appeared suddenly in the Andromeda nebula, attracting the attention of the astronomical world and furnishing an estimate of the distance of the nebula. To confuse matters, this was a supernova of tremendous, indeed unbelievable luminosity, far exceeding that of any known star. The general conclusion was that the nova was a relatively near star of reasonable luminosity rather than a very distant star of incredible luminosity. Were the Andromeda nebula a galaxy beyond our own and comparable in size, it was calculated that the nova would have to have a luminosity nearly fifty million times greater than that of our sun.

The supernova of 1885 and also one of 1895 initially seemed to contradict the belief in the spiral nebulae as galaxies. Not until 1917 when novae of lower apparent luminosity (ordinary novae in the Andromeda nebula and supernovae in more distant spiral nebulae) were discovered did the novae begin to furnish support for believers in the island universe theory.

The observations of novae in 1917 were made possible by the new tool of astronomical photography which, along with astronomical spectroscopy, was one of the important new observational tools of astronomy emerging at the end of the nineteenth century. Fixing an image by chemical means began with the work of Louis Daguerre and others. A few photographs of the brightest stars were taken as early as 1850, but the field of astronomical photography really began to develop after 1880 and the introduction of sensitive dry plates. In 1880, Henry Draper in the United States took a photograph of the Orion nebula; Andrew Ainslie Common in England obtained a photograph of the same nebula two years later; and Isaac Roberts began in 1883 an ambitious program to photograph star clusters and nebulae. It is interesting to note that amateur astronomers pioneered the development of the new tool of astronomical photography, much as was also the case with astronomical spectroscopy. Perhaps professional astronomers were too occupied with already established research programs using traditional methods.

The photographic reflecting telescope was soon to become a basic instrument of astronomical research, thanks in no small measure to the efforts of George Willis Ritchey. His family fled Ireland during the potato famine of the mid-nineteenth century and settled in Cincinnati where Ritchey became an assistant at the observatory. He next moved to Chicago where he met George Ellery Hale and became superintendent of the optical shops and instrument construction at the Yerkes Observatory. Ritchey followed Hale to California where the Carnegie Institution established the Mount Wilson Observatory. Ritchey was in charge of the construction of the 60-inch reflecting telescope which was designed specifically for astronomical photography. This instrument, completed in 1908, and the 100-inch completed in 1919 made possible observations which were to revolutionize twentieth-century astronomy.

Although not the most important observation, nor even one which absolutely required the new photographic telescopes, the discovery in 1917 of more novae in spiral nebulae did play a role in debates over the structure of the universe. Ritchey found on a photograph taken of the spiral nebulae NGC 6946 on July 19, 1919, a faint star not present on a plate taken almost a month before, on June 25. Independently of Ritchey, Heber D. Curtis in the course of his nebular photography program at the Lick Observatory was comparing 1901 and 1915 plates of the spiral nebula NGC 4527, and he discovered in March of 1917 a new star. On a subsequent search of old plates, Curtis found two novae in the spiral NGC 4321. More novae were found by Francis Pease and by Harlow Shapley on Mount Wilson Observatory plates.

Curtis believed that the occurrence of novae in spiral nebulae favored the island universe theory because novae are a galactic phenomena (26 had been found in our galaxy). Excluding the novae of 1885 and 1895, Curtis found an average maximum brightness for galactic novae of 5 compared to 15 for novae in spiral nebulae (novae in the Andromeda nebula and supernovae in more distant galaxies). Assuming equality of absolute magnitude for galactic and spiral novae, the novae in spiral nebulae and hence the nebulae themselves were 100 times as distant as novae in our galaxy and thus far outside the bounds of our stellar system. Harlow Shapley came to a similar conclusion regarding the distances of novae and seemed almost willing to believe that novae attained a most extraordinary absolute brightness.

Controversy Over Internal Motions in Spiral Nebulae

Shapley's tentative acceptance of the island universe theory was very short lived, though, for within two years he was to become the champion of the other side. The observational report primarily responsible for his conversion was Adriaan van Maanen's claim (subsequently disproved) to have measured internal motion (which was generally interpreted as rotation) in a spiral nebula.

The rotation of spiral nebulae was widely accepted. In the eighteenth century Immanuel Kant had realized that the arrangement of stellar systems might well be similar to that of the solar system with its planets circling the center, and Simon Laplace's nebular hypothesis gave rotation, at least by implication, to the nebulae. In the nineteenth century, the Earl of Rosse's observation of the spiral structure of M 51 led to his conclusion "that such a system should exist without internal movement, seems to be in the highest degree improbable: we may possibly aid our conceptions by coupling with the idea of motion that of a resisting medium; but we cannot regard such a system in any way as a case of mere statical equilibrium." Early in the twentieth century, T.C.Chamberlin's planetesimal hypothesis called for rotation of spiral nebulae, a phenomenon also suggested by the appearance of the spirals in photographs taken by James Keeler at the Lick Observatory. Finally, the new tool of spectroscopy in the hands of Slipher and others furnished conclusive proof of the rotation of spiral nebulae, though not necessarily large enough motion to detect from a comparison of photographs taken several years apart.

The new tool of photography also held out promise of confirmation of the speculations concerning rotation. Late in the nineteenth century, Isaac Roberts compared a photograph with earlier visual reports and briefly thought he had detected rotation of M 51, before he was corrected. In the twentieth century, astronomers at all three of the new mountain observatories in the American West, Mount Wilson, Lick, and Lowell, attempted to detect the rotation of spiral nebulae from a comparison of photographic plates. Chamberlin had visited Mount Wilson in 1915, and he may have had some influence on the research program there. Whatever the motivation, Ritchey asked van Maanen to compare for signs of rotation photographs of M 101 taken by Ritchey. At the same time, Curtis at the Lick Observatory and Carl Lampland at the Lowell Observatory were engaged in similar enterprises. The independent and simultaneous efforts emphasize that the time was ripe early in the twentieth century for the attempted photographic detection of the rotation of spiral nebulae.

Ritchey had devised a moveable plate holder and fast-acting camera shutters which made it possible to adjust for tracking errors and to build up many short exposures taken at moments of little atmospheric fluctuations. Using Ritchey's plates as well as some from the Lick Observatory, van Maanen reported the detection of internal motions in the spiral nebula M 101. His result was seemingly confirmed at other observatories (though van Maanen improperly claimed corroboration in some instances), and he seemed able to repeat his success on other spiral nebulae. Many astronomers accepted his work, more so in Britain than in the United States. Van Maanen's observations initially seemed compatible with British theoretical work, while observational evidence incompatible with his results was beginning to be found at the new American observatories.

The major questioning of van Maanen's results came from astronomers objecting to the import of the claimed observations upon the island universe theory. Shapley pointed out that if M 101 were comparable in size with our galaxy and at a great distance beyond the boundaries of our galaxy, then its outer edge would have to be moving at an impossibly great speed, greater than that of light. Curtis replied with doubts of van Maanen's measurement. There was some private criticism of van Maanen's work, but it was well accepted by many astronomers and stood until 1924 as the major discrepancy in an otherwise consistent picture of the spiral nebulae as island universes.

The Period-Luminosity Relation and Distance Determinations

The long debate over whether the spiral nebulae are or are not galaxies finally came to an end in 1924. The decisive observation was the discovery of Cepheid variable stars in spiral nebulae and the subsequent distance calculation using the period-luminosity relation.

The period-luminosity relation for Cepheid variables was established largely by the work of Harlow Shapley. Early in the twentieth century, distances to stars were known for only a few nearby stars, and direct parallax measurements were limited by instrumental capability to a few hundred of the closest stars. Statistical parallaxes extended distance measuring techniques far enough to reach approximately a dozen Cepheid variable stars in our galaxy. Henrietta Leavitt studying variable stars at the Harvard College Observatory noticed that the brighter variables in the Small Magellanic Cloud have longer periods. Ejnar Hertzsprung at the Potsdam Observatory made the first attempt to calibrate the relationship on nearby Cepheids with distances estimated from statistical parallaxes. Shapley repeated Hertzsprung's determination and combined the results with Leavitt's data for the Small Magellanic Cloud and with all other available data. Astronomy now had a most powerful tool; one could measure the easily observed period of a distant Cepheid variable star, assume it is similar to nearby Cepheids (there is a difference, changing the calculated distance by a factor of two), calculate its luminosity from the period-luminosity relation, and then derive its distance by comparing the calculated luminosity with the apparent brightness (which is reduced proportionally as the distance increases).

The first application of the new tool was by Shapley at the Mount Wilson Observatory, and his subject of investigation was the size of our galaxy. He found Cepheids in globular clusters, assumed that the globular clusters form the skeleton of our galaxy, and thus had the size of the galaxy. His figure was approximately ten times larger than previous estimates. Somewhat ironically, the results of the first use of the period-luminosity relation seemed to argue against the theory of the spiral nebulae as island universes. With our galaxy so large, the spiral nebulae to be outside it would have to be at even greater distances than previously estimated. And if they were comparable in size to our galaxy, then van Maanen's rotation would be impossibly fast.

Shapley left Mount Wilson for the directorship of the Harvard Observatory, and seems to have lost the opportunity to extend his distance determinations to the spiral nebulae. Maybe he did not believe that individual stars were being resolved in photographs of the nebulae, because

Ritchey had told him that he thought the observed condensations were not actual stars but maybe stars in the process of formation. The opportunity was left for Edwin Hubble to grasp.

At the Mount Wilson Observatory in 1921, Hubble had John Duncan take photographs of the irregular nebula NGC 6822 with the 100-inch telescope. By 1924 they had more than 50 plates and 11 Cepheids. The distance was estimated to be 700,000 light years, far beyond anyone's limits of our galaxy. At the same time that NGC 6822 was providing a successful test of the use of the period-luminosity relation, Hubble moved on to the spiral nebulae. He found that they too are clearly at great distances outside our galaxy.

THE EVOLUTION OF THE UNIVERSE

Hubble's discovery of Cepheid variables in spiral nebulae very quickly settled the centuries old debate over the extragalactic nature of the spiral nebulae. At the same time that he was answering one question, Hubble was also rendering a whole new field of astronomy susceptible to empirical investigation. It is with Hubble's work, particularly his demonstration of the expansion of the universe, that cosmology became an observational science.

The Establishment of the Velocity-Distance Relation

Once Hubble had determined the distances of several spiral nebulae, he was able to test and then demonstrate the existence of a most important relation between the velocity of objects and their distances.

In 1917, the Dutch astronomer Willem de Sitter had discovered a second static solution to Einstein's field equations which gave a model of the universe in which there was a predicted increase of the apparent radial velocity of objects at greater distances from the observer. By 1920, de Sitter knew of radial velocities for some twenty-five spiral nebulae, mostly determined by Slipher, but he did not have the other half of the relationship: the distances. An ill-considered attempt by another astronomer in 1924 to test the relation on globular clusters for which distances were known was done improperly, and the claimed positive results were quickly rejected by others. A conclusive test awaited the determination of distances of spiral nebulae, because the globular clusters might not be distant enough to display a measurable velocity-distance relation.

Hubble's distance estimates for some 24 extragalactic nebulae in 1929 made possible a re-examination of the question of a velocity-distance relation. Hubble had distances for five nebulae (and a sixth companion) from Cepheid variables. From these nebulae, he calibrated the magnitude of the brightest star in a nebula and used that to estimate the distances of fourteen more nebulae. Then he used these twenty to calibrate the average magnitude of nebulae, and used that figure to estimate the distances of four more nebulae. With distances for 24 of the 46 extragalatic nebulae for which radial velocities had been determined, Hubble was able to show that the data indicated a linear correlation between distances and apparent velocity displacements. Hubble had already had Milton Humason measuring radial velocities of faint nebulae, and in 1929 the two embarked on a plan to test the velocity-distance relation over as great a range in distance as the 100-inch telescope at Mount Wilson would allow. (For the nearer spiral nebulae with extended surfaces, the size of the telescope had not been a factor; but once the radial velocities of nearby spiral nebulae had been measured, the lead in this endeavor passed from Slipher to Hubble and Humason at the Mount Wilson Observatory where they measured spectra of nebulae which were little more than point sources.) A 1931 paper by Hubble and Humason further established the existence of an empirical velocity-distance relation.

The Interpretation of the Velocity-Distance Relation

Hubble's estimate of an empirical velocity-distance relation coupled with J.H.Oort's estimate of the mass of our galaxy led to a crisis in astronomical theory. Prior to 1930, there had been two possible cosmological models, Einstein's system A and de Sitter's system B. Hubble's observations ruled out system A, and Oort's work, extended by de Sitter, ruled out system B. Arthur Eddington realized that the problem was that scientists had been only looking for static solutions. Actually, a few expanding models had been proposed during the 1920's, but not until 1930 and the crisis created by the realization that neither Einstein's nor de Sitter's models were compatible with observations, did astronomers recognize the necessity of finding an alternative model of the universe.

There were soon three alternative cosmological models to choose from, all with an observed relation between velocity and red shift. One was that of the Belgian George Lemaitre, who proposed an expanding universe in which the apparent Doppler effect due to the variation of the radius of the universe was approximately proportional to the distance of the object. A simpler alternative to Lemaitre's expanding space was proposed by the English theoretician E.A.Milne. His purely kinematic theory began with a concentrated density distribution and random motions. After some time, the faster moving particles would be at greater distances from the central core, at distances approximately proportional to their velocities. A very different interpretation of the observed red shifts was also possible, as Fritz Zwicky at the California Institute of Technology pointed out. A gravitational drag on light, vaguely analogous to the Compton effect, would give an increasing interaction for light quanta from galaxies at increasingly greater distances, and thus the appearance of a velocity-distance relation.

Given the proliferation of new cosmological models, Hubble determined to choose between the different models on the basis of observational data. He saw mathematicians dealing with an infinite number of logically consistent worlds, theoreticians segregating out those worlds compatible with known observations, and observers like himself then developing new factual information to further reduce the number of possible worlds.

In collaboration with Richard C. Tolman at the California Institute of Technology, Hubble devised observational criteria to distinguish between the models of Lemaitre, Milne, and Zwicky. The predicted number of nebulae to successive limits of apparent magnitude was different for each model, and a choice between the three models was quite simple in principle. There were, though, practical difficulties, such as accurately determining the limiting magnitude and the threshold of identification, of counting every nebular image on a photographic plate at and above the threshold of identification, and of correcting for differing observing conditions.

Hubble's main observational work during the 1930's consisted of surveys of nebulae to increasingly fainter limiting magnitudes made with the 100-inch telescope at Mount Wilson. With this data, he hoped to show that Lemaitre's theory was to be preferred. When the initial data pointed instead towards Zwicky's model, Hubble concluded that more data was needed before one could confidently choose between the different cosmological models. The search for new data at increasingly greater distances characterizes twentieth-century observational cosmology.

The Advent of Radio Astronomy

Hubble could confidently expect new data from greater and greater distances when the 200-inch telescope under construction at Mount Wilson was completed in the late 1940's. The unexpected development in observational capability was the development of radio astronomy following from war-time work on radar. Radio observations opened up a new part of the electro-magnetic spectrum and made possible surveys to greater limiting magnitudes than is possible with telescopes using the optical portion of the spectrum.

Radio waves were demonstrated by Heinrich Hertz in 1888, and there were several early attempts to detect radio waves from the sun, by Thomas Edison in 1890, by Sir Oliver Lodge in 1894, and by Karl Jansky with Bell Telephone in 1932. Jansky did detect radio waves, not from the sun but from the center of our galaxy, as did Grote Reber, also in the United States. War-time research on radar set the stage for the rapid development of radio astronomy in Britain and in Australia immediately after the war. In the Netherlands, reports of Reber's work reached Leiden during the war, and Oort's interest in the structure of the galaxy and H.C.van de Hulst's prediction of the 21 centimeter hydrogen line gave the Netherlands also a prominent place in the new field of radio astronomy. The development of radio astronomy seems similar to the earlier advances in astronomical spectroscopy and photography in that amateurs played prominent roles in establishing the observational techniques and demonstrating their worth, and in that scientific predominance in astronomy shifted among nations with the development of new fields and techniques of research.

Radio astronomy has contributed importantly to the choice between cosmological models, which became in the late 1940's and after a choice between the expanding model and the newly-proposed steady-state model. The number of radio sources at different flux levels and the existence of a radiation field are two most important datum of modern cosmology, both of which seem to favor the expanding model.

Twentieth-century cosmology has become very complex in terms of theory and in the interpretation of observations in the light of different theories, and the traditional distinction between theory and observation seems less clear than formerly. In a most stimulating paper, Hermann Bondi has even argued that an examination of the relative reliabilities of theory and observation shows that errors in theory are less frequent than in observational work. With this in mind, and with a reluctance to move from the history of observational cosmology in the first half of the twentieth century to predictions for the second half of the century, let us turn to Professor Whitrow's discussion of theoretical cosmology in the twentieth century.

BIBLIOGRAPHIC NOTE

The works of Michael Hoskin constitute a very good history of modern cosmology. The best single sources are G.J.Whitrow's The Structure and Evolution of the Universe and Man Discovers the Galaxies by Richard Berendzen, Richard Hart, and Daniel Seeley. The latter volume appeared

after this review was written. Papers by Berendzen and Hart, by J.D.
Fernie, and by Deborah Jean Warner provide valuable supplements on a
variety of topics.

ACKNOWLEDGMENT

This paper was written while holding Faculty Senate Research Award
3740-x038 from the University of Kansas. The University subsequently
refused to honor the grant.

Professor G.J. Whitrow

THEORETICAL COSMOLOGY IN THE TWENTIETH CENTURY

> .... Or if they list to try
> Conjecture, he his Fabric of the Heav'ns
> Hath left to their dispute, perhaps to move
> His laughter at their quaint opinions wide
> Hereafter, when they come to model Heav'n
> And calculate the Starrs ...

Milton's famous lines (1) referred to the dispute between those who believed in the Ptolemaic and Copernican systems respectively. No doubt if he had lived this century he would have felt mutatis mutandis that they would be no less appropriate when applied to modern theoretical cosmologists. For, although as described by Dr. Hetherington great advances in observational technique have been made this century, particularly since the end of the First World War, we are still obliged to resort to speculative hypotheses and extrapolations when we try to construct world-models compatible with the available data. For our direct knowledge of the universe is necessarily confined to a limited region of space and time, and even to interpret the data available from this limited region a long chain of theoretical assumptions is required. Once we pass beyond this region we only have our own general ideas about the nature of the physical world to support our speculations. Margaret Fuller, the celebrated nineteenth-century mystic, once declared in her usual over-enthusiastic manner "I accept the universe!" Her contemporary, Thomas Carlyle, commented as a dour Scot, "By God, she'd better!" That was in the middle of last century. Today we should feel like making a different retort: "which universe?"

In the nineteenth century, as in the Middle Ages, men thought that they knew the broad outlines on which the physical universe had been constructed. In the Middle Ages the world was a system of concentric spheres with the Earth at the centre. In the nineteenth century the solar system was regarded as the appendage of an otherwise typical star, moving in a vast system of stars, the limits of which were unfathomable. Due to the genius of Newton and his successors, it was thought certain that the solar system was governed by the inverse square law of gravitation. Indeed, in 1846 this law received a triumphant vindication in the discovery of the planet Neptune, predicted by mathematical calculation by Adams and Leverrier. Moreover, evidence from the study of binary stellar systems was accumulating that the stars as well as the planets were governed by the same iron law, and there seemed to be an overwhelming probability that its range was universal throughout the whole realm of infinite space.

Nevertheless, as the century wore on one or two small clouds appeared in this otherwise blue sky. At the time the achievement of Adams and Leverrier seemed to be one of the greatest in the history of astronomy, but from the point of view of the general advancement of knowledge more importance must be attached to another, and apparently far less successful, calculation made by Leverrier some years later on the motion of the planet Mercury. The theory of its motion had long given trouble, so that, as the historian Agnes Clerke has said, "the planet seemed to exist for no other purpose than to throw discredit on astronomers." (2) Even to Leverrier's powers of analysis it long proved recalcitrant, but in September 1859 he announced to the Academy of Sciences in Paris that a body of about the size of Mercury itself could produce the required effect of advancing the perihelion of Mercury's orbit by the 38 seconds of arc a century not otherwise explained. Strangely enough, from time to time claims were made that this planet, to which Leverrier gave the name 'Vulcan', had been seen. All proved to be illusory, but the reality of the perihelion effect could not be denied. In 1884 Simon Newcomb corrected Leverrier's 38 seconds to 43 seconds of arc. Writing in 1885, Agnes Clerke said of this anomaly in Mercury's motion, "Its elucidation constitutes one of 'the pending problems' of astronomy." (3)

Meanwhile in 1874 the physicist Carl Neumann had raised a specifically cosmological difficulty associated with Newton's theory of gravitation. (4) Just over twenty years later essentially the same difficulty was discussed by the stellar astronomer and head of the Munich observatory Hugo Seeliger. (5) As is well known, Newton in 1692 had argued that a finite material universe in infinite space would condense under its self-gravitation into one massive lump and that as this had not happened the universe must consist of an infinite number of masses, or 'stars', distributed at great distances from one another throughout infinite space. Seeliger pointed out that, if Newton's law of gravitation acted throughout such an infinite universe, the number of stars inside a sphere of large radius R would tend to be proportional to R^3 and hence the gravitational intensity at any point on the surface of such a sphere would be proportional to $R^3/R^2 = R$. Consequently, the gravitational field thereat would become indefinitely large for increasing values of R. Since the centre of the sphere can be anywhere we choose, there should be an infinite gravitational field throughout the universe, which is impossible. Seeliger suggested that this difficulty could be avoided by modifying the Newtonian law of gravitation and introducing a new factor of the form $e^{-\lambda r}$, where r denotes radial distance and λ is what we would call nowadays a new 'constant of nature'. The value of λ would have to be so small as to make the new factor effectively unity except for very large values of r. Although Seeliger's suggestion was criticized as being purely ad hoc, a factor of this type had been suggested previously to account for the anomaly in the motion of Mercury. In order to account for this observed discrepancy between Newtonian theory and observation, however, the value of λ would have to be such that the corresponding perihelion motion of the Earth would be several times greater than that revealed by observation. (6)

Another alleged objection to the infinite Newtonian cosmos is the notorious Olbers's Paradox of infinite sky-brightness that has been much discussed in recent years. It is, however, not such a serious difficulty to contend with as Seeliger's Paradox. For, implicit in the traditional accounts, which can be traced back a century before Olbers to Halley, is the assumption that there is not only an infinite space in which stars are more or less evenly distributed but also an infinite past time during which they have been radiating on the average at the same rate. This was no doubt an excusable assumption in Olbers's day, when the general law of the conservation of energy had not been formulated and the source of stellar radiation was unknown. But in this thermonuclear age it is clear that this assumption must be abandoned, unless some speculative hypothesis about the continual creation of new stars is involved. Indeed, to interrupt the historical narrative for a moment, taking the currently accepted approximate value for the mean density ρ of matter in the universe $(10^{-29}$ to 10^{-31} grams per cc), it follows that if all this were converted into radiation the mean-radiation density of the universe ρc^2 would be only about 10^{-8} to 10^{-10} ergs per cc, which is about 10^9 to 10^{10} times less than the radiation-density at the surface of the Sun.

An interesting attempt to reconcile the conception of an infinite Newtonian universe with the criterion of finite gravitational intensity everywhere (and also to evade the usual criticism associated with the traditional view of Olbers's paradox) was made in 1922 by the Swedish astronomer C.V.L. Charlier (7) in terms of a hierarchical model originally suggested by the eighteenth century mathematician J.H. Lambert. In this model N_1 stars form a galaxy, or system of the first order G_1, N_2 galaxies form a meta-galaxy, or system of the second order G_2, and so on indefinitely With certain simplifying assumptions, Charlier found that, if the radius of a typical (spherical) system G_i is denoted by R_i, then the difficulties discussed by Seeliger and Olbers respectively, could be avoided if $R_{i+1} > \sqrt{(N_{i+1})}\, R_i$, for all positive integers i. Nevertheless, although galaxies and clusters of galaxies exist, most astronomers with one or two exceptions, notably de Vaucouleurs do not believe in hierachy of clusterings of the type envisaged by Charlier.

1. THE FIRST PHASE IN THE DEVELOPMENT OF MODERN THEORETICAL COSMOLOGY

A new era in the history of cosmology dawned in 1917 when Albert Einstein (1879-1955) applied his new theory of gravitation to the construction of a homogeneous isotropic world-model. A significant difference between Newton's law of gravitation and that embodied in the field-equations of

Einstein's General Theory of Relativity would only occur
when very large differences of gravitational potential were
involved. Such differences are likely to occur only in
astronomical situations. The three well-known tests of
general relativity that Einstein considered in his definitive paper of 1916, which included the advance of perihelion that automatically accounted for the anomaly in
Mercury's motion that Agnes Clerke had described thirty
years before as one of the 'pending problems' of astronomy,
were all very small effects in the solar system. To get a
more quantitatively significant difference between the
predictions of his theory of gravitation and Newton's,
Einstein had to turn to the large scale properties of the
universe.

At the time when Einstein first considered the cosmological problem the only astronomical velocities known were
very small compared with the velocity of light. He therefore had no reason to depart from the traditional belief
that the universe considered as a whole is static. Although
in the same year, 1917, as he constructed his famous world-
model he also published his great paper on the quantum
theory of radiation, he made no reference to Olbers's radiation paradox in his cosmological considerations. This has
sometimes been thought surprising, but as I have already
indicated it should rather be regarded as evidence of Einstein's penetrating physical intuition that he realized that
the more significant objection to a static Newtonian universe is the difficulty raised by Seeliger, of which Einstein took full account.

In the synopsis of the second of his Rhodes Lectures at
Oxford in 1931, he expressed his views on cosmology very
clearly: "If one describes a three-dimensional sphere large
enough to contain many stars about any point in the universe
and divides the enclosed mass by the volume of the sphere,
one obtains a mean density ρ which may be assumed to reach a
limiting value as the radius increases. It is to be assumed
that this value is not zero, since otherwise the universe
would be virtually empty. It is natural to assume that this
density ρ is the same everywhere in the universe." He
pointed out, in his 1917 paper (8), that difficulties arise
if it is assumed that ρ vanishes at infinity, so that the
universe is an island in infinite space. For, by applying
Boltzmann's law of distribution relating to gas molecules to
the stars regarded as forming a system in thermal equilibrium, he showed that, since a finite ratio of densities
corresponds to a finite ratio of potential, zero density
at the boundary would necessitate zero density everywhere.

A further argument for assuming that, on the average, ρ
must have the same non-zero value everywhere was based on
his rejection of the possibility that ρ might have a very
high but finite limiting value at infinity. He regarded
this as unacceptable because differences of gravitational
potential in the universe must be low enough for "the
stellar velocities generated by them not to exceed the velo-

ocities actually observed." He concluded that a static
universe should be one of uniform non-vanishing density and
be, as we now say, homogeneous and isotropic but that it was
impossible on the basis of Newtonian theory for such a universe to be either infinite (because of Seeliger's Paradox)
or finite in infinite Euclidean space. Since Newtonian
theory seemed to be inconsistent with the observed properties of the universe in the large, he therefore asked
whether the assumption that the cosmic density ρ is the
same everywhere is compatible instead with the field equations of general relativity?

The assumption of constant non-zero density implies,
according to general relativity, a constant non-zero spatial
curvature. Einstein found, however, that his original field
equations could not be solved to produce a satisfactory
static world-model. To escape from this dilemma, Einstein
first considered how Newton's theory might be modified to
yield a universe of constant non-zero density. He did not
regard this as something to be taken seriously, but "as a
foil for what is to follow". He pointed out that, if in
place of Poisson's equation we write

$$\nabla^2 \phi - \wedge \phi = -4\pi G \rho,$$ (1)

where $\wedge$ denotes a new universal constant, a uniform finite
potential ϕ is compatible with a uniform finite density ρ.
The gravitational theory of general relativity corresponds
to Newtonian theory based on Poisson's equation in its
original form (Newtonian solutions can often be deduced
from general relativistic ones by expanding in powers of
G/c^2 and taking the zero-order terms in this parameter).
Corresponding to the modified Poisson-equation, Einstein
obtained modified field equations containing new terms
$-\wedge g_{rs}$, where g_{rs} are the coefficients appearing in the metric.
Using these equations, Einstein found a new solution to the
cosmological problem. The resulting world-model had a
finite radius R and mass M, related to $\wedge$ by the formulae

$$\wedge = \frac{4\pi G}{c^2} = \frac{1}{R^2}, \qquad M = 2\pi^2 \rho R^3,$$ (2)

and hence

$$GM = \frac{1}{2}\pi c^2 R.$$ (3)

From these equations we see that the 'cosmical constant' $\wedge$
must be positive, and that zero $\wedge$ corresponds in this model
to zero density ρ.

Although Einstein made no quantitative predictions about
the universe, his achievement should not be underestimated.
He had succeeded in constructing an apparently satisfactory
world-model, which did not seem to be possible on Newtonian
theory. His model occupied the whole of a non-Euclidean
spherical space. It was therefore finite but unbounded and
free from the difficulties previously encountered in cosmology. Moreover, as with so many of Einstein's previous
achievements, he had succeeded in relating particular concepts in a new way. For, not only was $\wedge$ related to the
mean density but the radius R depended on the mass M, so
that the total extent of space was determined by the
quantity of matter in the world. Indeed, Einstein believed

that his model satisfied what he called 'Mach's Principle'.
This much-disputed metaphysical principle, which Einstein
named after the anti-metaphysical philosopher Ernst Mach,
requires that, in some sense or other, the laws and general
properties of the universe are dependent on its material
contents.

At this time, early in 1917, Einstein was in frequent
contact by letter with Willem de Sitter (1872-1934), Director of the Leiden Observatory and, unlike Einstein, a professional astronomer. The purpose of this contact was to
get the General Theory of Relativity known in England, since
there was no direct communication possible between Germany
and England owing to the war. Consequently, de Sitter was
able to send in the years 1916-17 three papers from neutral
Holland for publication in the Monthly Notices of the Royal
Astronomical Society, all bearing the title "Einstein's
theory of gravitation and its astronomical consequences".
As has been pointed out by McCrea, the title is revealing
because, whereas Einstein had thought of astronomy as
helping to test his theory of general relativity, de Sitter
wanted to use the theory to help elucidate phenomena in
astronomy. (9) In preparing the first two of these papers -
in which he presented Einstein's theory afresh - de Sitter
was stimulated to think about the subject himself and make
his own remarkable contribution to theoretical cosmology in
the third paper. All three papers eventually made a great
impression on A.S. Eddington (1882-1944), at that time one
of the Honorary Secretaries of the R.A.S. Eddington's
initial reaction was rather cool."You understand, he wrote,
"that every paper published by the R.A.S. has to be sent to
a referee". Also, "we are having trouble with the printers",
"the paper is longer than we normally print", and so on.
But in the end de Sitter's papers were accepted, Eddington
himself became greatly interested, and as a result he and
Dyson, the Astronomer-Royal, started planning, in the darkest
times of the war, the crucial eclipse expeditions of May 1919
to test Einstein's quantitative prediction of light deflec-
tion by the Sun. The success of these expeditions made
Einstein's name a household word for the rest of his life.

In the third of the papers submitted by de Sitter in 1917
to the R.A.S. he presented his own uniform cosmological model based on the same field-equations, with the cosmical
constant $\wedge$, used by Einstein. In Einstein's model universe,
the smaller the density the greater the radius. In de
Sitter's model, unlike Einstein's, space-time has an intrinsic structure of its own, independent of the density of
matter. Indeed, de Sitter's model was in fact empty. Since
its properties did not therefore depend on the presence of
matter, it was not in accord with Mach's Principle and its
discovery was a great embarrassment to Einstein. In a
letter that he wrote to de Sitter on 1917 March 24, Einstein
did not approve of de Sitter's solution of the cosmological
problem. (10) In particular, he thought that it contained
a singularity at a finite distance from the observer, but

de Sitter's assistant Kluyver pointed out that Einstein occasionally made elementary mistakes in calculus and his objection was in fact invalid. In a letter to de Sitter, dated
1917 June 14, Einstein wrote, "Your metric does not make
sense to me. It could only apply without 'world-material',
that is stars". Actually, at that time Einstein assumed a
value of about 10^{-22} grammes per c.c.for the mean-density of
the universe, from which he derived for his model a radius R
of 10^7 light-years - values that now seem ludicrously bad
estimates. As de Sitter wrote in another context some years
later, although Einstein was a very great scientist, "his
knowledge of astronomy was minimal." (11)

In Newtonian analogy, the additional terms involving a
positive Λ introduced into the field equations of general
relativity by Einstein corresponded to a force of cosmical
repulsion. In Einstein's static world-model this just balances world-gravitation. In de Sitter's universe, however,
world-gravitation is everywhere zero, but cosmical repulsion
is not. Consequently, whereas a test-particle introduced
into Einstein's universe will remain at rest if it has no
initial motion, a similar particle introduced into de Sitter's
universe will, in general, automatically acquire an ever-
increasing velocity of recession from the observer. Consequently, de Sitter suggested that in the actual universe
the most distant celestial objects might prove to be moving
away from us. This may be regarded as the first prediction
of the expansion of the universe. His suggestion was linked
with another important characteristic in which his world-
model differed from Einstein's. In the latter time is
featureless, but in the original form in which de Sitter
presented his universe each observer perceives an horizon at
which time apparently stands still. (Of course, this is not
an absolute effect. At any point on the relative horizon of
a given observer the time-flux experienced by an observer
there would be the same as that experienced by the given observer.) Nevertheless, to draw from the apparent slowing-
down of distant clocks and periodic phenomena such as atomic
oscillations any conclusion about the predominence of red-
shifts in the de Sitter universe proved to be a very tricky
business, as has been made clear by North. (12) Indeed, for
some years the subject was beset by confusion, although the
possibility of systematic spectral shifts for remote objects
meant that de Sitter's universe greatly interested astronomers, notably Eddington. (13) We now realize that much of
this confusion was due to the mathematical accident that the
de Sitter metric was expressed in a time-independent, i.e.
quasi-static, form. What de Sitter had in fact discovered in
1917 was one of the simplest models of an expanding universe.
With a more physically appropriate choice of coordinates, as
was first shown by G. Lemaitre in 1925 (14), and independently by H.P. Robertson in 1928 (15), the metric of the de
Sitter universe can be expressed as

$$ds^2 = dt^2 - \{R^2(t)/c^2\}(dr^2 + r^2 d\theta^2 + r^2 \sin^2\theta \, d\phi^2), \quad (4)$$

where

$$R(t) = \exp\{\sqrt{(\Lambda/3)}\,ct\}. \quad (5)$$

It is in this form that we now normally study it, namely as the limiting case of an expanding universe as the mean density everywhere tends to zero. We therefore no longer consider de Sitter's as a static universe, its apparent changelessness being a mathematical fiction.

2. THE SECOND PHASE

The deliberate investigation of non-static solutions of the cosmological problem was initiated by a scientist who was by profession neither a theoretical physicist like Einstein nor an astronomer like de Sitter. Alexander Alexandrovich Friedmann (1888-1925) was a Russian meteorologist who had studied under Bjerknes at Leipzig before the first World War. In 1920 he was appointed to a chair in hydrodynamics at Petrograd. His contact with the talented physicists whom he met there led to his interest in relativity, and before long he gave a course on tensor analysis. Many years later, in recalling the discussions on relativity that took place at that time in the seminar of the Physical Institute of the University V.A. Fock remarked "Friedmann put the emphasis not on the physics but on the mathematics. He strove for mathematical rigor and gave great importance to the complete and precise formulation of the initial hypotheses." (16) Friedmann wrote his two main papers on relativity in 1922 and 1923. Although only the first was publised in Russian both were published in German, in the Zeitschrift für Physik in 1922 and 1924, respectively. (17). The translations were made by Fock. The year after the second paper appeared Friedmann died of typhoid fever, some months after he had been made Director of the main geophysical institute in Russia.

Friedmann's work was based on Einstein's field equations with the cosmical constant Λ. When it later became clear that a static world-model was incompatible with the observations of the extragalactic red-shifts Einstein abandoned this constant, but Friedmann retained it in his researches although he investigated non-static models. His retention of Λ has been defended by Fock, who has argued that, although in non-cosmological astronomical problems we can require that at 'infinity', i.e. at great distances from bodies under consideration, space is Euclidean and the energy-momentum tensor zero , so it is necessary that $\Lambda = 0$, in the cosmological problem we cannot automatically impose these conditions 'at infinity' and any particular choice of Λ is a special cosmological hypothesis. Consequently Fock maintained that "contrary to a general impression", Friedmann's use of the equations with non-zero Λ was "entirely justified". (18)

Friedmann's specifically cosmological postulates involved a 3-space coordinate system relative to which the bulk distribution of matter was assumed to be stationary (co-moving coordinates). Also, he postulated that time is orthogonal to this 3-space and that the latter is of constant (positive, zero or negative) curvature, although he did not explain

clearly why these assumptions were made. He reduced the
metric used to the mathematical form

$$ds^2 = dt^2 - \left\{ R^2(t)/c^2 \right\} d\sigma^2, \qquad (6)$$

where $d\sigma^2$ is the metric of the 3-space. He considered both
the stationary case, in which R = constant, and the non-
stationary case, in which R varies with time. His most remarkable results were obtained in the latter case. He not
only considered positive curvature, but in his second paper
he established for the first time the possibility of a non-
stationary world with negative curvature of the co-moving
3-space. His main results can be summarized in the following
equation for R(t):-

$$\dot{R}^2 = A/R - kc^2 + \tfrac{1}{3} \Lambda c^2 R^2, \qquad (7)$$

where the constant A is given by A = $(8\pi G/3)\, \rho\, R^3$, and
k is the 'curvature index' (1 for positive curvature and
-1 for negative) of the 3-space. Friedmann also made the
point that world-space could be curved without necessarily
being of finite extent.

How did Einstein re-act to Friedmann's unexpected discovery that general relativity (with or without Λ) admitted
non-static solutions of the cosmological problem? Soon
after the publication of Friedmann's first paper he published a note (19) in which he said that he had found a
mistake in Friedmann's work and that when it was corrected
his solution reduced to a static one. One of the physicists
of Petrograd who was a colleague of Friedmann's, Y.A. Krut-
kov, visited Einstein in Berlin at Friedmann's request (he
was travelling abroad at the time) and with great difficulty,
as he later told Fock, he convinced Einstein that he was
wrong (20). As a result, there appeared a second note by
Einstein in which he fully acknowledged his error and
praised Friedmann's work. (21)

According to Fock, Friedmann's concern was to indicate the
different possible cosmological solutions of Einstein's
equations and leave the physicists and astronomers to decide
which, if any, could apply to the actual universe. For this
reason his work did not arouse much interest and was neglected for some years. In the meantime the problem was attacked
without knowledge of Friedmann's work by the Belgian mathematician and physicist Georges Lemaitre (1894-1966). After
graduating in mathematics and physics at the University of
Louvain in 1920 with a thesis on approximations to functions
of many variables, Lemaitre was ordained Abbé in 1923 and
eventually he was made a Canon. Simultaneously with his
religious studies he became interested in Einstein's theory
and he prepared a thesis on relativity and gravitation that
won him a travelling scholarship. He thus spent the year
1923-4 at Cambridge, where he came under the influence of
Eddington. After a further year abroad, in the United States,
he returned to Louvain, where he became a full professor in
1927. In April of that year he published his famous paper
"Un univers homogène de masse constante et de rayon crois-

sant rendant compte de la vitesse radiale des nebuleuses extragalactiques."(22) This paper, like those of Friedmann, at first attracted little attention, no doubt partly due to its publication in a journal that was not widely read. G.C. McVittie (23) recalls that nearly three years later when he was a research student at Cambridge Eddington suggested to him that he work on the redshift problem. "I well remember the day," he writes, "when Eddington, rather shamefacedly, showed me a letter from Lemaitre which reminded Eddington of the solution to the problem which Lemaitre had already given". Eddington confessed that, although he had seen Lemaitre's paper in 1927, he had completely forgotten about it. He quickly made amends by a letter to "Nature", published on 7 June 1930, in which he drew attention to Lemaitre's brilliant paper and moreover persuaded the Royal Astronomical Society to republish it in English translation in 1931. (24) What had transformed the situation and made astronomers throughout the world turn eagerly to Lemaitre's work was Hubble's announcement in 1929 of the systematic recession of the galaxies subject to his famous law of redshifts.

Lemaitre's work was not only more physically oriented than Friedmann's but, although he concentrated on positive curvature, in another respect his work was more general. Like Friedmann he considered a homogeneous fluid model, but initially he made allowance for a uniform pressure p in the basic equations derived from Einstein's general relativity with the Λ term. In units with c = 1, he wrote these as

$$3\dot{R}^2/R^2 + 3/R^2 = \Lambda + K\rho, \tag{8}$$

$$2\ddot{R}/R + \dot{R}^2/R^2 + 1/R^2 = \Lambda - K p, \tag{9}$$

where K = 8πG. If ρ and p are regarded as functions of time alone, then the solution of the problem depends on the determination of the scale-function R. Lemaitre made the simplifying assumption that 'the mass of the universe is constant', and this led him to the equation

$$t = \int \frac{dR}{\sqrt{(\Lambda R^2/3 - 1 + \alpha/3R + \beta/R^2)}}, \tag{10}$$

where α is a constant connected with the total mass and β is a constant connected with the pressure. If the total pressure of both matter and radiation is taken to be zero, then β = 0. From this equation de Sitter's solution is obtained by taking α as well as and β to be zero, and Einstein's by taking β to be zero and R to be constant. In the latter case there is a relation between α and Λ, namely α = 2/$\sqrt{\Lambda}$. Lemaitre also considered this relation to hold more generally with R not a constant; in other words, he considered a model of constant mass and zero pressure that expanded from an initial state given by Einstein's 1917 model. Consequently, he presupposed that both Λ and the curvature of 3-space were positive. He then obtained a solution for R that increases steadily as t increases from the initial value (in the infinite past) R = R_o, where Λ = 1/R_o^2. Le-

maitre's model has the remarkable property that as t runs
from $-\infty$ to $+\infty$, the model tends asymptotically from Einstein's
universe to de Sitter's.

Lemaitre went beyond Friedmann in using his model to interpret the observed redshift in the spectra of galaxies regarded as a Doppler effect. As McVittie has remarked, "With
this step, cosmology, as it is known today, was launched."
(25) Lemaitre's model leads to the formula

$$\frac{\delta\lambda}{\lambda} = \frac{v}{c} = \frac{\dot{R}}{R}\,r = \frac{\sqrt{(1 - 3/y^2 + 2/y^3)}}{R_o\sqrt{3}}\,r, \qquad (11)$$

for the Doppler shift $\delta\lambda/\lambda$ and the associated velocity of
recession v at distance r, where $y = R/R_o$. This gives
Hubble's velocity-distance law, its limiting form as R increases indefinitely being $v = cr/R_o\sqrt{3}$. Using Hubble's value
for his 'constant' (H in the empirical relation $v = Hr$) and
the value of R_o depending on the estimated density of matter
in the universe, Lemaitre calculated that currently $y = 21.5$.
He thus obtained a model according to which the universe
began in the infinite past with a radius of about 1/20 of
its present radius and subsequently expanded, at first very
slowly and then at an accelerating rate. According to
Hubble's observational data, the initial radius appeared to
be 10^9 light-years and the total mass came out to be about
2×10^{55} grammes, equivalent to about 10^{22} times the mass of
the Sun, or roughly 10^{11} times the mass of our Galaxy.

As previously mentioned, Lemaitre's model, which was also
adopted by Eddington, who showed in 1930 that the Einstein
universe was unstable (26) and might therefore be expected to
expand (or contract), depends on the hypothesis that the
cosmical constant is non-zero and positive, a view which both
of them defended for the rest of their lives. In Newtonian
terms, the associated idea of a universal force of repulsion
between large aggregates of matter could be regarded as the
reason for the expansion. From 1945 onwards, Lemaitre developed the idea of a pre-Einstein state of the universe in his
notion of the "Primeval Atom". (27) According to this, the
universe originated in the explosion of a highly condensed
radioactive super-atom. Later it passed slowly through the
unstable Einstein state and then continued to expand ever
faster as comical repulsion overcame gravitational attraction.

In 1931 Einstein, who as early as 1919 wished to discard
the cosmical constant Λ since in his view it was a purely
ad hoc addition to the field equations of general relativity,
discussed in his Rhodes Lectures at Oxford the solution of the
Friedmann-Lemaitre equations with positive curvature and zero
Λ. (28) These assumptions lead to an oscillating universe.
Shortly afterwards, O. Heckmann of Göttingen pointed out that
there were 9 possible sets of conditions for a uniform world-
model according as Λ and the curvature of 3-space were taken
to be positive, zero or negative, respectively. Already in
1929, H.P. Robertson had deduced the general form of space-
time metric, far more clearly than Friedmann, given by
formula (6) above, applicable to all world-models that are
homogeneous and isotropic. (29) In so doing he drew attention to the assumption introduced by Weyl in 1923 (30), and

later made use of by others, that the world-lines of all matter in the universe form a coherent pencil of geodesics with a common origin in the past.

Following Heckmann's recognition of the possibility that not only Λ but the curvature (of 3-space) might also be zero, Einstein and de Sitter in 1932 published a celebrated short paper on what must be regarded as mathematically the simplest world-model based on the field equations of general relativity. (31) As they remarked, since there was no direct observational evidence for the 3-space curvature, "the question arises whether it is possible to represent the observed facts without introducing a curvature at all." The Einstein-de Sitter model as it came to be called, is based on zero curvature as well as zero Λ. It has an expansion factor $R(t)$ which is proportional to $t^{2/3}$ and an age (present epoch) given by $\frac{2}{3}\left(\frac{1}{H}\right)$, where H is the current value of Hubble's constant in the velocity-distance law. The density at the present epoch is given by $3H^2/8\pi G$. With Hubble's value for $1/H$ of approximately 2×10^9 years, this came out to be about 4×10^{-28} gm/cc and the age about 1.2×10^9 years.

Finally, in January 1933 this second phase in the development of 20th century theoretical cosmology culminated in the publication of Robertson's famous and definitive report on "Relativistic cosmology" in "Reviews of Modern Physics".

3. THE THIRD PHASE

In the same month as Robertson's comprehensive report appeared on what may be called the 'orthodox theory of the expanding universe', the astrophysicist E.A. Milne (1896-1950), who had been appointed in 1929 to the newly established Rouse Ball Chair of Mathematics at Oxford, published a long paper in the "Zeitschrift für Astrophysik" (32) in which he challenged the assumption that world-models must be based on general relativity, with or without a cosmical constant. Instead, he constructed an expanding world-model based on special relativity, the relative velocities of recession of the galaxies being assumed to be uniform, whereas in each of the non-static models previously considered the velocities were necesarily non-uniform. For example, in the Lemaitre model the velocities increase with time whereas in the Einstein-de Sitter model they decrease.

Milne's active interest in cosmology was first aroused in May 1932 by a series of letters in "The Times" following the publication of a report on the Ludwig Mond Lecture given in Manchester by Sir James Jeans, in which he had said that we shall never know why the universe is expanding. Milne objected to this as "scientific mysticism". In tackling the question himself he felt that it was necessary to return to first principles. He pointed out that, in general, any system of uniformly moving particles initially distributed at random in a finite region of space would eventually be found to be receding from one another according to Hubble's law. Although in this simple model gravitation was neglected, Milne felt that his simple explanation of Hubble's law indicated that the properties of a swarm of freely moving par-

ticles expanding uniformly from an initial configuration of
maximum congestion, (ideally, an initial point-singularity)
should be investigated. Since the motions concerned were
uniform, he used the method of Special Relativity, or rather
a cosmological modification of it that he gradually developed and eventually called 'Kinematic Relativity'. He used
this method in preference to the technique of expanding 3-
space and co-moving coordinates employed in the orthodox
theory of relativistic cosmology.

Milne's world-model consisted of a system of uniformly
moving particles (idealized galaxies) receding from a common
origin at a particular instant, which he regarded as the
origin of time. The relative motion of each particle at
distance r at epoch t was expressed by a law of the form
r = vt, so that comparison with Hubble's empirical law,

$$v = Hr, \tag{12}$$

implied that the reciprocal of H yielded the current value of
t, approximately 2×10^9 years according to Hubble's data.
At any epoch t, subsequent to the initial epoch zero, the
system fills the interior of a Euclidean sphere of radius
ct, the most distant particles appearing to recede with
speeds arbitrarily close to c.

All the world-models so far considered were based on the
principle of homogeneity, each region having the same properties as any other region. Milne adopted the same hypothesis and introduced the useful term 'cosmological principle'
to denote it. His model, like previous models, was also
isotropic about each constitutent particle. The number of
particles was infinite, there being an inpenetrable barrier
of apparently infinite density at distance ct. Otherwise, if
there were only a finite number of particles inside a Euclidean sphere of radius ct, there would exist a definite rim
of particles, from each of which the general appearance of
the system could not be the same as from an interior particle.

The Milne universe depends on the assumption that, since
each galaxy is assumed to be at a centre of symmetry the
total gravitational pull on it is zero, and therefore it is
possible to regard the recessional motion of the galaxies as
uniform or unaccelerated. In other words, the model can be
regarded as self-consistent without neglecting gravitation,
although since it is of non-zero density it is not compatible
with the field equations of general relativity. Instead,
Milne attempted to reformulate and deduce the fundamental
laws of physics from his cosmological hypotheses. He found
that in his relativistic model the universal constant of
gravitation is not a secular invariant, as had always previously been assumed, but increases linearly with time. This
was the first suggestion that the temporal behaviour of fundamental physical constants was a problem that should be
investigated. Another novel consequence of Milne's cosmological researches was the idea that different uniform scales
of time might be implied by different kinds of physical phenomena, e.g. gravitational time in his view is not the same
as atomic time and the time of radioactive decay. (33)

Besides these ideas associated with his uniformly expanding world-model, one of Milne's most important achievements was to resuscitate Newtonian cosmology by introducing in 1934, partly with the aid of W.H. McCrea, the study of Newtonian expanding universes. (34) For once the possibility was admitted that the universe as a whole is not static, the previous objections to a Newtonian universe in Euclidean space no longer apply. Milne and McCrea found that equation (7) above with $\Lambda = 0$ also applies in Newtonian cosmology and that the local properties of the relativistic models with zero cosmical constant and zero pressure in expanding spaces of positive, zero or negative curvature are observationally the same as those of the respective Newtonian universes in which every constituent particle has a velocity either less than, equal to, or greater than, the velocity of escape from the world-gravitational field in the neighbourhood of the observer.

One of the most important consequences of Milne's work in cosmology was that it stimulated A.G. Walker (35) to develop a rigorous derivation of the metric form (6) above characterizing any homogeneous isotropic world-model. This metric is now known as the Robertson-Walker metric.

In 1937 Dirac (36) introduced a completely different type of uniform cosmological model to explain the occurrence of very large dimensionless numbers in physics. For example, the ratio of the electrical to the gravitational force between a proton and an electron is of order 10^{39}. Dirac noticed that, if one defines an 'atomic' unit of time as the time light takes to travel a distance equal to the radius of an electron, then the number of such units in the reciprocal of Hubble's constant H (which according to various world-models is of the order of magnitude of the age of the universe) is also of order 10^{39}. This remarkable coincidence suggested to Dirac what he has since called 'The Large Numbers hypothesis'. This asserts that all the large dimensionless numbers occurring in Nature are connected with the present epoch expressed in atomic units and thus vary in time. As a result, if we assume that the masses and charges of elementary particles are constants, it easily follows that the universal constant of gravitation G varies inversely with the epoch. Moreover, if the dimensionless combination $G\rho/H^2$ is taken to be a constant of order unity, as in effect Dirac assumed, then ρ must vary inversely with epoch t and hence the expansion rate of the model is proportional to the cube root of t, and so it is decelerating. The current age of the universe according to Dirac's model is given by $\frac{1}{3}\left(\frac{1}{H}\right)$, i.e. only about 6×10^8 years according to Hubble's original estimate of his 'constant'. Dirac's argument was later used by P. Jordan to develop a more complicated theory of cosmology.

Meanwhile, a much more elaborate and far-reaching attempt to account not only for the occurrence of large dimensionless numbers in physics but for the exact values of all the pure numbers that are physical constants, e.g. the fine structure constant, was developed between 1928 and 1944 by Sir Arthur

Eddington, culminating in his posthumous treatise "Fundamental Theory" (37). Throughout this work, in which unlike Dirac he assumed that all the constants, however large, are secular invariants, he retained his belief that the world originated in an Einstein universe of mass 2×10^{55} grammes, containing 10^{79} protons and electrons in equal numbers. Irrespective of the fate of his detailed work in this subject, we can salute Eddington's achievement in drawing attention to the need to explain the various fundamental dimensionless constants in terms of a satisfactory theory of cosmology.

4. THE FOURTH PHASE

In the early nineteen-thirties when models of the expanding universe were first widely discussed, little was known about the stellar evolution and the possible age of the solar system. Estimates of the age of the Earth based on study of the radioactivity of the rocks were of the order of two thousand million years, about the same as the age of the universe indicated by Milne's model, but rather greater than that indicated by the Einstein-de Sitter universe. Moreover, as de Sitter emphasized in his Lowell Lectures at Boston in 1932, theories of the expanding universe seemed to restrict the time available for stellar evolution to a limit that was difficult to accept. (38) He considered the possibility that the extragalactic red-shifts were not due to motion after all but to some other cause, e.g. that light automatically loses energy as it travels through space, but he dismissed all such ad hoc hypotheses. Instead, he concluded that the hypotheses on which the concept of the expanding universe were then based "may at some future stage of the development of science have to be given up, or modified, or at least differently interpreted."

In the late nineteen-thirties the thermonuclear origin of stellar energy came to be generally accepted following the researches of Bethe and Weizsäcker, and by the late nineteen-forties it was becoming clear that the age of the Sun is probably of the order of five thousand million years. As a result, it seemed to many that current theories of the expansion of the universe allowed insufficient time for stellar evolution - and even for the past history of the Earth, since geophysicists were becoming convinced that this exceeded three thousand million years. Only for those like Lemaitre and Eddington (who died in 1944) who believed in the existence of the cosmical constant, and so could extend the age of the universe significantly beyond the value indicated by the reciprocal of Hubble's constant, was there no serious embarrassment. But many theoretical workers agreed with Einstein's view that the cosmical constant should be rejected.

In 1948 a radically different challenge from that of Milne to the orthodox theory of the expanding universe was advocated by three Cambridge astrophysicists F. Hoyle, H. Bondi and T. Gold. Although Hoyle's views (39) differed in some respects from those of Bondi and Gold (40), they all

argued that the universe is in a steady state, but instead
of rejecting the Doppler interpretation of the redshifts
they invoked the hypothesis of continual creation. As a
result, although galaxies recede from each other and evolve,
there was claimed to be no temporal evolution of the universe as a whole. As old galaxies and clusters stream away
from each other, new galaxies and clusters are continually
being formed to fill up the gaps in space that would otherwise appear. Since the age of the universe on this hypothesis was infinite, it was obviously compatible with any
estimates of the ages of the stars and planets. Great
stress was laid by the supporters of the new theory on the
apparent time-scale difficulty of most previous theories
from which their theory was free.

This argument began to be undermined in 1952 when Walter
Baade of the Mount Wilson and Palomar Observatories pointed
out a mistake made by astronomers in evaluating the zero-
point in the period-luminosity relation for Cepheid variables which had led Hubble to underestimate the distances of
the galaxies and as a result to overestimate H in the law
$\mathcal{U} = rH$. Consequently, the time-parameter given by the reciprocal of H had to be increased to a value nearer 5×10^9
years. Then, in 1958, as a result of further observational
work by Alan Sandage, it was clear that this parameter had
to be increased to be a value between 10 and 20 thousand
million years.

Nevertheless, argument continued for many years between
the supporters of the steady-state with continual creation
hypothesis and those who favoured an evolutionary view of
the universe. Although evidence from radio astronomy from
the early nineteen-fifties onwards, particularly that bearing on the properties of radio galaxies and their distribution in space, and also the discovery of quasars tended to
favour the idea of cosmic evolution, decisive observational
evidence against steady-state cosmology came only in 1965.
In that year, quite unexpectedly, A.A. Penzias and R.W.
Wilson, of the Bell Telephone Laboratories in New Jersey,
discovered what is now known as the cosmic black-blody microwave radiation . This is most readily interpreted as the
relic of the primeval fireball predicted by George Gamow
and others in the late nineteen-forties on the basis of an
explosive origin of the universe. (41)

5. THE FIFTH PHASE
    In the fifth and current phase of modern theoretical
cosmology the dominant belief is that the universe has
expanded and evolved from an initial very high-temperature
state. Physicists have developed in great detail theories
of the initial stages of the universe and the creation
first of elementary particles and then of stars and galaxies.
Mathematical interest has concentrated on the existence of
singularities in space-time. In particular, it has been
shown by highly sophisticated mathematical techniques including topological considerations that, allowing for some
relaxation of strict adherence to the principles of homo-

geneity and isotropy, theoretical world-models based on general relativity must contain singularities, an initial singularity corresponding to an epoch of world-creation at a finite epoch in the past. (42)

REFERENCES

(1) Milton, J., Paradise Lost, viii, 75-80
(2) Clerke, A.M., History of Astronomy during the Nineteenth Century, A. and C. Black, Edinburgh, 1885, p. 293.
(3) Clerke, A.M., ibid., p. 296.
(4) Neumann, C., Leipzig Abh. Math. Phys., 1897, xxvi, 97.
(5) Seeliger, H. von, Astron. Nachrichten, 1895, cxxxvii,129.
(6) Clemence, G., Reviews of Modern Physics,1947, 19, 261.
(7) Charlier, C.V.L., Arkiv. für Matematik, Astronomi och Fysik, 16, 1922, No. 22, 1.
(8) Einstein, A. Sitz. der Preuss. Akad. d. Wiss., 1917,142.
(9) McCrea, W.H. J. Brit. Astron. Assoc., 1972, 82, 179.
(10) Kahn, C. and Kahn F., Nature, 1975, 257, 452.
(11) Kahn C. and Kahn, F., op cit.,454.
(12) North J.D., The Measure of the Universe, Clarendon Press, Oxford, 1965, pp. 92 et seq.
(13) Eddington, A.S., The Mathematical Theory of Relativity, University Press, Cambridge, 1923, pp.161-6
(14) Lemaitre, G., J. Math. Phys.1925, 4, 188.
(15) Robertson, H.P., Phil. Mag., 1928, 7, 835.
(16) Fock, V.A., Soviet Physics Uspekhi, 1964, 6, 473.
(17) Friedmann, A.A.,Zeit f. Phys. 1922, 10, 377; 1924, 21, 326.
(18) Fock, V.A., op. cit., 474.
(19) Einstein, A., Zeit f. Phys., 1922, 11, 326
(20) Fock, V.A., ibid.
(21) Einstein, A., Zeit. f. Phys., 1923, 16,228.
(22) Lemaitre, G., Ann.Soc. Sci..Bruxelles, 1927, 47A, 49.
(23) McVittie, G.C., Q.J.R.A.S.,1967, 8, 295.
(24) Lemaitre, G.,M.N.R.A.S., 1931, 95, 483
(25) McVittie, G.C., op.cit., 296.
(26) Eddington, A.S., M.N.R.A.S.,1930, 90, 668.
(27) Lemaitre, G., L'Hypothèse de l'Atome Primitif,Editions du Griffon, Neuchatel, 1946.
(28) Einstein, A., Sitz. Preuss Akad, Wiss.,1931, 235.
(29) Robertson, H.P., Proc. Nat. Acad. Sci. (Washington), 1929, 15, 822.
(30) Weyl, H., Phys. Zeit., 1923, 29, 230.
(31) Einstein A. and de Sitter W.,P.Nat.Ac.Sci.1932,18,213.
(32) Milne, E.A.,Zeit. f. Astrophys., 1933, 6, 1.
(33) Milne, E.A., Proc. Roy. Soc., 1937, 158A, 324; Milne, E.A. and Whitrow, G.J., Zeit. f. Astrophys., 1938, 15, 342.
(34) Milne, E.A.,Q.J. Math. (Oxford), 1934, 5, 64; McCrea, W.H. and Milne, E.A., ibid, 1934, 5, 73.
(35) Walker A.G., Proc. Lond. Math. Soc., 1937, 42, 90.
(36) Dirac, P.A.M., Nature, 1937, 139, 323; Proc. Roy. Soc. A, 1938, 165, 199.

(37) Eddington, A.S., <u>Fundamental Theory</u>, University Press, Cambridge, 1946.
(38) de Sitter W., <u>Kosmos</u>, Harvard U.P., Cambridge (Mass.) 1932, 131 et seq.
(39) Hoyle, F., <u>M.N.R.A.S.</u>, 1948, 108, 372.
(40) Bondi, H. and Gold, T., <u>M.N.R.A.S</u>., 1948, 108, 252.
(41) Alpher, R.A.,Bethe,H. & Gamow,G. Phys.Rev.1948, 73,803.
(42) Hawking, S.W. and Ellis, G.F.R., <u>The Large Scale Structure of Space-Time</u>, The University Press, Cambridge, 1973.

CHURCHILL, Prof. F.B.
Goodbody Hall 130, Bloomington,
Indiana 47401, U.S.A.

LIST OF CONTRIBUTORS

COSTABEL, P.
Académie Internationale d'Histoire
des Sciences, 12 Rue Colbert,
F-75002 Paris, France

ALBURY, Dr W.R.
School of History and Philosophy
of Science, University of
New South Wales, Australia

CROSLAND, Prof. M.P.
Unit for the History, Philosophy
and Social Relations of Science,
Physics Building, University of
Kent, Canterbury, Kent CT2 7NH, U.K.

d'ALVERNY, Mme M.T.
58 Rue de Vaugirard,
Paris 75006, France

BEAUJOUAN, Prof. G.
3 Rue Louis Rolland,
F-92120 Montrouge, France

DAUB, Prof. E.E.
General Engineering Department,
University of Wisconsin-Madison,
Madison, Wisconsin 53706, U.S.A.

BEN-DAVID, Prof. T.
Department of Sociology, Hebrew
University, Jerusalem, Israel

DOBROV, Prof. G.M.
Rue Vorovskogo 22, V-Kiev, U.S.S.R.

BENNETT, Dr J.A.
National Maritime Museum,
London SE10 9NF, U.K.

DRAKE, Prof. S.
219 Glen Road, Toronto, Ontario
M4W 2W2, Canada

BIERMANN, Prof. K-R.
Akademie der Wissenschaften der
D.D.R., Alexander-von-Humboldt
Forschungsstelle, Zimmerstrasse 94,
108 Berlin, D.D.R.

FISHER, Dr N.W.
Department of History and
Philosophy of Science, Kings
College, Aberdeen, U.K.

BYNUM, Dr W.F.
Sub-department of History of
Medicine, University College,
London WC1, U.K.

GOODSTEIN, Dr J.R.
California Institute of Tech-
nology, Institute Archives 1-32,
Pasadena, California 91125, U.S.A.

CARDWELL, Prof. D.S.L.
Department of History of Science
and Technology, University of
Manchester, P.O. Box 88,
Manchester M60 1QD, U.K.

GOWING, Prof. M.M.
History Faculty, Indian Institute,
Broad Street, Oxford OX1 3BQ, U.K.

CASINI, Prof. P.
Via del Bargellino 8,
50014 Fiesole, Italy

GUERLAC, Prof. H.E.
Society for the Humanities,
Cornell University, Ithaca, N.Y.
14853, U.S.A.

CAWOOD, Dr J.A.
6 Brook Terrace, Slaithwaite,
Huddersfield, U.K.

HALL, Prof. A.R.
Department of History of Science
and Technology, Imperial College,
London SW7, U.K.

HEIMANS, Prof. J.
Quinten Massÿsstraat 4,
Amsterdam, The Netherlands

HERMANN, Prof. A.F.
Ganghoferstr.16, 8204 Brannenburg,
Degerndorf, F.R.G.

HESSE, Prof. M.B.
Department of History and Philo-
sophy of Science, University of
Cambridge, Cambridge, U.K.

HETHERINGTON, Prof. N.S.
Department of History, University
of Kansas, Lawrence, Kansas
66075, U.S.A.

HIEBERT, Prof. E.
Department of History of Science,
Science Center 235, Harvard Uni-
versity, Cambridge, Mass. 02138,
U.S.A.

HORNIX, Dr W.J.
Faculty of Science, Katholieke
Universiteit, Nijmegen,
The Netherlands

HOSKIN, Dr M.A.
16 Rutherford Road, Cambridge
CB2 2HH, U.K.

HOYER, Prof. U.
Göttendorferweg 32,
4403 Drensteinfurt 3, F.R.G.

KLEIN, Prof. M.J.
Department of History of Medicine
and Science, Yale University,
New Haven, Conn. 06520, U.S.A.

KNIGHT, Dr D.M.
Philosophy Department, Durham
University, 50 Old Elvet, Durham
DH1 3HN, U.K.

KROBER, Prof. G.
Institut für Theorie, Geschichte
und Organisation der Wissenschaft
der DDR, 108 Berlin, D.D.R.

LERVIG, Dr P.
Institute of Physics, University
of Aarhus, DK8000 Aarhus C,
Denmark

MERLAU-PONTY, Prof. J.H.P.
5 Rue du General Castelnau,
75015 Paris, France

MIKULINSKY, Prof. S.R.
Institute of the History of
Science and Technology, Staropansky
per. 1/5, Moscow K-12, U.S.S.R.

NAKAYAMA, Mr S.
3-7-11 Chuo, Nakano, Tokyo, Japan

NEEDHAM, Dr J.
Gonville and Caius College,
Cambridge, U.K.

NORBERG, Dr A.L.
The Bancroft Library, University
of California, Berkeley,
California 94720, U.S.A.

NORTON, Mr B.J.
Department of History of Medicine,
University College, London WC1 6BT,
U.K.

OLBY, Dr R.C.
Department of Philosophy, Leeds
University, Leeds LS2 9JT, U.K.

RASHED, Mr R.R.
10 Allee des Mouilleboeufs,
92350 Le Plessis-Robinson, France

RAVETZ, Dr J.R.
5 Moor Park Avenue, Leeds LS6 4BT,
U.K.

REINGOLD, Dr N.
Joseph Henry Papers (S1-319),
Smithsonian Institution,
Washington D.C. 20560, U.S.A.

RICHTA, Prof. R.
Ustav pro Filosofii a Sociologi
CSAV, Praha 1, Jilsta 1, PSC
110 00, Czechoslovakia

RIGHINI-BONELLI, Prof. M.L.
Instituto di Storia delle Scienze,
Piazza dei Giudici 1, 50122
Florence, Italy

ROSSI, Prof. P.
Instituto di Fisica, Universita
di Lecce, Italy

SALOMON, Prof. J-J.
19 Rue Decamps, 75116 Paris,
France

SCHROEDER-GUDEHUS, Dr B.
Institut d'Histoire et de Socio-
politiques des Sciences, Université de Montreal, C.P.6128,
Canada

SLOAN, Dr P.R.
Program in History and Philo-
sophy of Science, University of
Notre Dame, Indiana 46556, U.S.A.

THACKRAY, Prof. A.W.
History and Sociology of Science,
University of Pennsylvania, Phil.
19104, U.S.A.

VOLKOV, Dr V.A.
Institute for the History of
Science and Technology, Academy
of Sciences, Staropansky per. 1/5,
Moscow K-12, U.S.S.R.

WARDESKA, Dr Z.
Maklakiewicza 13/32, 02-642 Warsaw,
Poland

WATANABE, Prof. M.
16-3 Higashi-cho, 1-chome,
Koganei-shi, Tokyo 184, Japan

WEBSTER, Mr C.
Wellcome Unit for History of
Medicine, 47 Banbury Road,
Oxford, U.K.

WEINER, Prof. C.
Technology Studies Program, 200-224
M.I.T., Cambridge, Mass. 02139,
U.S.A.

WHITROW, Prof. G.J.
Department of Mathematics, Imperial
College, Huxley Building, Queen's
Gate, London SW7 2BZ, U.K.

WHITROW, Mrs M.
41 Home Park Road, Wimbledon,
London SW19, U.K.

WRIGLEY, Dr E.A.
SSRC Cambridge Group for the
History of Population and Social
Structures, 27 Trumpington Street,
Cambridge CB2 1QA, U.K.

YAROSHEVSKY, Prof. M.G.
Institute of the History of Science
and Technology, Staropansky 1/5,
Moscow K-12, U.S.S.R.